SMALL ANIMAL
CARE & MANAGEMENT

SECOND EDITION

SMALL ANIMAL CARE & MANAGEMENT

SECOND EDITION

DEAN M. WARREN

DELMAR

THOMSON LEARNING™   Australia   Canada   Mexico   Singapore   Spain   United Kingdom   United States

DELMAR

THOMSON LEARNING

Small Animal Care & Management, Second Edition
by Dean M. Warren

Business Unit Director:
Susan L. Simphenderfer

Executive Editor:
Marlene McHugh Pratt

Acquisitions Editor:
Zina M. Lawrence

Developmental Editor:
Andrea Edwards Myers

Editorial Assistant:
Elizabeth Gallagher

Executive Production Manager:
Wendy A. Troeger

Project Editor:
Amy E. Tucker

Production Manager:
Carolyn Miller

Executive Marketing Manager:
Donna J. Lewis

Channel Manager:
Nigar Hale

Cover Design:
Dutton & Sherman Design

For permission to use material from this text or product, contact us by
Tel   (800) 730-2214
Fax   (800) 730-2215
www.thomsonrights.com

Library of Congress Cataloging-in-Publication Data
Warren, Dean M. (Dean Marvin), 1944-
    Small animal care & management/Dean M. Warren. -- 2nd ed.
        p. cm
    Includes bibliographical references.
    ISBN 0-7668-1424-6
        1. Small animal culture. 2. Pets. I. Title: Small animal care and management. II. Title.

SF409.W37 2001
636.08'3--dc21   00-020911

NOTICE TO THE READER

Publisher does not warrant or guarantee any of the products described herein or perform any independent analysis in connection with any of the product information contained herein. Publisher does not assume, and expressly disclaims, any obligation to obtain and include information other than that provided to it by the manufacturer.

The reader is expressly warned to consider and adopt all safety precautions that might be indicated by the activities herein and to avoid all potential hazards. By following the instructions contained herein, the reader willingly assumes all risks in connection with such instructions.

The Publisher makes no representation or warranties of any kind, including but not limited to, the warranties of fitness for particular purpose or merchantability, nor are any such representations implied with respect to the material set forth herein, and the publisher takes no responsibility with respect to such material. The publisher shall not be liable for any special, consequential, or exemplary damages resulting, in whole or part, from the readers' use of, or reliance upon, this material.

Contents

Preface

When I started a class on small animal care several years ago, it soon became apparent that there weren't any textbooks available in the subject area. I purchased numerous books from pet shops and book stores to use in developing my course and spent many hours creating teaching materials, quizzes, and tests. Because some educators may be reluctant to start a course in small animal care, because of their lack of knowledge and teaching materials in this subject area, I proceeded to write a text and an accompanying instructor's guide.

I tried to write a book that would be flexible enough to be used by the general reader, pet enthusiast, and students with varying abilities at varying grade levels. If it is used as a text for an animal science class, the instructor can decide how difficult or easy to make the course. Numerous scientific names are used; the instructor must decide whether the student should learn them. This book also encourages the reader to research and explore subject areas. In a text of this type, it is impossible to thoroughly cover each area. Readers may wish to seek additional information, and for this reason, Internet sites have been included in this second edition. The author and Delmar made every attempt to ensure that all Web site URLs were accurate at the time of printing. However, due to the fluid nature of the Internet, we cannot guarantee they will remain current for the life of the edition.

The instructor's guide contains the same discussion questions that are found at the end of each chapter and the answers to the discussion questions. Also available through the publisher is a Classmaster CD-ROM, which contains the Teacher's Resource Guide, the instructor's guide, and a computerized test bank. With these supplements, in addition to this text, I believe that an instructor can create a course in small animal care and have the basic materials needed.

The text contains a glossary of terms that might be unfamiliar to the reader. An index is also provided to assist one in quickly finding subject areas and topics. A strong point of the text is that it has many excellent photographs and illustrations, and the second edition has been improved by many full-color photographs and illustrations.

The pet and companion animal industry is a rapidly growing segment of our economy. Employees with knowledge and skills to work with small animals are constantly needed.

It is the sincere hope of this author that this text meets the needs of those who wish to learn about the proper care and management of small animals.

Acknowledgments

I wish to thank the many persons who have helped in putting this text together. I especially want to thank Mary Martha Scott, the librarian at Columbia City High School, for her assistance in checking the manuscript for grammatical errors and for assisting in locating reference materials.

I want to thank Isabelle Francais for her excellent photographs of many of the animals found in the book. Without her generous assistance, this book wouldn't have the exceptional visual element of illustrations that it contains. Aaron Norman provided the excellent photographs of fish. Chapter 20 is greatly enhanced because of his photographs.

I wish to thank the United States Department of Agriculture for providing photographs and resource material on diseases.

Dr. Jack L. Albright of Purdue University and William McVay, retired agriculture teacher at Whitko High School, South Whitley, Indiana, provided materials and input on Animal Rights and Animal Welfare. Cindy Raker, agriculture educator at Carroll High School, Fort Wayne, Indiana, provided me with her class outline and materials on dogs and cats. It was her material that I used in my early classes on small animal care.

I wish to thank my former students; the staff of Columbia City High School, Columbia City, Indiana; and the agriculture educators in Indiana for their support and encouragement throughout this project. I particularly want to thank Jack Caster, Computer Coordinator, for his assistance with word processing and other computer problems. His assistance helped me tremendously during the writing of the first edition.

A special thanks goes to Delmar for seeing a need for a text on small animal care and giving me an opportunity to provide one. And, also to Andrea Myers and all the staff at Delmar who had a hand in putting this text together. I truly appreciate their assistance and patience throughout this project.

I especially wish to thank Suzanne Copple, Production Editor, and her staff at Graphic World Publishing Services. Her careful reading, formatting, and editing skills enhanced the readability of this edition.

The author and Delmar also wish to thank the following reviewers for their time and content expertise:

Rachel A. Zabel
New Richmond High School
New Richmond, Wisconsin

Harold Johnson
El Molino High School
Forestville, California

L. William Shrum
Seminole Vocational Education Center
Seminole, Florida

Charles Speck
Warren Central High School
Bowling Green, Kentucky

Small Animal Care
and Management Second Edition

Dean M. Warren

Small Animal Care & Management

ADDED FEATURES

- Pedagogical use of FULL COLOR to accelerate learning through highly illustrative material.

- Competency-based level of achievement needed for success is spelled out in front of each chapter.

- FULL-COLOR species identification.

44
SECTION ONE

PET CARE WORKERS

Pet care workers provide many services for the owners of small animals. They are employed in animal hospitals, boarding kennels, animal shelters, pet grooming parlors, pet training schools, and pet shops. Some pet care workers have their own businesses. See Figure 5–1.

Kennel attendants care for animals and keep their quarters clean. They watch the animals closely to make sure that they are in good health, and they often exercise animals and keep records on the animals' health, feeding, and breeding. At times they may also have to dispose of diseased or unwanted animals. See Figure 5–2.

Animal groomers bathe pets and use special solutions to keep them free of ticks, fleas, and other pests. They often brush the pets and trim their hair and nails.

Dog trainers teach dogs to hunt, herd, or track; to obey signals or commands; to guard lives or property; to run races; or to lead the blind. Handlers command trained animals during shows, sporting matches, or hunts. See Figure 5–3.

Animal breeders arrange for the mating of animals and care for the mother and young. They usually train or sell animals when they reach the appropriate age. They often specialize in one breed, such as German Shepherd dogs or Siamese cats.

Pet shop owners care for the birds, fish, cats, dogs, and other animals they offer for sale, as well as deal with customers. In addition, they must take care of the details involved in running a business. De-

pending on the size of the shop, the owner may employ a manager, animal care persons, and pet counselors. Stores usually employ an average of three to five workers. See Figure 5–4.

Figure 5–2 *Kennel attendants feed and care for animals and keep the facilities clean. Courtesy of Brian Yacur and Guilderland Animal Hospital.*

Figure 5–1 *High school students can get employment as animal hospital workers. Responsibilities usually include cleaning and disinfecting cages and equipment, feeding animals, and performing other duties assigned by the veterinarians. Courtesy of USDA.*

Figure 5–3 *Dogs can be trained for many different jobs and skills depending on what they will be used for (i.e., herding, hunting and tracking, police work, or assisting people with disabilities). Courtesy of PhotoDisc.*

CHAPTER 18 Reptiles

OBJECTIVES

After reading this chapter, one should be able to:
- describe the characteristics of reptiles.
- compare the four orders of reptiles.
- describe the methods used in handling and feeding reptiles.
- describe the housing and equipment needs of reptiles kept in captivity.

TERMS TO KNOW

arboreal	ectotherms	scutes
brille	hemipenes	terrapins
brood	lamellae	terrarium
carapace	oviparous	tympanum
casque	ovoviviparous	vivarium
crepuscular	plastron	viviparous
dimorphism		

CHARACTERISTICS

Reptiles are cold-blooded vertebrates that possess lungs and breathe air. They also have a bony skeleton, scales, or horny plates covering the body, and a heart that has two auricles and in most species one ventricle.

There are approximately 6,500 species of reptiles belonging in four different orders: Testudines (called Chelonia in some references), which are made up of turtles, tortoises, and terrapins; Serpentes, which includes snakes, pythons, and boas; Squamata, which includes of iguanas and lizards; and Crocodilia, which includes crocodiles, alligators, caimans, and gharials.

Fossil remains of reptiles date back to the Carboniferous period 340 million years ago. Other fossil remains dating back 315 million years are of small agile, lizard-like animals. The evolution of reptiles continued through the Permian and into the Triassic periods. During the Triassic period, reptiles became the dominant group of terrestrial vertebrates inhabiting the earth; they continued their dominance for about 160 million years. The reptiles reached their greatest number during the Cretaceous period; then many became extinct. Today, only four orders of the fifteen that existed during the Cretaceous period have survived.

258

The Italian Greyhound. The Italian Greyhound is a very old breed, descending from the Egyptian Greyhound of over 6,000 years ago. This is the smallest of the family of dogs that hunt by sight. It is believed to have been brought into Europe by the Phoenician traders and developed in Italy, thus the Italian Greyhound name. There is some disagreement as to whether the breed was developed for hunting or for a pet and companion. The breed is very affectionate and thrives best when this affection is returned. The coat is short and smooth and requires very little grooming. The Italian Greyhound is odorless and sheds very little. The breed is sensitive, alert, intelligent, and playful. It may be timid, meek, and wary of strangers. See Figure 7–49.

CHAPTER 7 ■ Dogs 93

Figure 7–49 *Italian Greyhound. Photo by Isabelle Francais.*

The Manchester Terrier. The Manchester Terrier breed was developed in the Manchester district of England during the nineteenth century by crossings between the Black and Tan Terrier, the Whippet, the Greyhound, and the Italian Greyhound. Its nickname is the "rat terrier" because it has long been recognized for its ability to hunt rats and other rodents. It is considered intelligent, active, and alert. Today, it is primarily a small companion dog. In 1959, two varieties were recognized: the toy and the standard. See Figure 7–50.

Figure 7–50 *Manchester Terrier. Photo by Isabelle Francais.*

The Pekingese. In ancient times, the Pekingese was held sacred in China, the land of its origin. The exact date of origin is not known, but its existence traces back to the Tang dynasty of the eighth century. The very old strains (held only by the imperial family) were kept pure, and the theft of one of these sacred dogs was punishable by death. Introduction of the Pekingese into the Western world occurred as a result of the looting of the Imperial Palace at Peking by the British in 1860. The breed is sensitive, extremely affectionate with its owner, loyal, somewhat aloof, reserved, and quiet. Although Pekingese are not aggressive, they are courageous and will stand their ground against larger adversaries. They are primarily an apartment lapdog, but they make good watchdogs as well. The long, fine coat needs frequent care, and the teeth need to be cleaned to prevent early decay. See Figure 7–51.

Figure 7–51 *Pekingese. Photo by Isabelle Francais.*

ENHANCED CONTENT

The updated and enhanced content addresses the evolving small animal curriculum:

- The physical restraint of animals and general guidelines for risk and safety management for small animals is discussed in depth.

- Expanded coverage of all species. This coverage provides the most up-to-date information available on small animal anatomy, handling, feeding, housing and equipment, history, breeds, uses, grooming, training, diseases and ailments, and reproduction.

- Extensive coverage of the skills and educational requirements needed for small animal students to move on to a specialized field of study.

EXTENSIVE TEACHING/LEARNING PACKAGE

Instructor's Guide to Text

The Instructor's Guide provides answers to the end-of-chapter questions and additional material to assist the instructor in the preparation of lesson plans.

Classmaster CD-ROM

Order #0-7668-1426-2

This new supplement provides the instructor with valuable resources to simplify the planning and implementation of the instructional program. It includes transparency masters, motivational questions and activities, answers to questions in the text, and lesson plans to provide the instructor with a cohesive plan for presenting each topic. Also included is a computerized testbank with more than 500 questions that will give the instructor an expanded capability to create tests.

SECTION

I

Introduction to Small Animal Care

OBJECTIVES

After reading this chapter, one should be able to:

- briefly describe the history of animals on the earth.
- list the important time periods in the evolution of animals.
- describe how animals probably became domesticated.
- describe the importance of the small animal industry.
- describe how organisms are classified.

TERMS TO KNOW

Animalia	Monera	Reptilia
Aves	notochord	taxa
binomial nomenclature	Osteichthyes	taxonomy
Chordata	pharyngeal	trinomial nomenclature
Fungi	placental mammals	vertebrate
invertebrates	Plantae	
Mammalia	Protista	

A BRIEF HISTORY OF ANIMALS

The oldest direct traces of life on earth date back 3.4 to 3.5 billion years. In rocks that age in Australia and Southern Africa, geologists have found stromatolites, layered structures created through the activity of primitive algae and bacteria. Other Australian rocks of similar age provide even more direct evidence of ancient life. Sections of these rocks, known as cherts, show fossilized remains of blue-green algae themselves.

Rocks also reveal even more distant, indirect traces of life. Living things use particular isotopes (atomic forms) of the element carbon preferentially. The mix of carbon isotopes detected in rocks from Greenland more than 3.8 billion years old show evidence of life on earth.[1]

Some 2.2 billion years ago, free oxygen was present in the atmosphere. Living things used this reactive substance in the biochemical functions of their cells. The free oxygen in the atmosphere also produced a layer of ozone, which filters out the ultraviolet light from the sun that is harmful to life. See Figure 1–1.

Era	Period	Epoch	Biologic events	Years before present (B.P.)	Geologic events — Events refer especially to North America and do not reflect great worldwide variations
CENOZOIC	Quaternary	Recent	Modern man	11 thousand	Ice ages; then warmer
		Pleistocene	Early man	0.5 to 3 million	
	Tertiary	Pliocene	Large carnivores	7 million	Continental elevation; cool
		Miocene	Abundant grazing mammals	25 million	Plains and grasslands; moderate
		Oligocene	Apes, monkeys, whales	40 million	Mountain erosion; mild
		Eocene	Radiation of placentals	60 million	Mountain erosion; rain and mild
		Paleocene	First placental mammals	70 million	Mountain building; subtropical
MESOZOIC	Cretaceous		Climax of giant land and marine reptiles, followed by extinction; flowering plants; decline of gymnosperms	135 million	Spread of inland seas and swamps; building of Andes, Himalayas, Rockies; mild to cool
	Jurassic		First birds; first mammals; dinosaurs abundant	180 million	Continents with shallow seas, building of Sierra Nevada mountains; cool then mild
	Triassic		First dinosaurs; mammal-like reptiles; conifers dominate plants	230 million	Continents elevated; widespread deserts; cool and dry
PALEOZOIC	Permian		Radiation of reptiles; displacement of amphibians; extinction of many marine invertebrates	280 million	Continents elevated; building of Appalachians; cold and dry
	Carboniferous	Pennsylvanian	First reptiles; giant insects; great conifer forests	310 million	Shallow inland seas; extensive coal deposits; warm and moist
		Mississippian	Radiation of amphibians; abundant sharks; scale trees and seed ferns	345 million	Inland seas; warm to hot; swamplands
	Devonian		First amphibians; freshwater fishes abundant; bryozoans and corals	405 million	Inland seas; first forests; mild
	Silurian		First jawed fishes	425 million	Continental seas and reefs; mild
	Ordovician		Ostracoderms (first vertebrates); abundant marine invertebrates; first land plants	500 million	Submergence of land; warm
	Cambrian		Origin of many invertebrate phyla and classes; trilobites dominant; marine algae	600 million	Three periods of land submergence; mild
PRE-CAMBRIAN	Proterozoic Archeon Hadean		Fossil algae; other fossils extremely rare; evidence of sponges and worm burrows	2.5 billion 3.8 billion 4.5 billion	Volcanic activity; mountain building; glaciations; variable climate

Figure 1–1 The history of the earth is divided into periods of time. This Geologic Time Chart shows the geologic events that occurred in North America and the living organisms that were present during the geologic events. By permission of McGraw-Hill Company.

The first land plants were believed to have become established about 420 million years ago during the Ordovician period. This is also the period of time when invertebrates, such as arthropods and worms, appeared on land. These first terrestrial invertebrates fed on decaying plant material.

Vertebrate animals, in the form of amphibians and reptiles, appeared during the Devonian period about 370 million years ago. The bony fish appeared during the Ordovician period and became abundant during the Devonian period. See Figure 1–2.

During the Triassic period, 180 million years ago, dinosaurs appeared along with mammal-like reptiles. The first birds and mammals appeared 135 million years ago during the Jurassic period, when dinosaurs became abundant. See Figures 1–3 and 1–4.

Figure 1–2   Coelacanth, a primitive, lobe-finned, bony fish that some believe may be the ancestor of land animals. Coelacanth was thought to be extinct until some living specimens were discovered off the East Coast of South Africa. Negative #325025 (Photo by Rota). Courtesy Department of Library Services, American Museum of Natural History.

During the Cretaceous period 135 to 70 million years ago, dinosaurs and marine reptiles reached their period of greatest abundance and then disappeared. The reason for their demise is still debated by scientists.

With the extinction of the dinosaurs, an opportunity arose for smaller creatures to evolve and become abundant. During the Paleocene and Eocene epoch 70 to 40 million years ago, **placental mammals** evolved, dispersed, and adapted to new environments. Placental mammals have a placenta through which the embryo and fetus are nourished while in the uterus. Many of the small animals discussed in this book are placental mammals and evolved during this period.

Figure 1–3   Archaeopteryx is one of the ancestors of modern birds. It lived 150 million years ago during the Jurassic period. This fossil was discovered in a Bavarian stone quarry. Negative #34711. Courtesy Department of Library Services, American Museum of Natural History.

Figure 1–4   A reconstruction of Archaeopteryx from fossil remains. Illustrated by Ron Ervin.

DOMESTICATION OF ANIMALS

Domestication of animals probably began when human hunters brought back the young of the adult animals they killed. Dogs are believed to be the first animals domesticated. Wolves may have hung around villages looking for meals, and wolf cubs were probably domesticated and became companion animals. In addition, they may also have warned the villages of approaching predators or enemies. The discovery of a dog's jawbone found in a cave in Iraq dates back 12,000 years. This is the earliest indication of a domesticated animal.

The cat's association with humans dates back about 3,500 years ago. Pictures of cats have been found in the Egyptian tombs. The domestic cat probably provided pleasure and companionship, as well as controlled rats and mice.

THE SMALL ANIMAL INDUSTRY

Pet ownership is at an all-time high—58.9 percent of all American families, about 58.2 million households, include at least one companion animal. See Figure 1–5. Cats outnumber dogs as the most common household pet, see Figure 1–6; there are more than 59.1 million cats in the United States compared with 52.9 million dogs. Dogs are found in 31.6 percent of the nation's households and cats in 27.3 percent, see Table 1–1. Most cat owners have an average of two cats.

American households also contain 12.6 million birds, 4.9 million rabbits, 1.8 million hamsters, and 55 million fish.[2] See Table 1–2.

Altogether, an estimated 10,500 to 11,000 retail pet stores nationwide produce some $21 billion in sales, See Figure 1–7.[3] Veterinary expenditures in 1996 were nearly $7.008 billion for dogs, $3.971 billion for cats, and $91.2 million for birds. These figures represent increases in veterinary expenses of 42.2 percent for dogs, 96.8 percent for cats, and 141.9 percent for birds since 1991.[4] This could put total expenditures for pet care in the United States at well over $32 billion, an increase of more than 50 percent since 1991.

Figures from the United States Department of Agriculture in the Animal Welfare Enforcement Report for fiscal year 1997 showed a total of 1,267,828 animals used in laboratory experiments. Of this total, there were 75,429 dogs, 26,091 cats, 56,381 primates, 272,797 guinea pigs, 217,079 hamsters, and 309,322 rabbits. Since the use of birds, laboratory rats, and laboratory mice are currently excluded from Animal Welfare Act (AWA) regulation, no accurate figure is available, although the Animal Welfare Enforcement

Figure 1–5   Pets are an important part of the family. Courtesy of Carolyn Miller.

Figure 1–6   Animals can teach children many things such as responsibility, compassion, and trust. Courtesy of Tracey A. Carter.

Table 1–1 United States Pet Ownership—The Percentage and Number of Pet-Owning Households and Pet Population Estimates, December 31, 1987, 1991, and 1996.

	Percentage of Households (%)			Number of Households (Millions)			Average (Mean) Number Owned Per Household			Total Population (Millions)		
	1987	1991	1996	1987	1991	1996	1987	1991	1996	1987	1991	1996
Dogs	38.2	36.5	31.6	34.7	34.6	31.2	1.51	1.52	1.69	52.4	52.5	52.9
Cats	30.5	30.9	27.3	27.7	29.2	27.0	2.04	1.95	2.19	54.6	57.0	59.1
Birds	5.7	5.7	4.6	5.2	5.4	4.6	2.48	2.16	2.74	12.9	11.0	12.6
Horses	2.8	2.0	1.5	2.6	1.9	1.5	2.63	2.54	2.67	6.6	4.9	4.0

Source: Reprinted with permission from "U.S. Pet Ownership & Demographics Sourcebook," Center for Information Management, American Veterinary Medical Association, 1997.

Table 1–2 United States Pet Ownership—The Number of Specialty and Exotic Pet-Owning Households and Pet Population Estimates, December 31, 1991 and 1996.

Type of Pet	Number of Pets Per Household		Number of Households (1,000)		Population of Pets (1,000)	
	1991	1996	1991	1996	1991	1996
Fish*	9.05	8.92	2,652	6,228	23,997	55,554
Ferrets*	1.45	2.00	189	395	275	791
Rabbits*	3.22	2.63	1,420	1,878	4,574	4,940
Hamsters	1.39	1.86	947	1,008	1,316	1,876
Guinea pigs	1.77	1.87	473	583	838	1,091
Gerbils	2.18	2.76	284	277	619	764
Other rodents	2.31	2.42	379	435	875	1,053
Turtles	1.87	1.78	379	534	708	950
Snakes	3.88	4.14	189	217	735	900
Lizards	1.66	1.55	189	455	314	705
Other reptiles	2.97	2.75	95	336	281	924
Other birds (pigeons and poultry)	13.78	13.16	379	336	5,220	4,423
Livestock	7.12	11.61	473	524	3,371	6,083
All others	3.37	3.26	189	376	638	1,225

*Fish, ferrets, and rabbits were specifically listed on the 1996 survey, but they were not listed in 1991.

Source: Reprinted with permission from "U.S. Pet Ownership & Demographics Sourcebook," Center for Information Management, American Veterinary Medical Association, 1997.

Figure 1–7 A variety of pet foods are available on the market today. Courtesy of Brian Yacur and Guilderland Animal Hospital.

Report lists 150,987 other animals used in experiments. See Figure 1–8.

Registered with the AWA for fiscal year 1997 are 1,243 research facilities, 2,098 exhibitors, 4,043 dealers, 96 carriers, and 309 intermediate handlers. Research facilities include hospitals, colleges and universities, diagnostic laboratories, and many private firms in the pharmaceutical and biotechnology industries. Exhibitors operate animal acts, carnivals, circuses, public zoos, roadside zoos, and marine mammal displays. Carriers include airlines, motor freight lines, railroads, and other shipping lines. Intermediate handlers usually provide services for animals between

Figure 1–8 *Animals are also used in research.*
Courtesy of Agricultural Communications Department,
University of Georgia.

consignor and carrier; they also care for animals delayed in transit. Dealers are persons selling lab animals for research and teaching, wild animals for exhibition, or pet animals at the wholesale level.

Animal care is a very large and rapidly growing industry. Pet care workers, zoo administrators, laboratory animal care workers, small animal breeders, animal trainers, animal groomers, veterinarians, veterinary technicians, and biologists will continue to be needed for this important industry.

CLASSIFICATION OF ORGANISMS

Taxonomy is the science concerned with the naming and classification of organisms. The Greek philosopher and biologist Aristotle is credited with the first attempts at classifying organisms based on their structural similarities. An English naturalist, John Ray (1627–1705), developed a more comprehensive system of classification, but Carolus Linnaeus (1707–1778), a Swedish botanist, is credited with developing our modern method of classification.

Linnaeus's system has been expanded since then and today is made up of seven categories, or **taxa**. These categories are kingdom, phylum, class, order, family, genus, and species. These categories can be subdivided into finer categories, such as subphylum or subclass. Today about 30 different taxa are recognized.

All organisms are divided into one of five kingdoms:

1. **Monera** are the bacterial organisms that lack a true nucleus in the cell and reproduce by fission.

2. **Protista** are single-celled or microscopic animals, including algae.

3. **Plantae** include multicellular photosynthesizing organisms, higher plants, and multicellular algae.

4. **Fungi** are molds, yeasts, and fungi.

5. **Animalia** is composed of the invertebrates and vertebrates.

All animals discussed in this book belong to the phylum **Chordata**. All members of this phylum have the following four distinctive characteristics:

1. an embryonic **notochord,** which is usually replaced by the spinal cord

2. a dorsal tubular nerve cord running down the back side

3. **pharyngeal** or throat area gill slits

4. a rear area tail

Although they possess these structures at some embryonic stage, these characteristics may be altered or disappear in later stages of life.

All animals discussed in this book then belong to one of the following four classes:

1. **Mammalia** is the class of vertebrates that possess mammary glands, have a body that is more or less covered with hair, and possess a well-developed brain.

2. **Reptilia** is the class that includes snakes, lizards, and turtles. This class comprises cold-blooded vertebrates having lungs, an entirely bony skeleton, a body covered with scales or horny plates, and a heart with two atria and (usually) a single ventricle.

3. **Aves** is the class for birds. These are warm-blooded vertebrate animals with two legs, wings, and feathers and that lay eggs.

4. **Osteichthyes** are the bony fish. This is a large group of vertebrate animals that live in the water and have permanent gills for breathing, fins, and a body usually covered with scales.

Linnaeus's system of naming species is called **binomial nomenclature** and uses the Latin language. Each species has a name composed of two words. The first word is the genus and is written with the first letter capitalized. The second name is the species and is written with a small-cased first letter. The domesticated dog, for example, is named *Canis familiaria. Canis* is the genus and *familiaria* is the species. In some cases, an organism may have a subspecies name and would then use a **trinomial nomenclature.** An example of a complete classification for the domestic dog would be as follows:

1. Kingdom — Animalia — Animal kingdom

2. Phylum — Chordata — Animals with vertebrae

3. Class — Mammalia — Animals that suckle young

4. Order — Carnivora — Flesh or meat eaters

5. Family — Canidae — The dog family (includes wolves, fox, and coyotes)

6. Genus — Canis — dogs

7. Species — familiaria — domestic dog

SUMMARY

Most of the small animals used as pets or in laboratories evolved during the Paleocene and Eocene epochs of the Tertiary period 70 to 40 million years ago. Modern birds evolved earlier during the Cretaceous and early Tertiary periods.

The Cretaceous period, 135 to 70 million years ago, was the period of the dinosaurs. After the demise of the dinosaurs, smaller reptiles evolved. These small reptiles included snakes, lizards, and turtles, and they could compete with the large number of mammals that were also evolving.

The bony fish are the oldest vertebrate group. Some of them remain virtually unchanged since their evolution during the Devonian period 400 million years ago.

Domestication of animals probably began with the wolf about 12,000 years ago. As humans moved from being hunters to farmers, other animals were domesticated. Cats were thought to have been domesticated some 3,500 years ago. These domesticated animals served as pets and companion animals.

Today, the pet industry is large and growing. Fully 58.9 percent of American households have a pet or companion animal. A large industry exists to supply food, equipment, and health care for all of these animals. Millions of other small animals, birds, reptiles, and fish are used in exhibitions and laboratories. Together these industries provide jobs, careers, and opportunities for a significant segment of our population.

DISCUSSION QUESTIONS

1. When did life on earth begin according to scientific data? Please support your response.

2. When did plants first appear on earth? How do we know this?

3. Why weren't plants the first life to appear on earth?

4. During which time period were dinosaurs most abundant?

5. During which time period did the placental mammals appear on earth?

6. Which animals were first to be domesticated? Why?

7. What agency has responsibility for regulating the care and management of small animals?

SUGGESTED ACTIVITIES

1. Visit the library and use reference materials to trace the evolution of various classes of animals through time.

2. From references in the library, research the various theories as to why the dinosaurs became extinct.

3. Obtain a copy of the Animal Welfare Act. What are the regulations covered under the act?

ADDITIONAL RESOURCES

1. Dinosaurs and Birds
   www.enchantedlearning.com/subjects/dinosaurs/Dinobirds.html

2. Geologic Time Scale
   www.geosociety.org/pubs/public/geotime1.htm

3. Geologic Time—Enchanted Learning Software
   www.enchantedlearning.com

4. Geologic Time Chart
   www.littleexplorers.com/subjects/Geologictime.html

END NOTES

1. Excerpt from *Atlas of the Living World* by David Attenborough. Copyright © 1989 by Marshall Editions Ltd. Reprinted by permission of Marshall Editions Ltd. All rights reserved.

2. Excerpt from "U.S. Pet Ownership & Demographics Sourcebook," Center for Information Management, American Veterinary Medical Association, 1997.

3. Excerpt from "Pet Industry Fact Sheet," Pet Industry Joint Advisory Council, 1998.

4. Excerpt from "U.S. Pet Ownership & Demographics Sourcebook," Center for Information Management, American Veterinary Medical Association, 1997.

CHAPTER 2

Safety

OBJECTIVES

After reading this chapter, one should be able to:

- explain the importance of safety when working and playing with small animals.
- list ten types of diseases that can be transmitted from animals to humans.
- describe how to prevent becoming infected.
- describe proper restraint procedures when working with small animals.
- list guidelines for safety when handling dangerous chemicals and when working with small animals.

TERMS TO KNOW

evulsions

immune gamma globulin

intermediate hosts

intradermal

intramuscular

parasites

reservoir

sustenance

zoonoses

RISKS WITH SMALL ANIMALS

Owning and working with small animals can be very rewarding, but one needs to realize that there are a growing number of health risks. These risks are real, and everyone should take precautions.

Zoonoses are diseases that can be transmitted from animals to humans. Rabies is an example of a zoonosis that has been around for a long time. Ancient Greeks recognized that rabies can be transmitted by dog bites. The bubonic plague of the 1400s was caused by rats.

Ascarids (*Toxocara* species) and hookworms (*Ancylostoma* species and *Uncinaria stenocephala*) are common roundworms of dogs and cats. These parasites of dogs and cats can be passed on to humans. **Parasites** are organisms that live on or within another organism or host, which derives its **sustenance** (food or nourishment) from the host. Children are most at risk because they play with dogs and cats and play in the soil where these animals have been. The parasite is transmitted through contact with the animal's feces or contaminated soil. Symptoms of roundworms in humans are fever, headache, cough, and poor appetite. Play areas and sand boxes should be kept clean or covered and free of animal feces. Dogs and cats should be treated for roundworms. Deworming of dogs and cats is the most effective means of preventing contamination. See Figure 2–1.

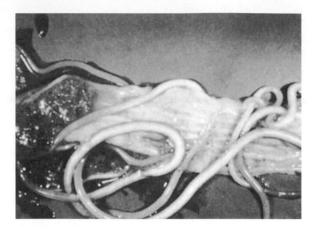

Figure 2–1 Common roundworms of dogs, Toxicana canis, *can be transmitted to humans. Transmission is usually through contact with feces or contaminated soil. Courtesy of North Dakota Department of Education.*

Toxoplasmosis is a disease produced by infection of the parasite *Toxoplasma gondii.* It can be carried by several different animal species, but it is usually spread to humans through cat feces or contaminated litter. Cats usually become infected by ingesting contaminated mice.

Most humans infected with *Toxoplasma* organisms will not develop symptoms because most people carry antibodies against the disease. People with immune system defects or those receiving immunosuppressive therapy can develop symptoms, however. Symptoms in humans include fever, headache, swollen lymph glands, cough, sore throat, nasal congestion, loss of appetite, and skin rash. Pregnant women also are at a greater risk. Toxoplasmosis can cause miscarriage, premature births, and blindness in the unborn child. Pregnant women should avoid cleaning litter boxes. Daily cleaning of litter boxes is important because the organism in the feces of an infected cat becomes infective after 36 to 48 hours. Rubber, disposable gloves should be worn when cleaning the litter box, and hands should be washed thoroughly afterward. Children's play areas and sand boxes should not be allowed to be used as litter boxes. See Figure 2–2.

Ringworm is a skin disease caused by a fungus; it is not caused by a parasitic worm as the name would imply. Dogs, cats, chinchillas, guinea pigs, rats, mice, and rabbits can be infected with ringworm and then transmit the infection to humans. Ringworm is spread by direct contact with a person or animal infected with the fungus. It can also be spread indirectly through contact with articles (such as combs or clothing) or surfaces that have been contaminated with the fungus. Symptoms of ringworm are the appearance of round, scaly, or encrusted lesions on the skin; hair is usually absent from these areas. The ringworm is infectious as long as the fungus remains present in the

Figure 2–2 Toxoplasmosis is usually spread to humans through cat feces and cat litter or contaminated soil. Caution should be taken when disposing of cat litter and cleaning cat litter boxes.

skin lesion. The fungus is no longer present when the lesion starts to shrink. Animals with rashes should be evaluated by a veterinarian. If the animal's rash is caused by a fungus, no one should be allowed to come in contact with the animal until the rash has been treated and heals and the animal has been given a bath. Iodine soap or antifungal drugs are used to cure the problem in humans. See Figure 2–3.

Psittacosis, or parrot fever, is a disease transmitted by caged birds of the Psittacine subfamily (parrots, budgerigars, and other related birds).

Humans can be infected by contact with the feces of contaminated birds or from fecal dust. Symptoms in humans include coughing, chest pains, fever, chills, weakness, vomiting, and muscular pain.

Preventive measures include the use of spray disinfectants to wet the feathers of birds and to eliminate lice and mites. In handling birds and cleaning cages, one should wear dust masks or protective face shields. A blood test can be used to confirm psittacosis, and antibiotics are used to treat the disease in both birds and humans. See Figure 2–4.

Cat-scratch fever is associated with cat scratches or bites. The affected area becomes swollen and is slow to heal, and the lymph nodes may swell and become tender and painful as well. The disease is

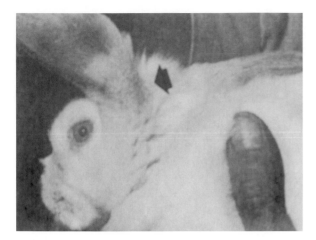

Figure 2–3   Ringworm infection appears as round, scaly, or encrusted patches on the skin. The fungus that causes ringworm can be transmitted to humans. Reprinted from Raising Rabbits, *copyright 1977 by Ann Kanable. Permission granted by Rodale Press, Inc., Emmaus, PA 18098.*

Figure 2–5   Cat scratches and injuries from cats can be avoided by properly handling the animals. The animal should be held close to the body and supported on the arm so that the animal feels secure. Courtesy of Brian Yacur and Guilderland Animal Hospital.

usually not serious and can be treated with antibiotics. Learning to properly handle cats can help prevent scratches and bites. See Figure 2–5.

Salmonellosis is a disease caused by infection from *Salmonella* bacteria. The bacteria can be transmitted to humans and animals; children and the elderly are especially at risk. Inflammation of the stomach and intestines results in vomiting, diarrhea, and abdominal pain. Symptoms usually appear 12 to 72 hours after infection, and the illness usually lasts four to seven days.

Salmonella bacteria can be carried by dogs, cats, birds, guinea pigs, hamsters, rats, mice, and rabbits. Pet turtles and other reptiles are a common source of infection in humans; the U.S. Food and Drug Administration (FDA) has banned the sale of pet turtles with a shell of less than 4 inches because small turtles can be carriers of *Salmonella.* Proper sanitation and husbandry practices will greatly reduce the risk of infection. See Figure 2–6.

Streptococcal bacteria infections cause sore throats, especially in children, and can be transmitted by dogs. Penicillin can eliminate the infection in both humans and dogs.

"Injuries from animal attacks, especially animal bites, are a major community health problem in the United States. While many bite injuries are minor, a few are fatal. National estimates range from 300 to 700 bites per 100,000 population each year. Since many bites go unreported, the actual number of animal bites is probably considerably higher than the nearly two million reported annually.

"Children five to nine years old are the primary victims of animal bites. Better than five percent of that age group report a bite every year; this is greater than the combined annual reports of measles, mumps,

Figure 2–4   Spray disinfectants used to wet feathers can eliminate lice and mites. Courtesy of Brian Yacur.

Figure 2–6   Proper sanitation is important in controlling Salmonellosis. Children should be taught to wash their hands after handling or playing with animals.

chicken pox, and whooping cough. Although children in this age group comprise less than nine percent of the population, they are the victims of nearly 30 percent of animal bites. Children nine to fourteen years old are the group with the next most bites.

"While injuries from animal attacks can become infected and fatalities do occur, the injury is more frequently psychological rather than surgical. Bite wounds can consist of lacerations, **evulsions** (a tear or pulled out wound), punctures, and scratches.

"The potential for infection varies; although less than five percent of dog bites become infected, up to 50 percent of cat bites do. This increased rate is attributed to the difficulty in effectively irrigating the puncture wounds that typically occur from cat bites.

"Generally, animal bites have not been shown to be a high risk for tetanus contaminations; very few cases are reported each year in the United States. However, this may be due in part to the high level of tetanus immunity in the general population.

"The most important consideration in the treatment of the patient bitten or scratched is whether rabies treatment should be initiated. If rabies is a consideration and the animal cannot be identified with certainty, immunization is recommended.

"Except in certain localities near the Mexican border and along the Atlantic Coast where rabies is especially prevalent, domestic dogs and cats in the United States have a low likelihood of rabies infection. Moreover, domestic animals can generally be identified and quarantined for observation or testing.

Bats and feral carnivores, particularly foxes, coyotes, skunks, bobcats, ferrets, and raccoons, are frequently infected with rabies in the wild and are the most common source of human rabies in the United States."[1] Successful vaccination programs that began in the 1940s caused a decline in rabies among domestic animals; however, as the number of cases of rabies among wild animals has increased, so has the number of cases among domestic animals.

In 1997, forty-nine states, the District of Columbia, and Puerto Rico reported 8,509 cases of rabies in animals and 4 cases in humans to the Centers for Disease Control and Prevention (CDC). Hawaii has never reported a case of rabies. The number of cases reported is an increase of 19.4 percent since 1976. See Figures 2–7 and 2–8.

Wild animals accounted for 93 percent of the reported cases; 50.5 percent of the rabies were in raccoons, 24.9 percent in skunks, 11.3 percent in bats, and 5.3 percent in foxes. The number of cases reported represents a 23.3 percent increase in rabies in

Reported Cases of Rabies, 1997

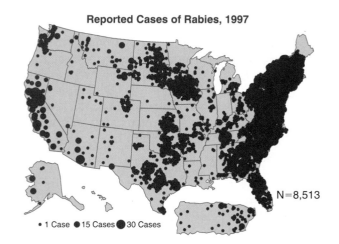

N=8,513

• 1 Case ● 15 Cases ⬤ 30 Cases

Figure 2-7   The number and location of reported cases of rabies in 1997.

Cases of Animal Rabies, 1955-1997

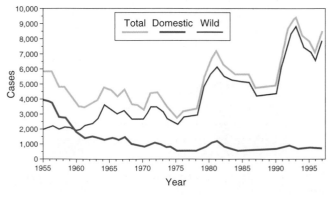

Figure 2-8   Cases of rabies have increased since 1955.

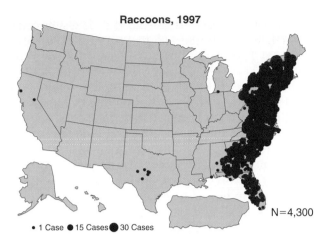

Raccoons, 1997

N=4,300

• 1 Case ● 15 Cases ● 30 Cases

Figure 2-9   The number and location of reported cases of rabies caused by raccoons in 1997.

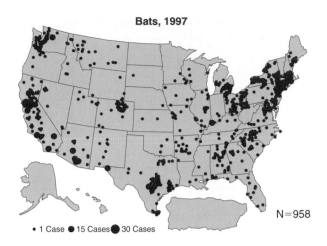

Bats, 1997

N=958

• 1 Case ● 15 Cases ● 30 Cases

Figure 2-11   The number and location of reported cases of rabies caused by bats in 1997.

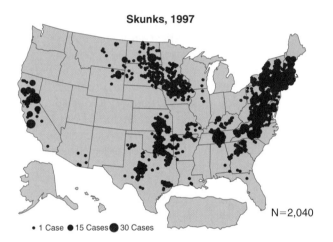

Skunks, 1997

N=2,040

• 1 Case ● 15 Cases ● 30 Cases

Figure 2-10   The number and location of reported cases of rabies caused by skunks in 1997.

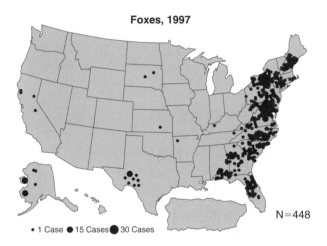

Foxes, 1997

N=448

• 1 Case ● 15 Cases ● 30 Cases

Figure 2-12   The number and location of reported cases of rabies caused by foxes in 1997.

raccoons, a 19.6 percent increase in skunks, an 8.7 percent increase in bats, and a 29.3 percent increase in foxes since 1976. See Figures 2–9 through 2–12.

Domestic animals accounted for 7.2 percent of all rabid animals reported in 1997. This was a 6.3 percent increase in cases since 1976. See Figures 2–13 through 2–15.

During the twentieth century, the number of human deaths attributed to rabies declined from 100 or more each year to an average of 1 to 2 each year. Four deaths in humans as a result of rabies were reported in 1997. (Refer to Chapter 7 for more information on rabies.)

If one is exposed to a potentially rabid animal, one should wash the wound with soap and water and seek medical attention immediately. Whether to treat a patient bitten by an animal for rabies will depend on the following criteria:

■ the species of the animal that caused the bite

■ the frequency of rabies in the community

■ the circumstances surrounding the bite (Was the animal provoked or was it an unprovoked attack?)

■ the behavior of the biting animal

■ whether the animal can be quarantined and observed

■ whether the animal's head (in the case of a wild animal) can be sent in for laboratory examination of the brain[2]

Animal care workers should be alert to animals that are vicious or potentially hazardous. These workers can protect themselves against rabies with a series of three **intradermal** (within the skin) pre-exposure injections over a three-week period. Workers who have received these injections would then need only two **intramuscular** (within the muscle) boosters if they become exposed. Nonprotected workers will require a much more expensive series of five intramuscular injections of rabies vaccine and one rabies

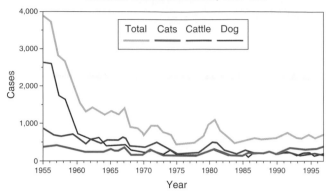

Rabies in Domestic Animals, 1955-1997

Figure 2-13   Cases of rabies in domestic animals have decreased since 1955.

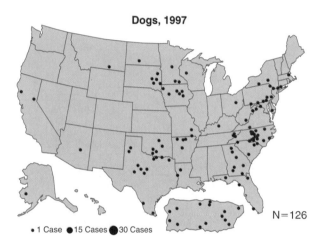

Dogs, 1997

N=126

• 1 Case ●15 Cases ●30 Cases

Figure 2-14   The number and location of reported cases of rabies caused by dogs in 1997.

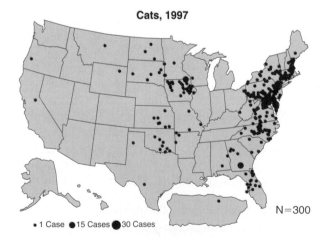

Cats, 1997

N=300

• 1 Case ●15 Cases ●30 Cases

Figure 2-15   The number and location of reported cases of rabies caused by cats in 1997.

immune gamma globulin (antibodies) injection over a twenty-eight-day period. Current vaccines are relatively painless and are given in the arm.

In general, proper treatment of animal bites and scratches is very important in preventing infection and more serious complications. First-aid procedures for bites and scratches include immediately and thoroughly washing the area with soap and water. This should be followed by medical attention; most doctors will give a tetanus shot if none has been received within the past five years. Proper bandaging of the wound and appropriate antibiotics will also be given. Proper handling of animals is also important to prevent injury to workers; all animal care workers should be instructed in proper techniques. To prevent animal bites, children should be taught to avoid unfamiliar animals and that any animal may bite if frightened, ill, or injured. Familiar animals may bite when startled or disturbed when sleeping or eating, or if they are handled roughly. Small children should always be supervised when handling or playing with animals. The proper handling of animals is covered in each chapter; for information on a specific species of animals, refer to that chapter.

Rocky Mountain spotted fever is found in all areas of the country and is primarily transmitted by the American dog tick, although six other species of tick have also been found to be carriers. The organism causing the disease, *Rickettsia rickettsii,* can be transmitted to humans by the bite of ticks. Ticks can be found in wild settings or can be brought into the household by dogs and cats.

The *Rickettsia* organism multiplies in the cells of the small peripheral blood vessels, causing coagulation of the blood, which is then forced into surrounding tissues. Fever, headache, nausea, vomiting, and skin rash are symptoms of Rocky Mountain spotted fever. Early diagnosis and treatment with antibiotics is important. Anyone who experiences these symptoms within two weeks of a possible tick bite or exposure should see a doctor immediately. If not treated, Rocky Mountain spotted fever can cause death. Normal grooming of animals after they have been outside will help locate and eliminate ticks. Attached ticks should be removed carefully; one must avoid breaking off the mouth parts in the skin because they may cause infection. Forceps and gloves should be used to avoid contamination of the fingers, and the area should be treated with antiseptic.

Lyme disease is a tick-transmitted disease affecting both humans and animals. In 1969, a grouse hunter in Wisconsin developed a strange, red rash around a tick bite; this is believed to be the first documented case of a rash that is often the first symptom of the disease.

In 1977, doctors observed arthritis-like symptoms in a group of children in the town of Lyme, Connecticut. These doctors are credited with giving the disease its present name. See Figures 2–16 and 2–17.

Lyme Disease: A Growing Threat

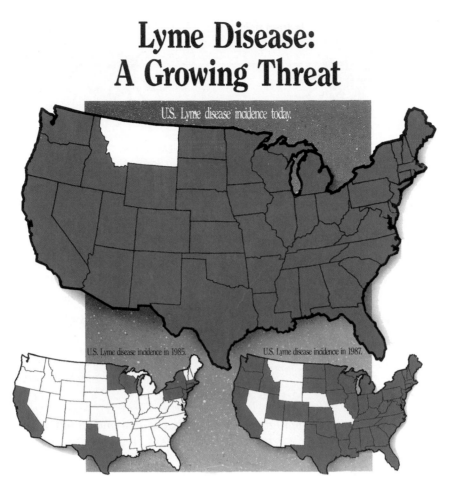

Figure 2–16 *Lyme disease is a growing threat. These maps show the spread of the disease since 1985. Courtesy of Fort Dodge Laboratories.*

Lyme disease was reported in thirty-one states in 1989 and in forty-seven states in 1992. In 1982, 492 cases were reported to the CDC; and more than 7,000 cases were reported in 1989. By 1998, forty-nine states and the District of Columbia had reported cases of Lyme disease. (See Appendix C for the reported cases of Lyme disease by each state in 1997.) Montana has never reported any cases. Almost 16,000 cases were reported in 1998. Since 1982, 125,000 cases have been reported to the CDC. Based on reported cases, during the past ten years, 90 percent of Lyme disease cases occurred in ten states: New York, Connecticut, Pennsylvania, New Jersey, Wisconsin, Rhode Island, Maryland, Massachusetts, Minnesota, and Delaware.

Five species of ticks have now been recognized as carriers of the infectious bacteria *Borrelia burgdurferi* and can bite and infect humans. See Figures 2–18 and 2–19. In the Northeast and Midwest, the pinpoint-sized Deer tick, *Ixodes dammini,* and in the Midwest and Southeast, the Black-legged tick, *Ixodes scapularis,* are the recognized carriers. White-tailed deer serve as hosts for these two species of ticks. With no hosts present, they feed on other animals, including humans. See Figure 2–20.

On the East Coast, white-footed mice serve as a **reservoir** (an immune host) for the *Borrelia* organism. These ticks have a two-year life cycle. They lay

Reported cases of Lyme disease, United States, 1997*

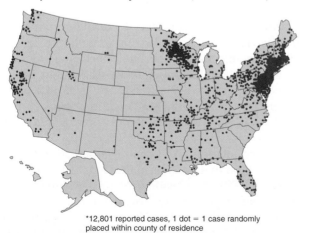

*12,801 reported cases, 1 dot = 1 case randomly placed within county of residence

Figure 2-17 *The number and location of reported cases of Lyme disease in the United States in 1997.*

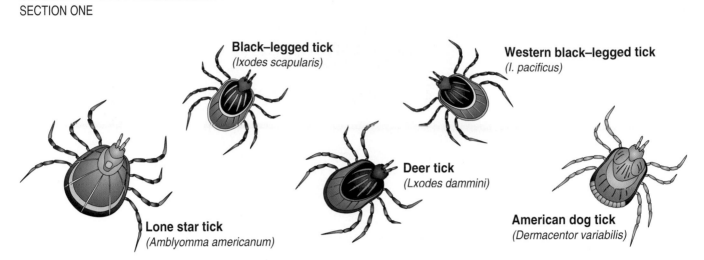

Figure 2–18 *These five ticks are most often associated with the transmission of the Lyme disease organism.*

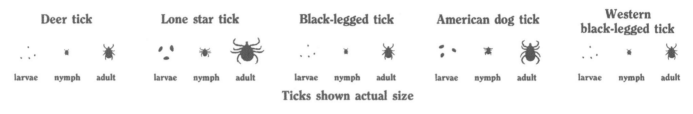

Figure 2–19 *The nymphs of ticks are very small and appear as specks on a human palm. Courtesy of Fort Dodge Laboratories.*

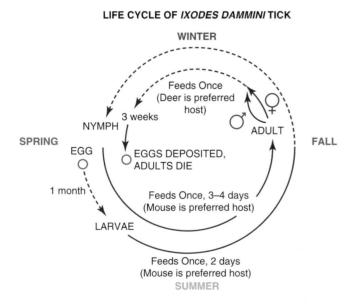

Figure 2–20 *The life cycle of the* Ixodes dammini *tick.*

their eggs in the spring, and the larvae that emerge feed on white-footed mice. The larvae then remain dormant during the winter and develop into nymphs the following spring. The tick larvae pick up the bacteria from the white-footed mouse and transmit it to humans.

In the Pacific coastal states of California, Oregon, and Washington, the Western black-legged tick, *Ixodes pacificus,* is recognized as the carrier of the *Borrelia* organism. The dusty-footed woodrat, a common California rodent, is the reservoir for the *Borrelia* organism in the West. Another species of tick, *Ixodes neotomae,* can also be a carrier. This tick does not bite humans, but it carries the organism that infects the woodrat. See Figure 2–21.

Two other species of ticks that may carry the organism are the American dog tick *(Dermacentor variabilis)* and the Lone star tick *(Amblyomma americanum).* The American dog tick is found in the Midwestern and Eastern United States, and the Lone star tick is common in the Southern United States. All six ticks have been found on ground-feeding birds and songbirds as well, which may explain why the disease has spread so rapidly over the United States.

The first symptom of Lyme disease is the distinctive skin lesion that appears in three to thirty-two days after the tick bite. The lesion begins as a small red spot that expands into a large irregular circular or oval-shaped area with a red outer border. Other lesions may appear as the bacteria spreads within the skin; these lesions may itch and burn. Variations of the distinctive symptoms may appear, including measle-like

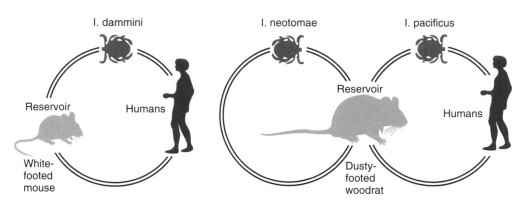

Figure 2–21 *In the Northeast and Upper Midwest, the white-footed mouse serves as a reservoir for the Lyme disease bacteria. In Western areas, the dusty-footed woodrat serves as one of the reservoirs for the bacteria.*

eruptions or hives; these lesions usually disappear in about three weeks. Flu-like symptoms, including aching muscles, stiff neck, fatigue, fever, chills, painful joints, nausea, coughing, sore throat, loss of appetite, swollen lymph glands, irritated eyes, and aversion to light, may also be present. These symptoms too may disappear within three weeks; however, the fatigue and muscle aches may last for several months. If treatment with antibiotics is not started, the bacteria can spread throughout the body and cause severe damage to the organs of the body.

Anyone experiencing Lyme disease–like symptoms should contact a doctor as soon as possible for diagnosis and treatment, which usually consists of the antibiotics tetracycline or doxycycline for adults and phenoxymethyl penicillin or amoxicillin for children. Two vaccines have been developed to prevent Lyme disease. One has recently been approved by the FDA and the second is still waiting approval. Both vaccines will require three injections over a twelve-month period. In regions where Lyme disease is prevalent, vaccines are recommended for at-risk persons between the ages of fifteen and seventy.

Ticks are found in grassy or wooded areas and can be brought into the home by dogs and cats that have been allowed outside. Adult ticks usually crawl around on a person for some time before they decide to attach themselves to the skin and are usually large enough to see and remove; however, the small larvae are so tiny that they may go unnoticed. Daily showers and personal hygiene are important in helping eliminate these small larvae. Pet owners should wash their hands and arms after grooming a pet that has been outside, and pet care workers should wear gloves when washing and grooming animals.

If ticks have attached themselves, they can be removed by grasping the tick as close to the head as possible and pulling it straight out, being careful not to squeeze or twist it. One must be careful not to break off the head or to crush the tick, because this may

serve as a means of infection. After the tick has been removed, the area should be disinfected with alcohol or another disinfectant.

Pet owners who allow their pets outside AND humane shelter workers are probably at greatest risk. Other persons at high risk are trappers and hunters who venture into grassy and wooded areas. Cases of Lyme disease have also been reported among taxidermists. All persons working in high-risk areas should wear rubber gloves and take necessary precautions.

The tapeworms *Echinococcus granuosus* and *Echinococcus multilocularis,* which are sometimes carried by dogs and cats, can cause a rare, but potentially fatal, disease, Alveolar Hydatid Disease (AHD). The tapeworm species *Echinococcus multilocularis* was originally confined to the Alaska coast, the cold tundras of Canada, and the eastern hemisphere; however, it has now been identified in the Dakotas, the Central Plains, and as far south as the Carolinas. Possible illegal shipments of animals have been linked to the spread of the disease. Fox, coyotes, and mice are the normal **intermediate hosts** (a host that the parasitic organism lives on or in during an immature stage) for this species of tapeworm, but the parasite is now appearing in domestic dogs and cats. The tapeworm species *Echinococcus granuosus* is found in California, Utah, Arizona, and New Mexico, where it normally uses sheep as the intermediate host. See Figure 2–22.

Infective *Echinococcus multilocularis* and *Echinococcus granuosus* eggs, accidentally ingested by humans, hatch in the small intestine, penetrate the intestinal wall, and migrate to the liver where they produce parasitic tumors or cysts. See Figures 2–23 and 2–24. Clinical signs may not appear until the growth of the cyst has progressed for several years. Surgical removal of the cyst is the preferred treatment, although in many cases the cyst has affected other organs and may not be removable. AHD has a fatality rate of 50 to 75 percent with or without surgery.

Echinococcus multilocularis and *Echinococcus granuosus* are difficult to identify because the eggs are identical to the *Taenia* tapeworm species that is common to dogs and cats.

Preventive measures need to be taken to minimize the risk of human infection. Pet owners and pet care workers in rural areas are at higher risk and should take precautions through personal protection and hygiene to prevent hand-to-mouth transfer of eggs.

PHYSICAL RESTRAINT OF ANIMALS

Animals may need to be restrained from normal movement for examination, collection of blood or other samples, administration of drugs, or therapy. Prolonged restraint should be avoided. Sick or injured animals may become frightened and confused. Special care may need to be taken to prevent injury to the animal and to the handler.

Figure 2-22   The spread of E. multilocularis *tapeworm in North America. Since 1992, the disease appears to have been confined to the cooler areas and doesn't show much further spread.*

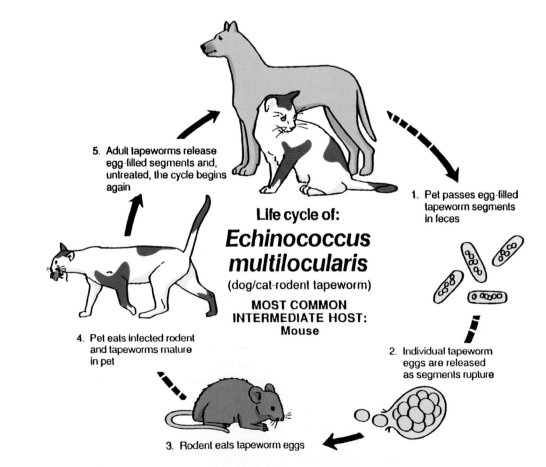

5. Adult tapeworms release egg-filled segments and, untreated, the cycle begins again

Life cycle of:

Echinococcus multilocularis

(dog/cat-rodent tapeworm)

MOST COMMON INTERMEDIATE HOST: Mouse

1. Pet passes egg-filled tapeworm segments in feces

2. Individual tapeworm eggs are released as segments rupture

4. Pet eats infected rodent and tapeworms mature in pet

3. Rodent eats tapeworm eggs

Figure 2-23   Life cycle of E. multilocularis *tapeworm. Courtesy of Miles Inc.*

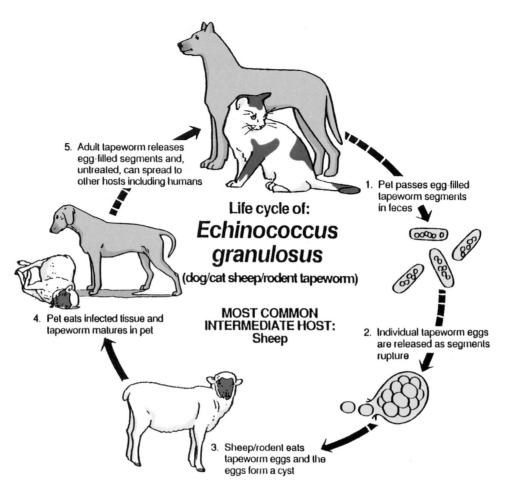

5. Adult tapeworm releases egg-filled segments and, untreated, can spread to other hosts including humans

1. Pet passes egg-filled tapeworm segments in feces

Life cycle of:
Echinococcus granulosus
(dog/cat sheep/rodent tapeworm)

**MOST COMMON INTERMEDIATE HOST:
Sheep**

4. Pet eats infected tissue and tapeworm matures in pet

2. Individual tapeworm eggs are released as segments rupture

3. Sheep/rodent eats tapeworm eggs and the eggs form a cyst

Figure 2-24   Life cycle of E. granulosus *tapeworm. Courtesy of Miles Inc.*

Cats are nervous animals, and when subjected to new situations or introduced to strange people, the normally good-natured family pet can bite and scratch.

When working around the head or neck of a cat, the animal can be wrapped in a blanket. This controls the legs of the animal. Cats can also be placed in a zippered-type canvas bag. The animal can be placed in the bag with its head exposed. To hold the cat's head, place the palm of the hand on the back of the head and grasp the head between the thumb and fingers.

Cats can be transported in small cages or "cat carriers." Care may need to be taken when removing the cat from the carrier to prevent injury to the cat or the handler. One may need to use gloves when reaching into the carrier. The cat should be grabbed by the scruff of the neck with one hand and then grabbed by the rear legs with the other hand. The cat can then be lifted up and removed from the carrier.

Cats can be caught and restrained by a catch pole, which is a device that consists of a five- or six-foot pole with a rope and noose attached. The noose should be slipped over the animal's head and front legs and around the chest. This tool should be used only when all other methods of restraint have failed.

Dogs can be restrained while the animal is in a standing or sitting position by placing one arm under the dog's neck with the forearm holding the head. The other arm is placed around the animal's body, and the animal is held close to the handler's body.

Dogs can be placed in a lying position by reaching over the back of the animal and grabbing the front legs with one hand and the rear legs with the other hand. The animal should be gently lifted and the animal's body allowed to slide down into a lying position. The handler's forearms can be lowered to rest against the animal's head and body to restrain the animal.

Muzzles can be used to prevent bites. Muzzles are placed over the dog's mouth. Different sizes are available, and care must be taken to ensure that the muzzle fits snugly and comfortably. A muzzle can be made from a narrow strip of gauze or cloth.

1. Make a loop in the material and slip the loop over the dog's nose and mouth.

2. Then tighten the loop by pulling on the ends.

3. Then cross the ends under the dog's jaw and bring them up behind the ears and tie them in a bow (see Figures 2–25 and 2–26).

4. The bow can be untied quickly by pulling on the ends.

An "Elizabethan collar" can be made from cardboard or plywood, or it can be purchased (see Figure 2–27). These collars prevent the dog from biting while being handled.

Catch poles can also be used to restrain dogs. The noose can be placed over the head of the animal and around the neck. The dog should be restrained in this manner only long enough to be tranquilized or subdued.

Rabbits can be picked up by grabbing the scruff of the neck with one hand and lifting up while placing the other hand under the rump. Rabbits can also be held with one hand holding the scruff of the neck and the other hand supporting the abdomen. Remember to support the hindquarters of the animal at all times. Rabbits seldom bite, but they can inflict injury with their hind legs.

Rabbits should not be placed on a smooth surface. They may injure themselves while trying to hop or move. Rabbits should be placed on a piece of carpet or on a towel.

Restraint boxes are also available that the animal can be placed in. The box should be small enough to prevent the rabbit from struggling or kicking. Usually, the head is extended from this type of restraint box.

Rats and mice that have not been handled may bite, and special care may be needed when handling them. Gloves can be used to place the animals on a piece of carpet or on a piece of screening. Once they are placed on the carpet or screening, they can be picked up with bare hands. The animal should be grabbed gently by the tail. Don't grab the end of the tail, but grab as close to the body as possible. The animal will try to pull away, and while it is in a stretched-out position, use your other hand to grab all the loose skin on the neck and shoulders that you can and lift the animal up. The animal should be restrained in this manner and unable to bite. Special restraining boxes are also available for rats and mice.

OTHER RISKS

Many of the chemicals used in working with small animals are potentially dangerous. All workers should be instructed in the safe use, storage, and disposal of chemicals and containers. Chemicals such as pesticides can enter the body in several ways; the skin is the most common portal of entry into the body. To prevent absorption through the skin, the worker must wear protective clothing and equipment.

Gloves should be worn by workers when handling chemicals. Plastic or rubber unlined gloves that

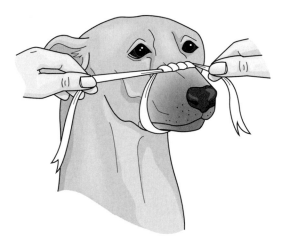

Figure 2-25   Applying muzzle material around dog's nose.

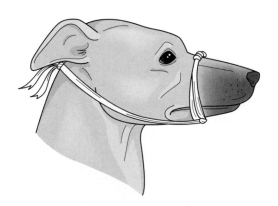

Figure 2-26   Securing muzzle behind the ears.

Figure 2-27   Elizabethan collar.

are liquid-proof are commonly used when handling chemicals. Canvas, cotton, or leather gloves should not be worn because they can be soaked with the chemicals and provide a continuous source of exposure. Leather gloves should be worn when handling animals that could possibly bite or scratch. Various length gloves are available for different animals and situations.

Unlined, lightweight, rubber boots should be worn when handling pesticides; leather or canvas shoes should not be used. Boots with nonslip soles should be worn when working on wet floors. Shoes with steel toes may be beneficial to workers handling large cages and heavy equipment.

Workers' eyes should be protected at all times when working with chemicals. Tight-fitting goggles with antifog lenses and ventilation holes are recommended as the best protection; however, full-face shields also provide protection against chemicals splashed into the face. Safety glasses protect the face and eyes from possible scratches and bites.

Workers in areas of high noise levels should wear ear protectors; damage to hearing can result from prolonged exposure to loud noises.

In some situations, the head and neck area may need to be protected from chemical exposure. Plastic safety helmets, broad-brimmed hats, or baseball caps with plastic sweat bands are recommended; cloth sweat bands are not recommended.

Uniforms, coveralls, or lab coats should be worn when working with pesticides and toxic substances and will protect workers from animal scratches. Protective clothing should be laundered daily and not worn outside the facilities.

Respirators provide protection from inhaling toxic substances; workers should read labels to determine proper use of respirators.

GENERAL GUIDELINES FOR SAFETY

No safety guidelines can cover all situations, so it is extremely important for workers and students to follow all instructions from supervisors and teachers and on all labels when handling chemicals. The proper procedures in treating and reporting accidents should be taught. The following are some general guidelines to follow for safety in the workplace or school lab.

1. Always wear protective clothing and equipment when the job requires it.

2. Always wash protective clothing and equipment after use to prevent contamination.

3. Wash hands and face after completing a job to make sure all chemical residue is removed.

4. If required, shower after completing a job so that chemical residue is completely removed from the body.

5. Wash hands frequently while working with animals, especially if working with different species and in different areas. This will prevent contamination to other animals and will prevent self-contamination.

6. Keep hands away from the mouth, eyes, and face when working with chemicals and animals to prevent self-contamination.

7. Do not consume food or drinks in areas where contamination could occur, and do not store these items in areas where contamination could occur.

8. Remove uniforms, lab coats, and coveralls when leaving an area that could be contaminated.

9. Never wash uniforms, lab coats, or coveralls with regular clothing.

10. Make sure all containers are correctly labeled to prevent the misuse of a chemical.

11. Dispose of all chemicals and their containers according to proper procedure or instructions on the label.

12. Students and small animal workers should be instructed in the proper methods of handling small animals.

13. First-aid kits should be kept in the work area or instructional area. Workers and students should be made aware of the location of first-aid kits.

SUMMARY

Working with small animals is an interesting and rewarding career; however, there are risks that everyone should be aware of.

Several diseases can be transmitted from animals to humans. The potential for infection in caged, domestic animals is low, but when dealing with animals that have been exposed to the outdoors, the risks increase. Several of these diseases can be life-threatening, so it is important that workers and pet owners are knowledgeable about the prevention and treatment of them.

Several chemicals and pesticides may be used when working with small animals. Proper safety is important to prevent accidents.

Safe handling techniques for all small animals are important in preventing scratches and bites to handlers, as well as preventing injury to the animal.

Most injuries and accidents dealing with small animals can be prevented by being aware of safety guidelines and proper procedures.

DISCUSSION QUESTIONS

1. What is a parasite?

2. What is a host? What is an intermediate host?

3. What is a zoonosis?

4. What is the life cycle of the Deer tick and the Black-legged tick?

5. What is the life cycle of the Western black-legged tick?

6. Why did Lyme disease spread so rapidly throughout the United States?

7. Why are the tapeworm species discussed in this chapter so hard to identify?

8. Why are children and the elderly more at risk from some of the diseases and injuries covered in this chapter?

9. Why shouldn't a pregnant woman handle a cat litter box?

10. Why is early diagnosis and treatment important in Rocky Mountain spotted fever?

11. What determines whether someone should be given immune gamma globulin injections for rabies?

12. What are some safety guidelines to follow when working with small animals?

SUGGESTED ACTIVITIES

1. Visit the school or community library and do some further research concerning Lyme disease. How can Lyme disease be prevented? Can the disease be eliminated? What are some difficulties in dealing with Lyme disease?

2. Check with local health officials and determine the incidence of animal bites in your area.

3. Research the tapeworm species covered in this chapter and the diseases caused by them. How can they be prevented or eliminated?

4. Make a list of proper methods for storing and disposing of dangerous chemicals.

5. Bring in labels from pesticide containers. What is the proper mixture of the material? What safety practices are listed on the label? How should the container be disposed of?

ADDITIONAL RESOURCES

1. Occupational Safety and Health Administration (OSHA)
www.osha.gov

2. McGill Laboratory Safety Manual, 3rd Edition 1994
www.mcgill.ca/eso/labman/1.htm

3. Bowling Green State University, Laboratory Safety and Chemical Hygiene Plan
www.bgsu.edu/offices/envhs/labsafe.htm

4. Centers for Disease Control and Prevention
www.cdc.gov

5. Division of Vector-Borne Infectious Diseases
National Center for Infectious Diseases
Centers for Disease Control and Prevention
P.O. Box 2087
Fort Collins, Colorado 80522
(970) 221-6400
www.cdc.gov/ncidod/dvbid/lymeepi.htm

6. Guide for the Care and Use of Laboratory Animals
www.nap.edu/readingroom/books/labrats/

END NOTES

1. Excerpt from Wishon, P. M., and A. Huang. 1989. "Pet associated injuries: The trouble with children's best friends." *Children Today* 18 (May/June): 24–27.

2. Kreisle, M.D., Ph.D., "Dealing with Bites and Other Occupational Hazards," Seminar of the Societal Issues in Animal Management, Purdue University, May 19, 1991.

Small Animals as Pets

OBJECTIVES

After reading this chapter, one should be able to:

- list the questions a person should ask before buying a pet.
- compare the different options available for obtaining a pet.
- describe what is meant by responsible pet ownership.
- list the important lessons children can learn from their pets.
- describe the benefits pets can have for the elderly.
- define euthanasia.
- describe the importance of euthanasia to a pet owner.

TERMS TO KNOW

allergies   neutered
euthanasia   spayed
entropion

CHOOSING A PET

People have kept domesticated animals for more than 10,000 years; dogs were probably the first animals domesticated, followed by cats. Dogs have been called man's "best friend," and until recently, more dogs were kept as pets than any other animal; cats now outnumber dogs, however.

Before World War II, many pets were hunting, farm, or show animals. Space and poverty limited animal ownership in cities. Following World War II, more people were moving from the farms to the cities, and people enjoyed increased incomes and prosperity. The demographics of the family changed. People were marrying and having families later in life and having fewer children. More people were electing to remain single. As children grew up and left home, older couples were experiencing the loneliness of the "empty nest." Pets, therefore, provided companionship in these fast-paced, changing times. Pets offer nonjudgmental acceptance, love, and companionship no matter what the age of a person. See Figure 3–1.

Unfortunately, many pets are abandoned each year because they don't fit in with the lifestyle of the family. Many pets lose their "cuteness" as they get older,

Figure 3–1   Pets provide companionship in today's fast-paced society. Courtesy of Tom Schin.

cost too much to feed, or put too much time demand on the family. Millions of unwanted animals pass through animal shelters each year, a result of uncontrolled breeding and abandonment. Many of these animals are eventually put to sleep (euthanasia, see Glossary). Experts estimate that each year more than 10 million pets are put to death nationwide.[1] Many more pets are abandoned to fend for themselves, and stray animals become problems in many areas.

No matter how cute an animal appears, careful thought should precede the purchase of a pet, because a hasty decision may lead to later disappointment.

When selecting a pet, a person should consider the following questions carefully before making a purchase:

1. Where should I obtain a pet?

2. How much space do I have for a pet?

3. What kind of animal does my lifestyle allow?

4. How much will the animal cost?

5. What will the future bring?

6. Does everyone in the family want a pet?

7. What kind of personality do I have?

8. Is this animal a fad or status symbol?

9. What am I going to use the animal for?

Where Should I Obtain a Pet?

The most obvious answer is a pet shop; however, other options are available.

Pet shops are usually conveniently located where consumers can purchase a pet and pet products. Pet shops serve as intermediaries between the breeder and consumer and make a profit on the difference between the buying price and the selling price of an animal. Animals in pet shops may come from anywhere in the country. Consumers have little knowledge of where the animals come from, the conditions under which they were raised and shipped, or any knowledge of pedigree. Dogs and cats may be registered, but were the parents healthy and bred by reputable breeders?

Purebred breeders usually are in business for the long haul. Their success depends on their reputation. If they produce animals that possess inherited defects or are ill-mannered, their business will suffer. Most breeders will guarantee satisfaction and are willing to work with the customer.

Purchasing direct from the breeder will save money by eliminating the middle-person. When visiting the breeder, the buyer can observe the parents and determine traits that can be passed on to the offspring.

Animal and humane shelters are where many abandoned animals end up. Many of these animals would make excellent pets if given the opportunity. A person would not only get a wonderful pet, but would also save the animal's life. The cost of such an animal will be much less than one purchased from a pet shop or breeder. Shelters usually charge only enough to cover expenses. There may be a charge to have the animal **spayed** or **neutered,** and most shelters require this procedure. Information on the animal will not be available unless the previous owner can be contacted. See Figure 3–2.

Friends and neighbors offer one of the easiest and cheapest sources to obtain a pet. Dogs and cats running loose often result in unwanted pregnancies and animals; friends and neighbors may give these animals away free. The sire of these litters may be unknown, but some of the traits of the dam can be observed to get an idea of what the animal will be like when grown.

How Much Space Do I Have For a Pet?

A small one-room apartment doesn't allow much space for active animals. Living in rural areas, on the other hand, may provide large acreage for an animal to roam. A person needs to be familiar with the space requirements of an animal before investing money in it.

Figure 3–2 Dogs do not have to be purebred to provide loving companionship. "Mutts" can make excellent family pets as well as companions. Courtesy of Tracey A. Carter.

What Kind of Animal Does My Lifestyle Allow?

Animals require different amounts of space, time, and care. Dogs need daily exercise and grooming; this varies considerably among the different breeds. Longhaired dogs require more grooming whereas active dogs, like the greyhound, require more exercise. It is important that people be familiar with the breeds they plan to obtain and are sure the pets will fit into their lifestyle. Cats, on the other hand, require less care and time.

If a person's lifestyle doesn't allow for a dog or cat, then a hamster or gerbil may make a more desirable pet. Provided with enough food and water, they can be left alone for several days and will fit well into small apartments.

How Much Will the Animal Cost?

Many people don't look past the initial purchase price of their pet. Puppies and kittens don't cost much to care for, but as they grow to adult size, food can easily cost from $25 to $40 per month for dogs and $7 to $16 for cats.[2] In addition to feeding, there may be expenses in the form of toys, grooming, and veterinary bills. Many cities and counties also have taxes on animals. Because of high veterinary costs, many people are opting for pet health insurance. Insurance plans vary but cost from $100 to $400 per year.

According to the American Kennel Club and *USA Today* research, the cost of keeping a pet dog for an average of 11 years is an average of $14,000: broken down, it translates into $4,020 for food; $3,930 for veterinary services; $2,960 for grooming, toys, equipment, and housing; $1,070 for flea and tick treatments; $1,220 for training; and $1,400 for miscellaneous expenses.[3]

Experts say pet owners should expect to pay about $550 a year caring for their dogs, about $390 for their cats, $244 for guinea pigs, $188 for rabbits, $130 for birds, $96.47 for hamsters, and $89.30 for fish. First-year costs will be higher because owners will need to purchase new equipment. Yearly costs include such basics as food, shelter, and routine visits to the veterinarian.[4]

What Will the Future Bring?

A person should plan ahead. Will that cute, cuddly animal still be wanted when it matures? Will the family continue to be able to afford a pet? If the animal was obtained for the children, will they be around or will Mom and Dad be the ones to care for it? Can the son or daughter take the animal to college when the time comes?

Does Everyone in the Family Want a Pet?

A pet should not be brought into the family unless every member agrees to it and is willing to share in the responsibility for the animal's care. An animal brought into the family as a gift or surprise may become an unwanted and uncared for pet.

Another factor to consider is that of **allergies.** Allergies to certain animals and animal hair or dust are fairly common. It is important to determine whether any member of the family is allergic to a particular species of animal. Finding this out can prevent the disappointment of having to give up a pet.

What Kind of Personality Do I Have?

Animals have personalities just like people. It is important that the pet one decides to purchase complements his or her personality. Dogs vary greatly in their personalities; some are docile, whereas others can be active and playful. These personality traits are characteristics of the breeds; therefore, it is important to be familiar with the breed of animal you intend to obtain. Cats are less active than dogs; other animals are usually kept in pens or cages and thus may be ideal for less active, quiet personalities.

Is This Animal a Fad or Status Symbol?

Many times, animals or special breeds of animals have been fads or status symbols. When the fad fades away, they are unwanted or abandoned animals. An example is the wrinkled-skinned Shar-Pei. See Figure 3–3. The following appeared in *Atlantic Monthly* magazine: "Twelve of these Chinese guard and fighting dogs were brought into the United States in the late 1960's. Offspring sold for as much as $10,000. Sloppily bred from a limited gene pool, these dogs have developed autoimmune deficiencies leading to severe skin disorders, and faulty bites that sometimes make it impossible for them to eat. Their hips are bad; their eye lids need to be surgically cut from over the eyes. Many display the foul temper of a punch-drunk fighter. Since the collapse of the market on Shar-Pei, many are appearing in animal shelters, abandoned by owners no longer interested in supporting a dog that has fallen from vogue."[5]

Shar-Pei can develop a condition called **entropion,** which causes the eyelid to roll in toward the eye, rub against the cornea, and cause irritation to the cornea. Watery eyes, infection, and corneal ulcers can occur. Surgical correction is required, and animals with this condition should not be bred because the condition is believed to be hereditary.

Another example of a fad breed is the Dalmatian. The popularity of Dalmatians increased rapidly since Disney released the animated *101 Dalmations* in 1984 and reached a peak in 1993 after the re-release of the animated film in 1991. The rank of registrations for Dalmatians rose from thirty-fifth in 1984 to ninth in 1993. Since 1993, the number of registrations has declined, and the Dalmatian now ranks thirtieth. Unfortunately, many of the animals purchased during the peak population of their breed found their way to animal shelters, were abandoned, or were euthanized once they grew up and were no longer cute little puppies.

Many organizations exist around the country to help rescue Shar-Pei, Dalmatians, and other breeds from animal shelters and to help find homes for unwanted animals.

People need to take some time and give serious thought to the decision to obtain a certain breed of dog. People should check out the different breeds, become familiar with their characteristics and needs, and ask the questions posed in this unit before obtaining an animal.

What am I Going to Use the Animal For?

This consideration is especially important with dogs. If the animals are going to be used for a particular purpose, they should be obtained from a breeder who raises them for that purpose. Several breeds, long noted for their hunting abilities, have been bred only for show. After several generations, the breed may no longer have the characteristics to hunt. Good examples of this are the Irish Setter, Cocker Spaniel, and Poodle breeds. It is important to be familiar with the breeder and the pedigree of his or her animals.

CHILDREN AND PETS

Almost all children sooner or later will beg their parents for a pet; most will want a dog because of its playfulness. Bringing a pet into the family should be based on the age and maturity of the child. Many professionals believe that a child must be at least three years of age to be considered mature enough to have a pet. Prior to age three, children are probably too young for the responsibility a pet requires. Children younger than three would treat an animal like a stuffed toy and not be able to handle an animal gently or with the care needed. See Figure 3–4.

Kittens, hamsters, gerbils, rabbits, or mice are good pets for younger children because they are less demanding. As the child learns the basic needs and behavior of these pets, and as the child learns to be consistent in his or her behavior, then the child may be ready for a more demanding pet.

Children can learn many things from their pets, such as responsibility, social skills, respect and compassion, and grief and coping with loss.

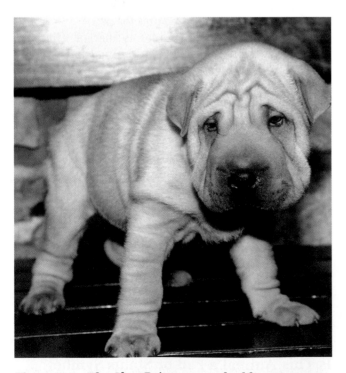

Figure 3–3 The Shar-Pei, once a valuable status symbol, are appearing in shelters since the collapse of the market on these wrinkled-skin dogs. Courtesy of Stacey Riggert.

Responsibility

Responsibility can be learned from taking care of a pet. This in turn develops one's self-esteem and self-

Figure 3–4 Children can learn many things from caring for a pet. Courtesy of Carolyn Miller.

worth. By caring for an animal, children learn to be competent individuals. These characteristics can be reinforced by praising children for the way they care for their pets.

Social Skills

All children will experience the feeling of being unwanted or unloved. A pet will always accept children and provide them with a sense of being loved and wanted.

By learning to understand a pet's body language, emotions, and needs, children develop a stronger sense of responsibility and personal worth.

Children can develop skills that are important for when they become adults. Many of these same skills are important in caring for children, the elderly, and the disabled.

Respect and Compassion

Respect and compassion can be developed by the daily caring for a pet. A child will recognize changes in the animal's behavior, such as illness, depression, or annoyance at the child. By recognizing these changes and having consideration, a child can transfer these traits to others. The child also learns to appreciate the effect his or her actions have on others. Erika Friedman, a professor of health and nutrition sciences at Brooklyn College of the City University of New York, in an article for the *Courier*, states, "In early childhood pets provide a connection to the natural world and are instrumental in teaching respect for other living things. For slightly older children, the responsibility of a pet may be instrumental in developing self-esteem, self confidence, and independence."[6]

Grief and Coping with Loss

Grief is the normal response to the loss of a pet, especially if the pet had become an important part of one's life. For many children, the loss of a pet may be the first time they experience death. The feelings and emotions that children experience should be discussed. Discussing the loss, and perhaps performing a ceremony or burial, may make children feel better.

In the grieving process, most people go through five stages. During stage one, there is shock, denial, and isolation. The loss may not be accepted, and one may block the loss out of one's mind or deny that the death has occurred. During the second stage, there is anger. The anger may be aimed at family members, friends, or the veterinarian who couldn't help the pet or performed the euthanasia. There may also be feelings of guilt or fear during this stage. During the third stage, there are feelings of helplessness. One may question whether he or she fed and cared for the pet correctly, whether the pet would have lived if taken to the veterinarian earlier, or what could have been done differently. During the fourth stage, there are feelings of depression, sadness, hopelessness, and regret. There may be worry about what to do with the pet and what will be the costs involved. The final stage is the acceptance of the loss.

Support of friends and family is important in dealing with a loss. Talking and discussing one's feelings are important. Discussions with the veterinarian may be helpful so that one understands the facts and procedures that were done to help the pet.

Everyone deals with loss differently, and the time involved in the grieving process varies with individuals. Children usually get over a loss more quickly than adults, but that does not mean they grieve any less. When dealing with the loss of a pet, children need to understand that most animals do not live as long as humans and that they do not fear death. Children need to focus on other activities to get their mind off the loss. After the child has accepted the loss, it may be time to replace the pet.

PETS AND PET THERAPY

Researchers are beginning to realize that pets are important in helping people overcome illness. Pets can help the elderly live longer, happier lives by helping to lower their blood pressure and helping them to recover from heart attacks and other serious illnesses. "Heart rate, for example, is lower when people sit quietly or read aloud in the presence of a friendly animal, than when doing so alone, according to Erika Friedmann, associate professor of Health Sciences at Brooklyn College. A year after their heart surgery, Friedmann found higher survival rates for patients with pets in their homes than for those who had none, even after accounting for overall health differences among patients. Elderly people who own pets also make fewer visits to doctors than do those who are without animal companions, possibly because the animals mitigate loneliness."[7] See Figure 3–5.

When University of Pennsylvania researchers studied a group of seriously ill heart patients, they found that pet owners had much better survival records. In the year the study lasted, the death rate for patients who did not own pets was 28 percent, whereas pet owners had a death rate of 6 percent.[8] A study in the December 15, 1995, issue of the *American Journal of Cardiology* found that dog owners, in particular, were significantly less likely to die within 1 year of a heart attack, compared with heart attack survivors who did not own dogs.[9]

Other studies have documented that petting a dog or cat lowers blood pressure in hypertensive patients and that older people show improved alertness when caring for a pet. Pets often seem to be especially valuable for some people experiencing major life changes, such as unemployment, illness, or the loss of a loved one.[10] A study published in the April 1999 edition of the journal *AIDS Care* found that people with AIDS who have pets are less likely to suffer from depression than people with AIDS who do not own pets.[11] Several studies have also demonstrated that pet owners tend to have lower cholesterol and blood pressure levels than non–pet owners and are therefore at a reduced risk for cardiovascular disease.[12] A study pub-

lished in the March 1999 *Journal of the American Geriatrics Society* showed that senior citizens who own pets are less likely to be depressed, are better able to tolerate social isolation, and are more active than those who do not own pets.[13]

H. Ellen Whiteley, DVM, in an article in the *Saturday Evening Post* states, "Pets help keep us young: they decrease loneliness, and they give us the opportunity to be needed. Pets also offer a healthier lifestyle by stimulating us to exercise while we take care of them."

Dr. Whiteley continues, "the elderly may benefit the most. At a time in their lives when they have returned to depending on others, they need to feel a sense of responsibility. Pets fulfill this need because they depend on their owners for care and attention. In return, the pets offer love and unqualified approval."[14]

Studies also indicate that pets can have a beneficial effect on the disabled. Dogs can be trained to perform numerous tasks, including opening and closing doors, turning switches on and off, pulling wheelchairs, carrying packages, and assisting individuals who are blind. See Figure 3–6. Not only can they perform these tasks, but they contribute to a person's self-esteem and psychological well-being. Dogs and other animals have been used in programs in nursing homes to lift the spirits and encourage activity and involvement by patients. Severely withdrawn patients, especially children, have shown responses when pets have been introduced to their therapy. Pets are also beneficial for children coming home from school before their parents get home. The children have their pets to care for and enjoy until their parents get home.

A study in the April 3, 1996, issue of the *Journal of the American Medical Association* concluded that trained service dogs can be highly beneficial and po-

Figure 3–5   Pets can be valuable in helping the elderly lead longer, happier lives. Courtesy of Deborah Rhodes (left) and Amy Tucker (right).

Figure 3–6 Some companion animals, especially dogs, can be very valuable to people with disabilities. Dogs can be trained to provide assistance to people with blindness, deafness, and to many who are confined to wheelchairs. Courtesy of the Guide Dog Foundation for the Blind, Inc.

tentially cost-effective components of independent living for people with physical disabilities.[15]

EUTHANASIA

Perhaps the kindest thing you can do for a pet that is severely sick or injured is to have your veterinarian induce death quickly and humanely; this is called euthanasia. The decision to have your pet euthanatized is a serious one and never easy to make.

Your relationship with your pet is special. When you acquired your pet, you assumed responsibility for its health and welfare. Fortunately, few owners are faced with making life-or-death decisions for their pets; however, such a decision may become necessary.

Euthanasia may be the most difficult decision you will ever make regarding your pet. Your veterinarian, family, and friends can assist in the decision-making process. You should consider what is best for your pet, for you, and for your family. Quality of life is im-

portant for pets and people alike.

If your pet can no longer get around comfortably; if your pet isn't eating or willing to eat; if your pet is no longer aware of or appreciates your companionship; if your pet is apt to bite family members, strangers, or children; if your pet can no longer respond in normal ways; if there is more pain than pleasure in your pet's life; or if the care and treatment become too much financially, then euthanasia may be a necessary option.

Euthanasia is accomplished with an injection of a solution that stops nerve transmissions and causes muscle relaxation. The veterinarian may sometimes administer a tranquilizer to calm and relax your pet so that the euthanasia procedure can be carried out more humanely and peacefully. The euthanasia solution is usually injected into a vein in the foreleg. Six to twelve seconds after injection, the animal will lapse into a deep unconsciousness. Death will come quickly and painlessly.

OVERPOPULATION

Each year, approximately 27 million dogs and cats are born in the United States; 10 million are euthanized each year as unwanted or abandoned. This overpopulation is a result of natural matings, matings for commercial purposes, and parents who want their children to experience "the miracle of childbirth."

To solve the overpopulation and unwanted animal problem, the number of puppies and kittens born must be reduced. To accomplish this, three events need to occur:

1. People must be educated to the problem.

2. Animal control laws must be passed and existing laws enforced.

3. People must recognize the importance of spaying and neutering.

Many unwanted animals end up in animal shelters where they are euthanized; the shelters have taken the responsibility to eliminate the problem of unwanted animals. Adults and children need to be educated and take responsibility for controlling the breeding of their dogs and cats.

Licensing of animals, leash laws, fence laws, and other animal control laws need to be enforced. Other laws and ordinances controlling the breeding of animals may also need to be passed.

Spaying and neutering are surgical procedures that prevent animals from reproducing. Spaying, where the ovaries and uterus are removed, is performed on females. This prevents the female from coming in heat and from becoming pregnant. Neutering, where the testicles are removed, is performed on males. This prevents the male from producing sperm and impregnating a female.

The procedures have been controversial but, if performed at a young age, can cost less than many other surgeries. Spaying is recommended at three to five months of age for females; neutering of males is recommended at seven to nine months of age, just before they reach sexual maturity.

During the surgical procedure, the animals are fully anesthetized, so they feel no pain. Some pain or discomfort may be experienced later, but all signs of discomfort disappear within a few days.

Spaying eliminates potential problems and risks associated with pregnancy and birth. Females are also less susceptible to mammary cancers, and uterine infections are eliminated. Spaying also eliminates the discomfort and frustration experienced by a female in heat.

Neutering eliminates testicular tumors and reduces the risk of prostate disorders. Neutered males no longer roam the community and are less apt to be hit by cars or injured in fights. Neutered dogs no longer practice their "mounting" behavior, which at times can be annoying or embarrassing, and they no longer spray urine to mark their territory.

SUMMARY

Since World War II, we have experienced a pet explosion in the United States. Pets provide companionship in today's fast-paced society. Unfortunately, many of these pets get abandoned each year. Millions of unwanted animals pass through animal shelters each year, a result of uncontrolled breeding and abandonment.

Pets can have an important influence on their owners, both young and old. Children can learn important lessons that can help them in later years, and pets can have therapeutic benefits to the elderly.

Pets have also been of great value to people with disabilities. The deaf, blind, and those confined to wheelchairs or bed can receive practical and psychological benefits.

People must be educated to the fact that they are responsible for their pets and for the controlled breeding of their pets.

Approximately $1 billion is spent each year in euthanizing unwanted animals.

DISCUSSION QUESTIONS

1. Are there unwanted animals in your area? What happens to them? Does a problem exist with unwanted animals? What should be done to correct the problem?

2. What are the questions one should ask before obtaining a pet?

3. What are four sources from which you can obtain pets? What are the advantages and disadvantages of each source?

4. What are some of the various lifestyles that people have? What pets might best fit those lifestyles?

5. What are some other costs of owning a pet besides the initial purchase price?

6. What lessons can children learn from having a pet? Are there other lessons not listed in the text?

7. Do animals have personalities? What are some of the personalities you have observed?

8. What are some other animal fads or status animals of recent years?

9. What do we mean by responsible pet ownership?

10. What do we mean by the terms *spay* and *neuter*? Should we have our pets spayed and/or neutered?

11. What are the benefits of pets to the elderly?

12. What are some of the therapeutic uses of pets?

13. What is euthanasia and why should it be considered by a pet owner? How would you counsel someone who is dealing with the loss of a pet?

SUGGESTED ACTIVITIES

1. With the approval of the instructor, invite pet shop owners, various pet breeders, and animal shelter personnel to your class to make presentations.

2. Visit a veterinarian and discuss with him or her the process of euthanasia and how pet owners facing the loss of a pet are counseled. Report the information back to the class.

3. Get involved with groups that take animals into nursing homes or hospitals. If no groups are in your area, consider starting such a program.

ADDITIONAL RESOURCES

1. American Veterinary Medical Association (AVMA)
   www.avma.org

2. Mayo Foundation for Medical Education and Research
   www.mayohealth.org

3. Dog-Play
www.dog-play.com

4. Pet Life
www.petlifeweb.com

END NOTES

1. Doyle, K. L. 1991. "Once a pet peeve, now a crime." *Insight* (February 4): 62.

2. Miller, J. 1991. "Financing fido." *Changing Times* 45 (March): 68.

3. Copyright 1997, *USA Today.* Reprinted with permission.

4. Silverman, F. 1997. "Costly critters: try getting Fido and Fluffy to live within your means." *Hartford Courant* [Online, October 5], p. B1 (4 pp.). Available: http://proquest.umi.com/pqdweb.

5. Derr, M. 1990. "The politics of dogs." *The Atlantic Monthly* 265(3) (March): 70.

6. Friedmann, E. 1998. "People and pets." *Courier* 41 (February): 13.

7. Burke, S. 1991. "In the presence of animals." *U.S. News & World Report* 112(February 24): 7, 64–65.

8. Reprinted with permission from *Mother Earth News,* copyright 1985, Sussex Publishers, Inc.

9. Friedmann, E., and S. A. Thomas. 1995. "Pet ownership, social support, and one year survival after acute myocardial infarction in the Cardiac Arrhythmia Suppression Trial (CAST)." *American Journal of Cardiology* 76(17) (December 15) 1213.

10. Mayo Foundation for Medical Education and Research. 1997. "Pet therapy, medical specialists on four legs" [Online, January 16]. Available: www.mayohealth.org/mayo/9701/htm/pet_ther.htm [1999, July 13].

11. Siegel, J. M., F. J. Angulo, R. Detels, J. Wesch, and A. Mullen. 1999. "AIDS diagnosis and depression in the Multicenter AIDS Cohort Study: The ameliorating impact of pet ownership." *AIDS Care* 11(2): 157–170.

12. Mayo Foundation for Medical Education and Research. 1997. "Pet therapy, medical specialists on four legs."

13. Raina, P., D. Waltner-Toews, B. Bonnett, C. Woodward, and T. Abernathy. 1999. "Influence of companion animals on the physical and psychological health of older people: An analysis of a one-year longitudinal study." *Journal of the American Geriatrics Society,* 47(3) (March): 323–329.

14. Whiteley, H. E., DVM. 1986. "The healing power of pets." *Saturday Evening Post* 258(7) (October): 22.

15. Allen, K., Ph.D., and J. Blascovich, Ph.D. 1996. "The value of service dogs for people with severe ambulatory disability." *Journal of the American Medical Association* [Online, Volume 275, No. 13, pages 967–1060]. Available: www.ama-assn.org/sci-pubs/journals/archive/jama/vol_275/no_13/abstract.htm. [1996, April 31].

CHAPTER

4

Animal Rights and Animal Welfare

OBJECTIVES

After reading this chapter, one should be able to:

- define the terms *animal rights* and *animal welfare* and be able to compare them.
- identify important persons in the animal rights movement.
- identify important dates and acts of legislation associated with animal welfare.
- compare the issues concerning animal rights and animal welfare.

TERMS TO KNOW

animal rights

animal welfare

confinement systems

ecoterrorism

factory farming

humanize

intensive operations

specieism

unethical

vivisection

HISTORY

The terms *animal welfare* and *animal rights* are becoming more common today. Many people believe the two terms to be synonymous, and they are often used interchangeably by the public and the media.

Animal welfare is the position that animals should be treated humanely. This includes proper housing, nutrition, disease prevention and treatment, responsible care, proper handling, and humane euthanasia or slaughter. Animal welfare people believe that animals can be used for human purposes but that they should be treated so that discomfort is kept to a minimum.

Animal rights is the position that animals should not be exploited. Animal rights people believe that animals should not be used for food, clothing, entertainment, medical research, or product testing. This also includes the use of animals in rodeos, zoos, circuses, and even as pets. They believe it is ethically, morally, and inherently wrong to use animals for human purposes under any condition. Many are against **vivisection** and believe that living animals should not be used for surgical operations and experiments in which the structure and function of organs are studied. They further believe that animals should not be used in experiments in which diseases and various therapy methods are studied.

The modern animal rights movement, which began in the early 1970s, is composed largely of people from urban areas who are vegetarians. It draws heavily on philosophers and theologians; others who have been drawn into the movement have been activists in other areas such as human rights.[1]

As early as 1964, aroused public indignation over the welfare of animals in Great Britain resulted when a London housewife and vegetarian, Ruth Harrison, published a book titled *Animal Machines: The New Factory Farming Industry*. The main emphasis of the book centered on two areas: the use of antibiotics, hormones, and additives in animal feeds; and modern **factory farming,** where chickens are kept in cages and veal calves are kept in small crates. This book had a major impact in England. An investigation was conducted and laws were passed concerning the care and treatment of chickens, turkeys, pigs, cattle, sheep, and rabbits.

The animal rights movement continued to gain momentum in the latter 1970s with a publication by Peter Singer titled *Animal Liberation*. Peter Singer, an Oxford University–trained Australian philosopher and vegetarian, is often considered the founder of the modern animal rights movement. *Animal Liberation* condemned the use of animals produced for food and their use in research.

In 1990, animal rights groups staged a "March for the Animals" rally in Washington, D.C. Crowds estimated from 15,000 to 24,000 people marched down Pennsylvania Avenue. They carried banners and placards proclaiming "Animals Are Not For Wearing," "Fur is Dead," and "Animals Have Rights, Too."

America has the distinction of having the first laws on the books to protect farm animals from cruel treatment. This was a legal code titled "The Body of Liberties" passed by the Massachusetts Bay Colony in 1641.

The first anticruelty law was passed by the New York legislature in 1828. The legislation claimed: "Every person who shall maliciously kill, maim, or wound any horse, ox, or other cattle, or sheep, belonging to another, or shall maliciously and cruelly beat or torture such animal, whether belonging to himself or another, shall upon conviction, be adjudged guilty of a misdemeanor."

In 1866, the American Society for the Prevention of Cruelty to Animals (ASPCA) was formed in New York, largely to look after the welfare of disabled horses and mules and save them from abandonment. This was America's first humane society and was founded by Henry Bergh.

Since no one had been prosecuted under the New York State Anti-Cruelty Law, Henry Bergh drafted in 1867 "An Act for the more effectual prevention of cruelty to animals." This act had ten sections and has served as the example for drafting many succeeding anticruelty laws. Forty-one states and the District of Columbia have present laws based on this act.

In 1906, the Animal Transportation Act was passed to protect animals traveling long distances by rail. This act was passed to provide humane care and treatment of animals destined for slaughter. Today 95 percent of the animals going to markets travel by truck, but the law was never amended to include them.

In 1958, the Humane Slaughter Act was passed and later amended in the 1970s to include the humane handling of animals prior to and during slaughter.[2]

In 1966, Congress enacted Public Law 89–544, known as the Laboratory Animal Welfare Act (AWA). This law regulated dealers who handled dogs and cats, as well as laboratories that use dogs, cats, hamsters, guinea pigs, rabbits, and primates in research.

The first amendment to the AWA was passed in 1970 (P.L. 91–579). This amendment authorized the regulation of other warm-blooded animals when used in research, exhibition, or wholesale pet trade.

An amendment in 1976 (P.L. 94–279) prohibited animal fighting ventures and regulated the commercial transportation of animals. In 1985, the Improved Standards for Laboratory Animals Act was enacted as part of the Food Security Act and further amended the AWA. These amendments required the issue of additional standards for the use of animals in research. Standards for the exercise of dogs and for a physical environment adequate to promote the psychological well-being of nonhuman primates resulted. Additional requirements included the establishment of Institutional Animal Care and Use Committees (IACUC) at research facilities, which included standards to ensure that pain and distress are minimized; that anesthetics, analgesics, and tranquilizers are used appropriately; and that researchers consider alternatives to painful procedures.

In 1990, injunctive relief and pet protection provisions were passed as part of the Food, Agriculture, Conservation, and Trade Act of 1990. The injunctive relief provision authorizes the Secretary of Agriculture to seek an injunction stopping a licensed entity from continuing to violate the AWA while charges are pending. The pet protection provision mandates the Secretary to issue additional regulations pertaining to random-source dogs and cats—animals that have been obtained from animal and humane shelters for use in laboratories. The act covers four categories of institutions: (1) dog and cat breeders; (2) zoos, circuses, and roadside menageries; (3) transportation of animals; and (4) research facilities.

The United States Department of Agriculture is charged with developing and implementing regulations to support the AWA. These regulations, Parts 1, 2, and 3, provide minimum standards for the care and handling, housing, feeding, sanitation, ventilation, shelter from extreme weather, veterinary care, and separation of species when necessary. Birds, laboratory rats, and laboratory mice are currently excluded from these regulations. The AWA regulations require the licensing of animal dealers, exhibitors, and operators of animal auction sales where animals regulated under the AWA are sold.[3]

The Horse Protection Act was passed in 1970 and amended in 1976. This act protects horses and regulates the horse show business. At issue is the showing or sale of horses whose gait is altered by inducing pain in their legs.[4]

In October 1986, an animal rights group known as the Animal Liberation Front (ALF) broke into laboratories on the Oregon State University campus and caused several thousand dollars' worth of damage. Six months later, ALF set fire to a laboratory on the campus of the University of California at Davis, causing nearly $5 million in damage. On Independence Day, 1989, ALF struck a research facility on the campus of Texas Tech, where Dr. John Orem was working with cats in an effort to understand the cause of Sudden Infant Death Syndrome (SIDS). The damages exceeded $55,000, but the break-in not only resulted in damage to the facility and equipment, but also caused a delay in the important research work of Dr. Orem. SIDS causes the death of more than 8,000 infants every year in the United States.

In 1989, the Farm Animal and Research Facilities Protection Act was introduced by Rep. Charles Stenholm (D-Texas). This act would have applied to facilities or premises where an animal is kept for food, agricultural research, testing, and education. It would make it a federal crime to disrupt the activities of the premises; steal or damage property or animals; enter a facility with the intent to commit a violation; or remain in a facility despite notice that entry was forbidden. Violators could be fined up to $10,000 and jailed for up to three years or both. See Figure 4–1.

Although the bill had 235 co-sponsors, it did not see floor action in the house prior to adjournment of the 101st Congress. Stenholm's office had released a list of ninety-one incidents between January 1981 and January 1990, ranging from bomb threats and trespassing to break-ins and arson, that were linked to groups associated with the animal rights movement.

On August 4, 1992, a bill introduced by Stenholm passed the House of Representatives and was merged with a similar proposal sponsored by Sen. Howell Heflin (D-Alabama), which had passed the Senate in the fall of 1991. On August 26, 1992, this bill, known as The Animal Enterprise Protection Act of 1992 (P.L. 102–346), was signed into law by President George Bush.

Facilities protected by this bill include farms, zoos, aquariums, circuses, rodeos, fairs, auctions, packing

Figure 4–1   Animal rights activists have caused considerable damage to research facilities around the country. Fire was set to facilities on the campus of Michigan State University and sulfuric acid was doused throughout the lab. Photo by Dave Weinstock, courtesy of The Farmer's Exchange.

plants, and commercial or academic enterprises that use animals for food or fiber production, agricultural research, or testing. See Figures 4–2 and 4–3.

The law imposes up to one year in jail for causing damages of $10,000 or more on an animal enterprise;

Figure 4–2   On the wall behind the clean-up crew are the words, "We will be back for the Otters!" Research at this lab was being conducted with mink. Photo by Dave Weinstock, courtesy of The Farmer's Exchange.

Figure 4–3 Animal rights activists vandalized a research facility at Oregon State University and spray painted a message on the wall. Photo by American Farm Bureau Federation staffer Julie Brown, courtesy of Farm Bureau News.

up to ten years in jail if an intruder injures a person during an attack on an animal enterprise; or up to a life sentence in jail if an intruder kills another person during an attack on an animal enterprise.

Even with the passage of The Animal Enterprise Protection Act, many crimes are still reported today. Many of these crimes are being committed under the guise of saving nature and have been given the term ecoterrorism. An example of this type of crime occurred on May 30, 1997. Nearly 10,000 mink were released from a fur ranch near Mount Angel, Oregon. The owner of the ranch later reported that approximately half of the animals died from fighting with each other, and nearly 1,300 of the animals he was able to round up would probably die.

These crimes are hard for law enforcement officers to solve and to prosecute. Although Section 43 of the Animal Enterprise Protection Act covers disruption of animal enterprises, many people are advocating an amendment to the Animal Enterprise Protection Act that would include adding to the list of protected persons loggers, miners, fishermen, farmers, trappers, ranchers, food outlets, processors, and all resources subject to ecoterrorism.

CURRENT ISSUES

Do Animals Have Rights?

Animal rights activists believe that animals have the same rights as humans. They believe any use of animals by humans reflects a bias that humans are superior to animals. Specieism is the term they use for this belief. They believe that because animals have nervous systems and feel pain, it is wrong to use animals for food or experimentation. They contend that one cannot ignore the rights of animals while advocating the rights of humans, because humans are animals. They contend that humans are supposed to care for and protect animals and not exploit them. To believe that animals have the same rights as humans is to humanize animals.

Animal welfare is the stand that animals should be treated humanely and without cruelty. Animals should have proper housing, management, nutrition, disease prevention and treatment, responsible care, and handling. Proponents of this position believe that animals can be used in research as long as all measures are taken to ensure that the animals are cared for humanely. Animals should be used only when no other alternatives exist to achieve the objectives of the research.

Should Animals Be Used for Food?

Animal rights activists argue that modern farming is inhumane and that eating meat is unhealthy. They contend that animals are constantly housed, kept in small cages with wire floors, and never allowed to see the light of day. Factory farming, as they call it, reportedly results in severe physiological and behavioral problems, as well as causing a long list of diseases and ailments. They contend that modern farming is controlled by large corporations that care only about profits and not about the welfare of animals.

Farmers have led the way in animal welfare. It has always been in their best interest to care for animals humanely and to see that they are well-fed and free of diseases and ailments. See Figure 4–4. Modern farmers could not stay in business long if their animals weren't healthy. Modern farming involves intensive operations, in which farmers try to increase their output through better breeding, feeding, and management. Many of these operations are confinement systems, where the animals are confined to cages or pens in partially enclosed or totally enclosed buildings. Farmers can get improved production through closer control of the environment. These buildings may have slotted floors, automatic feeding and watering systems, and forced air ventilation. See Figure 4–5.

Figure 4–4 Sows are usually placed in farrowing crates one week before they give birth. She will remain in the crate for three to four weeks. The use of a crate makes delivery easier, makes it easier for the farmer and veterinarian to work with her if necessary, and prevents the sow from rolling over and crushing her litter. Courtesy of Agriscience.

Most farm animals (except beef cattle) are kept in barns or suitable housing to protect the health and welfare of the animal. These facilities protect the animals from extremes in weather, diseases, and preda-

tors. Modern livestock facilities are well-ventilated, temperature-controlled, and provide the animals with a constant supply of fresh water and feed. Beef cattle are usually kept in pastures or open ranges, where they consume primarily nongrain feedstuffs such as grass, roughage, food processing by-products, and crop residues. See Figure 4–6.

Farming in the United States is not controlled by large corporations. Of the 2.2 million farms in the United States, 97 percent are family owned and operated; only 7,000 are non–family-controlled corporations.[5]

Both the federal government and the American Heart Association contend that a diet containing meat, milk, and eggs is appropriate. No studies can substantiate that a vegetarian diet is healthier than a diet that includes meat, milk, and eggs. The American Dietetic Association, American Heart Association, and others recommend 5 to 7 ounces of lean, trimmed meat daily because it provides large amounts of essential nutrients.

Should Animals Be Used for Experimentation?

Animals are used in three areas of experimentation:

- research in biomedical and behavior sciences
- testing of products for their safety
- education where animals are used for demonstrations and dissection

Animal rights activists believe that animal experimentation is **unethical** and unnecessary and that al-

Figure 4–5 Laying hens are kept in cages under controlled environments that increase the efficiency of the laying operation. Each hen receives adequate feed and water. Eggs, a valuable part of the diet of many Americans, are available at affordable prices because of the efficiency of these operations. Courtesy of Dr. Bill Muir, Purdue University.

Figure 4–6 Veal calves are kept in individual stalls so that they can receive individual attention. Diseases are kept to a minimum and farmers are able to feed and care for the calves more efficiently. The economic welfare of farmers depends on the proper care and treatment of their animals. Courtesy of Utah Agricultural Experiment Station.

ternatives to animal experimentation should be used. Animal rights activists contend that the value of animal research is greatly exaggerated and that misleading animal tests can have a devastating effect on human health.

Much animal rights literature depicts maimed and tortured research animals. Those who believe that animals have the same rights as humans also believe that all forms of animal research should be eliminated. Animal rights activists also point out that biomedical research is a vast, lucrative industry, supported each year by $15 billion in taxes and charity, while killing 65 to 100 million animals.

Several areas of animal experimentation have been attacked by animal rights activists:

1. Pound seizures, where dogs and cats from animal shelters have been sold to laboratory animal dealers and to research and educational facilities for use in animal experimentation.

2. The Draize Eye Test and the Skin Irritancy Test, which are used to measure the effects of products such as cosmetics on the unprotected eyes and skin of restrained rabbits. Rabbits are usually used for this test because they do not have tear ducts to wash the products from their eyes. In the skin tests, a patch of fur is shaved from the back and side of the animal. Products are applied to the area and then observed for any reactions. See Figures 4–7 and 4–8.

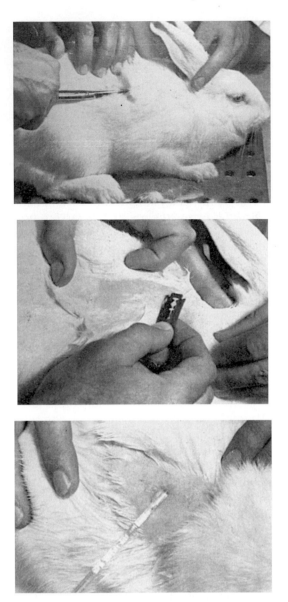

Figure 4–8   Skin Irritancy Tests are conducted on rabbits to test the effects of various products on their skin. A small patch of hair is shaved away and the product being tested is applied to the skin. Courtesy of The American Anti-Vivisection Society.

Figure 4–7   These rabbits are being used in a Draize Eye Test. The animals are restrained and various products being tested are placed in their eyes. Photo provided by People for the Ethical Treatment of Animals (PETA).

3. The approximately 6 million animals that are used in educational facilities for dissection. See Figure 4–9.

4. The Classical LD50 Test, which is used to determine the dosage required to kill 50 percent of a population of test animals within a specified period. Approximately 4 million animals are used in these tests each year. The animals are force-fed, injected, forced to inhale, or exposed to various substances.

5. Animals that are used in psychological research to discover the effects of pain, response

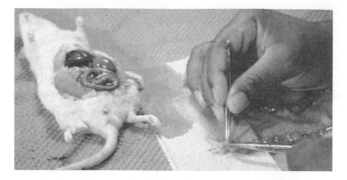

Figure 4–9   Many small animals are used in educational facilities across the country for dissection. Many alternatives to dissection now exist and many people question the validity of dissection as an educational tool. Photo provided by People for the Ethical Treatment of Animals (PETA).

to different stimuli, learning, and behavior. Estimates of animals used in this type of experimentation range from 900,000 to 1 million annually.

Most critics of the animal rights activists' views on experimentation believe that biomedical research would be limited or stopped completely if animals were not available for research. Many medical advances would not exist if animals were not used in research. Kidney, heart, liver, lung, and endocrine organ transplants are now possible because of prior research on animals. New drugs, vaccines, and surgical techniques were perfected with animal research. Vaccines against diphtheria, hepatitis, tetanus, and polio were developed after research on animals. Insulin is a result of research conducted with dogs.

The number of animals used in testing is declining. Approximately 90 percent of the animals still used in testing are rats and mice. The Animal Welfare Act of 1966, which regulates the care and treatment of laboratory animals, does not include rats and mice. Many companies no longer use animals for testing their products or have reduced animal testing programs. Several companies have offered grants for the research of "alternative" (replacement methods for animal testing) methods of testing products.

Some of these alternative methods now being used or suggested include the following:

1. The Bovine Corneal Opacity (BCOP) Test is being offered as an alternative to the Draize Eye Test. Cow eyes obtained from slaughter houses are used in this test. The cornea is removed, and the test substance is applied; the effects are then measured. This test does use the product of animals, but the eyes would have been discarded during the slaughter.

2. Computer videodisc simulations and three-dimensional computer programs are being used for teaching techniques without using animals.

3. Plastic anatomical models are being used for teaching in place of the actual dissection of animals.

4. "Cultured" skin and its use in testing irritation to cosmetics, household products, and other agents as well as the treatment of wounds shows promise. Cultured skin is grown *in vitro* (meaning in glass or in laboratory containers) in the laboratory. The cultured skin contains live cells that respond to stimuli similar to the skin of a live animal.

Primates make up about half of 1 percent of all animals used in experiments; however, their contributions have been significant. One example is the research into the cause and cure of poliomyelitis. The cause of poliomyelitis remained a mystery until 1908, when scientists were able to transmit the virus to monkeys. Years of research with primates followed until the early 1950s, when scientists were able to grow the virus in human cell cultures. This made the development of a vaccine possible. After a vaccine was developed, it was tested on monkeys to ensure that it was safe and effective. Kidney tissue from monkeys was also necessary to produce the vaccine. Today, the vaccine is produced from self-propagating cells.

Many people believe that monkeys, especially chimpanzees and the pigtail macaque, may be important in developing a cure for AIDS (Acquired Immune Deficiency Syndrome). The University of Washington reported in 1991 that it had successfully inoculated eight pigtail macaques with human AIDS and that the monkeys became critically ill with symptoms resembling those of human AIDS. Until then, the only other species successfully inoculated with the human AIDS virus was the chimpanzee, but no chimpanzee has yet developed clinical signs of the human AIDS.[6]

In mentioning AIDS research, it is also important to note that the United States Centers for Disease Control and Prevention (CDC) *Morbidity and Mortality Weekly Report* for September 11, 1992, announced that the blood of two laboratory workers working with monkeys had tested positive for simian AIDS. The two workers tested positive but had not become ill. This disease is highly fatal to monkeys. Laboratory workers need to be aware of guidelines published by the CDC in November 1988 to reduce the risk of transmission of "Monkey AIDS."[7]

Animal agriculture and medical research have combined to make significant contributions to the health and well-being of people. From the adrenal glands of cattle, epinephrine is used to relieve some symptoms of hay fever, asthma, and some allergies. It is also used as a heart stimulant in some crisis situations and by dentists to prolong the effect of local anesthetics.

Thrombin from cattle blood helps blood clotting and is valuable in treating wounds in inaccessible parts of the body; it is also used in skin grafting. Liver extract is combined with folic acid and injected to treat various types of anemia. Insulin derived from cattle pancreas is used to treat diabetes. Glucagon helps counteract insulin shock.

Heart valves from hogs are used in heart valve replacement surgery in humans. They are in some cases superior to mechanical valves because they don't stick and do not need the same level of anticoagulant infusion. In the last twelve years, 250,000 lives have been saved through implantation of swine heart valves.

Because of its similarity to human skin, pig skin is used to treat massive burns and large accidental skin removal. Gelatin made from pig skin is used for capsules and pills.

Thyroid extracts from swine are used to regulate the rate of metabolism in humans. Another extract is used to treat low calcium and phosphate levels and to regulate heartbeat. Extracts from swine pancreas are a source of insulin. Even with synthetic insulin, an estimated 5 percent of all people with diabetes are allergic to all but insulin from hogs.[8]

It is important that all animals used for research be cared for humanely. Facilities, researchers, and lab technicians must learn how to correctly care for, handle, and meet the needs of their animals. Animals should be used only when no other alternatives exist to achieve the objectives of the research. Many people believe that animal experimentation and research hold the key to the successful cure of AIDS, cancer, and heart disease, as well as to improve the health of all humans.

Should Hunting and Trapping Be Allowed?

Many animal rights activists believe that regulated recreational hunting and trapping is inhumane and unnatural. They also believe that hunters and trappers kill merely to inflict suffering and for the joy of killing.

Animal rights activists would also like to see a stop to the use of hunting dogs. They contend that using dogs to hunt is unnatural and inhumane treatment.

Animal rights activists attack state game agencies, claiming that they manage wildlife numbers not to maintain the balance of nature, but to provide large numbers of animals so hunters will have continued hunting. They claim that hunting and trapping does not prevent overpopulation, that hunting for food is no longer necessary, and that hunting should be banned completely.

Hunting advocates note that they are the prime revenue source for preserving nonendangered wildlife. In 1996, hunters spent a total of more than $20.6 billion; $651 million was spent for licenses, stamps, tags, and permits. These figures compare to $12.3 billion in total expenses and $532 million spent for licenses, stamps, tags, and permits in 1991. A portion of licenses, fees, stamps, tags, permits, and excise taxes on ammunition, firearms, and equipment is used to finance game research and management and to help purchase habitat that benefits all creatures. There were 13.769 million hunters in 1996[9] compared with 13.771 million in 1991.[10] Hunters make up about 7 percent of our population.

Throughout history, hunting has been a tradition of humans. Until recent times, humans hunted through necessity for food. As civilization progressed, humans' need to hunt for survival was reduced; most hunting today is recreational. See Figure 4–10.

Hunters themselves have long been concerned for the welfare of wildlife. Through their actions, hunters have identified problems such as pollution and its effect on wildlife, habitat loss, and abuse. They helped focus public attention on the plight of rare and endangered species. Hunters have been largely responsible for initiating wildlife laws and their enforcement, wildlife research and management, and establishment of parks and wildlife preserves.[11]

Today, wildlife populations are controlled by game management practices, habitat improvement, and carefully regulated seasons and bag limits. Wildlife management programs have been very successful in improving the populations of game species such as the whitetail deer, wild turkey, wood duck, and

Figure 4–10 *Hunting is an opportunity to enjoy the companionship of other hunters, to enjoy the outdoors, and to enjoy the tradition of hunting. Courtesy of Jeff Jackson.*

beaver, as well as populations of nongame species such as the trumpeter swan, egret, and heron.

At the turn of the century, the total population of whitetail deer in North America was estimated at about 500,000; in several states, whitetail deer had been completely eliminated. Nearly every state in the nation had closed its deer hunting season. After many years of closed hunting seasons, restocking programs, and management programs, the population had reached levels that were higher than ever before. By the 1960s, almost all states allowed some form of deer hunting. It is estimated that there are more than 18 million whitetail deer in the United States today. Many states have such large populations that the deer are causing destruction to the available vegetation and farmers' crops and are becoming hazards on our highways. See Figure 4–11.

The stories of the wild turkey and the wood duck are similar. By 1930, the wild turkey had been eliminated from almost all of our Northern states. A few states in the Southern United States still had small populations. Today, thanks to our nation's hunters, game agencies, and wildlife conservation organizations such as the National Wild Turkey Federation, there are more than 5 million wild turkeys roaming the continent in huntable populations in every state except Alaska, and they are even hunted in Ontario, Canada, and Mexico. Today, turkey hunting is one of the largest types of hunting in the United States, with close to 2.5 million sportsmen considering themselves turkey hunters.

The wood duck, after being on the verge of extinction by 1915, is now one of our most common breeding waterfowl in the Eastern United States. Unregulated hunting and destruction of hollow tree nesting sites almost eliminated this beautiful bird. See Figure 4–12.

Hunting programs are designed to harvest only the surplus animals. Wildlife cannot be stockpiled; many people do not realize the high natural mortality of wildlife. Nearly 75 to 80 percent of the bobwhite quail and dove populations will die annually from natural causes even if the birds are not hunted. Research shows that a good deer population can be maintained if hunters take 40 percent each year. Normally, hunters take only about 15 percent of the populations during annual hunting seasons; in many areas, hunters don't

A

B

Figure 4–11   There are more whitetail deer in the United States now than at any time in history. Courtesy of Brian Yacur.

Figure 4–12   Successful restocking programs and careful management programs have brought the (a) wood duck and (b) wild turkey back from the verge of extinction. Courtesy of PhotoDisc.

harvest enough deer. This leads to overpopulated areas where deer can severely damage the habitat.

SUMMARY

The terms *animal rights* and *animal welfare* are not synonymous. They represent two different viewpoints. Animal rights is the position that animals should have the same rights as humans and should not be exploited in any way. Activists for animal rights believe that animals should never be used for food, clothing, entertainment, medical research, or product testing. This also includes the use of animals in rodeos, zoos, and circuses, and even as pets. Animal rights activists believe that it is ethically, morally, and inherently wrong to use animals for human purposes under any condition.

Animal welfare is the position that animals should be treated humanely. Animal welfare people believe that animals can be used for human purposes but that they should be treated so that discomfort is minimized.

Four basic issues are being debated by animal rights and animal welfare groups:

1. Do animals have rights?
2. Should animals be used for food?
3. Should animals be used for experimentation?
4. Should hunting and trapping be allowed?

Students studying these issues need to be able to separate facts from opinions. The issues are much greater and involve more information than what can be presented here. Students need to research the organizations involved, review the facts or opinions they present, and then form their own opinions. Students need to be wary and not let emotions cloud the facts.

More than 7,000 groups are involved in animal protection; 400 of these are considered animal rights groups. Many of the groups do not consider themselves as animal rights or animal welfare groups but are dedicated to the preservation of a specific species of animal and its habitat.

DISCUSSION QUESTIONS

1. What was the title and who was the author of the book that aroused public indignation over the welfare of animals in Great Britain?
2. What is a factory farm?
3. Who is considered the founder of the modern animal rights movement and what was the name of his publication?
4. What do the terms *animal rights* and *animal welfare* mean?
5. Do animals have rights?
6. Should animals be used for food?
7. Should animals be used for experimentation?
8. Should animals be used for other purposes, that is, for hunting or trapping, for entertainment purposes in zoos and exhibits, for rodeos, or even for pets?

SUGGESTED ACTIVITIES

1. Visit the library and research the history of animal rights and welfare.
2. Research the various animal rights and welfare groups. Write letters to them requesting information, and then discuss the materials that were sent.
3. Research modern farming methods. Are they humane? Research and become familiar with the terms *factory farm, intensive farming,* and *confinement farming.*
4. Research each of the four issues presented in this chapter. Divide into groups and carry out a debate on the issues.
5. What is the ultimate goal of the animal rights movement? Is there one? Thoroughly study the animal rights movement, the groups, and the issues to answer these two questions.

SUGGESTED READING

1. *Animal Rights—Opposing Viewpoints,* David L. Bender and Bruno Leone, Series Editors, Janelle Rohr, Book Editor, Greenhaven Press, Inc., P.O. Box 28909, San Diego, CA 92198-0009, Copyright 1989.
2. *The Animal Rights Crusade—The Growth of a Moral Protest,* by James M. Jasper and Dorothy Nelkin, 1992, The Free Press, A division of Macmillan, Inc. New York.
3. *The Animal Rights Handbook,* 1990, Living Planet Press, 558 Rose Ave., Venice, CA 90291.
4. *Animals Have Rights, Too,* by Michael W. Fox, 1991, The Continuum Publishing Company, 370 Lexington Avenue, New York, NY 10017.
5. *Animals In Society,* by Zoe Weil, 1991, ANIMALEARN, a division of the American Anti-Vivisection Society, Suite 204, Noble Plaza, 801 Old York Road, Jenkintown, PA 19046-1685.

6. "Beyond Cruelty," by Katie McCabe, 1990, *The Washingtonian Magazine,* Volume 25 (February), Number 5.

7. *Congressional Research Service Report for Congress, Humane Treatment of Farm Animals: Overview and Selected Issues,* Geoffrey S. Becker, Specialist, Environmental and Natural Resources Policy Division, May 1, 1992.

8. *Guide for the Care and Use of Agricultural Animals in Agricultural Research and Teaching,* a booklet published by the Consortium for Developing a Guide for the Care and Use of Agricultural Animals in Agricultural Research and Teaching, March 1988, Editorial and Production Services, Association Headquarters, 309 West Clark Street, Champaign, IL 61820, (217) 356-3182.

9. *Animal Agriculture—Myths & Facts,* copyright 1988, Animal Industry Foundation, P.O. Box 9522, Arlington, VA 22209-0522, (703) 524-0810.

10. *Swine Care Handbook,* National Pork Producers Council, P.O. Box 10383, Des Moines, Iowa 50306, (515) 223-2600.

11. *Guide for the Care and Production of Veal Calves,* 3rd Edition, June 1, 1990, American Veal Association.

12. *Our Farmers Care,* Wisconsin Agribusiness Foundation, 2317 International Lane, Suite 109, Madison, Wisconsin 53704-3129.

ADDITIONAL RESOURCES

1. The National Fish and Wildlife Service www.fws.gov

2. The National Wild Turkey Federation www.nwtf.org

3. Ducks Unlimited www.ducks.org

4. Environmental Health Perspectives http://ehis.niehs.nih.gov/docs/1995/103-11/innovations.html

END NOTES

1. Gillespie, J. R. 1997. *Modern Livestock and Poultry Production.* 5th ed. Albany, NY: Delmar.

2. Excerpts from *Animal Welfare and Animal Well-being* by Dr. Jack L. Albright, Purdue University. Presented to the Indiana High School Agriculture Teachers' Workshop, June 25, 1991.

3. Excerpts from *Animal Welfare Enforcement,* Fiscal Year 1991, Animal and Plant Health Inspection Service, United States Department of Agriculture.

4. Excerpt from *Animal Welfare and Animal Well-being* by Dr. Jack L. Albright.

5. Excerpt from *Animal Agriculture: Myths and Facts,* Animal Industry Foundation, Arlington, Virginia, 1988.

6. Excerpts from International Primate Protection League. 1992. "The hit of the year." *News,* 19(3) (December).

7. Excerpt from International Primate Protection League. 1992. "Lab workers test positive for monkey AIDS," *News,* 19(3) (December).

8. Excerpts from *Animal Agriculture: Myths and Facts.*

9. National Fish and Wildlife Service, National Survey of Fishing, Hunting, and Wildlife-Associated Recreation [Online, 1997, November], page 74, Table 17, Expenditures for Hunting: 1996. Available: http://fa.r9.fws.gov/surveys.surveys.html. [1999, July 17].

10. National Fish and Wildlife Service, National Survey of Fishing, Hunting, and Wildlife-Associated Recreation [Online, 1993, March], page 83, Table 21, Expenditures for Hunting: 1991. Available: http://fa.r9.fws.gov/surveys.surveys.html. [1999, July 17].

11. Turkey/Hunting Information, History of Turkey Hunting and Role it Plays Today. [Online]. Available: www.nwtf.org/hunting/index.html. [1999, June 11].

Careers in Small Animal Care

OBJECTIVES

After reading this chapter, one should be able to:

■ compare opportunities in small animal care and management.

■ be able to describe the nature of the work, salaries, and requirements necessary for obtaining a job in the areas of small animal care.

TERMS TO KNOW

anatomists	geneticists	pharmacologists
animal trainers	laboratory animal care workers	physiologists
biochemists	laboratory animal technicians	small animal breeders
biologists	laboratory animal technologists	veterinarians
biophysicists	nutritionists	veterinary technicians
botanists	pathologists	zoo administrators
ecologists	pet care workers	zoologists
embryologists	pet groomers	

INTRODUCTION

The pet industry in the United States is a rapidly growing segment of today's business world. It is estimated that 31.6 percent of American households have dogs and 27.3 percent have cats. Cats outnumber dogs 59.1 million to 52.9 million, because most households have an average of two cats. Approximately 58.9 percent of U.S. households owned pets in 1996. Included are an estimated 55.0 million fish, 4.9 million rabbits, 1.8 million hamsters, 764,000 gerbils, and several million other household pets of various species.

Americans spend in excess of $20.3 billion per year on their pets. Veterinary expenses exceed $11 billion annually, and pet food manufacturers produce some $9.0 billion in sales.[1]

The pet care industry is just one area of small animal care. This chapter explores the jobs and careers that exist in the care and management of small animals.

PET CARE WORKERS

Pet care workers[2] provide many services for the owners of small animals. They are employed in animal hospitals, boarding kennels, animal shelters, pet grooming parlors, pet training schools, and pet shops. Some pet care workers have their own businesses. See Figure 5–1.

Kennel attendants care for animals and keep their quarters clean. They watch the animals closely to make sure that they are in good health, and they often exercise animals and keep records on the animals' health, feeding, and breeding. At times they may also have to dispose of diseased or unwanted animals. See Figure 5–2.

Animal groomers bathe pets and use special solutions to keep them free of ticks, fleas, and other pests. They often brush the pets and trim their hair and nails.

Dog trainers teach dogs to hunt, herd, or track; to obey signals or commands; to guard lives or property; to run races; or to lead the blind. Handlers command trained animals during shows, sporting matches, or hunts. See Figure 5–3.

Animal breeders arrange for the mating of animals and care for the mother and young. They usually train or sell animals when they reach the appropriate age. They often specialize in one breed, such as German Shepherd dogs or Siamese cats.

Pet shop owners care for the birds, fish, cats, dogs, and other animals they offer for sale, as well as deal with customers. In addition, they must take care of the details involved in running a business. De-

Figure 5–2   Kennel attendants feed and care for animals and keep the facilities clean. Courtesy of Brian Yacur and Guilderland Animal Hospital.

pending on the size of the shop, the owner may employ a manager, animal care persons, and pet counselors. Stores usually employ an average of three to five workers. See Figure 5–4.

Figure 5–1   High school students can get employment as animal hospital workers. Responsibilities usually include cleaning and disinfecting cages and equipment, feeding animals, and performing other duties assigned by the veterinarians. Courtesy of USDA.

Figure 5–3   Dogs can be trained for many different jobs and skills depending on what they will be used for (i.e., herding, hunting and tracking, police work, or assisting people with disabilities). Courtesy of PhotoDisc.

Figure 5–4 The duties of the pet shop owner will vary considerably depending on the size of the store and the amount of business that the store handles.

Pet shop managers normally schedule staff hours, set sales quotas, plan promotions, and recruit, hire, and train personnel. Managers would also place merchandise and animal orders, see that inspection standards are maintained, and be involved in customer and community relations. See Figure 5–5.

Pet shop counselors are encouraged to work closely with customers to decide which pet is best for them and to encourage them to purchase the products they need. Other tasks involve assisting customers with heavy items, answering telephone inquiries, operating the cash register, cleaning and straightening product displays, and restocking shelves and displays. See Figure 5–6.

Figure 5–5 Pet shop managers are in charge of the daily operations of the shop. Courtesy of Brian Yacur and The Feedbag Plus.

Education and Training Requirements

Many of these positions are entry level and require no special educational requirements. Many employers hire high school students and high school graduates. Experience with animals is helpful, and knowing the needs and habits of animals is an asset.

Figure 5–6 Some pet shops have counselors who can work with customers to determine which pet is best for them. Courtesy of Brian Yacur and The Feedbag Plus.

Courses in biology and animal husbandry help prepare students for pet care jobs. Business courses such as bookkeeping help prepare students for running a business.

Advancement Possibilities and Employment Outlook

Pet care workers can advance by becoming supervisors in large animal hospitals, kennels, or pet shops. Some experienced workers start their own kennels or pet shops. Pet care workers can increase their earnings by developing skills in special fields such as grooming, training, or breeding animals.

Because more households are keeping pets, the number of jobs in pet care should continue to increase. The best jobs will go to those who have experience in caring for animals.

Working Conditions

Pet care workers may work indoors or outdoors. They must get used to the animals and should not mind cleaning up after them. Their work may involve lifting heavy animals and equipment. They may have to drive a station wagon, van, or light delivery truck to pick up or deliver pets. Their work involves dealing with people, so workers should be friendly and courteous.

Many pet care workers work 40 hours per week. Some must also work or be on call evenings and weekends. There are some jobs for part-time workers to care for animals at night or on weekends.

Earnings and Benefits

Earnings vary depending on the location, type of job, and worker experience. Many pet care workers begin at minimum wage; workers are usually eligible for increases with experience. Pet shop workers and experienced groomers may earn a commission or percentage of sales for the grooming of an animal. Experienced kennel attendents earn about $200 to $300 per week.

Pet shop owners may have pretax profits of $35,000 to $200,000 depending on location and volume of business. Start-up investments in a pet shop may run from $70,000 to more than $200,000. Annual revenues may run from $175,000 to more than $1 million.[3]

Where to Go for More Information

American Animal Hospital Association
P.O. Box 15899
Denver, CO 80215-0899
(303) 986-2800
www.healthypet.com

American Veterinary Medical Association
1931 North Meacham Road, Suite 100
Schaumburg, IL 60173-4360
(847) 925-8070
www.avma.org

National Dog Groomers Association of America
P.O. Box 101
Clark, PA 16133
(412) 962-2711
www.nauticom.net/www/ndga

Pet Industry Joint Advisory Council
1220 Nineteenth Street, NW, Suite 400
Washington, DC 20036
(202) 452-1525
www.petsforum.com/pijac

Professional Handlers Association
15810 Mt. Everest Lane
Silver Spring, MD 20906
(301) 924-0089

ZOO ADMINISTRATORS

Zoo administrators[4] are employed at zoos throughout the United States. There are zoo directors, animal curators, veterinarians, and resident zoologists. Their job is to provide services for the entertainment and education of the public.

Zoo directors are at the center of a great circle of needs, demands, and financial restrictions. Directors are decision makers. They weigh conflicting needs and establish priorities.

Zoo directors deal with people and money. They rarely deal directly with the animals. Directors answer correspondence, hold interviews, and supervise a large staff of curators, keepers, veterinarians, maintenance people, and educators. Zoo directors are experienced, learned people who have gone through years of training as curators, veterinarians, or zoologists.

Animal curators are scientists in charge of the animals. They also do paperwork, often at night. During the day, they tour the zoo, looking after the animals' health and safety. They encourage breeding, apply medicines, study animal behavior, and read journals to keep well informed. See Figure 5–7.

To keep the animals healthy, zoo veterinarians spend much time practicing preventive medicine. They run routine blood, urine, and fecal analysis; control the animals' diets; and watch for parasites. They are faced with very special problems because little is known about diseases of some species.

Resident zoologists are also animal scientists. Their work often overlaps that of curators and vet-

Figure 5–7 *Zoo administrators are responsible for the daily care and feeding of animals.*

erinarians. Some zoo zoologists are in charge of the educational aspects of the zoo, such as preparing descriptions of animals or designing habitats that simulate those found in nature.

Education and Training Requirements

Curators are expected to have at least a master's degree in some branch of zoology. Directors are usually required to have degrees in zoology or life sciences and in business. Zoologists must have a doctorate. To prepare for a career as a zoologist, one can volunteer to help at the local zoo by feeding animals, cleaning cages, nurturing baby animals, and otherwise assisting wherever you are needed. One may be able to get a paying position while he or she completes his or her education.

Zoo veterinarians have the same training that other veterinarians have.

Applicants should apply directly to the personnel departments of various zoos. Check with your school placement office for openings.

Advancement Possibilities and Employment Outlook

Animal curators, veterinarians, and zoologists may be promoted to zoo director after years of experience. Zoo directors are already at the top of their profession; however, the director of a small zoo may become a director of a larger zoo.

The employment outlook for zoo administrators is fair. Persons seeking jobs in the field will encounter some difficulties through the year 2005 because of the anticipated slow growth in zoo capacity and low turnover rates for positions. Competition for all zoo positions will continue to be stiff.

Working Conditions

For directors, curators, veterinarians, and zoologists, the hours are long and exhausting. They may take paperwork home with them, stay at the zoo through the night nursing a sick animal, or get up in the middle of the night to meet a pair of rhinos arriving at the airport. They must pay close attention to detail and have great patience. Animal curators, veterinarians, and zoologists have the satisfaction of working with and learning from the animals. The director works under a great deal of pressure.

Earnings and Benefits

Curators earn an average of $38,000 per year, whereas zoo directors make about $30,000 to $80,000 per year. Benefits include paid vacations, accident and health insurance, and pension plans. Some staff members receive travel grants for expeditions to conduct research that will help them in their private studies and will eventually benefit the zoo.

Where to Go for More Information

American Zoo and Aquarium Association
7970-D Old Georgetown Road
Bethesda, MD 20814
(301) 907-7777
www.aza.org

LABORATORY ANIMAL CARE WORKERS

Laboratory animal care workers[5] take care of animals used in scientific research. Laboratory animal care workers include the positions of assistant laboratory animal technicians, laboratory animal technicians, and laboratory animal technologists. Scientists do experiments for drug companies, medical schools, and research centers. They learn about animal and human diseases and treatments. They also study animals to learn about behavior and intelligence. Animal care workers help scientists and carry out their instructions. See Figure 5–8.

Laboratory animal care workers may care for mice, guinea pigs, rats, rabbits, monkeys, dogs, or even insects or fish. Animal care workers provide food and water for the animals and clean their cages. They check for signs of illness and keep careful records. They may also order food and supplies. They sometimes help scientists or medical doctors do experiments.

Some laboratory animal care workers are veterinarian's assistants; they may administer medicine, treat minor wounds, and prepare animals for surgery.

Education and Training Requirements

A high school education is usually required for unskilled jobs in laboratories. Courses in science are also helpful. Some laboratories will give on-the-job training. Two-year animal care programs are available at several colleges and technical schools. Certification is available from the American Association for Laboratory Animal Science (AALAS) but is not required for

Figure 5–8 Laboratory animal care workers take care of animals being used in research. They usually work closely with the scientists conducting the research and carry out duties that are assigned to them by the scientists. Courtesy of PhotoDisc.

employment. There are three levels of certification: assistant laboratory animal technician (ALAT), laboratory animal technician (LAT), and laboratory animal technologist (LATG). Each level of certification has its own age, education, experience, and examination requirements.

Laboratory animal care workers need a knowledge of animal eating and sleeping habits, should enjoy working with animals, and must be able to follow directions carefully.

Advancement Possibilities and Employment Outlook

With training and experience, laboratory animal care workers can become supervisors, research assistants, or animal breeders. The employment outlook is good through the year 2006. Drug companies, medical schools, and research centers are employing increasing numbers of technicians to help them with experiments.

Working Conditions

Laboratory animal care workers usually work 40 hours per week and sometimes nights and weekends. Working areas are usually well lighted and pleasant; however, animal care workers are exposed to unpleasant smells. They spend most of their time working with animals rather than with people.

Earnings and Benefits

Salaries vary depending on education and experience. The average salary for beginning laboratory animal care workers is about $16,000 per year. Those with more experience earn, on the average, between $15,000 and $18,000 per year. Benefits include paid holidays and vacations, health insurance, and sometimes pension plans and paid tuition.

Where to Go for More Information

American Association for Laboratory Animal Science
70 Timber Creek Drive
Cordova, TN 38018-4233
(901) 754-8620
www.aalas.org

American College of Laboratory Animal Medicine
200 Summerwinds Drive
Cary, NC 27511
(919) 859-5985
www.aclam.org

SMALL ANIMAL BREEDERS

Small animal breeders[6] raise and market fur-bearing animals, laboratory animals, animals for pet shops, and animals for the general public. Breeders usually work in rural sections of the country where animals can be raised in outdoor cages. They may be self-employed or manage farms for owners. Laboratory animal breeders raise animals used in medical and research laboratories, such as mice, rats, guinea pigs, cats, monkeys, and dogs. See Figure 5–9.

Animals raised for laboratory use and experiments are closely monitored and controlled. They may be of all the same general genetic makeup. Correct feeding and good health are important.

Education and Training Requirements

There are no specific education requirements for becoming a small animal breeder. Some fur farmers start as laborers on fur farms and learn their skills on the job. Two- and four-year programs are available at technical schools and colleges. Courses in animal husbandry, biology, zoology, nutrition, animal hygiene, chemistry, and genetics are useful. One should also study bookkeeping, marketing, and business.

Advancement Possibilities and Employment Outlook

A few fur farmers acquire large herds and become wealthy. However, the profit margin is small and uncertain in fur farming; it takes a long time to develop a profitable business. Because imported furs are less expensive than American furs, the long-range outlook for fur farmers in this country is not good.

The success of the laboratory animal breeder depends largely on the size and health of the herd or stock. There will be a steady demand for laboratory animals, for increasing numbers of medical tests, and for behavior and scientific research.

Working Conditions

Self-employed small animal breeders work long hours every day of the week. A few who own large operations may work regular hours and have employees care for the animals at other times. Most people who work with animals must do some heavy lifting and are exposed to unpleasant odors.

Earnings and Benefits

Earnings vary widely among small animal breeders. The size of the operation, the type of animal produced, efficiency, and volume of business will influence the earnings. Some breeders earn modest profits; others make fairly large amounts of money. Workers will earn about $9,000 to $16,000 per year and generally receive benefits such as paid holidays and vacation. Self-employed breeders must provide their own insurance and vacations.

Where to Go for More Information

American Animal Hospital Association
P.O. Box 15899
Denver, CO 80215-0899
(303) 986-2800
www.healthypet.com

American Association for Laboratory Animal Science
70 Timber Creek Drive
Cordova, TN 38018
(901) 754-8620
www.aalas.org

National Association of Animal Breeders
P.O. Box 1033
Columbia, MO 65205
(314) 445-4406
www.naab-css.org

The Fur Commission U.S.A.
225 East Sixth Street, Suite 230
St. Paul, MN 55101
(612) 222-1080
www.furcommission.com

Figure 5–9 Animal breeders raise animals for sale to pet shops, laboratories, and the general public. Some may raise fur-bearing animals for their pelts, and others may raise rabbits for the fur and meat.

ANIMAL TRAINERS

Animal trainers[7] teach animals, including aquatic mammals and birds, to obey commands, compete in shows or races, or perform tricks to entertain audiences. They also may teach dogs to protect property or act as guards for the visually impaired. See Figure 5–10.

Many different animals are capable of being trained. Dogs, horses, lions, bears, elephants, parakeets, cockatoos, whales, porpoises, and seals can all be trained. The techniques used are basically the same, regardless of the type of animal. Animal trainers conduct a program consisting of repetition and reward to teach animals to behave in a particular manner and to do it consistently. Animal trainers may also be responsible for the feeding, exercising, grooming, and general care of the animals, either handling the duties themselves or supervising other workers.

Education and Training Requirements

Persons wanting to be animal trainers should like animals and have a genuine interest in working with them. Establishments that hire trainers often require previous experience as an animal keeper or aquarist, because proper care and feeding of the animals is an essential part of a trainer's responsibilities. Most trainers begin their careers as keepers and gain on-the-job experience in evaluating the disposition, intelligence, and trainability of the animals they look after. At the same time, they learn to develop a rapport with them.

Figure 5–10 Animal trainers teach animals to obey commands and to perform tricks for audiences. Courtesy of PhotoDisc.

Although previous training experience may give job applicants an advantage in being hired, they still will be expected to spend time caring for animals before advancing to the position of trainer.

Part-time or volunteer work in animal shelters, pet shops, or veterinary offices offers would-be trainers a chance to become accustomed to working with animals and to discover whether they have the aptitude for it. Experience can be acquired in summer jobs as caretakers at zoos, aquariums, museums that feature live animal shows, and amusement parks. For those with a special interest in horses, racing or riding stables can provide experience.

Race horse trainers must be licensed by the state in which they work. Otherwise, there are no special requirements for this occupation. Animal trainers who are themselves part of the entertainment act, however, should have a good stage presence, have a pleasing voice, and be able to speak well before an audience.

Advancement Possibilities and Employment Outlook

Most establishments have very small staffs of animal trainers, which means that the opportunities for advancement are limited; the progression is from animal keeper to animal trainer. A trainer who directs or supervises others may be designated head animal trainer or senior animal trainer.

Some animal trainers go into business for themselves and, if successful, hire other trainers to work for them. Others become agents for animal acts.

The demand for animal trainers is not great because most employers have no need for a large staff and tend to promote from within. The field is expected to decline through the year 2006. Criticism of animals used for purely entertainment purposes has reduced the number used for shows and performances. Some openings may be created as zoos and aquariums expand.

Working Conditions

The working hours for animal trainers vary considerably, depending on the type of animal, performance schedule, and whether travel is involved. For some trainers, such as those who work with show horses, hours can be long and irregular.

Except in warm climates, animal shows are seasonal, running from April or May through midautumn. During this time, much of the work is conducted outdoors. In winter, trainers work indoors, preparing for warm weather shows. Trainers of aquatic mammals, such as dolphins and seals, must feel at ease working around water. The physical strength required depends on the animal involved, and animal trainers usually need greater-than-average agility.

Earnings and Benefits

Animal trainers can earn widely varying amounts depending on their specialty and place of employment. Salaries can range from $10,000 to $100,000 per year. Animal training jobs are generally low paying, with average rates between minimum wage and $10 per hour. A few of the specialists, such as those working with dolphins in aquariums, can earn in the mid-$20,000 range, but few individuals earn more. Those who do earn the higher salaries are in upper management and spend more time running the business than working with animals. In the field of racehorse training, however, trainers are paid an average of $35 to $50 per day for each horse, plus 10 percent of any money their horses win in races. Show horse trainers may earn as much as $30,000 to $35,000 per year.

Where to Go for More Information

American Zoo and Aquarium Association
7970-D Old Georgetown Road
Bethesda, MD 20814
(301) 907-7777
www.aza.org

Animal Caretakers Information
Humane Society of the U.S.
Companion Animal Division
5430 Grosvenor Lane, Suite 100
Bethesda, MD 20814
www.hsus.org

Canadian Association of Zoological Parks and Aquariums
c/o Calgary Zoo
P.O. Box 3036
Calgary, AB T2M 4R8

PET GROOMERS

Pet groomers[8] comb, cut, trim, and shape the fur of dogs and cats. Although all dogs and cats benefit from regular grooming, shaggy, long-haired animals give dog groomers the bulk of their business. Some types of dogs need regular grooming for their expected appearance; among this group are poodles, schnauzers, cocker spaniels, and many types of terriers. More cats, especially long-haired breeds, are being taken to pet groomers; the procedure for grooming cats is very similar to that used for dogs. See Figure 5–11.

Education and Training Requirements

There are generally three ways that a person interested in dog grooming can be trained for the field: enrollment in a dog grooming school, working in a pet

Figure 5–11   Dog groomers specialize in grooming dogs. Courtesy of Tracy Fera, PET SPAS of America.

shop or kennel, or reading one of the many books on dog grooming and practicing.

Probably the best way to gain knowledge of dog grooming is to take an accredited dog grooming course. The National Dog Groomers Association provides a referral listing of approximately forty dog grooming schools throughout the United States. Three schools of dog grooming are recognized by the National Association of Trade and Technical Schools (NATTS): the Pedigree Professional School of Dog Grooming, The New York School of Dog Grooming (three branches), and the Nash Academy of Animal Arts. It is important for students to choose an accredited, licensed school, both for their employment opportunities and their own professional knowledge.

High school diplomas generally are not required for persons working as dog groomers. A diploma or GED certificate, however, can be a great help to persons who would like to advance within their present company or move to other careers in animal care that require more training, such as animal technicians.

To enroll in most dog grooming schools, a person has to be at least seventeen years old and be fond of dogs. Animals can sense when someone does not like them or is afraid of them; certain skills are needed to work with nervous, aggressive, or fidgety animals. A person must be patient with the animals, be able to gain the animals' respect, and like to give the animals a lot of love and attention. Persistence and endurance are also good traits for dog groomers because grooming one dog can take 3 hours or more of strenuous work. Groomers must be able to deal with pet owners tactfully and gain their trust. Repeat customers and referrals are extremely important, especially to self-employed groomers. Groomers should also enjoy working with their hands, have good eyesight, and possess manual dexterity to cut a clipping pattern accurately. Previous experience in dog grooming can sometimes be applied for course credits.

State licensing or certification is not required of dog groomers at this time. To start a grooming salon or other business, a person may need to get a city license.

Advancement Possibilities and Employment Outlook

The demand for skilled dog groomers has grown faster than average and is expected to continue through the year 2006. The National Dog Groomers Association estimates that there are more than 30,000 dog groomers and expects that more than 3,000 new groomers will be needed every year over the next decade.

Working Conditions

Working conditions can vary greatly, depending on the location and type of employment. Many salons and pet shops are clean and well lighted, with modern equipment and clean surroundings. Others may be cramped, dark, and smelly. Groomers need to be careful while on the job, especially when handling flea and tick killers, which are toxic to humans. When working with any sort of animal, a person may encounter bites, scratches, strong odors, fleas, and other insects. They may also have to deal with sick or bad-tempered animals.

Groomers who are self-employed can work out of their homes. Many people convert their garages into work areas. Some groomers buy vans and convert them into grooming shops. They can drive to the homes of the pets, which many owners find very convenient.

Groomers usually work a five-day, 35-hour week. If they work any overtime, they are compensated for it. Those who own their own shops or work out of their homes, like other self-employed people, work very long hours and can have irregular schedules. Other groomers may work only part-time. Groomers are on their feet much of the day, and their work can get very tiring when they have to lift and restrain large animals.

Earnings and Benefits

Groomers can charge either by the job or by the hour; generally they earn around $7.50 an hour. If they are on the staff of a salon or work for another groomer, they keep 50 or 60 percent of the fees they charge. For this reason, many groomers branch off to start their own businesses. Those who own and operate their own pet-grooming service can earn anywhere from $20,000 to $50,000 annually, depending on how hard they work and the type of business they attract.

Groomers generally buy their own clipping equipment, including barber's shears, brushes, and clippers. A new set of equipment costs around $275. Groomers who work at salons, grooming schools, pet shops, animal hospitals, and kennels often get a full range of benefits, including paid vacations and holidays, medical and dental insurance, and retirement pensions.

Where to Go for More Information

California School of Dog Grooming
727 West San Marcos Blvd. Suite 105A
San Marcos, CA 92069
(760) 471-0787 or (800) 949-3746
www.csdg.net/index1a.htm

National Dog Groomers Association of America
PO Box 101
Clark, PA 16113
(724) 962-2711
www.nauticom.net/www/ndga/index.html

New York School of Dog Grooming
248 East 34th Street
New York, NY 10016
(212) 685-3776
(800) 541-5541
www.collegeedge.com/details/college/1/67/d4_2067.asp

Nash Academy of Animal Arts
857 Lane Allen Plaza
Lexington, KY 40504
(606) 276-5301
www.collegeedge.com/details/college/1/5/d4_1905.stm

VETERINARIANS

Veterinarians,[9,10] also called doctors of veterinary medicine, treat and control animal injuries and diseases. They immunize healthy animals against disease and inspect animals and meat products to be used as food. Veterinarians also perform surgery, set broken bones, establish diet and exercise routines, and prescribe medicines.

Of the 55,000 veterinarians in the United States, about one-third treat small animals exclusively. Small animal veterinarians usually have private practices. See Figure 5–12.

Education and Training Requirements

To get the Doctor of Veterinary Medicine (DVM or VMD) degree, one must have graduated from one of the twenty-seven schools of veterinary medicine in the

Figure 5–12   One-third of the veterinarians in the United States work exclusively with small animals. Courtesy of Brian Yacur and Guilderland Animal Hospital.

United States (see Appendix A). To become licensed to practice veterinary medicine, one must also pass the state's oral and written licensing examination.

One must have at least two years undergraduate training at a college or university before one can apply for admission to a veterinary college. Many students earn a bachelor's degree before they apply for admission.

Advancement Possibilities and Employment Outlook

Veterinarians who are in private practice advance by expanding and by developing a good reputation in their community. Some veterinarians who work for large organizations are promoted to supervisory or management positions. The employment outlook for veterinarians looks good. Veterinarians will be needed to treat the increasing number of household pets and to care for and prevent diseases in animals raised for food.

Working Conditions

Veterinarians in private practice establish their own hours, although they may be called out in the middle of the night for emergencies. Private practitioners often work well over 40 hours a week, especially those

who treat large animals on farms. The chief risk for veterinarians is injury by animals; however, modern tranquilizers and technology have made it much easier to work on all types of animals.

Earnings and Benefits

Earnings vary depending on the years of experience, location, and whether the veterinarian is salaried or in private practice. Beginning veterinarians employed by the federal government earn $35,800 per year. The average yearly salary of veterinarians working for the federal government is about $57,600. For those in industry, the average yearly salary for veterinarians is about $44,500.

The earnings of veterinarians in private clinical practice varies according to the location, type of practice, and years of experience. New graduates employed in the established private practices of other veterinarians generally are paid an average of $29,000 per year. The average income of veterinarians in private clinical practice was $57,500 in 1995. Owners of private clinical practices must operate their practices as a small business. The average starting income for practice owners specializing in large animal care was $39,500, compared with an average income of $31,900 for veterinarians specializing in small animal care.

Where to Go for More Information

American Animal Health Association
P.O. Box 15899
Denver, CO 80215-0899
(303) 986-2800
www.healthypet.com

American Veterinary Medical Association
1931 North Meacham Road, Suite 100
Schaumburg, IL 60173-4360
(800) 248-2862 or (847) 925-8070
www.avma.com

Conference of Public Health Veterinarians
Office of Medical Services, Peace Corps
1900 K Street, NW
Washington, DC 20526
(202) 606-3512

U.S. Department of Agriculture
Animal and Plant Health Inspection Service
Federal Building
Butler Square West
4th Floor
100 North Sixth Street
Minneapolis, MN 55403
www.aphis.usda.gov

VETERINARY TECHNICIAN

Veterinary technicians[11] assist veterinarians and other members of the veterinary staff in diagnosing and treating animals for injuries, illness, and routine veterinary needs, such as standard inoculations and periodic checkups. See Figure 5–13.

Working under the supervision of the veterinarians, veterinary technicians perform many tasks, ranging from soothing and quieting animals under treatment to drawing blood, inserting catheters, and conducting laboratory tests. Oftentimes veterinary technicians have the closest contact to the clinic's animal population.

Education and Training Requirements

To become a veterinary technician, one needs to complete a two-year program leading to an associate degree from one of the sixty-five community or technical colleges whose program has been accredited by the American Veterinary Medical Association. A high school diploma is an essential requirement for admission to such a program. Courses in algebra, chemistry, biology, and English are essential.

Most programs offer students practical experience with live animals and field experiences under actual working conditions. Successful veterinary technicians combine a love for animals with good manual dexterity and an aptitude for record keeping, writing reports, and speaking effectively.

Advancement Possibilities and Employment Outlook

With experience and continuing education, veterinary technicians can expect to advance to greater levels of responsibility. In some settings, this simply means performing more difficult or sophisticated procedures, or it could mean that the technicians take on supervisory or other administrative responsibilities.

The employment outlook for veterinary technicians is very good. Contributing to this favorable outlook is the growth and interest in companion and recreational animals. Veterinarians are recognizing the value of qualified technicians in their practices, and more veterinary technicians are employed in recreational settings such as resorts, riding stables, and race tracks. Currently, there is a shortage of veterinary technicians.

Working Conditions

Veterinary technicians generally work 40-hour weeks. Hours may vary as veterinary technicians may need to adjust their schedules to allow for emergency work.

When working in the field, veterinary technicians may deal with unpleasant weather conditions while treating animals. Veterinary technicians may be at risk when treating injured animals.

People who become veterinary technicians care about animals, and they find caring for and treating animals very rewarding work. Veterinary technicians working in zoos may work with very interesting and sometimes endangered species.

Earnings and Benefits

About 70 percent of veterinary technicians are employed in private or clinical practice and research. Earnings for zoo veterinary technicians range from $17,000 to $35,000. Salaries in clinical or private practice range from $15,000 for recent graduates to

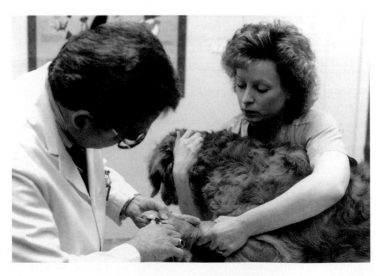

Figure 5–13 Veterinary technicians assist veterinarians. Courtesy of Brian Yacur and Guilderland Animal Hospital.

$40,000 for experienced graduates. Many factors affect earnings for veterinary technicians. Some factors include practical setting, geographic location, level of education, and years of experience. Benefits and vacation vary because most veterinary technicians work in private practice.

Where to Go for More Information

Association of Zoo Veterinary Technicians
(AZVT)
AZVT Office
c/o Louisville Zoo
PO Box 37250
Louisville, KY 40233
(502) 451-0440
www.worldzoo.org/azvt

American Veterinary Medical Association
1931 North Meacham Road, Suite 100
Schaumburg, IL 60173-4360
(800) 248-2862 or (847) 925-8070
www.avma.com

North American Veterinary Technician
Association
PO Box 224
Battleground, IN 47920
(765) 742-2216
www.avma.org/nvata

Association of Veterinary Technician Educators
Department of Veterinary Science
North Dakota University
Fargo, ND 58105

Canadian Veterinary Medical Association
339 Booth Street
Ottawa, ON K1R 7K1
(613) 236-1162
www.cvma-acmv.org

BIOLOGISTS

Biologists[12,13] study the origin, development, anatomy, function, and distribution of living organisms. They do research in many specialties that advance the fields of medicine, agriculture, and industry. See Figure 5–14.

Because of the breadth of the field, the nature of the work performed by individual biologists varies widely. The biologist may also be called a biological scientist, or life scientist, and may be identified by their specialties. **Botanists** study plants. **Zoologists** study animals. **Anatomists** study the structure and form of plants and animals. **Physiologists** study the life functions of plants and animals; these func-

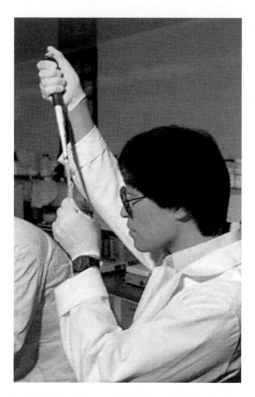

Figure 5–14 The jobs and duties of biologists vary considerably depending on the specific field in which they are employed. Courtesy of Utah Experimental Station.

tions include growth, respiration, and reproduction. **Embryologists** study the development of animals from their beginnings as fertilized eggs until the time they hatch or are born. **Geneticists** study heredity and how characteristics vary in all forms of life. **Pathologists** concentrate on the effects of diseases on the cells, tissues, and organs of plants and animals. **Nutritionists** study how food is used and changed into energy. **Pharmacologists** study the effects of drugs and other substances such as poisons on living organisms. **Ecologists** study the relationship of plants and animals to their environment. **Biochemists** study the chemistry of living matter. **Biophysicists** concentrate on the atomic structure and the electrical and mechanical energy of cells and organisms.

Education and Training

All biologists have earned undergraduate degrees in science. However, most go on to complete master's and doctorate degrees in their area of interest. A master's degree and a doctorate are essential for any biologist to conduct serious research, publish in scholarly journals, and obtain a faculty position at a college or university.

Advancement Possibilities and Employment Outlook

There are many possibilities for advancement in the field of biology, especially for those who have a doctorate degree. Job opportunities for biologists are expected to increase at a rate faster than the average for all occupations through the year 2006. Advances in genetic research leading to new drugs, improved plants, and medical discoveries should open up opportunities. Products now made by chemical or other means will soon be produced by biological methods. Efforts to preserve the environment may also result in an increased number of jobs.

Working Conditions

Working conditions for biologists vary considerably depending on the position and employer. Most will spend part of their time in laboratories, and these laboratories are usually clean, well lighted, and well equipped. Some may spend time in offices and classrooms. Others may do fieldwork and travel extensively.

The basic work week is 40 hours, but extra time may be spent working on research projects, writing up the results of projects, and reading to stay up with current information.

Earnings and Benefits

Earnings for biologists vary widely. The salaries depend on the education and experience of the scientist, the location, and the kind of job. Beginning salaries in private industry averaged $25,400 per year for persons with bachelor's degrees. Experienced biologists earn between $28,400 and $50,900 per year. Biologists in the federal government averaged $52,100 annually.

Where to Go for More Information

American Institute of Biological Sciences
1444 1 Street, NW, Suite 200
Washington, DC 20005
(202) 628-1500
www.aibs.org

Federation of American Societies for
Experimental Biology
9650 Rockville Pike
Bethesda, MD 20814-3998
(301) 530-7000
www.faseb.org

SUMMARY

There are few get-rich jobs working with animals. Salaries can be comfortable and increase with educa-

tion and experience. Most people who chose careers in animal care do so because of their love of animals.

Persons involved in animal care professions must realize that cleaning, feeding, and general care is a constant job. Overtime hours are common, especially when caring for sick or pregnant animals.

Employment opportunities look good for the future because 31.6 percent of American households have a dog, 27.3 percent have one or more cats, 1.9 percent have rabbits, and 1 percent have hamsters. Americans spend in excess of $20.3 billion a year on their pets; pet food manufacturers produce some $9.0 billion in sales, and veterinary care easily exceeds $11 billion annually.

The pet industry is just one area of small animal care. Numerous jobs exist in laboratory animal care, zoo animal care, animal health, animal training, and the many areas of the biological sciences.

DISCUSSION QUESTIONS

1. How large is the animal care industry? How important is it in your area?

2. What are the duties of the workers in the various small animal careers?

3. What are some of the different working conditions one might encounter?

4. What are the job requirements for the various animal care careers?

5. What are some background requirements that would be helpful when applying for a job in small animal care?

6. What requirements are necessary for advancement in the various careers?

7. What are some high school courses that would be helpful for students to take that might prepare them for a career in small animal care?

SUGGESTED ACTIVITIES

1. Make a list of small animal career workers in your area.

2. Visit some of the small animal career workers in your area and interview them to find out how they started in their profession, what education they needed, and what other requirements there were.

3. With the approval of the teacher, invite animal care workers to your classroom to present programs on their careers.

4. Make a list of desired characteristics, other than general education requirements, that are

helpful for a successful career in small animal care.

5. Do research and make a list of the various colleges and institutions that offer career preparation in the different areas of small animal care.

6. Write letters to the different organizations that are listed in this chapter to receive more information.

ADDITIONAL RESOURCES

1. 2000-01 Occupational Outlook Handbook
   http://stats.bls.gov/ocohome.htm

2. Career Consulting Corner
   www.careercc.com

END NOTES

1. Excerpt from *U.S. Pet Ownership and Demographics Sourcebook,* Center for Information Management, American Veterinary Medical Association, 1997.

2. *Career Information Center,* "Pet Care Workers": Macmillan Library Reference USA, 1633 Broadway, New York, NY, 10019, 7th Edition, Volume 5, pages 48–50, 1999.

3. Khan, S. and Philip Lief Group. 1988. *101 Best Businesses To Start,* New York: Doubleday, p. 433.

4. *Career Information Center,* "Zoo Administrators": Macmillan Library Reference USA, 1633 Broadway, New York, NY, 10019, 7th Edition, Volume 8, pages 129–130, 1999.

5. *Career Information Center,* "Laboratory Animal Care Workers": Macmillan Library Reference USA, 1633 Broadway, New York, NY, 10019, 7th Edition, Volume 7, pages 35–36, 1999.

6. *Career Information Center,* "Small Animal Breeders": Macmillan Library Reference USA, 1633 Broadway, New York, NY, 10019, 7th Edition, Volume 1, pages 93–95, 1999.

7. *Encyclopedia of Careers and Vocational Guidance,* "Animal Trainers": J.G. Ferguson Publishing Company, 200 West Jackson Boulevard, Chicago, IL, 60606, 11th Edition, Volume 2, pages 130–135, 2000.

8. *Encyclopedia of Careers and Vocational Guidance,* "Dog Groomers": J.G. Ferguson Publishing Company, 200 West Jackson Boulevard, Chicago, IL, 60606, 11th Edition, Volume 4, pages 71–75, 2000.

9. *Career Information Center,* "Veterinarians": Macmillan Library Reference USA, 1633 Broadway, New York, NY, 10019, 7th Edition, Volume 1, pages 146–148, 1999.

10. *Encyclopedia of Careers and Vocational Guidance,* "Veterinarians": J.G. Ferguson Publishing Company, 200 West Jackson Boulevard, Chicago, IL, 60606, 11th Edition, Volume 4, pages 772–774, 2000.

11. *Encyclopedia of Careers and Vocational Guidance,* "Veterinary Technician": J.G. Ferguson Publishing Company, 200 West Jackson Boulevard, Chicago, IL, 60606, 11th Edition, Volume 4, pages 774–777, 2000.

12. *Encyclopedia of Careers and Vocational Guidance,* "Biologists": J.G. Ferguson Publishing Company, 200 West Jackson Boulevard, Chicago, IL, 60606, 11th Edition, Volume 2, pages 298–303, 2000.

13. *Career Information Center,* "Biologists": Macmillan Library Reference USA, 1633 Broadway, New York, NY, 10019, 7th Edition, Volume 6, pages 92–95, 1999.

CHAPTER
6

Nutrition and Digestive Systems

OBJECTIVES

After reading this chapter, one should be able to:

- define the terms nutrition and nutrient.
- list the six basic nutrient groups.
- discuss the difference between the ruminant and nonruminant digestive system.
- discuss the importance of the various nutrients in the diets of animals.

TERMS TO KNOW

absorption	enzymes	nutrient
amino acid	hemoglobin	nutrition
antibodies	hormones	pH
assimilation	macrominerals	respiration
biochemical reaction	microminerals	ruminant animals
coprophagy	nitrogen-free extract	solubility
digestion	nonruminant animals	ventriculus
DNA		

NUTRITION

Adequate nutrition is important for pets. **Nutrition** refers to the animal receiving a proper and balanced food and water ration so that it can grow, maintain its body, reproduce, and supply or produce the things we expect from it. Animals can provide us with work, egg production, milk production, meat production, production of offspring for sale, fur or pelt production, and companionship. See Figure 6–1.

The term **nutrient** refers to a single food or group of foods of the same general chemical composition that supports animal life.

In discussing nutrients, we cover six basic nutrient groups or classes: water, proteins, carbohydrates, fats, vitamins, and minerals.

Water

Water is in every cell of the animal and is necessary for the following:

1. supporting **biochemical reactions** in the body, including **respiration, digestion,** and **assimilation** (see Glossary)

Figure 6–1   All animals must receive proper nutrition if they are to grow well and maintain their health. Courtesy of Catherine Wein.

2. transporting other nutrients

3. helping maintain body temperature

4. helping give the body its form

5. carrying waste from the body

Water is more important than any other nutrient group. An animal can go several days without food; it may become hungry and lose body weight, but it will not suffer significantly from the lack of food. However, without drinking water, there will be a loss of water from the blood, which results in a failure of proper circulation and decrease in the oxygen-carrying capacity. The tissues become dehydrated and the body becomes overheated. The cells of the body are starved for oxygen, and there is a slowdown in normal body functions. There is also a lowering of the body's resistance to disease.

Water is quickly absorbed into the body through the walls of the stomach. It is important that the animal receive a supply of good clean, fresh water. Any disease organisms in the water are also quickly absorbed into the animal's body, which creates a possibility of infection. Water is also a good carrier for drugs or chemicals for disease control.

Water makes up about 55 to 65 percent of an animal's body. A beef steer has about 55 percent of its body composition as water, whereas a mouse has about 65 percent. The composition of an animal's blood is 90 to 95 percent water, muscle is 72 to 78 percent water, and bone is 30 to 40 percent water.

Proteins

Proteins are complex nutrients composed of carbon, hydrogen, oxygen, and nitrogen. Proteins are needed in an animal's body for the following:

1. developing and repairing body organs and tissues, such as muscles, nerves, skin, hair, hooves, and feathers

2. producing milk, wool, and eggs

3. developing the fetus

4. generating enzymes and hormones

5. developing antibodies

6. transmitting DNA

Proteins found in feed materials are broken down into amino acids during the digestive process. These amino acids are then carried by the bloodstream to various parts of the body.

Twenty-five amino acids are found in animal feeds. Of these, ten or eleven are considered essential and the others are nonessential. Essential amino acids are those that cannot be produced by the animal's body and, therefore, must be supplied in the ration. The nonessential amino acids are those that are not needed or can be synthesized by the animal's body and do not need to be supplied in the diet.

Ruminant (see Glossary) animals are capable of manufacturing all the amino acids they require if any one or more amino acids are given in large enough quantity. In this process, nitrogen combines with carbohydrates to produce the various amino acids.

The essential amino acids are arginine, isoleucine, histidine, leucine, methionine, threonine, phenylalanine, lysine, tryptophan, and valine. Poultry also require the amino acid glycine. In animal diets, cystine can be used to replace insufficient amounts of methionine and thus is considered by some to be an essential amino acid.

The amount of the essential amino acids needed varies depending on the animal's function and stage of growth.

Carbohydrates

Carbohydrates contain the chemical elements carbon, hydrogen, and oxygen. Animals convert carbohydrates into energy that is needed for the following:

1. supporting bodily functions, such as breathing, digestion, and exercising

2. producing heat to keep the body warm

3. storing fat

Carbohydrates are made up of the group of chemicals called sugars, starches, and crude fiber. These groups are usually classified as **nitrogen-free extract** (N.F.E.) and crude fiber on most feed bags. Nitrogen-free extracts are the more easily and completely digested sugars and starches. Crude fiber is composed mostly of nondigestible bulk or roughage. Some animals, such as the ruminants, can handle larger amounts of crude fiber than the single-stomached animals.

Fats

Fats are made up of the same chemical elements as carbohydrates, but in different combinations. They contain 2.25 times as much energy as an equivalent amount of carbohydrates and proteins. Fats are essential in the diet for the following:

1. providing energy

2. aiding in the absorption of fat-soluble vitamins

3. providing fatty acids (Certain fatty acids such as linoleic, linolenic, and arachidonic acids are essential in the diet of animals. Most animal requirement for fat is less than 3 percent.)

Vitamins

Vitamins are organic substances required in very small amounts. Vitamins do not build body tissues but are necessary for specific biochemical reactions such as the following:

1. regulating digestion, absorption, and metabolism

2. developing normal vision, bone, and external body coverings such as hair and feathers

3. regulating body glands

4. forming new cells

5. protecting animals against diseases

6. developing and maintaining the nervous system

Vitamins are classified on the basis of their **solubility** such as being fat soluble or water soluble. Fat-soluble vitamins can be stored and accumulated in the liver, fatty tissues, and other parts of the body, whereas only limited amounts of the water-soluble vitamins are stored. Fat-soluble vitamins include A, D, E, and K. Water-soluble vitamins include C and the B-complex vitamins; the B-complex group includes thiamine, riboflavin, niacin (nicotine acid), pyrodoxine, inositol, cobalamin (B_{12}), pantothenic acid, folic acid, biotin, choline, and para-amino benzoic acid.

Vitamin A is required to prevent poor vision, respiratory ailments, digestive problems, and reproductive difficulties. Green leafy plants contain carotene (provitamin A), which animals convert to vitamin A in the body.

Vitamins D, D_2, and D_3 are associated with the use of calcium and phosphorus in the body. Animals deficient in vitamin D usually have weak trembling legs. Animals exposed to sunlight have plenty of vitamin D, whereas animals raised in confinement may need supplemental feeding of vitamins D, D_2, and D_3. Good-quality, sun-cured hay is a good source of these vitamins.

Vitamin E is required in female rats for normal gestation and in the male rat to prevent sterility; it is also considered important in other animals for successful reproduction. In ruminants, a lack of vitamin E is associated with a form of paralysis referred to as "white muscle disease" (and as "stiff lamb disease" in lambs). A lack of vitamin E in rabbits can cause a lack of interest in breeding, poor fertility, abortions, and miscarriages. Vitamin E has also been associated with muscle development. Most animals being fed proper rations are usually receiving sufficient amounts of vitamin E. See Figure 6–2.

Vitamin K is necessary for maintenance of normal blood coagulation and a lack of vitamin K has been shown to cause respiratory illnesses in rabbits. Green forages, seeds, and good hay should provide sufficient amounts of vitamin K.

Vitamin C, or ascorbic acid, is apparently produced in the digestive system of most animals and is not a consideration in feeding rations. However, guinea pigs and monkeys must receive vitamin C to prevent scurvy, which is characterized by swollen, painful joints and bleeding gums. Diarrhea, rough hair, and hair loss are also signs of vitamin C deficiency. Vitamin C is unstable and, when combined in feeds, begins to break down and be of little value. Bags of feed used for guinea pigs must be used

Figure 6–2 Vitamin E is important for successful reproduction. Courtesy of Catherine Wein.

Figure 6–3 Adequate nutrition is one of the most important factors in keeping and maintaining pets.

shortly after opening or the vitamin C dissipates. See Figure 6–3.

Vitamin B_1, or thiamine, was one of the first vitamins discovered; a deficiency of this vitamin was probably among the first known nutritional deficiency diseases. Symptoms of thiamine deficiency are decreased appetite, muscular weakness, and a characteristic paralysis known as polyneuritis or beri-beri. Thiamine is also required for normal metabolism of carbohydrates. Animals receiving raw, whole grains or their products do not usually develop thiamine deficiency. Freshwater fish contain an enzyme that destroys thiamine in the ration. Mink and fox fed raw, fresh fish develop thiamine deficiency; this condition can be overcome by cooking the fish.

Vitamin B_2 is also referred to as riboflavin. Lack of this vitamin leads to poor hatchability, deformed chicks, and poults. In swine and other animals, a lack of this vitamin results in crippled and deformed young. In rabbits, eye problems result. Riboflavin functions in several enzyme systems that are important in the metabolism of amino acids and carbohydrates. Whole or skimmed milk, green forages, and good-quality hay are sources of this vitamin. Riboflavin is available at low cost for the fortification and supplementation of breeding and growing rations.

Niacin was first recognized and designated as the vitamin preventing pellagra, or black tongue factor. Pellagra in humans shows up as a reddening of the skin and development of sores in the mouth and the intestinal tract, resulting in bloody diarrhea. In other animals, a lack of niacin causes digestive disorders and retarded growth. Ducks and pheasants appear to have a high niacin requirement. Niacin is a white powder that is available in adequate quantities at rather low cost. Most feeds contain some niacin, but usually not in sufficient amounts.

Pantothenic acid or panacid deficiency appears as abnormal skin condition around the beak and retarded growth in chicks and turkeys. In swine, a deficiency results in the development of a peculiar gait of the hind legs referred to as "goose stepping." This condition is caused by permanent damage to the sciatic nerve. Corn contains a low amount of pantothenic acid, but other feed materials have adequate amounts. Pantothenic acid is available from chemical sources at reasonable prices.

Vitamin B_{12}, or cobalamin, is essential as a coenzyme in several biochemical reactions. It is essential for normal growth, reproduction, and blood formation. This vitamin was originally referred to as the animal protein factor (APF). The most economical and reliable source of vitamin B_{12} is the chemically pure form. Animal protein feeds such as fish meal, fish solubles, liver meal, and dried milk products are also good sources of vitamin B_{12}.

Choline functions in the transportation and metabolism of fatty acids. Deficiencies of choline show up as slipped tendons or perosis in chicks and turkeys, and as kidney and liver damage in other animals. Animals lacking choline develop what is referred to as "fatty livers." Most feeds are good sources of choline; supplements of choline chloride ensure an adequate supply.

Folic acid, or folacin, is required for normal cell development and is essential in certain biochemical reactions; poor growth and various blood disorders are common symptoms. Most feeds are fair to good sources of folic acid, and supplementary rations are usually not required.

Biotin is related to carbon dioxide fixation and carboxylation. Dermatis, loss of hair, and retarded growth are common deficiency symptoms. Biotin usually is supplied in adequate amounts by common food materials and is usually not supplemented. When a diet contains raw egg white, avidin in the egg white combines with the biotin in the intestinal tract to make biotin unavailable to the animal.

Inositol is one of the B-complex vitamins that is widely found in feeds. Its exact function is not understood and deficiencies usually do not occur.

Para-amino benzoic acid is essential for growth of certain microorganisms. Its source is not well known, but it evidently is in adequate amounts because deficiencies do not occur.

Vitamin B$_6$, or pyridoxine, deals with fat metabolism and seems to be associated with the transportation and synthesis of unsaturated fatty acids such as linoleic, linolenic, and arachidonic acids. Deficiency symptoms may appear as poor growth, anemia, and convulsions in pigs. In rabbits, vitamin B$_6$ assists in digesting and making full use of the proteins in the diet. Most feed materials contain adequate amounts of vitamin B$_6$.

Minerals

Like the vitamins, minerals are essential to the support of the animal but do not contribute to tissue development. Minerals are components of the ash or noncombustible part of the ration that cannot be burned; it is the residue left after the material has been burned. The primary function of minerals is to supply the materials for building the skeleton and producing body regulators such as enzymes and hormones.

Of the twenty elements that function in animal nutrition, carbon, hydrogen, oxygen, and nitrogen are regarded as the nonmineral elements. There are sixteen minerals that are divided into two groups: the seven major or macrominerals that are needed in the largest quantity and are most likely to be lacking in the feed supply, and the nine trace or microminerals that are needed in trace amounts.

The macrominerals are calcium, phosphorus, potassium, sodium, sulfur, chlorine, and magnesium. The microminerals are iron, iodine, copper, cobalt, fluorine, manganese, zinc, molybdenum, and selenium.

Calcium is required by vertebrates in larger amounts than any other mineral. Ninety-nine percent of the calcium found in the body is deposited in the bones and teeth. Calcium is essential for bone, teeth, and eggshell formation; for normal blood coagulation; and for milk production.

A lack of calcium causes retarded growth and deformed bones in the young. A prolonged and extreme deficiency of calcium causes a condition known as "rickets," which is characterized by soft, flexible bones. In older animals, lack of calcium may lead to two common conditions: osteoporosis, a condition in which bones lose their density and become porous and brittle, and osteomalacia, a softening of the bones. Egg-producing birds and animals lay soft-shelled eggs when fed rations low in calcium. The most economical sources of calcium are limestone, oyster shell, bone meal, and defluorinated phosphates; crushed eggshells and cuttlefish are also good sources.

Phosphorus is closely associated with calcium in animal nutrition. Seventy-five percent of the phosphorus found in the body is deposited in the bones and teeth. Phosphorus is essential for the formation of bones, teeth, and body fluids and is required for metabolism, cell respiration, enzyme-based reactions, and normal reproduction. Symptoms of phosphorus deficiency are similar to those of calcium deficiency in growing animals.

The most economical sources of phosphorus are low fluorine phosphates and bone meal. Other good sources are wheat bran and middlings, high-protein feeds, bone-containing feeds, whole or skimmed milk, and most grains.

Potassium is required for many body functions, such as osmotic relations, acid-base balance, and digestion. Symptoms of a potassium deficiency are nonspecific but may appear as decreased feed consumption, slow growth, stiffness, and a loss of weight. Grains and most feeds are sufficient in potassium; supplements are usually not required.

Sodium and chlorine are necessary in the formation of digestive juices, control of body fluid concentration, control of body fluid pH, and in nerve and muscle activity.

Under most conditions, there are no deficiency symptoms. The only signs may be unthrifty-appearing animals and impaired performance. In heavy-working or heavy-producing animals, a lack of these two minerals may cause sudden death. Sodium is especially important in ruminant animals because it aids in the formation of sodium bicarbonate needed to neutralize acids produced by the fermentation process in the rumen. In birds, nervousness, poor feathering, and feather plucking are signs of deficiency.

Supplemental salt provided for animals normally provides them with a source of sodium and chlorine. Salt can be provided in the form of blocks, loose or free choice salt, spools that attach to cages, or a mineral mix.

Sulfur is utilized in the formation of sulfur-containing amino acids in the rumen and in the formation of various body compounds. Most feeds contain more-than-adequate supplies of sulfur; thus, deficiencies are usually not a concern. If a deficiency does appear, it expresses itself as a protein deficiency, which is a general unthrifty condition and poor performance.

Magnesium is necessary for many enzyme systems, in carbohydrate metabolism, and for proper functioning of the nervous system. Most common feeds contain enough magnesium so that a deficiency usually does not appear; however, if feed rations are low in magnesium, problems may occur. The most common problems are with cattle and sheep being fed low-quality forages, especially in late winter and early spring. Animals will develop a hypomagnesemia associated with severe toxic symptoms and frequently death. The condition is also referred to as "grass tetany" or "grass staggers."

Iron is necessary in the formation of hemoglobin. Deficiencies of iron usually show early in an animal's life because milk is low in iron content; thus, nursing animals are more susceptible. However, newborn animals normally have sufficient iron reserves in the liver and spleen to carry them through the early nursing period.

Anemia is the disease associated with iron deficiency. Animals with anemia show a loss of appetite and become weak. There will sometimes be a swelling around the head and shoulders; animals will also have trouble breathing and breathe with a characteristic dull "thumping" sound. Anemia can be prevented in young animals with concentrated ferrous sulfate administered orally, or with iron-based injections. See Figure 6–4.

Copper is necessary for proper iron absorption, for hemoglobin formation, in various enzyme systems, and in the synthesis of keratin for hair and wool growth. Most feeds contain sufficient supplies of copper; however, if the soil is lacking in copper, feeds grown in those areas may contain low amounts of copper. Copper is closely associated with iron, and a deficiency normally shows up as an anemic condition.

Iodine is important in the production of thyroxin by the thyroid gland. Most feeds and water found in the United States contain sufficient natural amounts of iodine; some areas around the Great Lakes and the far Northwest may be deficient. Common symptoms at birth or soon afterward are goiter (a swelling of the thyroid glands in the neck area), dead or weak animals, hairlessness, or infected navels.

Cobalt is an important component of the vitamin B_{12} molecule and is necessary in ruminant animals for proper synthesis of vitamin B_{12}. The microorganisms in the rumen require cobalt for growth. Symptoms of cobalt deficiency include poor appetite, unthriftiness, weakness, anemia, decreased fertility, slow growth, and decreased milk and wool production. Areas that have cobalt-deficient soils need to supplement this mineral in the diet.

Fluorine reduces the incidence of dental caries in humans and possibly other animals. This mineral is also associated with calcium and phosphorus utilization and is possibly important in retarding the breakdown of these elements in aging animals. Most feeds are usually sufficient in this nutrient, and fluorine is not a major concern in formulating diets.

Manganese is involved with the enzyme systems that influence estrus, ovulation, fetal development, udder development, milk production, growth, and skeletal development. A deficiency of manganese takes the form of delayed estrus, reduced ovulation, reduced fertility, deformed young, poor growth, lowered serum alkaline phosphates, abortions, resorptions, and lowered tissue manganese. Most feeds contain adequate amounts of manganese.

Zinc utilization is not fully understood, but it has long been recognized in laboratory animals that zinc is necessary in the diet to promote general thriftiness and growth, wound healing, and hair and wool growth.

In swine, a condition called "parakeratosis" is associated with zinc deficiency; calcium in the diet may render zinc unavailable or inadequate. Most feeds are adequate in zinc.

Molybdenum is a component of the enzyme xanthine oxidase, which is important in poultry for uric acid formation. This mineral is also important in stimulating action of rumen organisms. Molybdenum is usually adequate in most feeds.

Selenium is necessary for the absorption and utilization of vitamin E. Symptoms of selenium deficiency are similar to those of vitamin E deficiency and appear in the form of heart failure, paralysis, poor growth, low fertility, liver necrosis, and muscular dystrophy in cattle and lambs. Many areas of the country produce forages that are deficient in selenium, and supplementing diets is very important.

A lot of research has been conducted in recent years on pet nutrition. Proper feeding is easy today because of the variety of commercial pet foods available. Usually, a brand-name pet food contains a complete and balanced diet for pets, and recommendations for proper amounts to feed are included in the labeling.

Deficiencies usually do not occur among animals fed properly formulated diets; most problems occur from feeding animals table scraps or homemade diets. It is difficult to feed a homemade diet that contains all the nutrients required by animals.

Figure 6–4   Newborn animals should be watched carefully for signs of iron deficiency. Courtesy of Laurette Richen.

Supplemental feeding of vitamins and minerals is made easy with the use of trace mineralized salt, liquid vitamins, and minerals that are easily added to water.

ANIMAL DIGESTION

Digestion in animals is the process of breaking down food material into their various nutrient forms that can be absorbed into the animal's bloodstream.

Digestive systems in animals are basically of two types: ruminant and nonruminant. To understand the nutritional requirements of various animals, we must understand how they digest their foods. Ruminant animals are often referred to as "forage-consuming" or "multistomached animals," and include farm animals such as cattle, sheep, and goats and many animals commonly found in the zoo, such as elk, deer, giraffes, buffalo, camel, and antelope. The nonruminant animals are often referred to as "single-stomached" or "monogastric animals," and they include all of the small animals discussed in this book. Although the digestive systems of birds, rabbits, and horses differ somewhat, they are considered as nonruminant animals because they have a single digestive organ, or stomach.

Ruminant animals have four compartments to their stomach. These compartments are the rumen, reticulum, omasum, and abomasum. See Figure 6–5.

The rumen is the largest compartment and makes up about 80 percent of the total capacity of the stomach. A mature dairy cow may have a rumen capacity of 50 to 60 gallons. A ruminant animal chews food just

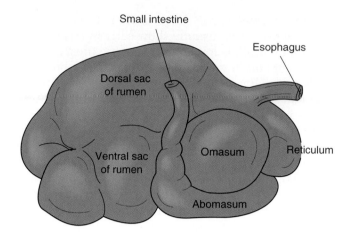

Figure 6–5 *A ruminant digestive system. Note the four compartments of the ruminant stomach.*

enough to aid in swallowing. When the food material reaches the rumen, it is worked upon by millions of bacteria and microorganisms. These bacteria and microorganisms transform low-quality protein and some nitrogen compounds into essential amino acids. They also manufacture many needed vitamins, including the B-complex group. Bacteria and microorganisms utilize some of these nutrients in food and, when they die, are digested and utilized by the animal.

Food material not fully digested in the rumen can be regurgitated in the form of cud. The animal chews on this cud and then swallows it back down into the rumen for further digestion.

Closely associated with the rumen is the reticulum. The reticulum works with the rumen in the formation of the cud for regurgitation. Foreign bod-

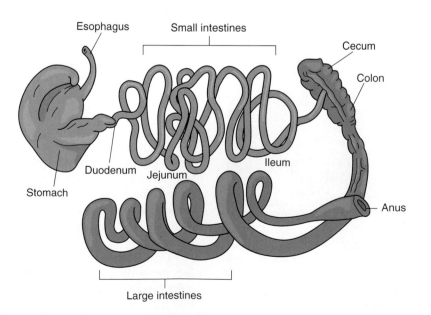

Figure 6–6 *A single stomach or monogastric digestive system.*

ies such as nails or pieces of wire can be held in the reticulum for long periods without causing serious injury. The reticulum is often referred to as the "hardware stomach" for this reason. Because of its beehive interior appearance, it is also referred to as the "honeycomb."

The third compartment is the omasum, which is made up of many layers of strong muscular tissue referred to as the "many plies." The function of this third compartment is to remove large amounts of water from food as it moves from the rumen to the fourth compartment.

The fourth compartment, or abomasum, is referred to as the "true stomach" because it functions very similarly to the stomach found in single-stomached animals.

The first step in the digestive process is the breaking, tearing, cutting, and grinding that takes place in chewing. Saliva is mixed with the food material in the mouth; saliva aids in the chewing process, helps the animal swallow, and aids in controlling the pH of the stomach. Saliva contains the first digestive enzymes in the digestive process. Salivary amylase works to change some starch into maltose, or malt sugar.

In **nonruminant animals,** food passes from the mouth to the stomach through the esophagus. Food is moved through the esophagus by a series of involuntary muscle contractions. See Figures 6–6, 6–7, and 6–8.

The functions of the stomach are to break down food material by muscular movement and to secrete digestive juices. These digestive juices begin to break down proteins and to break down fat into fatty acids and glycerol.

Food passes from the stomach to the small intestine, which is the primary site for the digestion and absorption of carbohydrates, fats, and proteins. The small intestine is divided into three sections: the duo-

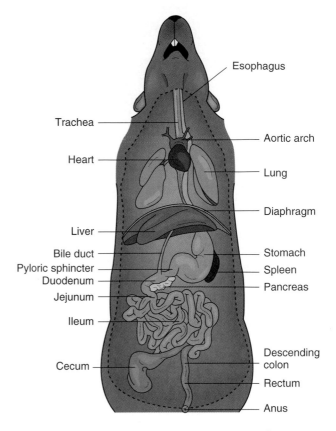

Figure 6–7 The rat is a monogastric animal.

denum, the jejunum, and the ileum. Enzymes from the pancreas enter into the duodenum and aid in the breakdown of fats; bile from the gallbladder enters the duodenum and aids in the breakdown of carbohydrates and proteins. The jejunum and ileum are active in the absorption of digestive nutrients.

Undigested food material passes from the small intestine into the large intestine. The large intestine is also divided into three sections: the cecum, the colon,

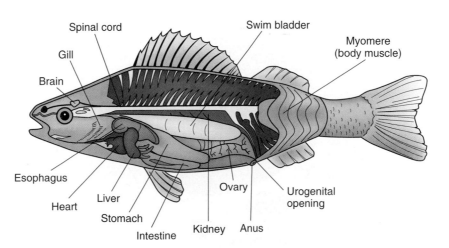

Figure 6–8 Fish are another representative of a monogastric animal.

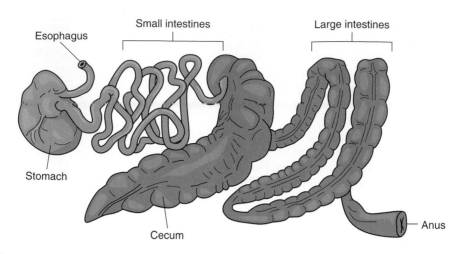

Figure 6–9   The digestive system of a horse. Note the large cecum that enables the animal to consume large amounts of roughage.

and the anus. The primary function of the large intestine is to absorb the water from the undigested material; to add lubricating mucus, which aids in the passage of the material through the large intestine; and to shape the material for easier passage. The rectum connects the large intestine with the anus.

The horse can consume large amounts of forages but is not considered a ruminant animal. It has a small, single stomach, but unlike other single-stomached animals, it has a large cecum and colon located between the small and large intestines. Bacterial action takes place in the cecum and allows the horse to digest roughage material, but not as efficiently as a true ruminant animal. Horses do not have a gallbladder. Bile is secreted into the duodenum directly from the liver. In feeding horses, one needs to reduce the amounts of

roughage and increase concentrates, which contain higher-quality proteins. See Figure 6–9.

The rabbit's digestive system is similar to that of the horse. It too has a large cecum that allows for the utilization of high-quality roughage material; bacteria are present in the cecum and help break down roughages. The maintenance of proper levels of bacteria in the cecum is very important in maintaining the health of rabbits.

Rabbits, unlike other animals, eat their feces. This is referred to as coprophagy and is usually done late at night or early in the morning. Feces that are eaten by the rabbit are usually light green, are soft, and have not been completely digested. Consumption of undigested feces allows the rabbit's digestive system to make full use of the bacterial action in the cecum. See Figure 6–10.

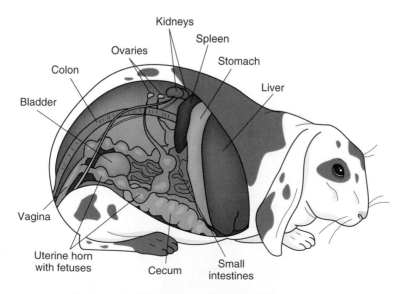

Figure 6–10   The digestive system of the rabbit is similar to that of the horse, because rabbits also have a large cecum. The rabbit consumes roughage materials that are broken down in the cecum.

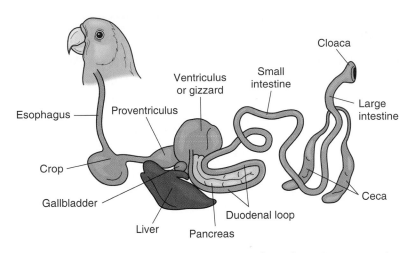

Figure 6–11 **Birds are considered monogastric animals; however, their digestive system is somewhat different from other monogastric animals.**

Birds, although considered as single-stomached animals, have several different organs in their digestive systems.

Birds do not have teeth, so no chewing or breaking down of food material takes place in the mouth, although saliva is added here to aid in swallowing. See Figure 6–11.

Food material passes down the esophagus, often referred to as the gullet, into the crop. The crop is an enlargement of the esophagus and serves primarily as a storage area for food material. Some softening of the food material also takes place in the crop.

The food material then passes to the ventriculus, commonly referred to as the "gizzard." This is the largest organ of the bird's digestive system, and its primary purpose is to grind and crush food before it enters the small intestine. The gizzard is composed of a horny, lined structure that is heavily muscled. Involuntary muscular action serves to break up the food material much like teeth action in other animals. Birds are often fed "grit" in the form of crushed granite, oyster shell, or other insoluble material that aids in the grinding and breaking down of coarse food material.

Food material then passes on to the small intestine. The first part of the small intestine is the duodenum, where enzymes are secreted from the pancreas to help break down the proteins, starches, and fats. Bile is secreted from the gallbladder to aid in the breakdown of fats. Most of the absorption takes place in the lower section of the small intestine.

Birds have two pouches where the small intestine connects into the large intestine. Although some bacteria are present in these pouches, very little digestion of the fiber is believed to occur here. Food material passes from the small intestine into the large intestine and then into the cloaca. The cloaca serves as a common junction for the bird's digestive, urinary, and genital systems.

SUMMARY

Nutrition refers to an animal receiving a proper and balanced food and water ration for proper health. The term *nutrient* refers to a single food or group of foods of the same general composition that supports animal life. There are six basic nutrient groups: water, proteins, carbohydrates, fats, vitamins, and minerals.

Digestion in animals is the process of breaking down food material into various nutrient forms that can be absorbed into the animal's bloodstream. Digestive systems in animals are basically of two types: ruminant and nonruminant. Nonruminant animals, such as swine, have a single digestive stomach. Ruminant animals, such as cows and sheep, have four compartments to their stomach. Poultry, although considered a single-stomached animal, have several different organs in their digestive system.

DISCUSSION QUESTIONS

1. Define the terms *nutrient* and *nutrition*.

2. Define the term *biochemical reaction.* What are the biochemical reactions that take place in the animal's body?

3. What are the nutrient groups, and what is their importance in the animal's diet?

4. List the various vitamins. What is their importance in the animal's body?

5. List the various minerals necessary in the diet of an animal. What is the importance each of them plays in the diet of the animal?

6. What is the difference between the digestive systems of a ruminant and a nonruminant animal?

7. List the four compartments of the ruminant stomach and explain the function of each.

8. How does the digestive system of a horse, rabbit, and chicken differ from a ruminant digestive system? Although they are considered nonruminants, how does the digestive system of a horse, rabbit, and chicken differ from most other nonruminants?

SUGGESTED ACTIVITIES

1. Visit a local feed processing plant and see how feeds are mixed, formulated, and produced.

2. Obtain ingredient tags from several different feeds. Compare the tags. How are the feed formulations different?

3. Conduct classroom tests using mice and rats. Feed some formulated feeds and some "junk" foods. Record and compare the results.

ADDITIONAL RESOURCES

1. Purina Pet Care Center
   http://www.purina.com/

2. Common Sense Guide to Feeding Your Dog and Cat
   http://www.dgp.toronto.edu/people/
   TabathaHoltz/cats/food1.html

3. Dog food and treat recipes
   http://soar.berkeley.edu/recipes/dog/
   indexall.html

4. Ruminant Nutrition
   http://www.hzn.com.au/Ruminant.htm

5. Hill's Health Center
   http://www.hillspet.com/public/
   health_center/index.html

CHAPTER 7

Dogs

OBJECTIVES

After reading this chapter, one should be able to:

- discuss the history of the dog.
- compare the seven major groups of dogs.
- list the breeds that make up the seven major groups of dogs.
- discuss the proper methods of feeding and exercising dogs.
- discuss methods of training dogs.
- discuss the grooming and health care of dogs.
- discuss the various diseases of dogs.

TERMS TO KNOW

anemia	estrus	placental membrane
colostrum	gestation	proestrus
conformation	heat period	rodenticide
congenital	pedigree	styptic

CLASSIFICATION

Kingdom — Animalia   Family — Canidae

Phylum — Chordata   Genus — *Canis*

Class — Mammalia   Species — *C. familiaria*

Order — Carnivora

HISTORY

The dog is probably the first animal domesticated by humans. The domestic dog originated about 12,000 to 14,000 years ago in Europe and Asia and was domesticated about 10,000 years ago. Many of our modern dogs probably descended directly from the wolf. These animals roamed in packs and probably gradually found it was easier to find food discarded by humans than to hunt for it. They probably became less wary of humans and gradually found their way into human encampments.

Figure 7–1 Miacis, *a small carnivorous mammal that lived in trees, is believed to be an early ancestor of several mammal species. Reprinted with permission of T.F.H. Publications, Inc.*

Humans found that they could depend on the dog to warn of danger, and the dog depended more on humans for food and shelter. The modern dog evolved as a result of selective breeding for specific purposes and as a result of the environment in which they lived.

The ancestor of the entire dog family is believed to have been a civet-like animal, *Miacis,* that lived 40 or 50 million years ago. See Figure 7–1. *Miacis* was a small animal with an elongated body and was probably arboreal, or at least seemed to have spent considerable time in the lush forests of its era. Next was the appearance, about 35 million years ago, of two apparently direct descendants of *Miacis.* These were *Daphaenus,* a large, heavy-boned animal with a long tail, and *Hesperocyon,* a small, slender animal who was the forerunner of the long-bodied, coyote-like "bear dogs" that eventually evolved into modern bears. *Hesperocyon* can be considered as the "grandfather" of the dog family. *Hesperocyon* retained the long body and short legs of the primitive carnivores, but unlike *Miacis, Hesperocyon* spent little time in the trees and began to hunt on the ground. Its claws were retractile, enabling it to walk on the ground and to climb trees.

From *Hesperocyon* there evolved two distinct dog types. The first, *Temnocyon,* was an important link in the evolutionary chain that led to the modern hunting dog of Africa, the Cape hunting dog. The second, *Cynodesmus,* is regarded as the ancestor of a large and diversified group of dogs that includes the modern Eurasian wolf and the American dogs, foxes, and wolves.

The animal considered to be the "father" of modern dogs, *Tomarctus,* was directly descended from *Hesperocyon. Tomarctus* had a body built for speed and endurance as well as for leaping and differed little in appearance from the modern dog. This was a hunter, an animal geared for the chase, that brought down prey with slashing teeth. The modern dog still retains much of *Tomarctus'* anatomical structure and is surpassed in speed only by the cheetah. See Figure 7–2.

As the evolutionary progress of the Canidae family continued, the progeny of *Tomarctus* developed into the modern dogs, wolves, foxes, coyotes, fennecs, and jackals. At the same time, descendants of *Temnocyon* gradually emerged as the Cape hunting dog of today. See Figure 7–3.

Figure 7–2 Tomarctus *is considered the ancestor of all canine-type mammals. Reprinted with permission of T.F.H. Publications, Inc.*

Descending in a direct line from *Tomarctus* are four major lines of dogs: the herd dogs, the hounds and terriers, the Northern and toy dogs, and the guard dogs. It is from these four lines, or groups, that modern dogs are descended.[1] Today, there are seven major groups, and more than 400 breeds that have been bred in the past 100 to 250 years.

The American Kennel Club recognizes 129 different breeds in seven groups. The seven groups are sporting dogs, hounds, working dogs, terriers, toys, herding dogs, and the nonsporting dogs. In addition to the seven groups, the American Kennel Club has established a miscellaneous class for breeds that have considerable interest but are not presently admitted to one of the regular groups.

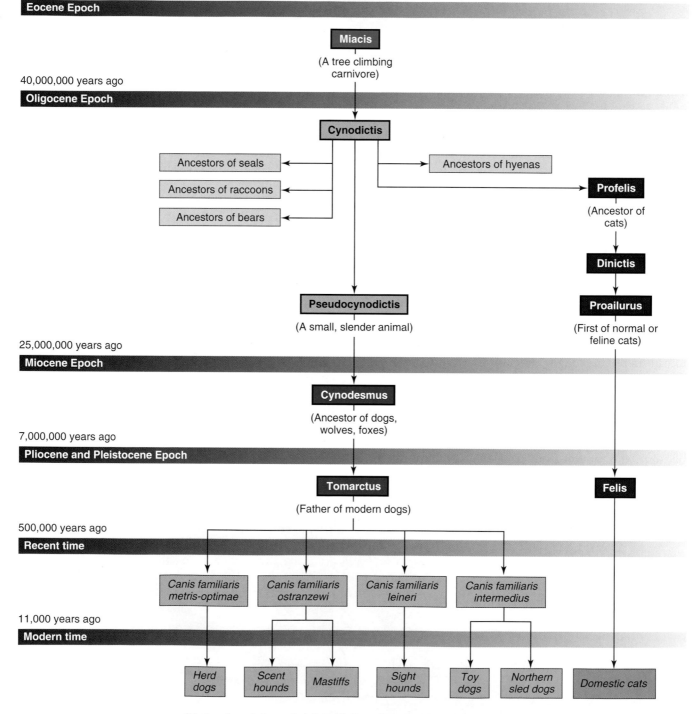

(Modern breeds have developed from various crossings of these varieties.)

Figure 7–3   This family tree traces the domestic dog's ancestors back about 40 million years to the tree-climbing mammal called Miacis. This tree also shows that domestic cats descended from this same tree-climbing mammal.

GROUPS AND BREEDS
The Sporting Group

This group of dogs was developed to assist the hunter in the pursuit of game. There are twenty-four breeds recognized in this group.

The Pointer. The Pointer is a breed that gets its name from the stance taken in the presence of game. The pointer was developed about 200 years ago in England from pointing breeds imported from Spain and Portugal. This is a powerful and agile animal with speed and endurance developed for hunting. Its short hair makes it neat and clean, and it requires less attention than some of the other hunting breeds. The Pointer possesses an even temperament and is very affectionate, intelligent, and patient with children and is a congenial companion. See Figure 7–4.

Figure 7–4   Pointer. Photo by Isabelle Francais.

The German Shorthaired Pointer. The German Shorthaired Pointer originated in Germany around 1600 from crossing the German Bird Dog, the Spanish Pointer, the English Pointer, and several other track and trail dogs. This versatile dog can be used for hunting almost all types of game. It is a rustic-looking dog with outstanding temperament and extraordinary hunting capabilities in all types of weather and terrain. The tail is usually docked to 40 percent of its length to prevent injury during the hunt. See Figure 7–5.

Figure 7–5   German Shorthair. Photo by Isabelle Francais.

The German Wirehaired Pointer. The German Wirehaired Pointer was developed from crossing the German Pointer, the Wirehaired Griffon, the Pointer, the Bloodhound, and the Airedale. The hair is hard and bristly like steel wool and is weather-resistant. The breed is noted for its keen nose, intelligence, staunchness at point, and tough constitution. The dog responds well to its master but can be somewhat unfriendly to strangers. See Figure 7–6.

Figure 7–6   German Wirehair. Photo by Isabelle Francais.

The Labrador Retriever.   The Labrador Retriever is native to New-foundland. It is a short coupled, strongly built, very active breed. It is an active water dog bred to assist the hunter in retrieving downed waterfowl, and it will readily plunge into the water in almost any weather. Its short, thick hair is water-resistant. The Labrador is noted for its excellent sense of smell and as being alert, friendly, and good-natured. Its excellent capabilities, fine temperament, and dependability have established it as one of the prime breeds for service as a guide dog for the blind or for search and rescue work. See Figure 7–7.

Figure 7–7   Labrador Retriever. Photo by Isabelle Francais.

The English Setter.   The English Setter is believed to have been produced from crossing the Spanish Pointer, the large Water Spaniel, and the Springer Spaniel. The breed has been trained as a bird dog in England for more than 400 years. It has gained popularity in the United States because of its usefulness and beauty. The breed is noted for having a mild, sweet disposition and aristocratic appearance. Because of its temperament and lovable disposition, it makes an excellent companion. See Figure 7–8.

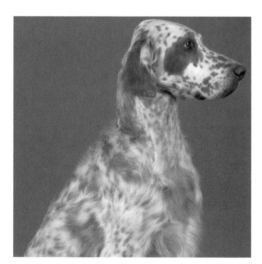

Figure 7–8   English Setter. Photo by Isabelle Francais.

The Irish Setter.   The Irish Setter is a solid red breed. It is probably older than the English Setter, but they may have common ancestry. Many believe that the breed is a result of crossing the Spanish Pointer, the English Setter, and perhaps the Gordon Setter. The big red breed can be used for almost all kinds of game. They are noted for their speed and toughness in all types of terrain. Other qualities include being independent and bold. The breed can also be gentle, lovable, and loyal. They may learn a little slower, but once trained, they will retain their abilities for a long time. The beauty, gaiety, courage, and personality of this breed has made it an ideal show dog. When it is bred for show, its field ability has been sacrificed. See Figure 7–9.

Figure 7–9   Irish Setter. Photo by Isabelle Francais.

The Brittany. Until 1982, the Brittany was registered by the American Kennel Club as Brittany Spaniel. Although called a spaniel, its method of working game is more like a setter. The breed is named for the French province in which it originated, and it is believed to be a cross of native spaniels of Brittany and English pointing dogs. The Brittany has steadily gained in popularity because of its merits as a shooting dog. It is smaller and works closer in. It is noted for its excellent sense of smell and its desire to please. It is adapted to all types of terrain and is always active, enthusiastic, and untiring. Because of its jovial character, it makes an excellent companion dog. See Figure 7–10.

Figure 7–10   Brittany. Photo by Isabelle Francais.

The Springer Spaniel. The Springer Spaniel is the foundation breed for all of the English hunting spaniels. During the 1600s, it was an ideal companion for a hunter in Europe. The breed is compact and strong and is noted for being cheerful, friendly, eager to please, courageous, and thorough. It is affectionate, good-natured, and sincere. It can be used in all types of terrain, especially one that is thick with brambles. The breed is hardy, long-legged, and more powerful and faster than other spaniels. See Figure 7–11.

Figure 7–11   Springer Spaniel. Photo by Isabelle Francais.

The American Cocker Spaniel. The American Cocker Spaniel first dates back to 1368. The Cocker Spaniel has retained its early classification as a sporting dog and is recognized as the smallest member of the sporting dog family. The breed gets its name from its proficiency at hunting woodcock. The breed is a cheerful, hardy, sporting dog with an elegant appearance. The Cocker Spaniel makes an excellent companion dog and is a lover of the home and family. This is a soft-hearted, sensitive dog that must be trained with a gentle hand. See Figure 7–12.

Figure 7–12   American Cocker Spaniel. Photo by Isabelle Francais.

Other breeds of sporting dogs include the following:

American Water Spaniel	Gordon Setter
Chesapeake Bay Retriever	Irish Water Spaniel
Clumber Spaniel	Sussex Spaniel
Curly-Coated Retriever	Vizsla
English Cocker Spaniel	Weimaraner
Field Spaniel	Welsh Springer Spaniel
Flat-Coated Retriever	Wirehaired Pointing Griffon
Golden Retriever	

(See Table 7–1 for color, size, and weight comparison of the sporting breeds.)

The Hound Group

This group of dogs is of two basic types: one type hunts by scent and the other by sight. The Greyhound and Afghan are two breeds of hound that hunt by sight; all others described hunt by smell.

The Afghan Hound. The Afghan Hound is a very old breed, and little is known about its origin and history. It is believed that the Afghan Hound existed thousands of years ago in Egypt and is native to the Sinai. The breed was discovered by the Western world in Afghanistan during the early 1800s. Originally, the breed was used for hunting. They would hunt in packs, pursuing game by sight and followed by huntsman on horseback. As a hunting hound, they were noted for their speed and their ability to travel rough terrain swiftly and sure-footedly. Today, the breed is popular as a show dog and companion. Because of their long, silky hair coat, Afghans require regular care and grooming. They are sensitive, intelligent, somewhat aloof, and dignified. Afghans respond to kindness and attention. See Figure 7–13.

Figure 7–13   Afghan Hound. Photo by Isabelle Francais.

The Basset Hound. The Basset Hound is an old breed developed in France. It is thought that the friars of the French Abbey of St. Hubert were instrumental in the development of this breed. Because of its excellent sense of smell, it is used for trailing rabbits, deer, raccoon, fox, and other game. It is a short-legged dog, heavier in bone (considering its size) than any other breed of dog. The Basset is gentle, affectionate, good with children, and devoted. In the field, it is capable of great endurance and determination. It may be stubborn at times and difficult to house-break. See Figure 7–14.

Figure 7–14   Basset Hound. Photo by Isabelle Francais.

		Average Weight in Pounds		Average Height at Shoulder in Inches	
Breed	**Color**	**Females**	**Males**	**Females**	**Males**
Pointer	Liver, lemon, black, orange; either in combination with white or solid-colored.	45–65	55–75	23–26	25–28
German Shorthaired Pointer	Solid liver or any other combination of liver and white, such as liver and white ticked, liver spotted and white ticked, or liver roan.	45–60	55–70	21–23	23–25
German Wirehaired Pointer	Liver and white, usually either liver and white spotted, liver roan, liver and white spotted with ticking and roaning, or sometimes solid liver.	45–60*	50–65*	22–24	24–26
Labrador Retriever	Black, yellow, or chocolate.	55–70	60–75	21½–23½	22½–24½
English Setter	Black, white, and tan; black and white; blue belton; lemon and white; lemon belton; orange and white; orange belton; liver and white; liver belton; and solid white.	50–60*	55–70*	24	25
Irish Setter	Mahogany or rich chestnut red with no trace of black.	60	70	25	27
The Brittany	Orange and white or liver and white in either clear or roan patterns. Some ticking desirable.	30–40	30–40	17½–20½	17½–20½
English Springer Spaniel	Black or liver with white markings or predominantly white with black or liver markings; tricolor: black and white or liver and white with tan markings.	49–55	49–55	19	20
English Cocker Spaniel	Many colors exist in solid and parti-colors.	26–32	28–34	15–16	16–17

Table 7–1   Color, Size, and Weight Comparison of the Sporting Breeds

*These figures are not given in the Official Standards and are thus approximate.

Information obtained from *"The Complete Dog Book,"* 17th edition, the official publication of the American Kennel Club, Howell Book House, 1986, Macmillan Publishing Company.

The Beagle.   The Beagle is one of the oldest breeds, but its actual origin is unknown. The modern-day Beagle was developed in the 1700s. The Beagle is noted as a rabbit hound and hunts in packs or singly. This little hound is cheerful, friendly, lively, and merry and makes an excellent family pet and companion. See Figure 7–15.

Figure 7–15   Beagle. Photo by Isabelle Francais.

The Black and Tan Coonhound. The Black and Tan Coonhound descended from the Talbot hound known in England during the eleventh century and was further developed through the bloodhound and the foxhound. The breed was developed in the United States and was officially recognized in 1945. It was specifically developed for hunting raccoons, but it can be used for hunting deer, bear, mountain lion, fox, and opossum. The breed is a passionate hunter and will work a scent with consummate skill and determination. The breed is intelligent, alert, friendly, eager, and aggressive. It responds well to its owner. See Figure 7–16.

Figure 7–16   Black and Tan Coonhound. Photo by Isabelle Francais.

The Bloodhound. The Bloodhound dates back to the third century AD. They were developed by the monks of St. Hubert in Belgium and were later brought to England by the Normans. The purebred Bloodhound is one of the most docile of all breeds. Its sense of smell is highly developed, and they have been used for tracking people and finding lost children, buried miners, and earthquake victims. Bloodhounds have been used by the police to track criminals, but unlike other breeds of police dogs, bloodhounds will not attack the criminal they are tracking. They are capable of following a scent that is several hours old. The Bloodhound is extremely affectionate, somewhat shy, and equally sensitive to kindness and correction by its owner. They are very good with children. See Figure 7–17.

Figure 7–17   Bloodhound. Photo by Isabelle Francais.

The Dachshund. The Dachshund is one of the older breeds of dog and dates back to the fifteenth century. In early Germany, it was known as the Teckel. The breed was perfected during the 1800s in Germany, where it was used for hunting badger. The dogs would trail the badger and follow it into its den. *Dachs* in German means badger, and *hund* in German means dog, thus the badgerdog or Dachshund. There are three varieties of the Dachshund: the shorthaired, the wirehaired, and the longhaired. Within the three varieties there are three sizes: normal, miniature, and toy. This is a short-legged, long-bodied, and muscular dog. It is considered clever, lively, courageous, bold, and confident. The Dachshund is good with children and is primarily used as a companion dog. It is considered a "barker" and makes a good watchdog. See Figure 7–18.

Figure 7–18   Shorthaired Dachshund. Photo by Isabelle Francais.

The American Foxhound. The American Foxhound descended from English hounds brought to America in 1650 by Robert Brooke. They were bred a century later to hounds imported from England in 1742 by Thomas Walker and to French hounds sent as a gift by Lafayette to George Washington. Although it is similar to the English hounds, the American Foxhound is smaller, possesses a keener sense of smell, and will work harder and faster. During the 1600s, these dogs were used to track Indians and later were used for hunting all types of animals. The breed is bold and aggressive in the field. They have been used primarily for hunting in packs and prefer the company of other dogs. They have not been used very much as house pets, although they are considered amiable and affectionate when not in the field. See Figure 7–19.

Figure 7–19   American Foxhound. Photo by Isabelle Francais.

The Greyhound. The Greyhound is a very ancient breed that traces its history back to the Tomb of Amten in the Valley of the Nile, which dates between 2900 and 2751 BC. The Greyhound has always had a cultural and aristocratic background and was a favorite of royalty in Egypt. The breed has been used for hunting practically all kinds of game, including deer, fox, wild boar, and rabbit. They hunted in packs and would catch up to their quarry and pull them down. Its greatest gift is speed, and it is primarily used today for racing, where it can reach speeds up to 40 miles per hour. The animal is also valued as a show dog. The Greyhound is very intelligent, sensitive, courageous, and loyal. Because it needs a lot of exercise, it is not suited to apartment life. It is not a recommended companion for children. If allowed to run loose in the country, it will kill small animals such as cats, geese, ducks, and chickens. See Figure 7–20.

Figure 7–20   Greyhound. Photo by Isabelle Francais.

The Norwegian Elkhound. The Norwegian Elkhound traces its history back to the early Vikings in 4000 to 5000 BC, when it was used for hunting elk. Over its history, it has been used to defend flocks from wolves and bear, to protect man from wild animals, and to hunt lynx, mountain lion, and raccoon. The animal is short and compact, and is noted for its stamina during the hunt and its eagerness to please. It is an exceedingly versatile breed also noted for its loyalty, affection, and intelligence. See Figure 7–21.

Figure 7–21   Norwegian Elkhound. Photo by Isabelle Francais.

Other breeds in the hound group include the following:

Basenji	Petit Basset Griffon Vendeen
Borzoi	Pharaoh Hound
English Foxhound	Rhodesian Ridgeback
Harrier	Saluki
Ibizan Hound	Scottish Deerhound
Irish Wolfhound	Whippet
Otterhound	

(See Table 7–2 for color, size, and weight comparison of the hound group.)

The Terrier Group

This group is divided into two subgroups: the long-legged, larger breeds such as the Fox Terriers, and the short-legged, small breeds such as the Dandie Dinmont.

The Airedale Terrier. The Airedale Terrier was developed in the county of York in England and takes its name from the river Aire that is located there. It descended from crossing the Old English Terrier or Broken Hair Terriers with the Otterhound. The breed was originally used for hunting bear, wolves, fox, badger, weasel, and other game. The breed is very docile, obedient, and lively, and it gets along well with children. It needs to feel loved and respected. During wartime, the dog was used to carry messages and was used as a police dog and bodyguard. See Figure 7–22.

Figure 7–22   Airedale Terrier. Photo by Isabelle Francais.

The Bedlington Terrier. The Bedlington Terrier takes its name from the small mining village of Bedlington, in the county of Northumberland in England, where it served as a watchdog. See Figure 7–23.

Figure 7–23   Bedlington Terrier. Photo by Isabelle Francais.

Breed	Color	Average Weight in Pounds		Average Height at Shoulder in Inches	
		Females	Males	Females	Males
Afghan Hound	All colors.	50	60	24–26	26–28
Basset Hound	Usually white with chestnut or sand-colored markings.	40–51	40–51	Not over 14	14
Beagle	Blue with black tigering, white, black, orange, or typical hound tricolor.	18–30	18–30	13–15	13–15
Black and Tan Coonhound	Black coat with tan markings on the muzzle, limbs, and chest.	60–70*	65–80*	23–25	25–27
Bloodhound	Black and tan, red and tan, and tawny.	80–100	90–100	23–25	25–27
Dachshund (normal)	Solid colors of red, tan, red yellow, and yellow. Bi-colors of black, chocolate, gray, and white. There are also speckled, streaked, and harlequin.	20	20	5–9 depending on the variety	
(minature)		9	9		
(toy)		8	8		
American Foxhound	Any color.	35–40*	40–45*	21–24	22–25
Greyhound	All colors.	60–65*	65–70*	20–25*	23–27*
Norwegian Elkhound	Gray with darker gray saddle. The muzzle, ears, and tail tip are black.	48	55	19½	20½

Table 7–2 Color, Size, and Weight Comparison of the Hound Group

*These figures are not given in the Official Standards and are thus approximate.

Information obtained from *"The Complete Dog Book,"* 17th edition, the official publication of the American Kennel Club, Howell Book House, 1986, Macmillan Publishing Company.

The Border Terrier. The Border Terrier was developed by the farmers and shepherds in the border area between Scotland and England for the purpose of helping drive out fox, which played havoc on sheep flocks. They needed a breed of dog with sufficient leg to follow a horse, but short enough to allow it to follow the fox into the den. The dogs also needed to be active, strong, and tireless and to have weather-resistant coats to withstand the prolonged exposure to drenching rains and mist in the area. The Border Terrier was thus developed and met all the necessary criteria. Because of its temperament and adaptability, it makes an excellent companion dog and is especially fond of children. See Figure 7–24.

Figure 7–24 Border Terrier. Photo by Isabelle Francais.

The Bull Terrier. The Bull Terrier dates back to the 1830s, when dogfights and bullfights were permitted and well attended. It is a cross of the Bulldog, the Old English Terrier, and the Spanish Pointer. This breed is noted for its great strength, agility, and courage. It can be exceedingly friendly and thrives on affection, but if provoked, it will readily fight to protect itself and its master. Over the years, the breed has been used as a guardian of livestock, a mouse and rat hunter, a companion dog, a guard dog, and a watchdog. There are two varieties of the breed: white and colored. See Figure 7–25.

Figure 7–25   Bull Terrier. Photo by Isabelle Francais.

The Dandie Dinmont Terrier. The Dandie Dinmont Terrier breed dates back to 1700 and is a cross of the Scottish and Skye Terriers. The Dandie Dinmonts were developed to hunt otters, badgers, weasels, martens, and skunks. Sir Walter Scott made the breed famous in his novel *Guy Mannering,* published in 1814. A character in the book, a farmer named Dandie Dinmont, owned six of the small dogs, and since the time of the novel, the breed has been known as "Dandie Dinmont Terriers." The breed is noted for its excellent hunting qualities and has retained those qualities. The Dandie Dinmont is intelligent, fond of children, and makes an excellent watchdog and companion. See Figure 7–26.

Figure 7–26   Dandie Dinmont Terrier. Photo by Isabelle Francais.

The Fox Terrier. The Fox Terrier has developed over many years into two distinct varieties: the Smooth and the Wire. Wires were crossed with the Smooths in the early days, but this practice has been almost universally discontinued. In 1984, the American Kennel Club approved separate breed standards for the Smooth Fox Terrier and the Wire Fox Terrier. It is believed that the Wire was developed from the old rough-coated black and tan working terriers and that the ancestors of the Smooth were the smooth-coated black and tans (the Bull Terriers, the Greyhound, and the Beagle). Fox Terriers are one of the best known and most widely distributed of purebred dogs. The breed is English in origin and dates back to the 1700s. Like many other terriers, the breed was developed to hunt den animals and was noted for its courage. Today, they are considered affectionate and trainable and make excellent companion dogs. See Figure 7–27.

Figure 7–27   Smooth-Coated Fox Terrier. Photo by Isabelle Francais.

The Miniature Schnauzer. The Miniature Schnauzer is of German origin and dates back to the fifteenth century. The miniature Schnauzer was developed from the Standard Schnauzer through crossings with the Affenpinschers and Poodles. The Miniature Schnauzer was recognized as a separate breed in 1899. As with the other breeds in the terrier group, it was developed to hunt badger, fox, weasel, and other vermin. The breed is characterized by its stocky build, wiry coat, and abundant whiskers and leg furnishings. The breed is hardy, healthy, and intelligent and is fond of children. It makes a charming and attractive companion. It is not as large or active as its larger cousin (the Giant Schnauzer) and makes a good apartment dog and family pet. See Figure 7–28.

Figure 7–28   Miniature Schnauzer. Photo by Isabelle Francais.

The Skye Terrier. The Skye Terrier traces its origin back to a Spanish shipwreck off the island of Skye in the Scottish Hebrides. Among the survivors were Maltese dogs, which mated with the native terriers and produced this small breed of terrier. Originally used for hunting in dens, the breed became a favorite at kennel shows because of its long, flowing, beautiful hair coat. The breed is characterized as being considerate, well mannered, and affectionate. It is friendly and lively with those it knows, but it is shy and cautious of strangers. See Figure 7–29.

Figure 7–29   Skye Terrier. Photo by Isabelle Francais.

The Welsh Terrier. The Welsh Terrier is a very old breed and in olden times was more commonly known as the Old English Terrier, or Black and Tan Wirehaired Terrier. The breed was recognized as Welsh Terriers in 1884–1885 and was developed to hunt in dens and work in packs with hounds. Today, it is considered an apartment dog and watchdog. It is friendly and outgoing with people and other dogs, showing spirit and courage. See Figure 7–30.

Figure 7–30   Welsh Terrier. Photo by Isabelle Francais.

Other breeds in the terrier group include the following:

American Staffordshire Terrier	Norfolk Terrier
Australian Terrier	Norwich Terrier
Cairn Terrier	Scottish Terrier
Irish Terrier	Sealyham Terrier
Kerry Blue Terrier	Soft-Coated Wheaten Terrier
Lakeland Terrier	Staffordshire Bull Terrier
Manchester Terrier	West Highland White Terrier
Miniature Bull Terrier	

(See Table 7–3 for color, size, and weight comparison of the terrier group.)

The Working Dog Group

This group of dogs has been developed to labor or work for humans. They may serve as guard dogs, sled dogs, police dogs, rescue dogs, and messenger dogs.

The Alaskan Malamute. The Alaskan Malamute is one of the oldest Artic sled dogs and was named for the native Innuit tribe called Mahlemuts (now spelled Malamute), who settled along the shores of Kolzbue Sound in the upper western part of Alaska. The origin of these dogs has never been ascertained, but we know they have been in Alaska for generations. When Alaska became settled by the white man, this Artic breed was bred with outside strains; with the increase in the popularity of sled racing as a sport, there was renewed interest in developing a pure strain of the Alaskan Malamute. The clean, odorless Malamutes, who do not bark, are loyal, intelligent, and affectionate toward their owner, although they are fighters among dogs. Therefore, firm handling and training are necessary. See Figure 7–31.

Figure 7–31   Alaskan Malamute. Photo by Isabelle Francais.

The Boxer. The Boxer is one of the many descendants of the old fighting dogs of the high valleys of Tibet. The breed showed up in many parts of Europe during the sixteenth and seventeenth centuries and reached its perfection in Germany within the past 100 years. The Boxer carries Bulldog, Terrier, and Bullenbeisser Mastiff blood. The breed was developed for two purposes: to fight bulls and to hunt bear. The breed is considered good-natured, loyal, playful, tolerant, and somewhat suspicious of strangers. It is a short-lived dog and often does not reach ten years of age. It is subject to rheumatism and must be exercised regularly. The Boxer is a watchdog, bodyguard, and guide dog for the blind and is used in police work. See Figure 7–32.

Figure 7–32   Boxer. Photo by Isabelle Francais.

		Average Weight in Pounds		Average Height at Shoulder in Inches	
Breed	**Color**	**Females**	**Males**	**Females**	**Males**
Airedale Terrier	The head and ears are tan, with the ears being a darker shade than the rest. The sides and upper parts of the body should be black or dark grizzle.	42	44	22	23
Bedlington Terrier	Two distinct colors, liver and blue.	15–21*	17–23	15½	16½
Border Terrier	Red, grizzle and tan, blue and tan, or wheaten.	11–14*	13–15	11–12*	12–13*
Bull Terrier	The White: all white, or white with black markings on the head. Colored: any color but white, with brindle the preferred color.	17–20*	18–21*	30–50*	40–60*
Dandie Dinmont Terrier	Two distinct colors: Pepper: blue gray to light silver with light tan or silver points and a very light gray or white topknot. Mustard: dark ocher color to cream with white points and topknot.	18–24	18–24	8–11	8–11
Fox Terrier	White should predominate; brindle, red, or liver are objectionable.	16	18	14½	15½
Schnauzer	Salt and pepper, black and silver, and solid black.	13–16*	15–19*	12–14	12–14
Skye Terrier	Coat is one overall color at the skin, but may be of varying shades of the same color in the full coat, which may be black, blue, dark or light gray, silver platinum, fawn, or cream.	23	25	9½	10
Welsh Terrier	The jacket is black spreading up onto the neck, down onto the tail, and into the upper thighs. The legs, quarters, and head are clear tan. The tan is deep reddish brown.	20	20	14	15

Table 7–3   Color, Size, and Weight Comparison of the Terrier Group

*These figures are not in the Official Standards and are thus approximate.

Information obtained from *"The Complete Dog Book,"* 17th edition, the official publication of the American Kennel Club, Howell Book House, 1986, Macmillan Publishing Company.

The Doberman Pinscher.   The Doberman Pinscher breed originated in Apolda, in Thuringia, Germany, around 1890 and takes its name from Louis Dobermann. The breed was developed by crossings among Great Danes, German Shepherds, Rottweilers, and Pinschers, with possibly some blood of the Beauceron and the English greyhound. At first, the Doberman was used as a guard dog and watchdog. Because of its intelligence and excellent nose, it has been used as a police dog, war dog, and hunting dog. The males and females have different temperaments. The females are calm, responsive, and affectionate with the family, but wary of strangers. The males are extremely intelligent, but impetuous, often aggressive, and must be trained by a dominant owner who is a good disciplinarian. It is a long-lived dog, often reaching fifteen to twenty years of age. See Figure 7–33.

Figure 7–33   Doberman Pinscher. Photo by Isabelle Francais.

The Great Dane. Drawings of dogs much like the Great Dane date back to Egyptian monuments of 3000 BC; early Chinese literature of 1121 BC describes a dog resembling the Great Dane also. It is believed that the Mastiff breeds of Asia and the Irish Wolfhounds are the principal ancestors of this breed. Its development into the modern type began in Germany during the 1800s. This giant dog was used for hunting wild boar and bear. During its history, it has seen service as a battle dog, a hunting dog, a cart dog, a watchdog, and a bodyguard. The Great Dane is spirited, courageous, friendly, and dependable. Today, it is used primarily as a watchdog and companion dog. See Figure 7–34.

Figure 7–34   Great Dane. Photo by Isabelle Francais.

The Great Pyrenees. The remains of this breed have been found in the fossil deposits of the Bronze Age, which dates between 1800 and 1000 BC in Europe. It is believed to originate from central Asia or Siberia. The Great Pyrenees were developed in the Pyrenees mountains, where they served as guardians to the shepherd's flocks. Their excellent sense of smell and sight proved invaluable to the shepherd, and because of their size and strength, they were almost unbeatable as guardian of the flock. Also known as the Great Dog of the Mountains, the breed is docile and makes a faithful, loving, and affectionate house pet. It does require considerable space, exercise, and grooming. Besides a guardian of sheep, the breed has been used as a guard dog for people and property, a guide and pack dog in ski areas, a rescue dog, and a messenger dog during World War I. See Figure 7–35.

Figure 7–35   Great Pyrenees. Photo by Isabelle Francais.

The Standard Schnauzer. The Standard Schnauzer is a German breed that dates back to the fifteenth and sixteenth centuries. It was developed from crossings of a black German poodle and a gray wolf spitz with wirehaired Pinscher stock. The dogs were originally used as rat catchers, guard dogs, and yard dogs. Before World War I, they were used in Germany to guard the carts of farm produce in the marketplace while farmers rested themselves and their teams at the inns. Today, the breed is used as a watchdog and a bodyguard as well as a family dog. It is feisty, intelligent, bold, and courageous. The breed is noted for being long-lived, clean, and almost odor-free. See Figure 7–36.

Figure 7–36   Standard Schnauzer. Photo by Isabelle Francais.

The Rottweiler. The Rottweiler is a descendant of the Drover dogs of Rome, which were a Mastiff type. Rottweilers were bred and developed in the German town of Rottweil, which is how they got their name. Drover dogs were brought into the area by Roman armies around 74 AD. The breed was used as a herding dog to drive cattle to and from the markets. This massive, powerful animal is considered peaceful, obedient, courageous, and intelligent. They are known to be very protective of their owners. Today, the breed is used as a herd dog, police dog, watchdog, and bodyguard. They also make excellent companion dogs. See Figure 7–37.

Figure 7–37 Rottweiler. Photo by Isabelle Francais.

The Saint Bernard. The Saint Bernard probably descended from large Mastiff dogs that were brought to the area that is now Switzerland by the Roman armies during the first two centuries AD. These dogs were crossed with native dogs that existed in this region and were used for a variety of herding, guarding, and drafting duties. The breed appeared around 1660–1670 in the Hospice of St. Bernard de Menthon near the Great St. Bernard Pass. The monks of the Hospice used the dogs for companionship and took them on trips of mercy, where they discovered the dogs were excellent pathfinders in the drifting snow, and the dogs' excellent sense of smell made them invaluable in locating helpless persons overcome during storms. In 1830, the breed was crossed with the Newfoundland, producing the first longhaired Saint Bernards. The longhaired variety proved unsuited for the cold weather condition because snow clung to their fur, obstructing the dog's movements. The Saint Bernard is obedient, peaceful, quiet, extremely loyal, and very gentle and loves the company of humans. See Figure 7–38.

Figure 7–38 Saint Bernard. Photo by Isabelle Francais.

The Samoyed Breed. The Samoyed breed was developed by and takes its name from the Samoyed tribe found in Siberia. The breed was used to herd reindeer and, as sled dogs, are noted for being able to pull heavy loads over long distances. It is a strong, dynamic, active dog that has proved itself as an excellent companion and watchdog. It is very gentle, kind, unusually intelligent, and affectionate and loves children. The Samoyed is considered by many the most beautiful breed in existence. Its only fault would be its frequent barking. See Figure 7–39.

Figure 7–39 Samoyed. Photo by Isabelle Francais.

The Siberian Husky. The Siberian Husky breed was originated by the Chukchi people of Northeastern Asia as an endurance sled dog. The breed was brought to Alaska in 1909 to compete in the All Alaska Sweepstakes Race. A team of Siberian Huskies won the race in 1910 and most of the racing titles in Alaska over the next decade as well. The breed is naturally friendly and gentle in temperament. The Husky possesses an independent nature at times, and although very alert, it lacks the aggressive or protective tendencies of a watchdog. It may be stubborn, easily bored, cranky, or ill-humored around children. See Figure 7–40.

Figure 7–40   Siberian Husky. Photo by Isabelle Francais.

The Komondor. The Komondor is a heavily coated dog that probably descended from the Aftscharka, which the Huns found on the southern steppes of Russia and brought with them into the area that is now Hungary. The breed is characterized by its great size, strength, courageous demeanor, and unusual, heavy, white coat. The breed makes an excellent guard dog and will readily fight wolves, bears, and other larger predators that would attack the flock. It is obedient, loyal, and devoted to its owner. See Figure 7–41.

Figure 7–41   Komondor. Photo by Isabelle Francais.

Other breeds in the working dog group include the following:

Akita	Great Swiss Mountain Dog
Bernese Mountain Dog	Kuvasz
Bullmastiff	Mastiff
Giant Schnauzer	Portuguese Water Dog

(See Table 7–4 for color, size, and weight comparison of the working dog group.)

The Herding Dog Group

This group was developed to aid the livestock herder with various species of livestock.

Breed	Color	Average Weight in Pounds		Average Height at Shoulder in Inches	
		Females	Males	Females	Males
Alaskan Malamute	The usual colors range from light gray through the intermediate shadings to black, always with white underbodies, parts of the legs, feet, and part of mask markings.	75	85	23	25
Boxer	Fawn and brindle. Fawn in various shades of light tan to stag red or mahogany. White markings may appear on the chest and on the face.	50–65*	60–70*	21–23½	22½–25
Doberman Pinscher	Black, red, blue, and fawn. Rust markings may appear above each eye and on the muzzle, throat and forechest; on all legs and feet; and below the tail. The nose will be solid black on black dogs, dark brown on red ones, dark gray on blue ones, and dark tan on fawns.	60–70*	65–80*	24–26	26–28
Great Dane	Five basic types: brindle, fawn, blue, black, and harlequin.	110–115*	120–150*	28+	30+
Great Pyrenees	White.	90–115	100–125	25–29	27–32
Standard Schnauzer	Pepper and salt or pure black.	60–70*	70–80*	17½–18½	18½–19½
Rottweiler	Always black with rust to mahogany markings.	80–90*	90–110*	22–25	24–27
Saint Bernard	Red with white markings or white with red markings.	90–120*	110–160*	27½+	27½+
Samoyed	Pure white.	50–60*	55–65*	19–21	21–23½
Siberian Husky	All colors from black to pure white. A variety of markings on the head are common.	35–50	45–60	20–22	21–23½
Komondor	White.	80–110*	90–125*	23½+	25½+

Table 7–4   Color, Size, and Weight Comparison of the Working Dog Group

*These figures are not given in the Official Standards and are thus approximate.

Information obtained from *"The Complete Dog Book,"* 17th edition, the official publication of the American Kennel Club, Howell Book House, 1986, Macmillan Publishing Company.

The Collie.   The Collie comes in two varieties: the more familiar rough-coated or longhaired Collie, and the smooth-coated Collie. The exact origin of the Collie is not known, but both varieties existed long ago as herding dogs of Scotland and northern England. Because sheepherding is one of the world's oldest occupations, the Collies date far back in history. Being no longer in demand as a herder, today's Collie serves as a devoted family dog. The Collie is kind, sensitive, intelligent, and loyal. It is very protective of its owner and family, but because of its sensitive nature, careful, gentle training is required. The Collie is wary of strangers and may also be lazy or stubborn on occasion. The breed is very versatile, being used as rescue dogs, as guide dogs for the blind, and in police work. Keeping burrs and knots out of the beautiful coat of the longhaired Collie requires frequent brushing. See Figure 7–42.

Figure 7–42   Collie. Photo by Isabelle Francais.

The German Shepherd. Several theories exist regarding the origin of this breed. Some believe that it is a result of crosses between various herding dogs and farm dogs existing in Germany, whereas others believe it is the result of matings between herding dogs, farm dogs, and wolves. The German Shepherd is courageous, obedient, steady, and loyal. It is affectionate with its owner and family, but wary of strangers. Originally used for sheepherding, the German Shepherd has, over its history, been used in time of war to carry messages, and as a rescue dog, police dog, guard dog, and guide dog for the blind. See Figure 7–43.

Figure 7–43   German Shepherd. Photo by Isabelle Francais.

The Old English Sheepdog. The Old English Sheepdog originated early in the nineteenth century in the counties of Devon, Somerset, and the duchy of Cornwall in England. Its exact ancestry is not known, but some believe its ancestors were the Scotch Bearded Collie, whereas others believe it descended from a longhaired Russian breed called the Owtchar. It is a steady, muscular dog with a characteristic gait much like that of a bear. It is an outstanding herd dog and has been used as a guard dog, sled dog, and retriever. Its friendly, affectionate disposition makes it an excellent companion dog and family pet. See Figure 7–44.

Figure 7–44   Old English Sheepdog. Photo by Isabelle Francais.

The Shetland Sheepdog. The Shetland Sheepdog is a breed that probably descended from the Collies that were brought to the Scottish island of Shetland. These Collies developed in miniature because of the environment and from being crossed with small Border Collies and other small breeds native to the islands. The breed resembles the larger, rough Collie in miniature. It is noted for its devoted, docile nature and its keen and all but human intelligence and understanding. It is considered very lively, perky, and affectionate. The Shetland Sheepdog makes an excellent family pet and companion. See Figure 7–45.

Figure 7–45   Shetland Sheepdog. Photo by Isabelle Francais.

The Welsh Corgi. There are two types of the Welsh Corgi: the Cardigan and the Pembroke. The ancestors of the Cardigan came into the hill country of Wales that is now known as Cardiganshire, in the British Isles around 1200 BC. The dog is a member of the same family that produced the Dachshund. The ancestors of the Pembroke came across the English Channel with Flemish weavers to Pembrokeshire in Wales in 1107. This dog sprang from the same family that includes the Keeshond, the Pomeranian, the Samoyed, the Chow Chow, the Norwegian Elkhound, and the Finnish Spitz. It has little or nothing of the Dachshund characteristics. See Figure 7–46.

Figure 7–46   Welsh Corgi, Cardigan. Photo by Isabelle Francais.

In comparing the two types, the Pembroke is shorter of body; the legs are straighter and lighter boned, and the coat is of finer texture. Cardigan ears are rounded, whereas the Pembroke's are pointed at the tip and stand erect. The Cardigan has a long tail, and the Pembroke a short one. The Pembroke is more restless and more easily excited. Both types were used to herd cattle, and they possessed unusual speed for having such short legs, as well as agility, balance, and endurance. The Corgi is noted for its intelligence and for being alert, lively, affectionate, friendly, and easy to train. See Figure 7–47.

Figure 7–47   Welsh Corgi, Pembroke. Photo by Isabelle Francais.

Other breeds in the herding dog group include the following:

Australian Cattle Dog	Belgian Tervuren
Australian Shepherd	Bouvier des Flandres
Bearded Collie	Briard
Belgian Malinois	Canaan Dog
Belgian Sheepdog	Puli

(See Table 7–5 for color, size, and weight comparison of the herding dog group.)

The Toy Dog Group

Certain dogs are called toy dogs because of their size; some are as small as 1.5 pounds. These dogs are very alert and popular as house pets and companions.

Breed	Color	Average Weight in Pounds		Average Height at Shoulder in Inches	
		Females	Males	Females	Males
Collie	Four recognized colors: sable and white, tricolored, blue merle, and white.	50–65	60–75	22–24	24–26
German Shepherd	Several colors exist. Black iron gray, ash gray, either uniform or regular shading of brown, yellow, or light gray. Nose is black.	60–75*	70–85*	22–24	24–26
Old English Sheepdog	Any shade of gray, grizzle, blue, or blue merle with or without white markings or in reverse.	70–100*	80–110*	20+	22+
Shetland Sheepdog	Black, blue merle, and sable; marked with varying amounts of white and/or tan.	18–25*	21–28*	13–16	13–16
Welsh Corgi	Cardigan: All shades of red, sable, and brindle. Blue merle with or without tan or brindle points. White flashings are usual on the neck, chest, legs, muzzle, underparts, tip of the tail, and as a blaze on the head.	25–34	30–38	10½–12½	10½–12½
	Pembroke: Outer coat of self colors of red, sable, fawn, black, and tan, with or without white markings.	28	30	10–12	10–12

Table 7–5   Color, Size, and Weight Comparison of the Herding Dog Group

*These figures are not given in the Official Standards and thus are approximate.

Information obtained from *"The Complete Dog Book,"* 17th edition, the official publication of the American Kennel Club, Howell Book House, 1986, Macmillan Publishing Company.

The Chihuahua. The Chihuahua is recognized as the smallest breed in the world. The Toltecs, a group of natives that date back to the ninth century AD in what is now Mexico, had a breed of dog called Techichi. This small breed indigenous to Central America is believed to be one of the ancestors of the Chihuahua. It is believed that the Techichi was crossed with a small hairless dog brought from Asia or Alaska over the land bridge where the Bering Strait now is. This hairless dog, similar to the one found in China, was responsible for the reduction in size. The name comes from the state of Chihuahua in Mexico. Because it dates back to the ninth century AD, it is considered the oldest breed of dog on the American continent. There are two varieties: a shorthaired variety and a wavy-haired variety. The Chihuahua is a companion dog and is noted for being courageous, energetic, lively, alert, proud, affectionate, loyal, and intelligent. It is clannish, recognizing and preferring its own kind, and as a rule not liking dogs of other breeds. See Figure 7–48.

Figure 7–48   Longhaired Chihuahua. Photo by Isabelle Francais.

The Italian Greyhound. The Italian Greyhound is a very old breed, descending from the Egyptian Greyhound of over 6,000 years ago. This is the smallest of the family of dogs that hunt by sight. It is believed to have been brought into Europe by the Phoenician traders and developed in Italy, thus the Italian Greyhound name. There is some disagreement as to whether the breed was developed for hunting or for a pet and companion. The breed is very affectionate and thrives best when this affection is returned. The coat is short and smooth and requires very little grooming. The Italian Greyhound is odorless and sheds very little. The breed is sensitive, alert, intelligent, and playful. It may be timid, meek, and wary of strangers. See Figure 7–49.

Figure 7–49   Italian Greyhound. Photo by Isabelle Francais.

The Manchester Terrier. The Manchester Terrier breed was developed in the Manchester district of England during the nineteenth century by crossings between the Black and Tan Terrier, the Whippet, the Greyhound, and the Italian Greyhound. Its nickname is the "rat terrier" because it has long been recognized for its ability to hunt rats and other rodents. It is considered intelligent, active, and alert. Today, it is primarily a small companion dog. In 1959, two varieties were recognized: the toy and the standard. See Figure 7–50.

Figure 7–50   Manchester Terrier. Photo by Isabelle Francais.

The Pekingese. In ancient times, the Pekingese was held sacred in China, the land of its origin. The exact date of origin is not known, but its existence traces back to the Tang dynasty of the eighth century. The very old strains (held only by the imperial family) were kept pure, and the theft of one of these sacred dogs was punishable by death. Introduction of the Pekingese into the Western world occurred as a result of the looting of the Imperial Palace at Peking by the British in 1860. The breed is sensitive, extremely affectionate with its owner, loyal, somewhat aloof, reserved, and quiet. Although Pekingese are not aggressive, they are courageous and will stand their ground against larger adversaries. They are primarily an apartment lapdog, but they make good watchdogs as well. The long, fine coat needs frequent care, and the teeth need to be cleaned to prevent early decay. See Figure 7–51.

Figure 7–51   Pekingese. Photo by Isabelle Francais.

The Pug.   The Pug is a very old breed that dates back to 400 BC. It is of Oriental origin, first appearing in the Buddhist monasteries of Tibet. From China it made its way to Japan and then to Europe. The Pug has a compact body and well-developed musculature. It has a round, massive head; a short, square muzzle; and a deep-wrinkled forehead. Its eyes are prominent with a sweet expression. The Pug is affectionate and loving, likes lots of attention, and loves children. Pugs are intelligent and easily trained. See Figure 7–52.

Figure 7–52   Pug. Photo by Isabelle Francais.

The Yorkshire Terrier.   The Yorkshire Terrier was developed in the 1800s by the working class in the English county of Yorkshire for the purpose of catching rats and other vermin. It is a crossing among the Waterside Terrier, the Black and Tan English Terrier, the Paisley Terrier, and the Clydesdale Terrier. Because of its long, silky coat, it has become a favorite lapdog and companion. The Yorkshire Terrier is lively, alert, affectionate with its owner, and wary of strangers. The Yorkshire Terriers' long, silky coat needs constant care and attention. They are best suited to adults who have plenty of time to devote to their required care. See Figure 7–53.

Figure 7–53   Yorkshire Terrier. Photo by Isabelle Francais.

The Shih Tzu.   The Shih Tzu breed dates back to 624 AD, when a pair was given to the Chinese court. They were said to have come from Fuh, which was believed to be in the Byzantine Empire. These dogs were bred in the Forbidden City of Peking. *Shih Tzu* means lion, and these small dogs resembled lions with their long, flowing hair. They were brought to England in 1930. This is a typical drawing-room dog. It is very lively and alert, with a distinctly arrogant carriage. Its long, silky hair needs daily attention to prevent matting and knotting. See Figure 7–54.

Figure 7–54   Shih Tzu. Photo by Isabelle Francais.

Other breeds in the toy dog group include the following:

Affenpinscher	Maltese
Brussels Griffon	Miniature Pinscher
Cavalier King Charles Spaniel	Papillon
Chinese Crested	Pomeranian
English Toy Spaniel	Silky Terrier
Japanese Chin	Toy Poodle

(See Table 7–6 for color, size, and weight comparison of the toy dog group.)

Table 7–6 Color, Size, and Weight Comparison of the Toy Dog Group					
Breed	**Color**	**Average Weight in Pounds**		**Average Height at Shoulder in Inches**	
		Females	**Males**	**Females**	**Males**
Chihuahua	Any color—solid, marked, or splashed.	1–6*	1–6*	6–9	6–9
Italian Greyhound	All colors and markings recognized.	11	11	13–15	13–15
Manchester Terrier (standard)	Jet black with rich mahogany tan about the face and lower legs.	12–22	12–22	9–13*	9–13*
(toy variety)		<12	<12	<9*	<9*
Pekingnese	Red, fawn, black, black and tan, sable, brindle, white, and parti-colored.	<14	<14	8–9*	8–10*
Pug	Silver or apricot fawn with black mask.	14–18	14–18	10–12*	11–13*
Yorkshire Terrier	Puppies are born black and tan. Adults are usually dark steel-blue or tan.	>7	>7	8–9*	8–10*
Shih Tzu	All colors, nose and eye rims black.	12–15	12–15	9–10½	9–10½

*These figures are not given in the Official Standards and are thus approximate.

Information obtained from *"The Complete Dog Book,"* 17th edition, the official publication of the American Kennel Club, Howell Book House, 1986, Macmillan Publishing Company.

Nonsporting Dogs

This group consists of miscellaneous breeds with a wide variety of sizes and characteristics. Basically, they are used as companion dogs.

The Boston Terrier. The Boston Terrier is an American breed that dates back to 1870 and that was developed in the Boston area. It originated from a crossing of an English Bulldog and a white English Terrier; later there was considerable inbreeding. The breed was recognized by the American Kennel Club in 1893. This is a clean-cut dog with a short head, snow-white markings, soft black eyes, and a body conformation of the terrier rather than the Bulldog. It is an excellent companion and house dog, noted for its gentle disposition and affection for its owner. The Boston Terrier is intelligent, alert, well mannered, and even-tempered, and it gets along well with children. See Figure 7–55.

Figure 7–55 Boston Terrier. Photo by Isabelle Francais.

The Bulldog. The Bulldog originated in England and was used for the sport of bull baiting. Its ancestry is probably that of the Asiatic Mastiff and other fighting dogs. The original Bulldogs were vicious, courageous, and almost insensitive to pain. In 1835, dog fighting and bull baiting became illegal in England. The Bulldog no longer had a use, but fanciers of the breed wanted to preserve it. Over the next generations, the ferocity was bred out of the dog and other desirable qualities were retained. The Bulldog we know today is good-natured, reserved, dignified, loyal, clean, and aristocratic. Over its history, it has been used as a war dog, guard dog, and police dog. Problems may occur at whelping because of the large size of the puppy's head. A veterinarian should be kept advised, and cesarean delivery may be necessary. See Figure 7–56.

Figure 7–56   Bulldog. Photo by Isabelle Francais.

The Dalmatian. The Dalmatian's origin, according to historians, is the center of much debate and disagreement. Models, engravings, paintings, and writings place the dog in ancient Egypt, Asia, and Africa. The dogs were frequently found in bands of Romanies, and like their gypsy masters, Dalmations have been well known but not located long in any one place. Authoritative writers place them first as a positive entry in Dalmatia, a region in west Yugoslavia. During their history, they have been used as a dog of war, a draft dog, and a herd dog. Dalmations are excellent for hunting rats and vermin. As sporting dogs, they have been used as bird dogs, trail hounds, and retrievers, as well as in packs for boar and bear hunting. But most important has been their use as a carriage or coach dog. The breed is noted for its intelligence and adaptability. The instinct for coaching is inbred and also trained. Dalmations will follow their master whether on foot, on horseback, or in a carriage. The Dalmatians are calm, quiet, sensitive, and loyal; they thrive on human companionship. They love children and make natural guardians. See Figure 7–57.

Figure 7–57   Dalmatian. Photo by Isabelle Francais.

The Lhasa Apso. The Lhasa Apso is a small dog with a heavy coat of hair. The breed originated in Tibet, where it was used as a guard dog because of its intelligence, quick hearing, and an instinct for distinguishing intimates and strangers. This breed is easily trained and responsive to kindness. To anyone the dogs trust, they are most obedient. They can be very affectionate to their owners and are more suited to adults who will pay lots of attention to them. See Figure 7–58.

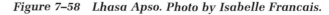

Figure 7–58   Lhasa Apso. Photo by Isabelle Francais.

The Poodle. The Poodle breed probably originated in Germany, where it was known as the Pudel. However, for years, it has been known as the national dog of France, where it was commonly used as a retriever and as a circus trick dog. In France, it was known as Canichi, which is derived from *chien canard,* or duck dog. The English word *poodle* comes from the German *pudel* or *pudelin,* meaning to splash in the water. There are three sizes: the standard, miniature, and the toy. All are of one breed, governed by the same standard of perfection. The breed is considered good-natured and very intelligent and responds well to their owners. Poodles are no longer used primarily as water dogs or retrievers, but they have gained fame as a show dog and companion. See Figures 7–59 and 7–60.

Figure 7–59   Standard Poodle. Photo by Isabelle Francais.

Other breeds in the nonsporting dog group include the following:

American Eskimo Dog	Keeshond
Bichon Frise	Schipperke
Chinese Shar-Pei	Shiba Inu
Chow Chow	Tibetan Spaniel
French Bulldog	Tibetan Terrier

(See Table 7–7 for color, size, and weight comparison of the nonsporting dog group.)

Figure 7–60   Toy Poodle. Photo by Isabelle Francais.

TABLE 7–7 Color, Size, and Weight Comparison of the Nonsporting Dog Group					
		Average Weight in Pounds		**Average Height at Shoulder in Inches**	
Breed	**Color**	**Females**	**Males**	**Females**	**Males**
Boston Terrier	Brindle with white markings and black with white markings.				
	(lightweight)	<15	<15	11–15	15–17
	(middleweight)	15–20	15–20	depending on weight	
	(heavyweight)	20–25	20–25		
Bulldog	Red brindle, all other brindles, solid white, solid red, fawn or fallow, and piebald.	40	50	13–15	14–16
Dalmatian	Two varieties: white with black spots and white with liver brown spots.	35–45*	40–50*	19–24	19–24
Lhasa Apso	All colors.	10–13*	13–16*	9–10	10–11
Poodle	All colors with the coat being an even and solid color at the skin.				
	(Standard)	35–45*	40–50*	>15	>15
	(Miniature)	16–35*	18–30*	10–15	10–15
	(Toy)	14–16*	15–18*	<10	<10

*These figures are not given in the Official Standards and are thus approximate.

Information obtained from *"The Complete Dog Book,"* 17th edition, the official publicaiton of the American Kennel Club, Howell Book House, 1986, Macmillan Publishing Company.

Miscellaneous Class

Currently, the American Kennel Club recognizes five breeds in the miscellaneous class. These are breeds where interest exists. Breeds in these classes may compete in obedience trials, earn obedience titles, and compete in conformation shows, but they are limited to the miscellaneous class and may not compete for championship points. If interest and growth of the breed continues, they may be admitted to one of the regular classes.

The Jack Russell Terrier is one of the breeds in the miscellaneous class. The Jack Russell Terrier was developed in Southern England about 200 years ago by an English clergyman, Parson Jack Russell. The breed is 12 to 15 inches tall and weighs 13 to 17 pounds. It was originally developed to hunt fox and shows an alert, confident, ready attitude. The coat is of two types: smooth and wirehaired. The coat colors are usually white with black or tan markings or a combination of the three. See Figure 7–61.

Figure 7–61 Jack Russell Terrier. Photo by Isabelle Francais.

Other breeds in the miscellaneous class include the following:

Anatolian Shepherd   Lowchen

Havanese   Spinone Italiano

ANATOMY

The anatomy and skeletal structure of the dog are idenfified in Figures 7–62 and 7–63.

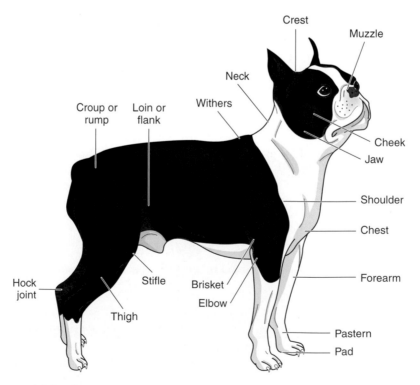

Figure 7–62 The anatomy of the dog.

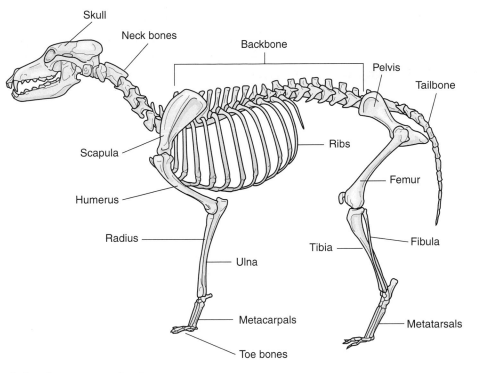

Figure 7–63   The skeletal structure of a dog.

CHOOSING A DOG

In selecting a dog, several major decisions need to be made:

1. what breed to buy

2. what source to buy from

3. what individual animal to buy

Within these decisions are many other questions that should be answered before making a final selection.

Bringing a puppy or dog home into the family can be an enjoyable experience for everyone, but unfortunately, many dogs end up in animal shelters, abandoned to run through the streets, dumped along the roadside, or destroyed. Time should be given to studying, researching, and fully considering the decision before bringing a dog home.

Selecting a Breed

In selecting a specific breed, the following questions should be considered:

1. Does one want a large breed or small breed of dog?

2. Is an active or quiet breed of dog wanted?

3. What type of hair coat should the dog have?

4. What is the purpose of the dog?

5. What price will have to be paid?

Dogs are available in all sizes, from the small Chihuahua that weighs less than 2 pounds, to the large Tibetan Mastiff that may reach a height of 31 inches and weigh up to 220 pounds. Where one lives and the environment in which the dog will be raised should also influence the decision. The Chihuahua and many toy dogs are fragile and are strictly apartment dogs. They require very little space and are more comfortable in the apartment environment. On the other hand, many of the large breeds require more space and lots of exercise. To be healthy and happy, they need an area to exercise in or a large area to roam freely.

Along with the size of the dog, consideration needs to be given to how active or quiet the breed is. The Fox Terrier, although weighing only about 18 pounds, is a very lively, active dog. The Old English Sheepdog and the Labrador, on the other hand, are fairly large breeds but can adapt well to apartment life. However, exercise should never be neglected with any breed.

Hair coats on dogs are mainly of three types: shorthaired, longhaired, and wirehaired or curly-haired. This decision should probably be based on the amount of time one has. Dogs with shorthair coats require very little grooming, although frequent brushing will remove old, dead hair and distribute the skin's oils. Longhaired dogs need regular brushing and combings, with extra attention given for mats and tangles. Some breeds, such as the Yorkshire Terrier, because of their long, silky hair, may require daily care. Dogs with wirehair and curly hair require regular plucking and special attention to remove dead hair and debris that may accumulate.

In making a selection of breed, one needs to consider what the dog will be used for. Over the years, dog breeds have been selected and bred for several purposes. Among the common uses are for hunting; as herd dogs, guard dogs, and show dogs; for killing rats, mice, and other "varmints"; as companions; or simply as pets. More specific uses would be for leader dogs for the blind, for smelling out and detecting drugs, as rescue dogs, and (in the case of the Greyhound) for racing.

Finally, one needs to consider the price. The common rules of supply and demand control the price of the various breeds. The more common and available the breed, the lower the price; the rarer the breed, the higher the price. There are exceptions to this rule, but this is usually the case. According to the number of registrations with the American Kennel Club, the most popular breeds today are as follows: Cocker Spaniel, Poodle, Labrador Retriever, German Shepherd, Golden Retriever, Beagle, Chow Chow, Miniature Schnauzer, Doberman Pinscher, and Shih Tzu.

Choosing a Source

After determining the breed to buy, the next step is to consider where and from what source to buy a dog. Several sources are available. These sources include your friends and neighbors, from newspaper and magazine advertisements, from purebred breeders, and from pet shops.

If one is just interested in obtaining a pet or companion dog and is not interested in a particular breed, dogs can be obtained from friends and neighbors. These "mutts" make lovable, friendly companions. Newspapers and magazines run advertisements for purebred as well as mixed-breed animals. A purebred breeder should be contacted for purebred dogs. In addition to newspaper and magazine advertisements, breeder names can be obtained from telephone directories; from local, state, or national kennel clubs; and from breed associations. Breeders of purebred animals are especially careful to maintain their reputations and will go to great lengths to see that future owners get a dog that meets their needs; satisfies them; is healthy, genetically sound, and free of any hereditary imperfections. Finally, dogs are available from pet shops. Pet shops obtain animals from local breeders; the dog may be sound and have registration papers, but many times they obtain their puppies from various "puppy mills" or "puppy factories." These places exist only to supply puppies to pet shops, and very little care is taken to promote sound breeding practices, sanitation, and good general care. Many of these "mills" and "factories" have been the subject of magazine articles and television shows. One should resist the impulse to buy an adorable puppy from a pet store and spend a little extra time in locating an animal that is right for one's family.

Selecting an Individual Animal

Once the decision has been made as to what breed is wanted and from what source the animal will be obtained, the prospective owner is ready to select an individual animal. Again, certain questions need to be answered:

1. What is the dog going to be used for?
2. How much does it cost?
3. What is its pedigree?
4. Is there a history of show winners or field trial champions in its ancestry?
5. Does its conformation meet standards?
6. Is it a male or a female?
7. How old is it?

The first four questions are closely related. Individual animals in a breed may be used for different purposes. The Irish Setter, for example, because of its beautiful, long, flowing hair coat, has long been admired as a show dog. Breeders who are primarily interested in show animals will breed for show quality and will pay very little attention to hunting instinct. As a result, it is possible to purchase an animal that has retained very little of its hunting instincts and would be of little use to a hunter. An animal that is of superior show quality or that has superior hunting instincts would be far more valuable than an ordinary animal sold as a pet or companion. **Pedigree,** or the ancestry of an animal, is very important in selection for show or for hunting. Many times animals are purchased by mail or phone, sight unseen, strictly because of the breeding that is used to produce the dog. The reputation of the breeder is very important in a purchase of this type because the purchaser is relying on the breeder to select an animal that is best suited. If the ancestry of an animal has a history of winning shows or field trials, the value of that animal is greatly increased. It comes from proven stock, and chances are it will be of superior show quality or a superior hunter.

Conformation. Conformation is the general structure, look, and makeup of the animal. In selecting animals for show, it is important that the animal meet the standards for the breed as set by the breed association. Again, animals that come from proven show stock usually meet these standards. Personal preference usually enters the decision when selecting specific color or markings of an individual.

Most puppies should not be taken from their mother until they are at least eight weeks old. It is often hard to determine at an early age the show potential of an individual, and one may want to wait until an animal is six months or older. When selecting animals, most people prefer young puppies because they are more adorable, usually are easier to train, have not been subjected to any other training, have not been subjected to long periods of abuse, and usually adapt easier to a family. However, one of the major problems with bringing young puppies into the home is the fact that they grow up to be dogs that are not as cute, require more time to feed and care for, and require more expense for food. When one buys an older animal, he or she knows exactly what the dog is like. Many older dogs that have been taken to shelters make excellent pets and companions for people interested in providing a good home for an animal.

FEEDING AND EXERCISE

To keep an animal healthy and in good condition, an adequate diet and clean water must be supplied. Beginning at about three weeks of age, fresh water should be available to puppies at all times. Even if they are still nursing, they may need additional liquid.

Types of Food

Several types of commercial foods are on the market for dog owners to choose from. Although one can mix his or her own rations, these commercial foods save time and are very convenient to use.

Three main types of commercial foods are available today: dry, semi-moist, and canned. These differ primarily in their ingredients and the amount of moisture they contain. The main ingredients in dry foods are corn, soybean meal, wheat millings, meat, and bone meal. Ingredients in semi-moist foods are corn, meat by-products, soybean meal, and corn syrup. Canned foods are of two types: (1) a ration-type that contains barley, meat by-products, wheat grain, and soy flour, and (2) a meat-type that contains meat by-products, meat, poultry, and soy flour. Dry foods

contain about 10 percent moisture, semi-moist contain 30 percent, and canned foods 75 percent moisture. See Figure 7–64.

Their nutrient composition also differs. Dry foods contain approximately 23 percent protein, 9 percent fat, and 6 percent fiber. Semi-moist foods have about 25 percent protein, 9 percent fat, and 4 percent fiber. The ration-type canned foods contain about 30 percent protein, 16 percent fat, and 8 percent fiber, whereas the meat-type canned foods contain 44 percent protein, 32 percent fat, and 4 percent fiber. Most foods contain about the same amount of energy. These are approximate figures and will vary with different brands and with different methods of preparation. Table 7–8 shows the daily ration necessary for dogs of various weights.

Dry foods have the advantage of being cheaper to purchase, are convenient to use, will not spoil, and help keep the dog's teeth clean.

Feeding Pregnant and Lactating Females

The amount of ration given to the pregnant female should increase as her weight increases. Her weight normally starts to increase about the fourth week of pregnancy, and just before giving birth, she may be consuming 35 to 50 percent more food than before. She should be fed three or four evenly spaced meals to avoid the discomfort that one large meal may cause. Modern commercial food provides an adequate and balanced ration during pregnancy, and supplemental feeding should not be needed.

Following whelping, the female needs two to three times as much food as before. This should be given in three or four meals per day. The ration should be the same as has been provided throughout the pregnancy.

Beginning at about three weeks of age, the young puppies will begin to lap food from their mother's food bowl, and the owner will need to consider weaning

---

GUARANTEED ANALYSIS

Crude Protein (Minimum) . . . . 21.00% Crude Fiber (Maximum) 5.00%
Crude Fat (Minimum) . . . . 12.00% Moisture (Maximum) 10.00%
Linoleic Acid (Minimum) . . . . .3.50% Vitamin E (Minimum) . . . . 30 IU/kg

INGREDIENTS

Lamb Meal, Ground Rice, Rice Bran, Rice Flour, Sunflower Oil (Preserved with mixed Tocopherols, a Source of Natural Vitamin E and Ascorbic Acid), Rice Gluten, Dried Whole Egg, Natural Flavors, Monosodium Phosphate, Dried Kelp (Source of Iodine), Choline Chloride, Vitamin E Supplement, Zinc Sulfate, Ascorbic Acid (Source of Vitamin C), Niacin, Ferrous Sulfate, Calcium Pantothenate, Vitamin A Supplement, Manganous Oxide, Thiamine Mononitrate (Source of Vitamin B_1), Vitamin D_3 Supplement, Riboflavin Supplement (Source of Vitamin B_2), Vitamin B_{12} Supplement, Pyridoxine Hydrochloride (Source of Vitamin B_6), Inositol, Folic Acid, Cobalt Carbonate, Biotin.

Figure 7–64 Always check the label when buying pet food.

Table 7–8 Daily Rations in Ounces Needed by Dogs of Various Weights*				
Dog's Weight (lb)	**Kcal Needed/lb Body Weight**	**Dry Type**	**Semi-moist**	**Canned**
2	65	1.3	1.7	3.4
5	52	2.5	3.2	7.00
10	44	4.3	5.5	10.75
15	39	6.1	7.8	15.25
20	37	7.5	9.7	18.75
30	33	10.0	12.8	25.0
45	30	14.0	18.0	35.0
75	26	20.4	26.2	51.0
110	24	25.0	32.2	62.5

*Caloric Requirements—Maintenance—and recommended daily food intakes of average adult dogs of various body weights. Individuals may require one-quarter more or less than these averages.

Source: Reprinted with permission of the American Kennel Club, Inc. *The Complete Dog Book,* 17th edition, copyright 1986, Howell Book House, Inc., New York, NY.

the pups. This can be done gradually; as the puppies begin to consume more solid food, the owner can start to decrease the amount of food given to the mother, remove her from the puppies during the day, and increase the amount of solid food given to the puppies. Weaning should be completed when the puppies are about six weeks of age. On the first day of weaning, all food should be removed from the mother; on the second day, about one-fourth of a normal maintenance ration should be given; on the third day, one-half ration; on the fourth day, three-fourths ration; and finally, a maintenance ration that will return her to her pre-breeding body condition.

Feeding Puppies

As mentioned in the previous paragraph, puppies begin to lap solid food at about three weeks of age. Milk, meat, eggs, and cottage cheese make ideal foods because they are easily digested and contain adequate proteins and nutrients. Be careful in feeding cow's milk because it may cause diarrhea, is not as concentrated, and does not contain as much protein, fat, and minerals as the mother's milk. The use of commercial substitutes or evaporated milk is recommended. Evaporated milk should be diluted at a rate of four parts milk to one part water. As the puppies grow, they should be started on one of the commercial foods especially formulated for puppies.

The amount to feed a puppy varies depending on its level of activity and metabolism. Usually, a puppy should be fed three or four meals a day; its digestive system is not yet developed to handle one large feeding per day. A young puppy must receive enough to provide its energy requirements and for adequate growth. One must be careful not to overfeed because the puppy may become overly fat. Placing one's fingers over the ribs and gently moving them forward and backward, a

thin layer of fat should be felt. If one cannot feel the ribs or the edges of the ribs, the animal has an excess of fat. The bones of a young, overweight puppy may not be developed to adequately carry excess weight, which could cause permanent injury to bones and joints.

Feeding Older Dogs

As dogs get older, their metabolism slows down and their need for calories decreases. Owners must feed a ration containing adequate protein and nutrients to maintain body condition. One should guard against overfeeding, which will lead to obesity and cut down on the expected life of the animal. Older dogs should receive smaller meals, which are easier to digest, several times a day.

Other Feeding Suggestions

Many people believe that feeding raw eggs to dogs improves the hair coat and produces more shine. Although the fat in egg yolks produces a shine to the hair, the egg white in raw form interferes with the absorption of biotin, a necessary vitamin. Therefore, eggs should be hard-boiled or cooked before serving to a dog.

Care must be taken when feeding bones to dogs. Large bones such as the knuckle and large marrow bones are all right for dogs to chew on; bones that splinter, such as chicken, turkey, or pork bones, should never be given to dogs. Splinters from these bones may get lodged in the digestive tract and cause disastrous results. Older, larger dogs have powerful jaws and can easily splinter most bones. It may be helpful to boil large bones to destroy any harmful bacteria before giving them to dogs.

Exercise

Exercise is necessary for most dogs if they are to remain healthy. Dogs are like people in that some need more exercise than others. The amount of exercise varies according to the size of the animal, age, breed, and what the dog is being used for. Even small apartment dogs should be allowed some freedom to move around and exercise. Care must be taken when exercising young puppies; their bones may not be developed, and risk of permanent injury could result. Older dogs need exercise also, but caution should be taken that they are not overworked. Short and more frequent walks are more suitable for older animals.

A few guidelines should be followed when exercising your dog:

1. Do not exercise any dog strenuously within 2 hours after it has eaten a large meal.

2. Dogs should receive a physical examination to rule out cardiac, circulatory, or skeletal and joint problems.

3. Before taking a dog out for strenuous exercise or strenuous hunting trips, the dog should be conditioned over time. Dogs are just like athletes and need proper conditioning to strengthen muscles, joints, and cardiac systems, as well as to toughen foot pads. This is especially important for obese dogs.

4. Dogs are eager to please their masters. They will not stop or give up a challenge, and one must be alert to signs of fatigue. Experience and working with the dog should allow the owner to note excessive panting, changes in color of the lips and face, vacant stares, and changes in behavior that might signal fatigue.

TRAINING

The amount and type of training a dog receives depends on what the dog is going to be used for. Almost all dogs should learn five basic commands: heel, sit, down, stay, and come. Most owners would like a dog that will respond and be under control. Dogs that are well trained can provide many hours of enjoyment and be great companions, whereas untrained dogs can be a nuisance.

Training should begin as soon as the new puppy arrives home. The puppy should learn its name, the basic corrective word *no,* and the praise words, *good boy* or *good girl.*

The puppy must learn what types of behavior are acceptable and what types are not. The owner's voice is the best corrective tool. A strong, authoritative "no" should be all that is necessary. "Good dog," "good boy," or "good girl" should be used when the puppy does something correctly. The puppy will soon learn that it will be corrected if it does something wrong and praised when it does something right.

The puppy should never be hit or threatened with a hand or an object. A puppy that gets hit will soon shy away from its owner for fear of being beaten. When this occurs, the owner has a nuisance dog. When the puppy does something wrong, it must be corrected immediately so that it will associate the mistake with correction. If the correction cannot be made immediately, it should not be made at all. Puppies are babies and are quick to forget. If they are punished for an inappropriate behavior that occurred several minutes earlier, the puppy may not realize why it is being corrected.

House-Breaking or House-Training

To be an enjoyable companion and member of the family, a dog should be house-trained. House-training can be accomplished by allowing the new puppy outside shortly after eating or several times during the day. The puppy will soon associate being allowed outside with relieving itself. If an occasional accident occurs, use the words *bad dog* and immediately take the puppy outside. Do not try to correct the puppy by sticking its nose in its waste. Again, it is important that the owner not hit or spank the puppy. Patience, correction, and praise will soon accomplish the objective. As the puppy gets older, it will make its way to the door and "ask" to be allowed outside.

Another method used is known as paper training. This involves laying several layers of newspaper down on the floor, usually in the area where the puppy will normally be fed. Place the puppy's food and water bowl on the newspaper; the puppy will usually relieve itself shortly after eating. Try to confine the puppy to the newspaper until it relieves itself. If a mistake occurs, use the words *bad dog* and place the puppy on the newspaper. The puppy soon will learn to associate the newspaper area with elimination. As the training progresses, the area covered by the newspaper can be made smaller and moved closer to the door. Eventually, the newspaper can be moved outside and the puppy allowed to go outside to relieve itself.

Serious Training

At about six months of age, the puppy is ready to start the more serious stages of training; dogs younger than this lack the ability to concentrate on more demanding training. Training periods should be held once or twice a day, be short in the beginning, and gradually increase to 15 to 30 minutes at a time. Longer training periods will tire the dog and may bore it. Do not forget to correct the dog immediately and be generous with praise when it performs well.

Equipment

Before beginning training, the trainer should purchase a choke collar and a leash. If used properly, the leash and choke collar make very good training tools. The trainer should put the collar on the dog and allow it to get used to it for a day or two. After the dog is used to the collar, the lead should be attached and the dog walked. No pressure should be applied on the collar. The trainer gradually increases control, and the dog learns that the collar and lead are used to restrain and not meant to harm it.

Basic Commands

The first command is to heel. To heel is to walk close to the side of the handler. The handler will want the dog to walk on his or her left side. As the handler starts to walk, he or she should use the dog's name and give the command to *heel*. The handler may have to give a slight pull on the lead to persuade the dog to follow. The handler should practice with the dog two or three times a day until the dog follows on its own accord.

Teaching the dog to sit is usually part of the process in teaching the dog to heel. When the handler stops walking, he or she would want the dog to sit on the left side. At first, when the handler stops walking, he or she should use the command *sit*. As the handler gives this command, he or she will pull up on the lead and, with the free hand, push down on the dog's rump. Gradually, the dog will learn to sit without command when the handler stops walking.

The next step is to have the dog remain in the "sit" position until it is called. When the handler has the dog in the sit position, he or she will give the command *stay*. As the command is given, the handler will place his or her left hand in front of the dog's muzzle and step forward with the right foot first. The handler will repeat the command and hold the dog if necessary until it gets the idea. Gradually, the handler will increase the time for the dog to stay and the distance walked away. A well-trained dog should stay on command for at least 3 minutes.

The handler may also want to teach the dog to stay in a standing position. While the dog is heeling, the handler will slow down, stop, and give the command *stand*. The dog may want to sit, because that is what it has been taught. The handler will not let it sit, but will simply start walking again and give the command *heel*. After a few steps, the handler will give the command *stand;* the handler will also give the command *stay*. Gradually, the dog will associate the two terms, *stand* and *stay*. The dog may get confused, so patience is necessary and the handler should remember to be generous with praise.

To teach the dog to lie down on command, the dog should be put in the sit position. The handler will kneel down beside the dog and reach over it with his or her left arm and grab the dog's left front leg. The handler will grab the dog's right front leg with the right hand and gently lift the dog's front feet off the ground and ease the dog down into a lying position. The handler will practice the *down* command along with the *stay* command until the dog will lie down and stay until it is called.

The command *come* is usually eagerly obeyed. By this stage, the dog should obey the commands *sit, stand,* and *stay,* so one simply calls the dog with the command *come*. In the beginning, if the dog hesitates, the handler should give a couple of tugs on the lead and the dog will come.

The handler should use all of the commands together during training so that the dog will obey them without the use of the choke collar and lead. If the dog hesitates, appears confused, or does not readily obey, the collar and lead will be put back on and the training routine continued.

The training steps listed here are brief. If one has questions about how to proceed, consult with a professional trainer or purchase training manuals that are readily available from pet shops and bookstores. Once the dog has mastered these basic commands, it is ready to go on to advanced training, depending on the plans for the dog. Specific training books on the various breeds are also readily available.

GROOMING AND CARE

Before purchasing a dog, it is important to consider the amount of time one has not only to train the dog but also to attend to its basic grooming and health care needs. Time and attention need to be given to the care of the dog's coat, nails, ears, eyes, and teeth.

Hair Coats

The amount of time spent on grooming depends on the type of hair coat; daily brushing is recommended to remove dead hair and distribute the skin's oils. Brushing down to the skin will help to remove flakes of dead skin and dandruff.

In addition to regular combing and brushing, longhaired dogs need to be checked for mats of hair, which commonly occur behind the ears and under the legs. These masses of hair can usually be teased out with a comb, but they occasionally need to be cut out. Care should be taken when cutting out mats so that one does not cut the skin. Longhaired dogs also have problems with burrs from plants. When cutting the hair, one should slide the comb down under the mat or burr and cut the hair on the outside of the comb. This method will be less likely to injure the dog.

The Terriers and wirehaired breeds need their coats plucked to remove dead hair. Plucking is accomplished by using a stripping knife, grasping a section of hair between knife and thumb, and pulling the knife away with a twisting motion. This will remove dead hair and trim live hair.

Many different kinds of equipment are available to groom a dog. The equipment needed depends on the breed, coat texture, and purpose of the dog. The local veterinarian, pet shop, or grooming parlor can give advice.

Bathing

Dogs should be bathed only when they become extremely dirty; frequent bathing removes natural oils and causes the coat to become dry and harsh. When bathing becomes necessary, a mild soap, baby shampoo, or coconut-oil shampoo should be used. Detergent soaps should never be used because some dogs may have skin reactions to them. Many of the shampoos for pets are medicated and help prevent parasites. When washing, one must be careful not to get shampoo in the dog's eyes. A few drops of ophthalmic ointment in the eyes may help prevent eye irritations from the shampoo. After bathing, it is important that the dog not become chilled. One should be careful of the temperature if bathing outside, and when bathing inside, one must make sure the dog is completely dry before letting it outside.

Nails

The dog's nails will need to be trimmed occasionally. How often depends on the surface the dog is kept on. Outside dogs usually wear their nails down naturally on the soil surface; inside dogs need to be checked more often. Nails can be clipped at home using clippers obtained from a veterinarian or pet shop.

Clippers should be sharp so that they cut the nail and not crush it; crushing the nail may cause pain to the nail bed. One must be extremely careful not to cut into the nail bed because this will cause bleeding. With clear or white nails, you can easily see the nail bed. Extreme caution needs to be taken when clipping dogs with black nails. Bleeding can be stopped using a **styptic** pencil. One should check the dew claws on the inside of the leg; these will not wear down and many times are overlooked.

Sharp, pointed scissors should never be used to trim nails because they may injure the dog if it should happen to move suddenly.

Ears

During routine grooming, the ears should be checked; a dog's ears need to be cleaned about once a month. A cotton swab or soft cloth soaked in mineral oil, hydrogen peroxide, or alcohol can be used. See Figure 7–65. Only the part of the ear that can be seen should be cleaned. One should use a finger and never poke any sharp or hard object into the ear; a quick head movement could cause serious injury. See Figure 7–66.

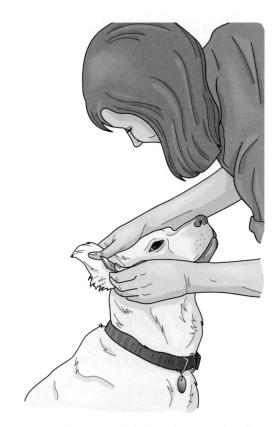

Figure 7-65   One should follow the veterinarian's instructions when applying ear drops or ointments in the ears of the pet.

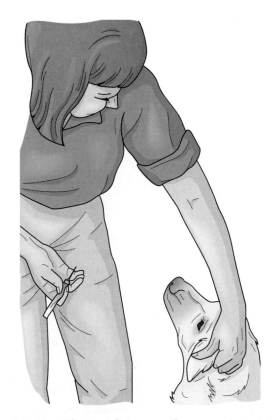

Figure 7–66   After applying ear drops or ointments, gently massage the ear.

Spaniels and other dogs with long hair need to be checked for mats and burrs that may block air movement and cause infections.

Eyes

Irritating substances can be removed from the eyes with boric acid solutions or other eyewash solutions. Dogs with large protruding eyes and long hair need special care to keep hair from rubbing the eyeball. Hunting dogs and other field dogs should be checked carefully after each outing for irritating substances. Serious irritations and injuries need to be brought to the attention of the veterinarian. See Figure 7–67.

Teeth

A dog's teeth are usually not prone to decay. However, dogs do develop plaque and tartar that can lead to periodontal disease. Periodontal disease is painful and may lead to premature tooth loss. A dog's teeth should be cleaned once or twice a week. This can be accomplished using a small toothbrush or a gauze pad. Tooth paste, salt water, or an equal mixture of salt water and baking soda can be used. The teeth should be scrubbed from gum to crown; regular cleaning will prevent many problems. If the plaque is not removed, it will develop into a hard yellow-brown or gray-white deposit on the teeth called tartar. Tartar is not easily removed and requires a visit to the veterinarian. See Figure 7–68.

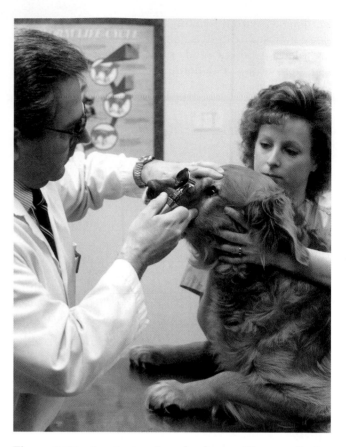

Figure 7-67   A veterinarian check-up will normally include checking the animal's eyes. Courtesy of Brian Yacur and Guilderland Animal Hospital.

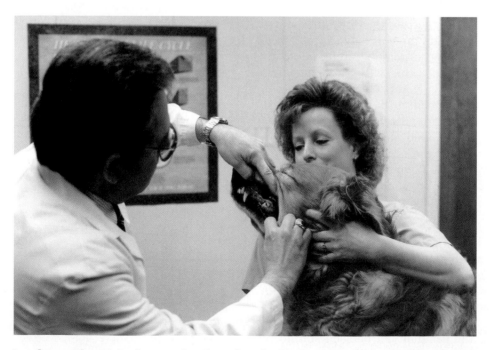

Figure 7–68   During the regular visit, the veterinarian also checks the animal's teeth and gums. Courtesy of Brian Yacur and Guilderland Animal Hospital.

Feeding the dog dry dog food and some of the various treats on the market also helps prevent the buildup of tartar. Dogs love to chew, and chewing will keep teeth clean and sharp. Rawhide, synthetic bones, and hard chew toys are pleasurable and beneficial for the dog.

COMMON DISEASES

Unfortunately, even the most well-cared-for dog may come down with some type of disease, parasite, or ailment. With regular grooming and care, an owner can be attuned to early warning signs. Some of these early signs are constipation or diarrhea, shivering, fever, watery eyes, runny nose, coughing, loss of appetite, vomiting, loss of weight, ravenous eating without gaining weight, increased urination, restlessness, straining to urinate, labored breathing, increased water intake, lameness, paralysis, obvious pain, or nervous symptoms. If any of these signs are observed, it is important that the dog be taken to a veterinarian as soon as possible. See Figures 7–69 and 7–70. Prompt diagnosis and treatment may prevent more serious conditions from developing.

This unit of the text is broken down into six areas: infectious diseases, noninfectious diseases, fungus

Figure 7-70 The veterinarian can easily check to see whether an animal is overweight.

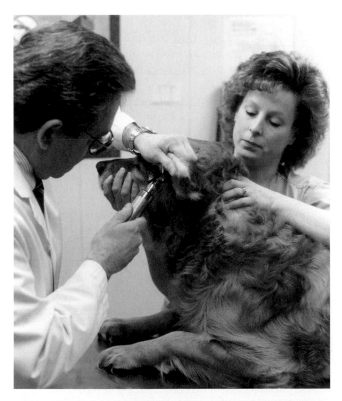

Figure 7-69 Pets should make regular visits to the veterinarian for examination and any vaccinations that may be needed. Courtesy of BrianYacur and Guilderland Animal Hospital.

diseases, internal parasites, external parasites, and poisonings. More emphasis is placed on the more common and serious problems that occur. The cause, symptoms, treatment, preventions, and control are listed when appropriate. This unit is intended as a general information unit and not as a diagnosis unit. It must be emphasized again that if any of the early warning signs appear, the dog should be taken to a veterinarian for diagnosis and treatment.

Infectious Diseases

This group of diseases is caused by pathogenic microorganisms such as viruses, bacteria, protozoa, and fungi that are capable of invading and growing in living tissues.

Canine distemper. Canine distemper is caused by inhalation of an airborne virus that spreads throughout the body. The condition most often occurs in young dogs between three and six months of age, although older dogs may be susceptible. Early symptoms are vomiting and diarrhea. Some improvement may be seen after the first symptoms appear, but weeks or months later, symptoms that show nervous system damage appear. These can range from tremors to full epileptic fits. The dog will eventually die or have permanent nervous system damage. Treatment is in the form of anti–canine-distemper serum, antibiotics to prevent secondary infection, cough suppressants, vomit and diarrhea treatments, and possible anticonvulsants. Canine distemper can easily be prevented with vaccination. Dogs should be vaccinated at eight weeks, twelve weeks, and sixteen weeks of age. Annual revaccination is recommended throughout the dog's life because of loss of antibody protection.

Infectious canine hepatitis. Infectious canine hepatitis is a highly contagious virus disease of dogs; the disease is spread from dog to dog through the urine. Infection usually is acquired by contact of urine or saliva in the mouth.

This disease is seen most often in dogs less than one year of age; affected animals appear dull and apathetic. They usually refuse to eat and may have an intense thirst. The body temperature increases, and swelling may occur in the head, neck, and lower portions of the abdomen; vomiting and diarrhea are common. The dog may moan in pain when eliminating. The gums may be pale, and sometimes hemorrhages appear on them; the tonsils are often enlarged and painful. Most dogs either recover or are dead within two weeks. Many die within a few days without warning. Vaccinations should be given at the same time as for canine distemper. Annual revaccination is also recommended.

Leptospirosis. Leptospirosis is caused by a bacteria and is a disease of major importance because dogs and other animals can act as reservoirs for human infections. The most common method of spread is through contact or ingestion of contaminated food and water. Outbreaks have occurred in humans and dogs after floods or after swimming in or drinking water contaminated by urine of rodents or other animals.

Early signs of the disease are high fever, loss of appetite, vomiting, and diarrhea; dehydration and depression may occur. Reddening of the membranes of the eyes and mouth are common. In some cases, yellowing of the gums and whites of the eyes may occur. The disease attacks the liver, kidneys, and gastrointestinal tract, causing pain and discomfort. Death usually occurs from severe damage to the liver and kidneys.

If any of the symptoms occur, your veterinarian should be consulted immediately. The bacteria responds to antibiotics, so if treatment is prompt, there is a chance for recovery. Vaccinations are available and should be given when the dog receives its distemper and hepatitis vaccinations. Annual revaccination is recommended for effective immunity.

Canine parvovirus infection. Canine parvovirus infection is a disease that appeared in the United States in 1977. The disease is caused by a small DNA-containing virus that requires rapidly dividing cells for growth to occur; thus, young pups are most susceptible. The main source of infection is ingestion of materials contaminated with feces from infected dogs.

There are two forms of the disease; one involves the intestines and the other the heart. In the intestinal form, the virus grows in the epithelial cells of the small intestine and rapidly destroys them. Early signs are vomiting followed by diarrhea, refusal to eat, and severe dehydration. The feces appear yellow-gray and are often streaked with or darkened by blood. Dogs may vomit repeatedly and have projectile and bloody diarrhea until they die. Others may have only loose feces and recover without complications. Most deaths occur within 48 to 72 hours following onset of clinical signs.

The heart form may be proceeded by the intestinal form or may occur without apparent previous disease. In this form, the virus rapidly attacks the muscle cells of the growing heart. Pups may act depressed and stop suckling shortly before they collapse, gasping for breath. There is no specific treatment. Pups that survive may have damaged hearts and die from heart failure weeks or months later.

Veterinarians recommend a series of vaccinations at eight, twelve, and sixteen weeks. Some veterinarians recommend a series at two-week intervals plus boosters at eighteen and twenty weeks. Annual revaccination is recommended for effective immunity. Dogs in high-risk areas or activities may need revaccinations at six-month intervals. Pregnant females should receive a vaccination two weeks before whelping so that protection can be passed on to the puppies.

Proper cleaning and disinfection of the kennel and other areas where dogs are housed are essential to controlling the spread of the virus. The virus is capable of existing in the environment for many months. A solution of 1 cup chlorine bleach in 1 gallon of water is an effective disinfectant.

Kennel cough. Kennel cough, more accurately called infectious tracheobronchitis, is an infectious respiratory disease of dogs marked by coughing and in some cases by fever, not eating, and pneumonia. It is caused by various viruses and bacteria alone or in combination.

This disease occurs primarily when dogs of varying ages and susceptibility are congregated under conditions that are less than ideal, such as pet shops, dog shows, kennels, dog pounds, animal shelters, or boarding and training kennels. The kennel cough agents are quickly spread when an infected dog coughs.

The most prominent sign of disease is a cough. The cough is usually dry and hacking, followed by gagging or expectoration of mucus. The most severe form of kennel cough is seen in dogs that are six weeks to six months of age. When there is a lack of appetite, nasal or eye discharges, and depression, the disease is usually severe and your veterinarian should be consulted. Left untreated, the disease could cause damage to the respiratory system in the form of bronchitis or pneumonia and eventually death. Antibiotics are usually an effective treatment.

Viral and bacterial vaccines are now available to control the principal agents involved in kennel cough. Dogs should be vaccinated when they receive their early puppy immunizations.

Rabies. Rabies is a viral disease of humans and other animals. The rabies virus attacks the central nervous system, and if the disease is not prevented, it causes death.

All warm-blooded animals can transmit rabies. The majority of all animal rabies cases in the United States are found in wildlife, particularly in skunks, raccoons, foxes, and bats. Among domestic pets, dogs and cats are the most commonly infected species. (Refer to Chapter 2 for more information.)

Rabies is usually transmitted from animals to humans by a bite from a rabid animal. The wound is contaminated with the virus found in the saliva of the infected animal. Signs of rabies develop within two weeks to three months after the bite has occurred. If the bite is on the head or face and is severe, symptoms may appear in ten days or less.

There are two types of clinical rabies in animals: "dumb" and "furious." During early stages of "furious" rabies, the animal may act strangely. It may be unnaturally withdrawn or abnormally affectionate. This stage usually lasts about two days and then is followed by the "mad" stage, where the animal wanders off and will attack and bite anything in its path. A common symptom of this stage is frothing at the mouth, which is caused by difficulty in swallowing. When the animal is over its wandering, it usually returns home, finds an isolated place to have its final spasms, lapses into a paralytic state, and dies.

"Dumb" rabies differs in that there is no wandering or "mad" stage. Paralysis of the lower jaw is usually one of the first signs, followed by paralysis of the limbs and vital organs, resulting in death.

When an animal is bitten, the virus is drawn to the nerves and follows the nerve fibers to the brain and the salivary glands. After the virus infects the brain, it reproduces rapidly, causing severe brain damage. The brain lesions lead to altered behavior, aggressiveness, progressive paralysis, and usually death.

In preventing and controlling rabies, all pets should be routinely vaccinated. Regular vaccinations every one to three years should be part of a pet's health program.

Coronavirus. Coronavirus is a disease that causes severe vomiting and diarrhea. It is a highly conta-

gious viral disease that spreads rapidly by way of contaminated feces. The disease was first isolated in 1971 from the feces of military dogs. The feces are usually orange, are foul smelling, and may be streaked with blood. If other viruses, bacteria, or parasites are present, treatment may be difficult. Early diagnosis is important so that early treatment may begin. A series of vaccinations at two-week intervals is now recommended for puppies at five to six weeks of age and continued until they are approximately eighteen weeks old.

Canine Brucellosis. Canine brucellosis is caused by the bacteria *Brucellosis canis.* The disease occurs worldwide and has been reported in every state in the United States.

Common signs of infection in females is abortion, failure to whelp, and enlargement of the lymph nodes. In male dogs, the scrotum and testicles swell, and one or both testes may also atrophy. Infected animals may show no sign of the disease; infection may occur readily across mucous membranes. Sexual intercourse is a common method by which the disease is spread; all breeding animals should be blood-tested. Infected animals should be neutered or spayed to reduce spread of the disease through breeding. No medicine is available against canine brucellosis. Treatment with antibiotics may be successful.

The potential of canine brucellosis being transmitted to humans should be considered, and all possible precautions should be taken when treating or handling infected dogs.

Canine herpesvirus. Canine herpesvirus causes severe illness and death in puppies less than six weeks old. The infectious organism is a virus, and only dogs are known to be susceptible.

Spread of the disease is by direct contact. Adult dogs show no signs of illness but can pass the virus in oral, nasal, or vaginal discharges. The organism can also be spread by way of saliva, feces, and urine. Unborn puppies may be infected in the mother's womb by primary infection of the mother or during passage through the birth canal.

Illness in young puppies usually occurs between the fifth and eighteenth day after birth. Symptoms include a change in the color of feces, difficulty in breathing, abdominal pain, ceasing to nurse, and constant crying. Affected puppies die shortly after the symptoms appear. No vaccines are available for this disease.

Pseudorabies. Pseudorabies, also known as Aujesky's disease or mad itch, is caused by a virus and occurs in areas where the disease is present in swine herds. Pseudorabies is primarily a disease of swine but can be transmitted to cattle, sheep, goats, dogs, cats, and wild animals. Dogs experience intense itching; self-mutilation may occur as the dog tries to relieve the itching. The disease progresses to convulsions, paralysis, and coma, followed by death within 24 to 72 hours. There is no effective drug for this disease.

Salmonellosis. Salmonellosis is caused by a bacteria organism that is found in both wild and domestic animals. It is usually spread by ingestion of the organism in food contaminated by infected feces.

Infected animals experience fever, diarrhea, and vomiting. Dehydration and severe inactivity usually accompany these signs of infection. Salmonellosis can be confused easily with other gastrointestinal diseases such as canine distemper and parvovirus infections.

People who handle animals may be susceptible to the infection from animals or may transmit the infection to animals because they are active carriers of the organisms.

Vaccination of dogs and cats is not done at present. Strict attention should be paid to disinfecting of facilities and to the personal hygiene of individuals working with animals. Infected animals can be treated effectively with antibiotics.

Haemobartonellosis. Haemobartonellosis is a bacterial infection in which the infective organisms attach tightly to the surface of red blood cells and eventually destroy those cells. Common signs of infection are fever, pale mucous membranes, weight loss, and depression. Animals may also be infected with other organisms. This disease is very responsive to antibiotics, and most animals survive without permanent damage.

Tuberculosis. Tuberculosis in dogs is caused by *Mycobacterium tuberculosis* and *Mycobacterium ovis* bacteria. Animals are infected by inhaling the organisms from respiratory excretions or by drinking infected milk. Symptoms usually occur as draining, nonhealing abscesses. Sometimes, local pain is noted, and diagnosis is made by isolating the organism from the lesions; surgical removal of the lesions is the preferred treatment. Response to antibiotics is usually poor.

Camylobacteriosis. Camylobacteriosis is caused by a bacterial organism spread by direct contact or consumption of contaminated water and food of animal origin. The organism has been isolated from human feces and has caused diarrhea in people. Origin of the human camylobacter organisms has been blamed on infected dogs and cats.

Symptoms are usually loss of appetite, inactivity, and mild diarrhea. The disease is responsive to antibiotics. Vaccination for the disease is not possible at

this time. Simple hygienic measures are important to prevent spread of the disease.

Noninfectious Diseases

This group of diseases is caused by physical injuries or genetic defects or are diseases that are not contagious.

Heart disease. Heart disease in dogs and cats occurs with moderate regularity. There are two types: **congenital** defects, which are present at birth, and acquired heart disease. The most common signs include coughing at night during sleep, coughing during exercise, inability to exercise, fainting spells during exercise, development of blue gums during exercise, open-mouth breathing at rest (not panting, which is the dog's normal way of perspiring), reluctance or inability to breathe when lying down, and development of a large pendulous abdomen because of accumulation of fluid.

There are several causes of acquired heart disease. These abnormalities usually begin slowly and increase in severity as the animal ages. Among the causes are degenerate disease of the heart valves, bacterial infections of heart valves and muscles, cancerous tumors, degenerate diseases of the heart muscle, and heartworms.

Cataracts. Cataracts in dogs cause a cloudy, white opacity of the lens. Cataracts may be hereditary or nonhereditary and usually cause blindness when fully developed. Removal of the cataract surgically is the only effective treatment.

Glaucoma. Glaucoma is a serious disease involving an increase in pressure within the eyeball caused by a variety of problems with the production, transport, and absorption of aqueous humor, the fluid within the eye. This pressure causes irreversible damage to the retina and optic nerve. Drops may be administered or surgery may be necessary.

Progressive Retinal Atrophy (PRA). PRA is a genetic disease in which the cells of the retina gradually degenerate, leading to the loss of sight. The age of onset of the disease varies with the breed and is usually breed specific. The first sign of the disease is loss of night vision; dogs undergo behavioral changes, especially when light is limited. There is no treatment for this disease.

Cherry eye. Cherry eye is caused by a prolapse and enlargement of the tear gland on the inner surface of the third eyelid. Cherry eye appears as a red, cherry-like growth. Antibiotics and anti-inflammatory drugs help relieve mild cases, but surgery is necessary to remove more serious cases.

Hip dysplasia. Congenital hip dysplasia (CHD) is an inherited trait. Environmental conditions such as too much exercise, especially at a young age; rough play; jumping; excess weight gain; and rapid growth can contribute to the age at which the condition occurs and to the severity of the condition.

The disease affects the hip joints and occurs more commonly in large-breed dogs. This widespread disease causes hind leg lameness as the result of pain originating in the malformed ball-and-socket joint. Affected dogs exhibit pain on attempting to rise and when moving, particularly in damp, cold weather.

Treatment is usually either nonsurgical or surgical. Nonsurgical treatment involves the use of drugs to relieve the pain. Surgical treatment may be necessary if the nonsurgical methods fail to reduce the pain or to improve the animal's condition. Before surgery one needs to consider the age and weight of the animal, how extensive is the surgery, and what is the function of the animal.

Arthritis. Arthritis is a degenerate joint disease that causes pain, lameness, and stiffness in the joints. The disease is usually associated with old age. Large dogs are more often victims, and the symptoms are more severe in obese dogs. Drug therapy can help relieve pain, and moderate activity is recommended.

Tetanus. Tetanus in dogs is caused by a bacteria whose spores are present in the soil and feces of various animals; most cases result from contamination of small puncture wounds and lacerations. The organisms produce toxins that cause overstimulation of the dog's nervous system.

Infected dogs usually experience spasms of the facial muscles. Spastic paralysis, lockjaw, inability to stand, prolapse of the third eyelid, and overextended head, neck, and legs may be seen as the disease progresses. Death usually occurs because the disease affects the respiratory muscles and other complications set in.

Botulism. Botulism in dogs results from ingestion of a toxin-producing bacterium. Animals usually acquire the toxin by ingesting contaminated material such as rotting carcasses or garbage. Botulism organisms have a paralyzing effect at the junctions of the muscles and nerves. The animal usually loses total muscle function.

Anal sacs. Anal sacs are located at the four o'clock and eight o'clock position on either side of the dog's anus. These sacs contain a substance that allows the dog to mark its territory; occasionally, these sacs become blocked and need emptying. They may become swollen and painful, and the dog may be seen licking the area and dragging his or her rear on the ground. A veterinarian should empty them the first time. After observing the procedure the veterinarian uses, the owner may want to attempt the procedure the next time the problem occurs.

Fungus Diseases

Ringworm. Fungus diseases are found throughout the world; ringworm is one of the most common. Ringworm in dogs is caused by three fungal organisms: *Microsporum canis, Microsporum gypseum,* and *Trichophyton mentagrophytes.*

Microsporum canis is most commonly found on cats but can infect dogs and humans as well. Infection is usually a result of direct contact with an infected animal.

Microsporum gypseum normally grows in the soil. Animals become infected by digging in or making contact with infected soil.

Trichophyton mentagrophytes are commonly found on wild rodents but can also be found on many species of animals. Dogs and cats become infected when they dig in the burrows of infected rodents.

Symptoms of ringworm begin as broken hairs around the face, ears, or feet. Reddened skin and scaly skin develop; in severe cases scaling and crusting accumulate and the skin becomes thickened and itchy. A characteristic redness develops around a healing center as the disease progresses. Lesions in dogs are usually more severe than in cats.

In treating ringworm, the skin in and around the lesions should be clipped. Local treatment with baths, dips, creams, or lotions containing antifungal agents should be applied. Oral medications may also be used. Treatment should continue for six weeks or longer.

Systemic fungal infections. Systemic fungal infections are so-called because they may involve one or all of the body systems.

Blastomycosis is caused by the fungal organism *Blastomyces dermatitidis.* The organism prefers moist soil enriched with bird or bat droppings; infection is caused by inhaling infected spores. Common symptoms are coughing, rapid breathing, pneumonia, and fever. In severe cases, there will be loss of weight, loss of appetite, inactivity, and even death. Nodular draining skin lesions are common with blastomycosis.

Histoplasmosis is caused by the organism *Histoplasma capsulatum.* This organism is also found in soil enriched with bird or bat droppings. Symptoms are similar to blastomycosis, except that the nodular draining skin lesions are not common; however, severe diarrhea is a common symptom.

Coccidioidomycosis is caused by the organism *Coccidioides immitis*. This organism prefers hot, dry alkaline soils. Symptoms are similar to blastomycosis and histoplasmosis, except that nodular draining lesions and severe diarrhea are not common. Coccidioidomycosis often affects the bones, resulting in swellings, lameness, and fractures.

Treatment of animals for all three fungal systemic diseases requires months of therapy with expensive and potentially toxic drugs.

Internal Parasites

Roundworms. Ascarids *(Toxocara canis)* of dogs, commonly called "roundworms," are large worms that reach 4 to 8 inches in length when mature. As egg-laying adults, they live in the small intestines of dogs less than six months old.

The major means of infection is by ingestion of eggs or by the transmission of infective larvae from mother dog to her unborn or nursing pups. The small larvae, which hatch in the dog's stomach, grow to an adult worm that begins to produce eggs in about four weeks. Older dogs develop a resistance to the adult worm, so if they ingest an infective egg, the hatched larvae stage does not grow to an adult worm but migrates to the muscle and remains inactive.

In male dogs, these larvae have no escape, but in females that become pregnant, the larvae become active and migrate via the umbilical cord to the liver of the fetuses. From the liver, the larvae make their way to the lungs, are coughed up, are swallowed, and make their way to the small intestine, where they grow to adult worms. The larvae can also be passed from the mother to the pups in the milk, although only about 5 percent are passed this way.

Clinical signs depend on the number of worms present. If only a few worms are present, there may be no evidence of infection, but if several hundred worms are growing to adults, the abdomen will distend, giving the pup a pot-bellied appearance. These worms deprive the pup of nutrients.

It is important to treat pups early so that the worms are expelled before they become egg-laying adults. Pups should be treated at two, four, six, and eight weeks of age. There are several drugs available that a veterinarian can prescribe. Treatment of the mother dog involves destroying the larvae ascarids in her musculature before the pups are born. Treatment must begin on about the fortieth day of pregnancy and continue daily until two weeks before whelping, or about thirty-seven consecutive days. Consult with a veterinarian for proper treatment.

Roundworms are a danger to children. Infected dogs should be kept away from children and their play areas. (See Chapter 2 for further information.)

Hookworms. Hookworms are bloodsucking parasites. The most common species is *Ancylostoma caninum.* The adult worm is about 1 inch long and has a large mouth that attaches to the small intestine and digests a plug of tissue; this causes a small bleeding site. When fifty to several hundred worms are present, the loss of blood can be dangerous.

Dogs heavily infected with hookworms develop anemia. The loss of blood into the intestine results in dark-colored feces. Gums of the animal will be white, and the animal will appear weak and listless.

Both young and adult dogs can have infections of adult hookworms, which shed eggs in the dog's feces. Once passed in the feces, hookworm eggs hatch into larvae and can infect dogs either by penetrating the skin when dogs lie down in damp, contaminated places, or by being ingested with contaminated food or water. Older dogs can harbor both adult hookworms and immature larvae in their muscles and mammary glands. Immature larvae can be passed to nursing pups by the lactating mother.

Larvae grow to egg-laying adults in about two weeks. If the number of hookworms is great, pups begin showing clinical signs of anemia at about two weeks of age and may die before three weeks of age.

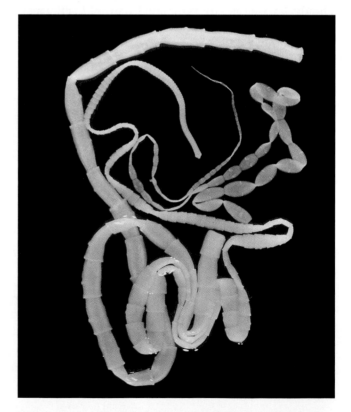

Figure 7–71 This type of tapeworm infects dogs and cats. Note the segmented body. These segments break off and pass out of the animal's body. These segments look like pieces of rice around the anal region and in the feces. Photo reprinted with permission of Miles, Inc., Animal Health Products.

Treatment for young pups and pregnant females is the same as that listed for ascarids. Older dogs should receive a single, one-day deworming once or twice per year as recommended by a veterinarian. Infected dogs should be kept away from children and their play areas. (See Chapter 2 for further information.)

Whipworms. Whipworms *(Trichuris vulpis)* live in the lower digestive tract of dogs, especially in the cecum. The adult worm is shaped like a whip, broad at one end and narrow at the other. It is about 2.5 inches long.

These worms thread their narrow end into the lining of the cecum to hold on. They produce football-shaped eggs that are passed in the feces of infected dogs. These eggs, protected by a thick shell, can live in the soil for several years.

Dogs with heavy infections will have irritation to the gut lining, causing loss of tissue fluids, which results in watery feces that is usually tinged with blood. Dogs will become dehydrated and, if not treated, will die. Consult a veterinarian for appropriate treatment.

Dogs should not be returned to the same yard pen if possible because they will become reinfected and require further treatments.

Tapeworms. Tapeworms are flat, segmented worms, usually 1 foot or more in length, that live in the small intestine. Four types infect dogs: *Diplylidium caninum, Taenia pisiformis, Echinococcus granulosus,* and *Echinococcus multilocularis.* See Figure 7–71.

Tapeworms shed their egg-filled terminal segment in feces of the infected animal. These segments are about the size of a grain of rice and may be seen crawling on the surface of feces.

An initial host is required for development of the young tapeworm to a stage that it will be infective. In the case of *Diplylidium,* the initial host is the flea; and in *Taenia,* wild rabbits serve as the initial host. See Figure 7–72.

Tapeworms are not harmful to the dog, but the two species *E. granulosus* and *E. multilocularis* can cause serious illness to humans (refer to Chapter 2). Drugs are available on the market that are effective in treating tapeworms. For complete control we must also control the fleas and prevent dogs from eating wild rabbits that serve as the initial hosts. Several products are on the market for control of fleas; these products consist of sprays, mists, oral tablets and liquids, cleansing and grooming shampoos, and dips. See Figure 7–73.

Heartworms. Heartworms, known as *Dirofilaria immitus,* live in the heart or in the major artery that carries blood from the heart to the lungs. Adult worms are thin and reach a length of 6 to 14 inches.

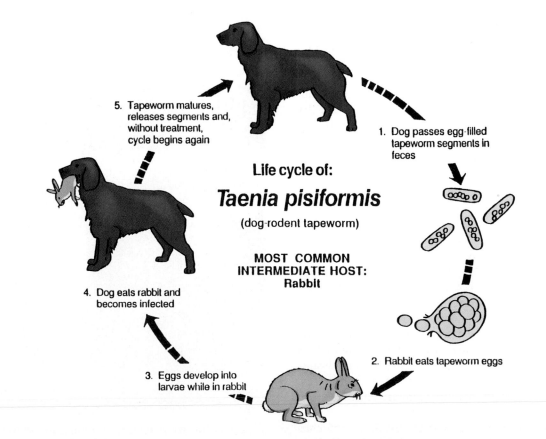

5. Tapeworm matures, releases segments and, without treatment, cycle begins again

1. Dog passes egg-filled tapeworm segments in feces

Life cycle of:

Taenia pisiformis

(dog-rodent tapeworm)

MOST COMMON INTERMEDIATE HOST: Rabbit

4. Dog eats rabbit and becomes infected

3. Eggs develop into larvae while in rabbit

2. Rabbit eats tapeworm eggs

Figure 7–72 Life cycle of Taenia pisiformis, *dog/cat tapeworm. Reprinted with permission of Miles, Inc., Animal Health Products.*

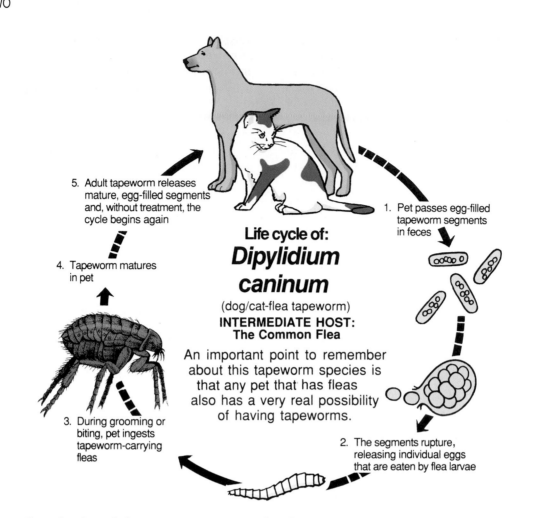

Life cycle of:
Dipylidium caninum
(dog/cat-flea tapeworm)
INTERMEDIATE HOST:
The Common Flea

An important point to remember about this tapeworm species is that any pet that has fleas also has a very real possibility of having tapeworms.

5. Adult tapeworm releases mature, egg-filled segments and, without treatment, the cycle begins again

4. Tapeworm matures in pet

3. During grooming or biting, pet ingests tapeworm-carrying fleas

1. Pet passes egg-filled tapeworm segments in feces

2. The segments rupture, releasing individual eggs that are eaten by flea larvae

Figure 7–73   Life cycle of **Dipylidium caninum.** *Reprinted with permission of Miles, Inc., Animal Health Products.*

Adult heartworms cause extensive damage to the vital organs; the lungs are the first affected. As the disease progresses, the heart becomes enlarged and does not function normally. Later, the liver and kidneys may be damaged. Considerable damage may occur before any outward signs are noticed. Signs include frequent coughing, sluggishness, rapid tiring, and labored breathing. These signs are easily noticed in hunting or working dogs. When the disease reaches critical stages, the animal is usually weak, has difficulty breathing, and may faint. See Figure 7–74.

Heartworm infection is transmitted by mosquitoes. When a mosquito bites an infected dog, it takes up blood that may contain microfilariae. These incubate in the mosquito for about two weeks, during which time they become infectious larvae. When the mosquito bites another dog, the infective larvae are passed on. These larvae migrate through the tissues of the body for two to three months and enter the heart, where they reach adult size in another three months.

Prevention of heartworm infection is more desirable than treatment. Three drugs are available from

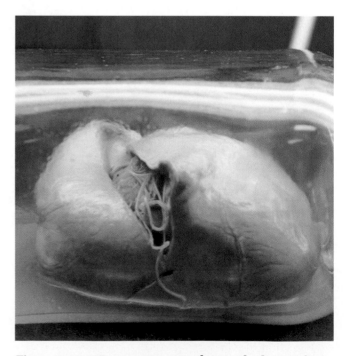

Figure 7–74   Heartworms can plug up the heart of an animal. Courtesy of Stacey Riggert.

veterinarians for prevention of heartworm. One involves a daily treatment, and the other two require a monthly treatment. All three are chewable tablets. The newest method of prevention is a minute dose of ivermectin given once a month. Tablets for this purpose are available in three sizes and are given according to the weight of the dog. See Figure 7–75.

Treatment for heartworm can be successful if the disease is detected early. Adult worms are killed with an organic arsenical drug given through a series of carefully administered injections.

External Parasites

Fleas. Fleas are wingless, brown, bloodsucking insects that infest warm-blooded animals. Young animals may die from severe blood loss.

Both the cat flea *(Ctenocephalides felis)* and the dog flea *(Ctenocephalides canis)* infest dogs. Fleas move rapidly on the skin and are most commonly found on the rump and groin area.

Flea eggs laid on the host are smooth and quickly fall off into the animal's environment. The eggs are oval, white, and glistening.

Small larvae hatch from the eggs and feed on the feces from adult fleas. After several molts, the last larvae stage forms a pupal case. The adult flea emerges from the pupal case and searches for an animal. Time required for the flea to develop from egg to adult may be as short as sixteen days.

Each bite a flea takes on the host animal allows a small amount of saliva to enter the skin and causes an irritating and allergic reaction that leads to severe itching. Biting and scratching around the rump and groin areas are the most common signs of flea allergy.

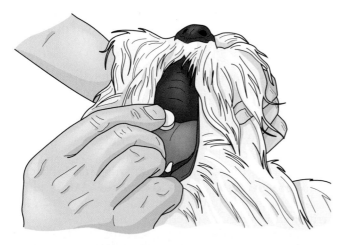

*Figure 7–75 **Heartworms can be prevented by giving heartworm pills (DEC) daily during the mosquito season and for two months after the mosquito season. Another drug, Ivermectin, can be given monthly.***

Hair loss, a red rash, and thickening of the skin are also commonly seen.

Flea feces can confirm the presence of fleas and consist of dried blood from the host animal. Mixed with a small amount of water, it will form a blood-red solution.

To control fleas, both the animal and its environment must be treated. Several products to control fleas are available. Among the products are flea shampoos, powders, rinses or dips, sprays, foams, pour-on products, oral insecticides, flea collars, foggers, and house and yard sprays. Read the labels on the products carefully and use them according to the directions on the labels.

Ticks. Ticks are bloodsucking arthropod parasites of the skin. Severe blood loss may result from heavy infestations. Ticks have a four-stage life cycle and, depending on the type, may require one, two, or three hosts to complete their life cycle.

Hard ticks *(Ixodidine)* and soft ticks *(Argasidae)* are the two main families of parasitic ticks. Hard ticks have hard shields on their backs distinguishing them from soft ticks. Most hard ticks require three different hosts to complete their life cycle; each stage feeds only once.

There are thirteen species of hard ticks that are of importance; two of these are major problems in dogs. The Brown Dog tick *(Rhipicephalus sanguineus)* can survive indoors and infest kennels and households, and the American Dog tick *(Dermacentor variabilis)* lives only in grasses and shrubs.

The Brown Dog tick is a three-host tick that attacks cats, horses, rabbits, and humans. The larval and nymph stage of the American Dog tick parasitizes field mice. Hard ticks are normally found securely attached by their head to the skin. Before removal, they should be sprayed with insecticide safe for use on animals or soaked with alcohol. Tweezers should be used to grasp the head to remove the tick. Cigarettes, lighters, gasoline, and kerosene can severely injure the skin and should not be used. In the case of the Brown Dog tick, the kennel or household premises where the dog stays may need to be treated.

Soft ticks have a leathery outer covering. The Spinose Ear tick *(Otobius megnini)* is the only medically important soft tick. The young stages (larvae and nymphs) live in the outer ear canal of dogs, cats, cattle, and horses. Adult soft ticks do not live on the animals. The larvae and nymphs can cause severe irritations to the ear canal and may occasionally cause paralysis and seizures in some animals. Care must be taken when removing these ticks because damage to the ear canal can occur. Control and treatment of ticks is similar to controlling fleas.

Lice.   Lice are wingless insects that are uncommon parasites of dogs. Lice spend their entire life cycle on the host. Lice are usually host specific.

The dog has one common biting louse *(Trichodectes canis)* and one common sucking louse *(Linognathus setosus)*. Female lice attach their eggs to the hair of the host. The young lice undergo several molts before becoming adults; development from egg to adult takes about nineteen to twenty-three days. Lice usually cause severe hair loss from scratching and rubbing. Dogs may be treated with dips, washes, sprays, or dusts. Two treatments about twelve days apart are usually necessary.

Mites.   Mites are tiny, sometimes microscopic, arachnids that are parasitic on many animals. There are five types of mites important to dog owners.

Demodectic mites *(Demodex* species) live in hair follicles. They are present on the skin of all normal animals and usually exist in small numbers that do not cause a problem. The problem occurs when the population of the mites on an animal increases dramatically. Because all animals have demodectic mites, the mites are not considered contagious.

Demodectic mites cause demodicosis, which is a potentially serious disease in the dog; there are two forms: localized and generalized. Patchy hair loss on the head, forelegs, and trunk is called localized demodicosis. In generalized demodicosis, hair loss, reddening, and crusting may involve the entire body. Dogs with generalized demodicosis may also develop bacterial infections in the skin.

The increase above normal numbers of mites is believed to be a deficiency in the immune system, and some breeds are more susceptible to the disease.

Sarcoptic mites *(Sarcoptic* species and *Notoedres* species) are highly contagious mites that affect many species of animals. This family of mites burrows within the outer layers of the skin, and the entire life cycle is spent on the host animal. The common skin disease sarcoptic mange or scabies is caused by sarcoptic mites. Sarcoptic mange is an intensely itchy disease that causes the animal to scratch, chew, and rub constantly. The constant scratching, chewing, and rubbing of the affected areas produces reddening, scaling, and crusting of the skin. The head, ears, and abdomen are most usually affected, but the entire body may become infected. Many times, animals will cause self-inflicted injuries as they scratch, chew, and rub the skin.

If sarcoptic mange is suspected, a veterinarian should be consulted. Weekly insecticidal treatments should be carried out over six weeks. All animals that have been in contact with the infected animal should be treated.

Ear mites *(Otodectes cyanotis)* are a common problem in dogs. These highly contagious mites are usually found in the outer ear but may be found on other areas of the animal. Ear mites feed on the wax and other materials found in the ear. Symptoms may vary from no signs to a severe irritation with thick, dry, black crusts in the ear canals. Dogs that are severely infected may shake their heads.

Ear drops containing an insecticide are usually applied in the ears. To be totally effective, the entire body should be treated with flea products, and all pets in the household should be treated.

Cheyletiella mites *(Cheyletiella yasguri)* are contagious mites that live on the surface of the skin. These mites cause cheyletiellosis, a condition referred to as walking dandruff, which produces severe scaling, usually on the back of the infected animal. Some itching may occur, but not nearly as bad as from the other types of mites that have been discussed. Treatment for cheyletiella mites is the same as for sarcoptic mange mites.

Chiggers.   Chiggers are larvae stages of the *Trombiculid* mites; only the larval stages are parasitic. The North American chigger *(Trombicula alfreddregesi)* is the most common chigger that affects animals and people. On animals, they cause an itchy, red rash on the belly, face, feet, and legs. Chiggers are orange-red and are usually picked up from contact with heavy underbrush. The larval mites remain attached to the skin for only a few hours, so the only treatment needed is something to stop the itching.

Poisonings

Several groups of chemicals can cause poisoning in dogs and cats. These groups include insecticides, plants, household products, rodenticides, herbicides, human and veterinary medicines, metals, and miscellaneous compounds.

Insecticides.   Insecticides such as the organophosphate and carbamate compounds produce most of the insecticidal poisonings in dogs and cats. Boric acid used in roach baits and arsenic used in ant traps also cause significant numbers of poisonings.

Plants.   Dogs that eat plants containing insoluble calcium oxalate crystals, such as in philodendron, dieffenbachia, pothos, and caladium, experience severe irritations of the mouth and intestinal tract.

Ingestion of poinsettia plants causes mild intestinal upsets, as do aloe vera, mistletoe, mushrooms, Japanese yew bushes, rhododendrons, azaleas, oleanders, lily of the valley, caster beans, and flower bulbs (iris, tulips, and daffodil).

Household chemicals. Cleaners and household chemicals containing bleach, ammonia, borates, hydroxides, pine oil, and phenol cause poisonings to pets.

Strychnine, a rodenticide used to kill rats and mice, is often ingested by pets. Rodenticides such as warfarin, brodifacoum, and diphacinone contain anticoagulants and are also ingested by pets. Anticoagulants cause bleeding in animals, usually more than two days after ingestion.

Herbicides. Herbicides used to kill weeds in lawns are also responsible for poisonings. Glyphosate, paraquat, and arsenic-based herbicides have been known to cause poisonings in dogs and cats.

Medications. Medications that are normally given to people and pets can be poisonous if left around and then ingested in large amounts or taken improperly. Many times, pet owners will give medications to pets that are not prescribed by veterinarians.

Metal poisonings. Metal poisonings are rare but can occur. The most common type is lead poisoning from animals ingesting old paint chips, fishing and drapery weights, roofing shingles, and used motor oil.

Antifreeze. Antifreeze used in radiators is a common source of poisoning in both dogs and cats; it has a sugary taste and is readily ingested. Even small amounts cause animals to appear drunk, depressed, and may even cause death.

Treating animals that have been poisoned must begin as soon as a pet owner recognizes a poison is involved; prompt diagnosis and treatment is extremely important. A veterinarian should be consulted immediately. If the cause of the poisoning is known, this can save valuable time in getting the correct treatment started.

REPRODUCTION

A female dog usually has her first **estrus** (**heat period** or season) between six and twelve months of age. Smaller toy dog breeds and terriers usually begin at six and nine months, but some of the larger breeds may not begin until two years of age.

The estrus period occurs at intervals of seven months on average and is preceded by a **proestrus** that lasts about nine days. During this period, the female is attractive to the male and has a bloody vaginal discharge and a swollen vulva. However, the female will not accept the male for mating during this time.

The estrus period follows and usually lasts nine days. Ovulation usually occurs in the first 48 hours.

Breeders differ in their opinion as to the best time to breed. From ten to twelve days after the beginning of discharge (proestrus) is the general period recommended. A good compromise is a double mating on the tenth and twelfth days, or on the eleventh and thirteenth days.

Females are usually taken to the stud for mating. After mating, the dogs may be locked together. This is called a "tie" and is perfectly normal. It results from a swelling of the bulbus glandis. During the tie, the male may move around so that they are positioned rear to rear. Do not try to separate them during the tie because doing so may injure both animals; they will separate naturally in 10 to 30 minutes.

Pregnancy

Pregnancy, or **gestation,** in dogs usually lasts an average of sixty-three days, with a variation from fifty-six to seventy days. Signs of pregnancy include increase in appetite, weight, and breast size. Some females may show very little weight and breast increase; a change may not be noticeable until around the twenty-eighth day.

A whelping box provides a nice place for the female to have her puppies. She should be introduced to her box about three weeks before whelping so that she can get comfortable with it. The bottom of the box should be covered with newspapers.

As the whelping date approaches, the female's temperature will be slightly below normal (100.4°F). About 24 hours before whelping, her temperature will drop further to about 96.8°F. Most females will refuse meals in the last 24 hours. Some may vomit a little shortly before whelping, and there will usually be a sudden discharge of green mucus a couple of hours before whelping as the placenta begins to separate.

A normal, healthy female usually gives birth easily without need of additional help; however, one should stand by and give assistance if necessary. Each puppy will be contained in its own **placental membrane,** which must be removed before the puppy can breathe. The mother will do this by eating the membrane and severing the umbilical cord. She will be busy licking the puppy to stimulate respiration, cleaning it, warming it, and allowing it to suckle.

A placenta will follow each puppy after a few minutes. Count them to make sure the number of placentas matches the number of puppies; a retained placenta can cause problems later.

One must be prepared to help the mother remove a puppy from its placental membrane, to clean it, and encourage it to nurse. When cleaning a membrane off a puppy, the owner may also have to sever the umbilical cord. A piece of cotton thread should be tied around the cord about ¾ to 1 inch away from the puppy's body;

the cord should then be cut ¼ to ½ inch beyond the thread, taking care not to pull on the cord and making sure hands are clean and sterile. An application of iodine solution to the naval cord is recommended. One must be careful of the mother also because she may not be too friendly.

The first milk produced by the mother is called **colostrum.** It is important that each puppy receive colostrum because it contains immunoglobulins that will help protect it from infectious disease.

Eclampsis is a calcium deficiency that may be observed shortly before whelping and during the first month of lactation. It usually occurs among the toy breeds and females with large litters that require a large milk production. Symptoms include restlessness, panting, whining, and getting up and down. As the problem worsens, the mother will become stiff, with twitching muscles and lack of coordination. If not treated, she will collapse into convulsions and die. A veterinarian must be consulted immediately to give her a calcium injection. Prevention is easy by ensuring that the pregnant female gets proper vitamin and mineral supplements, especially calcium.

SUMMARY

The dog is probably the first animal domesticated by humans. The domesticated dog originated about 12,000 to 14,000 years ago in Europe and Asia and was domesticated about 10,000 years ago.

The dog traces its ancestors back about 40 million years ago to a tree-climbing mammal called *Miacis*. *Miacis* was a small animal with an elongated body that probably spent most of its time in the lush forest areas.

Today, there are more than 400 breeds of dogs in the world. The American Kennel Club recognizes 129 different breeds in seven groups. The seven groups are sporting dogs, hounds, working dogs, terriers, toys, herding dogs, and nonsporting dogs.

Bringing a puppy or dog home into the family can be an enjoyable experience for everyone, but unfortunately, many dogs end up in animal shelters, abandoned to run through the streets, dumped along the roadside, or destroyed. Time should be given to studying, researching, and fully considering the decision before bringing a dog home.

To keep an animal healthy, an adequate diet and clean water are essential. Several types of commercial foods are available to choose from. Although one can mix his or her own rations, these commercial foods save time and are very convenient to use.

The amount and type of training a dog receives depends on what the dog is going to be used for. Almost all dogs should learn five basic commands: heel, sit, down, stay, and come. Most owners would like a dog that will respond and be under control. Dogs that are well trained can provide many hours of fun and enjoyment and can be great companions, whereas untrained dogs can be a basic nuisance.

Before purchasing a dog, it is important to consider not only the amount of time one has to train the dog but also the time needed to attend to its basic grooming and health care needs. Time and attention need to be given to the care of the dog's coat, nails, ears, eyes, and teeth.

Unfortunately, even the most well-cared-for dog may at times come down with some type of disease, parasite, or ailment. By regular grooming and care, an owner can be attuned to the early signs that may alert one to a problem. Some of these early signs are constipation or diarrhea, shivering, fever, watery eyes, runny nose, coughing, loss of appetite, vomiting, loss of weight, ravenous eating without gaining weight, increased urination, restlessness, straining to urinate, labored breathing, increased water intake, lameness, paralysis, obvious pain, or nervous symptoms. If any of these signs are observed, it is important that the dog be taken to a veterinarian as soon as possible. Prompt diagnosis and treatment may prevent more serious conditions from developing.

A female dog usually has her first estrus between six and twelve months of age. Smaller toy dog breeds and terriers usually begin at six and nine months, but some of the larger breeds may not begin until two years of age. This estrus period occurs at intervals of seven months on average.

Pregnancy, or gestation, in dogs usually lasts an average of sixty-three days with a variation from fifty-six to seventy days. Signs of pregnancy include increase in appetite, weight, and breast size. Some females may show very little weight and breast increase; a change may not be noticeable until around the twenty-eighth day.

The first milk produced by the mother is called "colostrum," and it is important that each puppy receive this because it contains immunoglobulins that help protect the puppy from infectious disease.

DISCUSSION QUESTIONS

1. What is the history of the dog? How has the modern dog evolved?

2. What are the seven major groups of dogs? How do they differ?

3. List the different breeds of dogs. What are the characteristics of the different breeds?

4. List the various terms that describe the behavior of the different breeds. What do these terms mean?

5. What things should be considered before selecting a breed of dog?

6. What are the various feeds available for dogs?

7. How does the feeding of pregnant and lactating females, of puppies, and of older dogs differ?

8. What are some of the general grooming and health care practices necessary to keep a dog clean and healthy?

9. What are the six groups of diseases that affect dogs?

10. List the diseases that go under each of the six groups. What are the symptoms of each disease?

11. What are some of the general care practices necessary for a pregnant female before, during, and after whelping? What are some practices for the general care of the new litter?

SUGGESTED ACTIVITIES

1. With the approval of the instructor, bring in dogs of various breeds and talk about the characteristics of the breeds.

2. Bring the labels from various dog foods and compare the ingredients that are used.

3. With the approval of the instructor, invite dog trainers into the class and ask them to explain the methods they use in training their animals.

4. With the approval of the instructor, invite local dog groomers to give demonstrations on the methods used to groom various breeds of dogs.

5. With the approval of the instructor, invite the local veterinarian into the class to talk about some of the common diseases of dogs.

6. Visit a dog show. Watch how the various groups of dogs are judged and watch what criteria are used to select the winners.

ADDITIONAL RESOURCES

1. American Kennel Club
www.akc.org/index.html

2. American Veterinary Medical Association
www.avma.org

3. Animal Clinic
www.animalclinic.com

4. Dog Owner's Guide
www.canismajor.com/dog

5. Imperial Shih Tzu by Jensen
myweb.rust.net/~shihtzu

6. Iowa State University
www.vetmed.iastate.edu

7. Net Vet Veterinary Resources
http://netvet.wustl.edu

8. United Kennel Club
www.ukcdogs.com

9. The Pet Center
www.thepetcenter.com

10. The Pet Zone
www.parentzone.com/petzone.htm

END NOTES

1. From the *Encyclopedia International,* 1979 edition. Copyright 1979 by Lexicon Publications, Inc. Reprinted by permission.

CHAPTER 8

Cats

OBJECTIVES

After reading this chapter, one should be able to:

- discuss the history of the cat.
- list the major groups and breeds.
- discuss the proper methods of feeding cats.
- discuss the general grooming and health care of cats.
- discuss the various diseases of cats.

TERMS TO KNOW

agouti	feral	occlusion
cochlea	Jacobson's organ	olfactory mucosa
colorpoint	jaundice	papillae
conjunctivitis	keratitis	points
coronavirus	nictitating membrane	

CLASSIFICATION

Kingdom — Animalia   Family — Felidae
Phylum — Chordata   Genus — *Felis*
Class — Mammalia   Species — *F. catus*
Order — Carnivora

HISTORY

Cats are descendants of the same carnivorous, tree-climbing mammal *Miacis* from which dogs descended. Cats, like dogs, trace their lineage through *Cynodictis.* The skeleton of these early primitive carnivores was much like that of a civet or a weasel. The body was long and flexible, the limbs short. All five toes were present, which appear to have been armed with retractile claws. These genera are customarily considered primitive dogs, but their general characteristics would make them ideal ancestors of the whole series of Carnivora.

Proailurus, the first normal or feline cat, appeared about 35 million years ago. The exact evolution of the cat from that time forward is not known, but modern cat types, like *Felis* and related genus groups with "normal" canine teeth, appeared about 7 million years ago. Zoologists do not know whether cats evolved from long-tusk or saber-tooth felines or whether they evolved as a separate and distinct genus.

In the last chapter, we learned that dogs were domesticated about 10,000 years ago. The cat, however, was not domesticated until much later; this was probably because of its independent, solitary nature. It is believed that domestication of the cat began around 4,000 years ago, although there is no clear documentation until around 2500 BC. Early Egyptian paintings and inscriptions suggest that the cat was kept in captivity, tamed, and eventually revered and protected. The cat was used in religious ceremonies, and many mummified cats have been found in Egyptian tombs.

As civilization spread from Egypt and the Middle East, domestication of the cat spread also. The Romans introduced the cat into Europe, and European explorers, travelers, and traders carried the cat to all parts of the world. Today, the domesticated cat can be found on every continent except Antarctica.

Cats have long been valued by farmers, ranchers, and home owners for their use in controlling rats and mice; they were also carried aboard ships for this purpose.

It is believed that the domesticated cat of today originated from the African Wild Cat *(Felis libyca)*. This cat was common to all areas of Asia and North Africa. The African Wild Cat had tabby or striped markings similar to the tabby today. It was larger than today's domestic cat but was a very quick, agile animal.

Two other cat species were the Jungle Cat *(Felis chaus)* and the European Wild Cat *(Felis silvestis)*. Both had varying tabby markings and were known to have been kept by the Egyptians and Romans.

Zoologists disagree on the exact number of cat species; there are about thirty-six or thirty-eight species depending on the method of classification.

Cats have erect ears and large eyes with vertical-slit pupils. They have twenty-eight to thirty teeth, five toes on each front foot, and four toes on each hind foot. All cats, except the cheetah, can retract their claws.

Cats are usually divided into three genus groups. One group is the large, roaring cats, *Panthera,* which includes the lion, leopard, tiger, and jaguar. Second are the cats that cannot roar, *Felis,* which include the small cats and the domesticated cat. The third group, *Acinonyx,* includes only one member, the cheetah. The cheetah belongs to a special group because it has claws that do not fully retract.

Domesticated cat breeds are commonly divided into two major groups: the shorthaired breeds and the longhaired breeds. There are thirteen longhaired breeds and twenty-six shorthaired breeds. Not all breeds are recognized by the popular cat clubs.

GROUPS AND BREEDS

Shorthaired Breeds

The Abyssinian. The Abyssinian is believed to be a direct descendant of the Sacred Cat of Egypt, which was worshiped and held sacred; however, its exact origin is not known. Modern-day Abyssinians come from several cats that arrived in England around 1868. It is believed that the cats were brought to England by soldiers returning to England from the Abyssinian War. (Abyssinia is now known as Ethiopia.) See Figure 8–1.

There are three recognized breed colors: ruddy, red, and blue. Ruddy is an orange-brown color, ticked with dark brown or black. The red is usually ticked with chocolate-brown, and the blue is a blue-gray, ticked with slate-blue. (Ticked refers to the darker colors that are found on the tips of each hair.)

Abyssinians are considered extremely affectionate, quiet, highly intelligent, and alert. The breed is very active, can learn to retrieve, and is fond of water.

The Abyssinian is medium sized with a firm, muscular, well-balanced body. Its head is slightly rounded and wedge-shaped with large ears and large, almond-shaped eyes.

Figure 8–1 Abyssinian. Photo by Isabelle Francais.

The American Shorthair. The American Shorthair is a descendant of shorthaired cats brought to the United States from England. They arrived with the early settlers and have been a part of our lives ever since. There are thirty-four recognized colors and patterns, with the classic tabby pattern the most common. The American Shorthair is extremely affectionate and makes a great lap cat; it is easily disci plined and trained. See Figure 8–2.

This cat is a medium- to large-sized cat. It is heavily built and well-muscled, with a broad chest, heavy shoulders, and thick hind legs. It has an oblong head, wide face, and wide-set ears. Females of the breed can reach 10 pounds, with males reaching 14 pounds.

Figure 8–2   American Shorthair. Photo by Isabelle Francais.

The American Wirehair. The American Wirehair originated from a litter discovered in a barn in upstate New York in 1966; the breed is a mutation of the American Shorthair. The general characteristics of this breed are the same as the American Shorthair; however, the hair coat is very dense, resilient, and coarse with a curly appearance. See Figure 8–3.

All the hair on the animal's body is curly, including the hair on the toes, ears, and whiskers. Extra care must be taken not to damage the wiry-hair coat. The coat should never be brushed or combed because this will rip out the hair. These hairs easily pick up dirt, dust, and lint. Regular bathing at three-week intervals may be necessary to keep the coat clean.

Figure 8–3   American Wirehair. Note the curly whiskers and rough, wiry hair coat that is a characteristic of the wirehair breed. Photo by Isabelle Francais.

The Bombay Breed. The Bombay breed was produced in 1958 by crossing a sable Burmese with a black American Shorthair; the result is a jet-black cat with gold to copper eyes. The Bombay is considered a graceful, charming cat that gets along well with others, even with strangers. See Figure 8–4.

The Bombay is a medium-sized cat, with a rounded head and broad forehead. The ears are rounded and tilt slightly forward, and there is a deep nose break. The Bombay may carry genetic abnormalities that are common to the breed.

Figure 8–4   Bombay. Photo by Isabelle Francais.

The Burmese. The Burmese breed developed in the United States in the 1930s; Dr. Joseph Thompson crossed a brown female Burma with a Siamese. The cross produced kittens with Siamese coloring, dark-brown kittens, and intermediate kittens. The dark-brown kittens were retained for breeding purposes. See Figure 8–5.

The Burmese is considered to have a sweet disposition and enjoys being held. However, the breed can become bossy, stubborn, and angry. The Burmese is a medium-sized cat with a solid, short body. The breed has a rounded head with a broad, short muzzle and a distinct nose break. They are sable-brown in color with gold eyes.

Upper respiratory diseases are common in kittens, and inbreeding has increased the incidence of heart defects, mutative malformations of the face and head, and other abnormalities.

Figure 8–5   Burmese. Photo by Isabelle Francais.

The British Shorthair. The British Shorthair is the oldest natural English breed and probably traces its lineage back to the domestic cats of Roman times. The British Shorthair is considered quiet and easygoing. It sleeps a lot and is content to lie around the house. See Figure 8–6.

There are eighteen recognized colors and patterns, with most colors having gold or copper eyes. Although it is related to the American Shorthair, the British Shorthair is larger and taller and has a large, rounded head with big, round eyes.

Figure 8–6   British Shorthair. Photo by Isabelle Francais.

The Chartreux. The Chartreux is among the oldest natural breeds. These cats are believed to have been brought to France from the Cape of Good Hope by Carthusian monks during the seventeenth century. It is not known whether the monks gave the cat its name or whether people associated the cat with the monastery and referred to them as Chartreux. The name of the monastery was known as "Le Grand Chartreux." See Figure 8–7.

The Chartreux is a large, robust cat with a broad chest and powerful hind legs. Males may reach a weight of 14 pounds. The head has a wide forehead, extremely full cheeks, and narrow but not pointed muzzle. The coat is gray-blue, and the eyes are gold to copper.

The Chartreux is considered warm, friendly, and highly tolerant of children and dogs. The breed is easy to train and is rarely heard. Males are considered clean, nonaggressive, and rarely spray or leave an odor.

Figure 8–7   Chartreux. Photo by Isabelle Francais.

The Colorpoint. The Colorpoint Shorthair was developed in England in 1947. **Colorpoint** gets its name from the color of the extremities: nose, ears, feet, and tail. Colorpoint cats have one general body color with darker extremities or **points.** Standard-point or regular-pattern Siamese was bred with red, domestic cats. This cross produced solid-red offspring, which were bred back to Siamese, producing tortoiseshell-pointed Siamese. When these cats were bred to one another and their offspring bred back to standard-point Siamese, the results were cats with red, cream, and tortoise points, as well as standard colorpointed offspring. Siamese breeders did not want to recognize any colors beyond the classic four point colors (seal, blue, chocolate, or lilac); thus, the new colors were given a separate breed identity. See Figure 8–8.

The breed is considered bold, very demanding, vocal, and full of nonsense. Colorpoints are considered very intelligent and loving. It is genetically Siamese and possesses the same characteristics, with an elongated, angular body and wedge-shaped head. The breed is medium in size, and the eyes, like those of the Siamese, are a vivid blue.

Figure 8–8 Colorpoint Shorthair. Photo by Isabelle Francais.

The Cornish Rex. The Cornish Rex originated in Cornwall, England, in 1950 as a result of mutation and inbreeding. The main characteristic of the Rex is its curved, rippling hair coat that is lacking in guard hairs. See Figure 8–9.

The Cornish Rex is very affectionate, extroverted, persistently inquisitive, and energetic. The breed is playful and excellent with children; it is considered a lap cat and should not be an outdoor cat.

The Cornish Rex is a small- to medium-sized cat with a long, narrow head, rounded forehead, and Roman nose. The ears are large, flared, erect, and set high on the head. The eyes are oval and give the cat a look of surprise. The body of the Cornish Rex arches like that of a greyhound; there are twenty-eight recognized colors and patterns.

Some care should be taken when brushing the cat, because the curly coat can be damaged easily. One should use a soft brush and avoid overbrushing.

The normal body temperature of the Cornish Rex is one degree higher than other breeds and may be confusing to veterinarians not familiar with the breed. Because Rex cats lack guard hairs, they need to be protected from extreme cold and heat.

Figure 8–9 Cornish Rex. Rex cats have extreme curl to their whiskers and curly hair coats. Photo by Isabelle Francais.

The Devon Rex. The Devon Rex is a result of mating a feral (wild) domestic cat to a cared-for stray cat; although the breed has a wavy hair coat similar to the Cornish Rex, the two breeds are different. See Figure 8–10.

The Devon Rex is a medium-sized cat, with a wavy coat that is longer and less curly than the Cornish Rex. The face is short and wedge-shaped with pronounced cheekbones, short nose, and prominent whisker pads. The large, wide-based ears set low on the head are an outstanding characteristic of this breed.

There are thirty-four recognized colors and patterns of the Devon Rex, with most having gold eyes. Like the Cornish Rex, grooming should be done carefully. This breed likes people, can be taught to retrieve, and is an excellent climber.

Figure 8–10   Devon Rex. Photo by Isabelle Francais.

The Egyptian Mau. The Egyptian Mau traces its ancestry back to 1400 BC; *Mau* is the Egyptian word for cat. The breed was first imported into the United States in 1953; the most important characteristic is its spotted pattern. There are three colors: silver, bronze, and smoke; the eyes are gooseberry-green. See Figure 8–11.

The character of the Mau varies considerably. Some make great pets, are patient with children, and enjoy being handled; others are excitable, unpredictable, and react adversely. They tend to become more friendly with one or two people. Bloodlines should be considered when purchasing an Egyptian Mau because of character trait differences.

The Egyptian Mau is a medium-sized cat. Its head is a gently contoured wedge with almond-shaped eyes; the ears are broad at the base and erect.

Figure 8–11   Egyptian Mau. Photo by Isabelle Francais.

The Exotic Shorthair. The Exotic Shorthair was developed by crossing Persians with American Shorthairs. This cross produced a breed that has a compact, short-legged build, with the short, pushed-in face characteristic of the Persians and the short, thick hair coat of the American Shorthair. See Figure 8–12.

This breed makes an ideal pet. It is calm, quiet, and good with children. The Exotic Shorthair is a medium- to large-sized cat and resembles the Persian, except for the short hair. It has a massive, round head, with tremendous skull and small, rounded ears. The eyes are large, round, and set apart; the nose is short and has a deep break.

Colors are the same as the American Shorthair with Persian color additions; the eye color is primarily copper.

Figure 8–12   Exotic Shorthair. Photo by Isabelle Francais.

The Havana Brown. The Havana Brown is a result of crossing a black domestic Shorthair with a seal-point Siamese carrying the brown gene. The resulting solid-colored brown cat was further developed into two separate breeds: the chestnut Oriental Shorthair and the Havana Brown. The name Havana Brown comes from the color, which is similar to rich Havana tobacco. See Figure 8–13.

The Havana Brown is an excellent companion cat that enjoys lots of attention. It is a medium-sized cat with a distinctive rectangular muzzle and is brown or mahogany in color. The ears are very large, stand erect, and are pointed slightly forward. There is very little hair on or in the ears; the eyes are a vivid green.

Figure 8–13   Havana. Photo by Isabelle Francais.

The Japanese Bobtail. The Japanese Bobtail cat is native to Japan. Paintings, wood carvings, and statues date it back to very early times. In Japan, this breed is known as Mi-Ke cats and is a symbol of good luck; *Mi-Ke* in Japanese means "three fur." See Figure 8–14.

The Japanese Bobtail is affectionate, is generally sweet-tempered, and enjoys human companionship. The Japanese Bobtail enjoys water and can be taught to retrieve.

The Japanese Bobtail is a medium- to large-sized cat; its distinctive, short tail is kinked. The hair on the tail is usually longer and forms a pompon; people sometimes refer to it as a bunny tail. The head forms an equilateral triangle with high cheekbones; the eyes are slanted and luminous.

The preferred color and pattern for the Bobtail is tricolored with a red, black, and white calico pattern. Bicolors of black and white or red and white are next in preference. There are twenty-three possible color combinations accepted for shows. The eyes may be gold, green, or blue. White cats may have one blue and one gold eye. Care must be taken in handling the Bobtail to prevent injury and pain to the sensitive bobtail.

Figure 8–14   Japanese Bobtail. Note the short "bobtail" of this breed. Photo by Isabelle Francais.

The Korat. The Korat is a centuries-old breed native to Thailand. It is rare even in Thailand, where they are highly prized and rarely sold. The Korat is considered intelligent and active. It makes an excellent companion cat, is protective of family members, and is reserved with strangers. See Figure 8–15.

The distinctive characteristic of this medium-sized cat is its silver-blue fur that is tipped with silver, producing a sheen or halo effect. The head is heart-shaped, and the eyes are very large, slanted, and luminous and appear oversized in proportion to the face. The preferred eye color is luminous green. Korat cats may be prone to upper respiratory viruses, and routine vaccinations are important.

Figure 8–15   Korat. Photo by Isabelle Francais.

The Malayan. The Malayan is an "other-colored" Burmese. The Burmese, as we learned earlier, is sable-brown in color. In the breeding that produced the Burmese, the colors blue, champagne, and platinum were also produced. These colors were rejected by Burmese breeders, but fanciers worked hard to establish them as a separate breed now recognized as the Malayan. The characteristics of the Malayan are basically the same as the Burmese; although stubborn and bossy, they can be outstanding companions. See Figure 8–16.

Figure 8–16   Malayan. Photo by Isabelle Francais.

The Manx. The Manx breed is native to the Isle of Man, an island between England and Ireland. The breed is the result of a genetic mutation and is among the earliest breeds registered in Europe, Great Britain, and North America. See Figure 8–17.

The Manx is affectionate but leans toward one-on-one relationships. The breed is alert, intelligent, healthy, and powerful; it can also be devilish and stubborn. The distinctive characteristic of the Manx is the complete lack of a tail. The perfect show animal is a rumpy and has a dimple where the tail should be. Two other types are the stumpy with a short tail and the longy with a complete tail. See Figure 8–18.

Figure 8–17   Manx "rumpy" cats do not have a tail; they have a dimple where the tail should be. Photo by Isabelle Francais.

The Manx is a solid, compact, medium-sized cat. It has a round head and muzzle, prominent cheeks, and a full face. The ears are of medium size and rounded at the tip. The rump of the Manx is considerably higher than the shoulders. There are many colors and patterns; therefore, this is not an important show characteristic. The eyes are round and primarily copper, but other eye colors are possible.

The Manx breed carries a lethal gene; its lack of a tail is a genetic defect and indicates possible spinal deformities and cartilage weaknesses. These may affect the legs and neck, causing paralysis.

Figure 8–18   Manx "stumpy" cats have a short tail. Manx "longy" cats have a complete tail.

The Ocicat. The Ocicat is a result of second-generation crossings of Siamese and Abyssinians; it was first developed in 1964. The first kittens resembled baby ocelots *Felis paradalis,* which are large, wild cats native to North and South America. They have a yellow or gray fur coat with black spots. The original breeders gave the cats the name Ocicat because of the similarity of the markings. See Figure 8–19.

The Ocicat is considered gentle, calm, friendly, and intelligent with a sweet temperament. It has a spotted, tabby-patterned coat. There are eight recognized color patterns. The hairs show bands of color in **agouti** pattern; where the bands fall together, a dot or dark area is formed. Agouti is a color pattern where one color is streaked or intermingled with another color.

The main characteristic of the Ocicat is its wild, feral appearance. Females of the breed are medium size, and the males are very large; males may weigh 12 to 15 pounds. The head is oval-shaped, with almond-shaped eyes that angle downward toward a broad nose. The ears are moderately large and sometimes tufted.

Figure 8–19   Ocicat. Photo by Isabelle Francais.

The Oriental Shorthair. The Oriental Shorthair resulted from the breeding that produced the Colorpoint Shorthair. The original crossings produced not only the colorpointed patterns of the Colorpoint Shorthairs but also solid and full-colored patterns. These cats became known as the Oriental Shorthair breed. See Figure 8–20.

The Oriental Shorthair is a talkative companion. The breed is considered intelligent, witty, graceful, and elegant. It is extremely affectionate and dependent on people; it is also playful and agile.

The Oriental Shorthair is a medium-sized breed and almost identical to the Siamese standard. The breed is long, slender, and fine-boned. The head is long, tapering, and slightly wedge-shaped. The ears are large and flared; the eyes are almond-shaped.

There are numerous colors of Oriental Shorthairs. They are found in solid, shaded, smoke, tabby, and multiple colors. The eyes of the Orientals are green. White cats may have blue or green but not odd eyes (one blue and one green). Oriental Shorthairs are usually of good health, but upper respiratory illnesses may occur.

Figure 8–20   Oriental Shorthair. Photo by Isabelle Francais.

The Russian Blue. The Russian Blue was brought into Europe from the White Sea port of Archangel in the 1860s. They have been called Archangel, Spanish Blue, Maltese Blue, and English Blue. This is a quiet, gentle, and somewhat reserved breed. The Russian Blue is fine with children and dogs; it is very affectionate with its owner but conservative with strangers. See Figure 8–21.

The main physical characteristic of the Russian Blue is its double plushy coat. The hair is short, dense, and fine. The Russian Blue is a medium-sized cat with a smooth, medium wedge-shaped head. The ears are large, wide at the base, and more pointed than round; the eyes are vivid green.

Figure 8–21   Russian Blue. Photo by Isabelle Francais.

The Scottish Fold. The Scottish Fold is a natural mutation. The first cat with folded ears was discovered on a Scotland farm in 1961. The Scottish Fold is an excellent family cat; it is reserved and rather shy, but affectionate. It has a sweet nature and is friendly. It is a quiet breed and gets along well with other cats and dogs. See Figure 8–22.

The main characteristic of the Scottish Fold is the ears that are folded forward and downward. This is a medium-sized cat with a rounded head, prominent cheeks, and prominent whisker pad. The eyes are large and round. There are twenty-three colors and patterns of the Scottish Fold. They are found in solid, shaded-tipped colors, tabby, and multiple colors.

Although generally healthy, the dominant gene that produces folded ears can cause crippling problems when two Scottish Folds are bred together.

Figure 8–22   Scottish Fold. Note the folded ears that are characteristic of this breed. Photo by Isabelle Francais.

The Siamese.   The actual origin of the Siamese cat is unknown. Pictures from Atutthaya, the old capital of Siam, date back to 1350 AD, which show Siamese cats kept by royalty. See Figure 8–23.

The Siamese is a one-on-one cat and usually expresses devotion to one particular family member. It is described as a prima donna: extraordinarily intelligent, precocious, talkative, loyal, fearless, and obstinate; it is also totally unpredictable.

The Siamese is a medium-sized cat with long, refined, elegant tapering lines. The head of the Siamese forms an equilateral triangle from its nose to the tip of the ears. The eyes are almond-shaped and slanted toward the nose. The bones are fine, the body is slender and tube-shaped, and the legs are long.

The general body color of the Siamese shows very little pigmentation. The pigmentation is darker at the points, and color is restricted to the points: mask, ears, tail, and feet. The point colors are seal, chocolate, blue, or lilac. The eye color is always deep, vivid blue.

Young Siamese may be subject to upper respiratory diseases; some also show evidence of heart problems. Owners need to be alert for sensitivity to vaccines and anesthetics.

Figure 8–23   Siamese. Photo by Isabelle Francais.

The Singapura.   The first Singapuras were imported into California in 1975 from Singapore. In Singapore, these were primarily street cats. Because of this background, it may show some shyness, but it warms readily to people who show it no harm. The Singapura is a quiet cat that adapts quickly, is very playful, and becomes very sociable around people. See Figure 8–24.

The Singapura is a small breed. Males weigh no more than 6 pounds, and females usually weigh less than 4 pounds. It has a rounded head with large eyes and ears. The nose is blunt on a broad muzzle with a slight stop below eye level.

The main characteristic of the Singapura is its ticked coat. The colors are usually yellow or ivory, with light brown hairs interspersed; there are darker bars on the inner front legs and back knee. The eyes are hazel-green or yellow and are described as being extremely beautiful and expressive.

Figure 8–24   Singapura. Photo by Isabelle Francais.

The Snowshoe Breed. The Snowshoe breed originated in the Kensing Cattery in Philadelphia, where two Siamese cats produced three females with white feet; the distinctive white feet led to its name. Snowshoe cats are considered people-oriented and will follow their owners everywhere. They are rather docile and make excellent house pets. See Figure 8–25.

The Snowshoe is medium to large in size. The body type is similar to the American Shorthair. It is medium-boned, well-muscled, and has a solid body with a medium neck in proportion to the body; the head forms a modified wedge with high cheekbones. The nose rises slightly at the bridge, and the ears are large, broad at the base, and pointed. The eyes are large, oval, and slant toward the nose.

The Snowshoe has white-tipped front paws and a white chest, chin, and muzzle; the hind legs are white to the hock. The mask, tail, ears, and legs are usually of defined seal or blue, with masking covering the entire face, except for an inverted white V-shaped pattern. The eyes are a bright, sparkling blue.

Figure 8–25   Snowshoe cats have distinctive white feet. Photo by Isabelle Francais.

The Sphynx. The Sphynx is a hairless cat that was first introduced by breeders in Canada in the late 1960s. Some evidence shows that hairless cats existed at the time of the Aztec Indians, and the Mexican Hairless was somewhat popular in the 1800s. See Figure 8–26.

The Sphynx breed, because of its lack of fur, is very dependent on people for survival. It cannot survive outdoors and may even catch a cold if exposed to drafty conditions. It is very affectionate, has a sweet nature, and adores body contact. It purrs constantly.

The main characteristic of the Sphynx is the tough, wrinkled, hairless skin. The breed has a sweet facial expression, expressive eyes, and large ears. The body type is similar to that of the Devon and Cornish Rex.

The Sphynx cat must be protected from the cold at all times; sweaters are recommended to keep the animal warm. The normal body temperature is four degrees higher than most other breeds. It is unable to store body fat, and thus requires frequent meals to maintain its body temperature.

Figure 8–26   Sphynx. Note the lack of normal fur and whiskers of this breed. Photo by Isabelle Francais.

The Tonkinese. The Tonkinese was developed during the 1950s through the 1970s and is a hybrid cross between Siamese and Burmese. The breed was recognized by the Canadian Cat Association in 1974 and by the Cat Fanciers Association in 1978. Tonkinese bred together usually produce a litter that is one-half Tonkinese, one-fourth Siamese, and one-fourth Burmese. See Figure 8–27.

The breed is said to be a combination of the best of Siamese and Burmese. It is considered clever, active, intelligent, fearless, and obstinate. It is also mischievous and a prankster. The Tonkinese is of medium size; the head is of modified wedge shape with a slight dip between the eyes. The body is of medium length, as is the tail and legs.

The Tonkinese are Siamese-patterned cats in five unique breed colors. The densely colored mask, ears, feet, and tail gradually lighten as you move from the points. The coat color, unlike the Siamese, is a dilution of the point color. The five breed colors are natural mink, blue mink, champagne mink, platinum mink, and honey mink; all colors have characteristic aqua eyes.

Like all Oriental breeds, young animals are susceptible to upper respiratory disease and may be sensitive to vaccines and anesthetics.

Figure 8–27   Tonkinese. Photo by Isabelle Francais.

Longhaired Breeds

The Balinese and Javanese. The Balinese and Javanese have been developed from litters of registered Siamese. Except for the coat length, the breeds maintain the same body style as the Siamese. The name Balinese is derived from the South Sea island of Bali, and the name Javanese is derived from Java. See Figures 8–28 and 8–29.

The Balinese and Javanese are considered warm, eager, inquisitive, and enthusiastic. They are also vocal and persistent. The Balinese and Javanese are longhaired cats with Siamese coat patterns. Balinese colors are seal point, chocolate point, blue point, and lilac point. Javanese colors include red, tortoise, and lynx point patterns. Color on the diamond-shaped mask, ears, legs, feet, and tail strongly contrasts with the pale, even body color. The eyes are deep, vivid blue.

Figure 8–28   Balinese. Photo by Isabelle Francais.

Although these two breeds are generally healthy, certain lines may have weak hind legs. Nasal obstruction or poor **occlusion** (upper and lower teeth not fitting together properly) can cause breathing through the mouth.

Figure 8–29   Javanese. Photo by Isabelle Francais.

The Birman.   The Birman is known as the sacred cat of Burma: the exact origin of the breed is not known. The breed first became known outside of Burma in 1898. The Birman is a very sociable cat that needs the company of others. It is considered a sweet-natured, charming, and intelligent cat. See Figure 8–30.

The breed is medium to large in size. The Birman is a sturdy, stocky cat with a long body. The head is strong, broad, and rounded, with a strong, rounded muzzle and low-set nostrils. The eyes are round, and the ears are medium in size with rounded tips.

The coat of the Birman is long and silky. The colors are seal point, chocolate point, blue point, and lilac point, all with white gloves (white paws). The front paws are white to the third joint, and the back feet are white up to the hocks. The long, silky coat of the Birman does not mat and requires little care.

Certain lines of the Birman may show genetic defects in the form of weak hind legs, nasal obstruction, and poor occlusion, which may cause breathing through the mouth.

Figure 8–30   Birman. Photo by Isabelle Francais.

The Cymric.   The Cymric (pronounced Kim-rik) is a longhaired Manx. During the original breeding of the Manx cats, some longhair cats occasionally resulted due to recessive genes. Both Shorthair Manx parents must have the longhair gene to produce Cymric offspring. Cymrics bred to each other produce only longhaired kittens. See Figure 8–31.

The Cymric is alert and sweet-natured. It is considered very quiet, extremely intelligent, and observant. It is excellent with other pets and with children.

The Cymric is a large, tailless breed with moderately long hair. The head is round with a broad muzzle and prominent cheeks. The front legs are short, balanced, substantial, and contrast with the higher, heavily muscled hindquarters. Color is not very important for show points. Taillessness in the Cymric can be associated with defects not always apparent in kittens. Be alert for signs of weakness in hindquarters such as unusual gait or limp.

Figure 8–31   Cymric. This is a longhaired "Manx." Note the lack of a tail. Photo by Isabelle Francais.

The Himalayan and Kashmir. The Himalayan and Kashmir were developed in Britain and North America in the 1930s. The crossing of Siamese, Persian, and Birman produced a Persian-type cat with a long, flowing coat and a Siamese pattern. During the development of the chocolate-point and lilac-point Himalayans, solid-chocolate and solid-lilac kittens often showed up in litters. These solid-colored Himalayans are now called Kashmirs. See Figures 8–32 and 8–33.

The Himalayan and Kashmirs are easygoing breeds that fit well into the family lifestyle. They like attention, purr loudly, and are considered intelligent and outgoing. Their voices are louder than that of the Persian but much quieter than that of the Siamese.

Figure 8–32   Himalayan. Photo by Isabelle Francais.

The Himalayan and Kashmirs are medium- to large-sized cats; the body is short and compact. The head is large on a short neck, and the nose is snub-like with good break. The ears are set low on the head; the eyes are large, round, and expressive. The eyes of the Himalayan are a vivid blue, whereas the eyes of the Kashmir are copper-colored. The very short nose of the Himalayan and Kashmir may be associated with tearing because of malformed tear ducts; breathing problems may also be possible in some lines.

Figure 8–33   Kashmir. Photo by Isabelle Francais.

The Maine Coon. The Maine Coon is considered to be the native American longhair cat that first appeared on the rocky coast of New England. It is the oldest natural breed in North America. The Maine Coon is considered lovable and friendly and enjoys the companionship of another pet or person. The breed has a tiny voice with squeaks and chirps. See Figure 8–34.

The Maine Coon is a large-sized cat, often reaching 12 to 18 pounds. It has a medium-wide head, medium-long nose, and large tufted, tall ears. The fur on the tail is especially long compared with the fur covering the body. The brown tabby is the best-known color and pattern for the Maine Coon, but a wide range of colors and patterns exists. Eye color varies from green to gold to copper; white cats have blue eyes.

Figure 8–34   Maine Coon. Photo by Isabelle Francais.

The Norwegian Forest Cat. The Norwegian Forest Cat is a breed native to Norway and is believed to be descended from numerous cats that lived in the Norwegian woodlands. This cat is considered exceedingly alert and intelligent. The breed is responsive and loving to its owner and must have companionship. See Figure 8–35.

The Norwegian Forest Cat is medium to large in size, with a moderately long body and powerful appearance due to a full chest and considerable girth. The tail is long and flowing. Almost all color combinations are acceptable.

Figure 8–35   Norwegian Forest Cat. Photo by Isabelle Francais.

The Persian Breed.   The Persian breed is one of the oldest and most popular breeds. The exact origin of the breed is not known, but the present Persian cat is believed to be descended from longhaired cats introduced into Italy from Asia and cats introduced into France from Ankara, Turkey. The Persian cat is well mannered, easygoing, and quiet. It is an excellent companion and adapts well to apartment living. See Figure 8–36.

The breed is medium to large in size and has a short, compact body. The head is large with ears that are small and round-tipped. The nose is short, broad, and stubbed with a deep break; the eyes are large and round, giving the cat a sweet expression.

The Persian breed is divided into six groups for show purposes: solid colors that usually have copper eyes, except for the white, which may have blue, copper, or odd eyes; shaded colors that have green, blue-green, or brilliant copper eyes; smokes that have copper eyes; tabby colors that have brilliant copper eyes, except for the silver tabby, which has green or hazel eyes; multicolors that have copper eyes; and the point-restricted colors that have vivid blue eyes.

Some white Persian cats are born deaf; this usually occurs in cats with blue or odd eyes. Odd-eyed cats, having one blue eye and one copper eye, may be deaf only on the blue-eyed side. Some lines with very flat faces may experience breathing problems or develop clogged tear ducts.

Persians usually require daily grooming. Cats that do not receive regular grooming develop tangles and knots in their fur. Regular bathing is necessary to remove excess oil from the coat.

Figure 8–36   Persian. Photo by Isabelle Francais.

The Ragdoll. The Ragdoll resulted from the mating of a white Persian and a seal-point Birman. The offspring of the cross were then mated to a sable Burmese; the breed was developed in the 1960s. It takes its name from the limp and floppy posture taken when being handled and is said to feel no pain or fear. The Ragdoll is considered docile, quiet, and composed. The breed is very intelligent and enjoys retrieving. See Figure 8–37.

The Ragdoll is exceptionally large and heavy. The head is moderately wedge-shaped, and the ears are of medium size. The eyes are large, oval, and blue in color. Four colors are common: seal point, chocolate point, blue point, and lilac point; these colors are found in three distinct patterns. In the first pattern, there can be no white. In the second, or mitten, pattern, there are white gloves or paws in addition to the color points and a white abdomen. The third, or bicolored, pattern has colored points with white on the head, body, and legs.

Figure 8–37   Ragdoll. Photo by Isabelle Francais.

The Somali. The Somali was produced in the 1960s when Abyssinian parents carrying the recessive gene for long hair were mated. The Somali is considered friendly, affectionate, companionable, playful, intelligent, and very active. See Figure 8–38.

The breed is medium in size. The body characteristics are basically the same as those for the Abyssinian, but with a medium-long coat and a fox-like, bushy tail. There are two special colors: ruddy, an orange-brown ticked with black; and red, ticked with chocolate-brown. Both breed types have gold or green eyes.

Figure 8–38   Somali. Photo by Isabelle Francais.

The Tiffany. The Tiffany was produced in England by crossing Himalayan and Burmese breeds during early attempts to develop solid-chocolate Himalayans (now called Kashmirs). The Himalayan-Burmese cross produced litters with long hair. See Figure 8–39.

The Tiffany is a very playful cat and makes a pleasant pet. It is relaxed, is devoted, and follows its owner around like a puppy. The breed is of medium size like the Burmese and has an Oriental body type. It is basically a longhaired version of the Burmese. The coat color is sable, and the breed has gold eyes. Very few exist in the United States.

Figure 8–39   Tiffany.

The Turkish Angora. The Turkish Angoras were originally known as Ankara cats because the breed originated in the province of Ankara, Turkey. They probably share the same distant ancestors as the Persian but developed from a different breeding population. See Figure 8–40.

This breed is considered polite, intelligent, and responsive. It is easily trained to retrieve and perform tricks. The Turkish Angora is fastidious and prefers a clean, spotless environment.

The Turkish Angora is a small- to medium-sized cat with a long, slender, graceful torso on a light frame. The ears are large and erect on a small- to medium-sized, wedge-shaped head.

The breed was originally bred only in white, but many colors and patterns exist today. The fur is medium in length and silky. The breed lacks the thick underfur that its Persian cousin has. The eyes are large, almond-shaped, and show radiant, deep color.

Figure 8–40   Turkish Angora. Photo by Isabelle Francais.

ANATOMY

The anatomy and skeletal structure of the cat are identified in Figures 8–41 and 8–42.

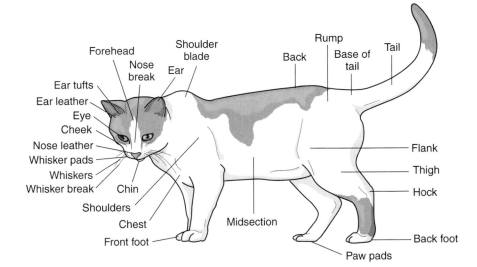

Figure 8–41 **The anatomy of the cat.**

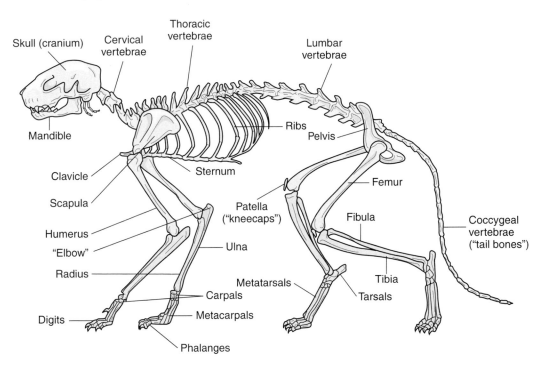

Figure 8–42 **The skeletal structure of the cat.**

CHARACTERISTICS

The body of the cat has developed for speed and flexibility. The bones, joints, and muscles form together to give the animal its size, shape, speed, and strength.

The skeleton of the cat contains 244 bones, which is about 40 more than a human; most of these extra bones are found in the spine and tail. Spread over this skeletal structure are approximately 500 separate muscles, compared with more than 650 in the human body. The largest muscles on the cat are found on the rear legs, which gives the animal its tremendous power for running and jumping.

One of the most obvious characteristics of the cat is its beautiful coat, which provides the cat with protection from excessive water loss, heat, cold, excessive sunlight, and physical injuries, and guards against the invasion of germs and parasites. Carnivores, like the cat, that use their forelegs for grasping prey also have whiskers and eyelashes that are sensitive to touch.

The sense of touch, although important in the cat, appears to be one of its less important senses. Specialized nerve endings that respond to touch are found throughout the skin. Petting the animal usually brings about a relaxing purr or, if not appreciated, results in the cat extending its claws and teeth.

The cat's nose and the paw pads are especially sensitive. Cats often extend a paw to investigate food, water, or unfamiliar objects. The nose and the tongue are then used to further determine smell and taste.

The ears of the cat are supersensitive. Being a hunter, the cat can detect the slight sounds of mice and other prey. The ears can detect sounds in the ultrasonic range, far higher in pitch than the human ear.

The organ within the ear that enables the cat to detect these sounds is the **cochlea;** its range of hearing is from 30 cycles per second up to 65,000 cycles per second (30 to 65 kHz). It can also detect and locate sound very efficiently.

Some white cats, especially the blue-eyed whites, may be deaf. This is caused by degenerative changes in the cochleas starting at about five days of age.

Another characteristic of the cat is its eyes, which point forward to provide for three-dimensional vision. The eyes of the cat function in almost total darkness as well as in bright daylight.

Cat's eyes are very similar to those of other animals but have some special features. They are designed to collect the maximum amount of light; the cornea and lens are large in relation to the overall size of the eyeball. They are highly curved, and the lens is farther back in the eyeball than in humans. These differences give the cat the ability to project an image on the retina five times brighter than the human eye. This, along with the greater sensitivity of the retina, gives the cat the ability to see in light six times dimmer than the human eye can see.

The cat eye has three eyelids; two close to protect the eye if anything touches the eyelashes, whiskers, or the eye itself. The third is known as the **nictitating membrane,** or haw, and moves diagonally across the eye under the eyelid to help lubricate the cornea.

The eyes of kittens remain closed for about ten to twelve days after birth. It takes about three months before the eye is completely developed.

Smell receptors located in the lining of the nose enable the cat to pick up the smell of airborne substances. Taste receptors located on the tongue allow the cat to detect and recognize different substances it ingests.

Cats differ from other animals in that they have a third set of receptors. These receptors, called **Jacobson's**

organ, are located in the roof of the mouth. The Jacobson's organ is stimulated by odors that are picked up by the tongue from the air and transferred to the organ when the cat presses its tongue against the roof of the mouth.

The **olfactory mucosa** is a special area of the nasal lining that picks up smells from the air. This area is almost twice the size of the olfactory mucosa of humans and indicates the importance of smell to cats.

The cat's tongue has several functions and is lined with knobs called **papillae.** The papillae in the center of the tongue form backward-facing hooks, which help hold food and are used to lick or scrape meat off bones. Papillae along the front and sides of the tongue carry the taste buds; large papillae located at the back of the tongue also carry taste buds.

The tongue is spoon-shaped and enables the cat to lap up liquids; the cat swallows after every four or five laps. The tongue is also used for grooming.

CHOOSING A CAT

Many people select cats for pets for their beauty and personality. Cats are independent and self-sufficient. They make excellent companions, especially for the elderly, and require less care than dogs. Cats fit well into apartment living and the less active lifestyle of the elderly person. The initial purchase price and the cost of feeding and keeping a cat is usually less than that for a dog. Cats are clean, generally requiring very little or no house-training. Depending on the circumstances, cats can be used to control mice and other rodents.

For the busy family or elderly person, it may be best to purchase an adult cat. It will not require as much time to care for as a young kitten. Adult cats should be taken to a veterinarian soon after obtaining them, because they may have parasites. This is especially true if the cat is a stray. Older cats may not live as long, which could be emotionally upsetting.

Young kittens adapt quickly to a new family; they can be fun to watch as they grow. The more attention a kitten receives, the closer it becomes attached to the family.

If one does not intend to breed the cat, it is best to have the animal spayed or neutered. It makes very little difference if the pet cat is male or female; both are equally intelligent, affectionate, and playful.

A male kitten reaches sexual maturity at about six or eight months of age; even if the owner lets it outside, it will start spraying areas to mark its territory. The smell is very strong and unpleasant; if one decides to keep a male cat, special housing may need to be provided.

A female kitten becomes sexually mature as early as five months of age. She usually comes in heat about every three weeks; unwanted litters may be expected by keeping an unspayed female cat.

If one is just looking for a pet, the common domestic, mixed-breed house cat will probably make a suitable selection. These animals have traits from several breeds and make affectionate, friendly pets.

Pedigree animals usually have both good and bad characteristics; one needs to be aware of these traits before purchasing a particular breed. Some breeds are basically inactive, solitary animals, whereas others are more active, playful, and demanding. Longhaired cats require more time for grooming. Also, the climate of the area is important; unless they have air conditioning, longhaired cats may become uncomfortable in high temperatures.

Mixed-breed animals can usually be obtained free from neighbors or friends or by checking advertisements in newspapers. To obtain some of the specialty breeds, one may have to contact local or state cat fancier clubs. Pet stores also offer cats, but it is important to make sure the animal has received vaccinations, because diseases may be a problem in some pet stores. Cats may also be obtained from animal shelters.

When selecting individual animals, one should be alert to signs of illness. Gums should be pale pink and the teeth white; older cats may show some yellowing of the teeth because of tartar buildup. The eyes should be bright and clear. Animals that have discharges emanating from the eyes or nose should be avoided. An extended, red third eyelid is usually a sign of illness. The nose should be cool and slightly damp. One should check the ears for a buildup of wax, which may indicate ear mites. The coat should be clean, glossy, and free of mats; one should check for traces of fleas and other parasites. A potbellied condition may indicate an animal with worms. One should feel the animal to check for lumps, abscesses, or other abnormalities. The rear area of the animal should be checked for signs of diarrhea or worms. The area and surrounding fur should be clean and free of mats.

Kittens are best purchased at eight to twelve weeks of age. One should select a kitten that is playful, alert, and lively and that does not shy away.

When selecting animals for breeding and show, one also needs to check for correct marking, eye color, coat color, and conformation. Be aware that adult colorings may not develop for several months.

FEEDING AND TRAINING

Cats are true carnivores and require almost twice as much protein in their diet as dogs. The best source of this is from animal products; 30 to 40 percent of the cat's diet should be animal-type proteins (meat, meat by-products, fish, eggs, and milk). See Figure 8–43.

About 10 percent of the diet should consist of fat; fat provides calories and the essential fatty acids.

A cat should be fed using one of the many commercial cat foods. If feeding fresh foods, it is important to provide a variety. Strictly feeding meat, chicken, fish, and other muscle meats may cause bone disease, stunted growth in kittens, poor eyesight, and other problems due to the lack of calcium and vitamin A. Too much liver can cause vitamin A poisoning. Milk is a good source of calcium, but it may cause gas and diarrhea in adult cats. Calcium can be supplemented by adding sterilized boneflour, calcium phosphate, calcium lactate, or calcium carbonate. Feeding too much of some oily fish such as tuna may through oxidation destroy vitamin E and lead to a deficiency called steatitis or yellow fat disease.

Raw egg white contains a substance that destroys the B vitamin biotin, but egg yolk and cooked egg white provide valuable protein, fat, and vitamins.

When feeding meat, it is important that all of the bone has been removed or chopped to prevent pieces

---

GUARANTEED ANALYSIS

Crude Protein . . . . .not less than 30.00%	Ash. . . . . .Not more than 6.50%
Crude Fat . . . . .not less than 14.00%	Magnesium . . . . .not more than 0.10%
Crude Fiber . . . . .not more than 2.00%	Taurine. .not less than 1500 ppm
Moisture . . . . .not more than 10.00%	Linoleic Acid . . . . .not less than 4.00%

INGREDIENTS
Chicken Meal, Ground Rice, Rice Gluten, Whole Dried Egg, Rice Flour, Poultry Fat (preserved with Tocopherols, a source of Natural Vitamin E), Rice Bran, Lamb Meal, Dried Brewers Yeast, Sunflower Oil (preserved with mixed Tocopherols, a source of Natural Vitamin E), Salt, Potassium Chloride, Natural Flavoring, DL-Methionine, Monosodium Phosphate, Taurine, Garlic, Ferrous Sulfate, Zinc Sulfate, Copper Sulfate, Cobalt Carbonate, Choline Chloride, Vitamin A Supplement, Vitamin D3 Supplement, Vitamin E Supplement, Niacin, Calcium Pantothenate, Riboflavin Supplement (source of Vitamin B2), Thiamine Mononitrate (source of Vitamin B1), Pyridoxine Hydrochloride (source of Vitamin B6), Folic Acid, Biotin, Inositol, Potassium Iodide, Vitamin B12 Supplement.

Figure 8–43 Cats require a high-protein diet. One should always check the label when purchasing pet food.

of bone from becoming lodged in the throat or digestive system. Cats should never be given chicken bones.

Canned foods contain more animal protein than the other commercial rations, have a higher fat content that makes them more palatable, and contain about 75 percent water. Because of the high water content, cats may not drink as much water, but water should be available at all times. Labels should be checked because some of the commercial rations may be nutritionally incomplete. Fresh and canned foods should not be fed straight from the refrigerator; these foods should be allowed to warm to room temperature before being served.

Semi-moist foods are usually less expensive because they contain some vegetable protein and are usually supplemented with nutrients to make them nutritionally complete. Semi-moist foods have chemicals added to keep them from drying out or spoiling; they contain about 30 percent water. Again, labels should be checked carefully because some of the semi-moist foods may not be nutritionally complete, especially for growing kittens.

Dry foods contain about 10 percent water and less fat and protein than semi-moist foods. Cats on dry diets should have plenty of water available. Some cats on dry diets may develop bladder problems. Milk, water, or gravy can be mixed with the food to improve palatability and to ensure that the cat gets adequate water intake. One may wish to feed canned foods occasionally to help prevent bladder problems, get the cat used to different types and textures of foods, and ensure that the cat gets a balanced diet. Dry foods do have the advantage of helping to clean the teeth and prevent the buildup of tartar.

The amount of food one gives depends on the cat's age, weight, breed, condition, and amount of activity it gets. Cats and young kittens will not consume enough food in one meal to last 24 hours. Two meals are recommended, and young kittens and females that are pregnant or nursing require more frequent feedings.

Cats should never be given a diet of dog food because it contains large amounts of cereals and vegetables. Because of this, the cat may not get enough animal protein. Dog food also lacks necessary amounts of vitamins A and B and some essential fatty acids. Many times cats will be seen eating grass. The exact reason for this is not known, but it may be an attempt to increase roughage in the diet or to eliminate a hairball.

The amount of exercise a cat needs varies considerably depending on the breed and where its home is. A cat living in a city apartment may get little exercise, whereas one living in rural areas may be allowed to run free. Apartment cats may need to be furnished with toys, cardboard tubes, or other play equipment to provide them with a means to exercise. Owners of valuable purebred or show cats may not want their animals to run free where they risk injury, loss, or unwanted litters. Outdoor cats get plenty of exercise; however, they run a greater risk of injury from fights, of death or injury on the roadways, and of contracting diseases and parasites. Cats living indoors should have a clean litter box and plenty of water. Owners should be aware of plants that may be poisonous to their pets and of other dangers if the cat is left alone for long periods. To protect the furniture, cats should be trained to use a scratching post.

Cats scratch to sharpen their claws, to remove loose scales and fragments of dry skin, and to leave a mark for other cats. Kittens should be trained to use a scratching post as soon as they are weaned. The kitten should be held near the post and its claws placed on the post. The kitten will soon learn what to do and will usually continue to use the same post. A cloth-covered post may offer the cat an alternative to clawing soft furniture and draperies.

Correct toilet training is easier with cats than with dogs because cats naturally cover their urine and feces. If a mother cat does a good job of raising her litter, she will probably train them to use a litter box. If a kitten does not know what the litter box is for, one can train it easily. Holding its front paws, the owner should show it how to scratch in the litter material. Every time the kitten appears to be looking for a place to urinate or defecate, it should be placed in the litter box.

Urine spraying is a natural part of a cat's behavior. The male cat will spray walls and furniture to mark its territory. If one can catch the cat in the act, spray it with a water pistol. Another alternative to try is to hang aluminum foil around the spraying area. Also, one can try feeding the cat in the target areas, because cats will usually not spray near feeding areas.

Some cats can be trained to sit, beg, jump through hoops, and do other tricks. Training should be similar to the methods used with a dog, with short training periods, firm verbal commands, and instant rewards. Whether a cat learns any tricks depends on its intelligence and whether it wants to oblige. In any case, training should not be forced on a cat that does not appear interested.

GROOMING AND CARE

A sleek, glossy hair coat is an indication of a cat's general health and care. All cats benefit from grooming, although the longhaired breeds require more attention. A cat's fur usually sheds in spring and fall, although some shedding may constantly occur. Grooming removes old, dead hair and lessens the risk of hairballs. Regular grooming gives the opportunity to check for parasites, skin disorders, and eye and ear problems. See Figure 8–44.

Figure 8–44 A large selection of combs, brushes, and pet care products is available for grooming pets. Courtesy of Brian Yacur and The Feedbag Plus.

Longhaired cats should receive daily care. If neglected, the hair will tangle and mat. Removal of these tangles and mats may be difficult and an unpleasant experience for the cat. See Figure 8–45.

Equipment for longhaired cats should include a comb with two sizes of teeth, a fine-tooth or flea comb, nail clippers, a grooming brush made with natural bristles (nylon may cause excessive static), and grooming powder (baby powder, talcum powder, or cornstarch).

Using a wide-toothed comb, comb all areas of the animal. One must be careful of the sensitive areas of the stomach, the insides of the legs, and under the tail. If the coat is free of tangles and mats, the fine-toothed part of the comb should be used. The skin and not just the outer fur should be combed; now, the fur should be brushed out. One should brush in the opposite direction to which the hair naturally lies and occasionally sprinkle grooming powder into the fur.

If the fur has become badly tangled and matted, scissors should be used to cut the mats out, being careful not to injure the animal. Blunt-ended scissors—never sharp-pointed scissors, knives, or razor blades—should be used. A sudden movement by the cat could cause serious injury.

Grooming shorthaired cats can usually be accomplished with a fine-toothed or flea comb. In many cases, hand grooming is sufficient to remove dead hair. A rubber grooming brush is also very effective, but it must be used carefully because good hair may also be removed. The use of a soft chamois, silk, or nylon pad causes some static in the coat and helps it cling tightly to the body. This is an especially effective way to complete the grooming for Siamese, Oriental Shorthairs, Colorpoint Shorthairs, and Burmese. Other breeds enjoy a light brushing with a soft brush.

The coat of the Cornish and the Devon Rex is delicate and easily damaged; a soft brush should be used occasionally, avoiding friction or overbrushing. Friction and overbrushing may cause the hair to break and may even cause bald spots.

Occasionally, it may be necessary to give a cat a bath. In the case of show animals, more frequent bathing is necessary. Cats should be accustomed to bathing at about four months of age. Cats get nervous from the sounds of the water, and unnecessary splashing should be avoided. The tub should be filled with about 4 inches of warm water. The cat should be gently lowered into the water and wet all over, being careful not to get water in the eyes or ears. If the cat will allow it, a piece of cotton should be placed in its ears. After the coat is thoroughly wet, one should apply shampoo and lather the coat, followed by a thorough rinse to remove all traces of the shampoo. The cat should be wrapped in a towel and patted dry or blown dry using a hair dryer on the warm setting. Excessive rubbing should be avoided because this causes the hair to tangle. The cat should be dry before letting it outside, including the insides of the ears. Many pet shampoos are on the market that are suitable; some of these are medicated to help control fleas

Figure 8–45 Regular grooming of one's pet will keep the pet looking nice and gives one the chance to examine the animal. Courtesy of Sandy Clark.

Figure 8–46 *A couple of a wide range of pet care products that are available. Photo by Dean Warren.*

and other parasites. Baby shampoo is also suitable. See Figure 8–46.

During regular grooming, the ears should be checked for mites; signs are a dark, crumbly residue inside the ear. Ear mites can be controlled with ear drops available at any pet store. See Figure 8–47.

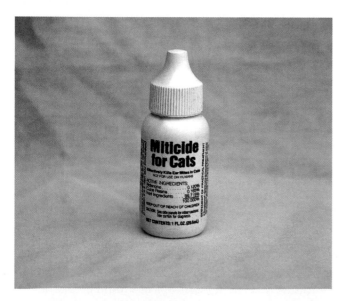

Figure 8–47 *Ear drop products for the control of mites are readily available at most pet stores and discount stores. Photo by Dean Warren.*

The eyes of the cat should be bright and clear; any discharge may indicate illness or infections. White cats, especially Persians, may show staining around the eyes; this staining can be removed by careful bathing. See Figure 8–48.

Teeth and gums should be observed during regular grooming and should be free of any soreness. Occasional use of dry food helps clean the teeth. Excessive amounts of tartar may need to be removed by a veterinarian to prevent tooth decay and gum disease. See Figure 8–49.

If claws require trimming, one should be careful not to cut into the pink area of the claw because this causes pain and bleeding. Specially designed clippers should be used—never use scissors. Cats that are using a scratching post will probably never need their claws trimmed. See Figure 8–50.

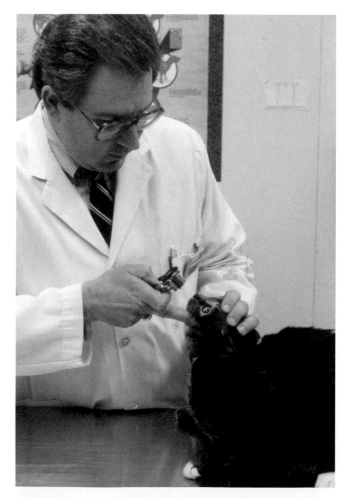

Figure 8–48 *Cats should be taken to a veterinarian for regular checkups and vaccinations. The veterinarian will examine the animal's eyes as part of a regular checkup. Courtesy of Brian Yacur and Guilderland Animal Hospital.*

Figure 8–50   The claws can easily be trimmed using special trimmers, or the veterinarian can trim the animal's claws during the regular checkup. Courtesy of Brian Yacur and Guilderland Animal Hospital.

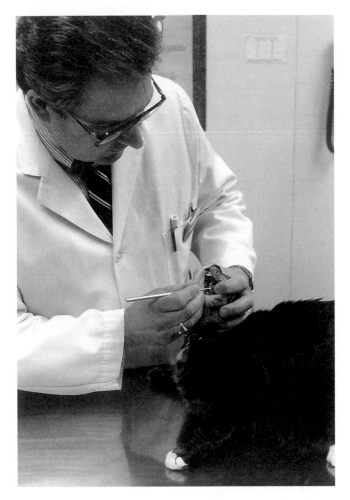

Figure 8–49   The veterinarian will also examine the animal's teeth and gums. Courtesy of Brian Yacur and Guilderland Animal Hospital.

COMMON DISEASES

This unit is broken down into six parts: infectious diseases, noninfectious diseases, internal parasites, external parasites, fungal diseases, and poisonings.

Common symptoms of illness that should alert the owner are loss of appetite, sudden loss of weight, changes in behavior, apathy, neglect of grooming, dull fur, loss of hair, itching or scratching, swelling of the body, constant vomiting, increased thirst, and diarrhea. If any of these signs are observed, the animal should be taken to a veterinarian for diagnosis and treatment.

Infectious Diseases

Feline Panleukopenia.   Feline panleukopenia, also referred to as feline infectious enteritis or cat distemper, is caused by a parvovirus or DNA virus. This is primarily a disease of young cats. Kittens less than

sixteen weeks of age may die at a rate of 75 percent. Older cats may have acquired some immunity and are less likely to die.

Feline panleukopenia is spread by direct contact among infected cats. Infection may also occur from infected food and water dishes, bedding, and litter boxes. Infected cats usually show signs of depression, loss of appetite, high fever, lethargy, vomiting, dehydration, and hanging over the water dish. Diarrhea may show large amounts of liquid, mucus, and blood.

The virus is capable of passing the placental barrier in pregnant cats. Infection of the fetus usually results in abortion, stillbirth, early fetal death, or permanent structural brain damage.

Recovery for kittens less than eight weeks of age is poor; there are no antibiotics that kill the virus. The veterinarian will try to combat the dehydration, provide nutrients, and prevent secondary infection with antibiotics. Immunity in kittens is possible from their mothers; this passive immunity is temporary and is usually ineffective after twelve weeks of age.

Vaccines offer the safest and most effective means of protection. Kittens should be vaccinated at nine and twelve weeks of age. Annual revaccinations are recommended throughout the cat's life.

Feline Herpesvirus and Feline Calicivirus.   Two respiratory virus infections are of primary importance in cats: feline herpesvirus (FHV) and feline calicivirus (FCV). The herpesvirus is a DNA virus, and the calicivirus is an RNA virus. The viruses of both can be shed in discharges from the nose, eyes, and throat. Transmission results mainly from direct contact between animals.

The virus may persist in cats for a long period of time after infection, and the cat can become a carrier of the virus. Early signs of infection are depression, sneezing, and coughing. As the infection progresses, severe eye and nasal discharges occur with an increase in temperature. Both viruses may cause ulcers in the mouth, but corneal ulceration appears only in feline herpesvirus infections. Vaccinations are available for both feline herpesvirus and feline calicivirus.

Feline Rhinotracheitis. When herpesvirus infection is confined to the respiratory tract, it is generally referred to as feline rhinotracheitis (PVR). Diagnosis of these diseases are based on symptoms, especially sneezing and discharges from eyes and nasal passages. Definite diagnosis depends on isolation and identification.

Vaccinations for respiratory viruses are usually included in combination with the panleukopenia virus vaccination. Two vaccinations are required for all ages. Kittens should be vaccinated at nine and twelve weeks. Annual revaccination is recommended.

Feline Infectious Peritonitis. Feline infectious peritonitis (FIP) was first observed in the early 1950s. The disease is caused by a coronavirus. Coronavirus infections are relatively common in domestic cats, but the majority of these do not produce signs of disease. Although a cat may show no signs of disease, it may serve as a carrier. The normal means of transmission is by direct contact.

The disease occurs in two forms. In one form, the abdomen and chest accumulate fluid. The cat experiences a fever, refuses to eat, appears depressed, and loses weight. As the disease progresses, the body cavities continue to fill with fluid, organ systems are affected, and signs of specific organ failure occur. Death may eventually occur as a result of organ failure.

The second form is usually associated with specific organ failure. Lesions affecting the cat's eyes are very common and may be the first indication of the disease. In this form, the disease may progress to the point that the cat dies from organ failure. Diagnosis is made from symptoms and laboratory results. Routine immunization against feline infectious peritonitis is not possible at this time. A vaccination for FIP was developed in 1991. One should consult a veterinarian about FIP vaccinations.

Feline Leukemia. Feline leukemia virus (FeLV) disease complex is caused by an RNA virus that occurs worldwide. The infection of cats is related to the density of the cat population in various areas. Infection in single-cat households is less likely than those in multiple-cat households and catteries.

The virus is excreted primarily in the infected cat's salivary secretions; however, the virus may also be present in respiratory secretions, feces, and urine. It is spread through direct contact and by sharing litter boxes, food, and water dishes.

Symptoms of the disease may not be present or may be very difficult to recognize. Fever, depression, loss of appetite, and enlargement of lymph nodes may be early signs. Kittens usually die as a result of pneumonia, accumulation of pus in the chest, intestinal infections, or blood infections. Those that survive the early phases of the disease may recover. Others develop a secondary phase of infections and become carriers or die. A vaccine for FeLV is now available.

Many disorders are associated with FeLV infections. It is estimated that about 30 percent of all severe illnesses in cats are caused by this virus.

Feline Enteric Coronavirus. Feline enteric coronavirus is the cause of infection in kittens between four and twelve weeks of age. The infection is spread by the ingestion of contaminated feces.

The virus affects the epithelial cells of the small intestine. Symptoms are low-grade fever, vomiting, and soft or watery diarrhea. Blood may be present in the feces, and dehydration is present in more severe cases. Vaccine for protection is not available at this time.

Feline Pneumonitis. Feline pneumonitis is caused by infection with *Chlamydia psittaci*. Runny eyes and noses are the most characteristic signs of infection. Early immunity is present in the colostrum, and vaccinations are available for further immunity. Annual revaccination is recommended.

Rabies. Rabies is a viral infection of all warm-blooded animals. Three hundred cases of rabies in cats were reported to the Centers for Disease Control and Prevention (CDC) in 1997. During 1997, only 126 cases of rabies in dogs were reported to the CDC. (Refer to Chapters 2 and 7 for more information on rabies.) Cats should be vaccinated against rabies. A first vaccination should be given at twelve weeks of age and a second vaccination at eighteen weeks of age. A booster should be given every three years.

Noninfectious Diseases

Feline Urologic Syndrome. Feline urologic syndrome (FUS), also referred to as feline lower urinary tract disease (FLUTD), is a urinary problem in cats. The condition may range from mild inflammation to a blockage of the urethra, uremic poisoning, and death. The condition is believed to be brought on by an improper diet, reduced water intake, and possibly a virus. If an animal does not have sufficient water intake, the urine can become concentrated and various salts will settle out and form stones. A diet high in magnesium and phosphorus may also be involved. If resulting stones block the urethra, prompt treatment is necessary to prevent uremic poisoning.

Entropion. Entropion is a condition in which the eyeball sinks into its socket or in which there is a spasm of the eyelid due to discomfort. The turning in of the eyelid allows hair to come in contact with the cornea, causing discomfort, watering, and possibly conjunctivitis (inflammation of the membrane lining the inner surface of the eyelids) and keratitis (an inflammation of the cornea). These conditions are more prevalent in the Persian and related breeds.

Wet Eyes. Wet eyes is a condition of excessive tear production or blockage of the canals that normally drain tears into the nasal cavity, causing tears to overflow at the inner corners of the eyes. This condition appears more frequently in the Persian and related breeds and is especially a problem with cats that are used for show.

Other conditions include tetanus, botulism, and the inflammation of the third eyelid. These are also problems associated with dogs.

Internal Parasites

Toxoplasmosis. Toxoplasmosis is caused by infection with *Toxoplasma gondii,* a single-celled protozoan parasite. This disease occurs throughout the world and has been observed in a wide range of birds and mammals. Cats, however, are the primary target for infection. The infectious agent may be picked up by eating raw meat or contaminated feces.

Infected cats may develop fever, jaundice, enlarged lymph nodes, difficulty breathing, anemia, eye inflammation, encephalitis, and intestinal disease. Abortion is also seen in pregnant cats that are infected. Cats may also develop stiff, painful muscles and have difficulty moving or be unable to move.

Accurate diagnosis is possible through laboratory tests; because symptoms are so widely varied, however, they are not reliable aids in diagnosis of the disease. Vaccination of animals against toxoplasmosis is not possible at this time.

Humans can become infected with *Toxoplasma gondii.* Transmission is usually through ingestion of infected raw or undercooked meat or by ingestion of feces. Ingestion is usually connected with handling cat litter boxes or contaminated soil.

Pregnant women should be especially concerned because unborn fetuses appear to be at great risk to toxoplasmosis during early pregnancy. Pregnant women should avoid handling litter boxes.

To avoid human infection, meat should be cooked thoroughly; children should be especially careful and wash their hands before eating. Gloves should be worn when working in areas where cats defecate. Children's sand boxes should be covered when not in use so that cats cannot use them as litter boxes. Litter boxes must be changed frequently and feces disposed of so that animals and humans cannot come in contact with the feces.

Ascarids and Hookworm. Cats can be infected with ascarids and hookworms. The common ascarid *(Toxocara cati)* and hookworm *(Ancylostoma tubaeforme)* of cats are different species from those that attack dogs; although the cat can be infected with ascarids and hookworms, the prevalence of the infection is not as great. This is probably related to the cat's burying its feces.

Cats acquire ascarid infections from ingesting eggs that have been passed in the feces of an infected animal. Nursing kittens may become infected with ascarid larvae through the milk of the mother cat.

Hookworm infection occurs when the larvae penetrate the skin or when the larvae are ingested in contaminated food or water. The larvae are not passed in the mother's milk as in the case of dogs.

Symptoms of infection by ascarids and hookworms are basically the same as those described in dogs. Symptoms may not occur in cats because cats do not become as heavily infected as dogs. The same drugs used for deworming dogs can be used to deworm cats. See Figure 8–51.

Tapeworms. Tapeworms are also found in cats. One species, *Dipylidium caninum,* infects both dogs and

Figure 8–51 Roundworms, hookworms, and tapeworms can be prevented with products available at pet stores and discount stores.

cats. Another species, *Taenia taeniaeformis,* is sometimes found in cats.

An initial host is required for development of the young tapeworm. The flea serves as the initial host of *Dipylidium,* and rats or mice serve as the initial host for *Taenia* in cats. See Figure 8–52.

Tapeworms do not cause major harm to cats; the same drugs used for control of tapeworm in dogs is used in cats. To prevent continued reinfection, one must also eliminate fleas and prevent the cat from eating rats and mice.

External Parasites

Lice. Cats are infected by one common species of biting louse *(Felicola subrostratus).* Healthy and well-cared-for cats are unlikely to be troubled by lice. When a cat is infected, the lice are usually found around the head. Many pesticides are available to control adult lice; treatment should be repeated at two-week intervals.

Mites. Mites are rare in cats; however, there are two types of demodectic mites: *Demodex cati* and an unnamed *Demodex* species. In demodicosis, there is patchy hair loss, reddening, and occasionally crusting on the neck, ears, and head. Severe cases may involve the entire body.

Feline Scabies. Feline scabies, or notoedric mange, caused by *Notoedres cati* is a skin disease that is common in certain local areas of the United States. These mites are highly contagious to other cats and may be transmitted to people and dogs. They cause hair loss and thickening and crusting of the skin, usually affecting the head and forelegs. Parasiticidal rinses are used to treat notoedric mange.

Feline Mites. Ear mites, walking dandruff mites, chiggers, fleas, and ticks that infect cats also infect dogs. These parasites were covered in Chapter 7; refer to that unit for more detailed information on external parasites. Remember that when treating with flea and tick powders, it is important that only powders recommended for cats be used. The powders used for dogs would be too strong for use on cats; cats lick their fur and would ingest the stronger powder, and possible serious injuries could result.

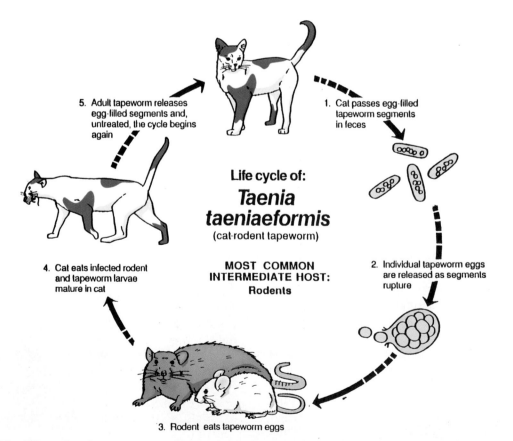

5. Adult tapeworm releases egg-filled segments and, untreated, the cycle begins again

1. Cat passes egg-filled tapeworm segments in feces

Life cycle of:
Taenia taeniaeformis
(cat-rodent tapeworm)

MOST COMMON INTERMEDIATE HOST:
Rodents

2. Individual tapeworm eggs are released as segments rupture

4. Cat eats infected rodent and tapeworm larvae mature in cat

3. Rodent eats tapeworm eggs

Figure 8–52   The life cycle of Taenia taeniaeformis, *cat-rodent tapeworm. Reprinted with permission of Miles, Inc., Animal Health Products.*

Fungal Diseases and Poisonings

These diseases and problems were also covered in Chapter 7; they are problems common to dogs also. Please refer to Chapter 7 for more detailed information.

REPRODUCTION

The age at which female cats (queens) first come into heat varies with breeds. The Siamese may come into heat as early as five months of age, whereas most longhaired breeds of the Persian type may not come into heat until ten months of age or more.

Estrus in cats is induced by lengthening daylight and rising temperatures of spring. The heat period lasts for nine or ten days. If the cat is not bred during the heat period, she will come back in heat in fifteen to twenty-one days.

A female cat in heat is restless and unusually friendly, rubbing up against objects, furniture, and the owner's legs. She usually has a decreased appetite and urinates more frequently. A repeated monotone howl or call is made, and she rolls around on the floor. Any touch to the back of the tail is likely to stimulate a crouch position, and further stroking at the back and base of the tail exaggerates this receptive position.

In response to the call of the female, receptive males (toms) also emit courtship cries and spray urine to leave scent marks to attract the female's attention. Immediately after mating occurs, the female releases a loud piercing cry, then turns and attacks the male. Ovulation (release of the egg cells) occurs about 24 hours after mating occurs.

Pregnancy in cats averages about sixty-five days but may vary from fifty-one to sixty-eight days. At three weeks into the gestation period, the nipples of the female become pink and the hair recedes slightly. At six weeks, pregnancy is obvious, as the female's abdomen shows an increase in size.

About two weeks before the kittens are due, the female becomes restless and searches for a quiet place to give birth. If one offers her a well-prepared box, she will usually take to it.

During the first stage of labor, the female's rate of breathing increases, and she may start to breathe through her mouth. As labor proceeds, there will be some discharge, usually colorless at first, but later becoming blood-stained.

The second stage of labor begins with the contractions of the abdominal muscles. At first, these contractions may occur once an hour; just before delivery they occur every 30 seconds.

The female repeatedly licks her genital area and may show signs of agitation. The vaginal opening di-

lates and the first amniotic sac appears. The sac and the enclosed kitten are usually expelled within 15 to 30 minutes. As soon as the kitten is expelled, the female ruptures the amniotic sac. She cleans the kitten; this persistent licking stimulates the kitten to breathe. Next, she severs the umbilical cord with her teeth. The placenta and the afterbirth are soon expelled, and the female promptly eats it. This natural instinct provides the female with a source of nutrients. The rest of the litter will follow at about 15-minute intervals.

The female may go out of labor for a period of time and then restart contractions, giving birth at intervals of perhaps 1 or 2 hours. In some cases, she may rest for a period of 24 hours, nursing her first kittens before going back into labor and delivering the rest of the litter.

Occasionally, a female does not attempt to remove the sac from the newborn kitten. She may not know what to do or may be too busy with the next kitten. In this case, one must remove the membrane. If the female still does not show interest, one must carefully sever the umbilical cord, rub the kitten dry, and stimulate it to breathe. Then, the kitten should be placed close to the mother's stomach. Hopefully, the kitten will begin to nurse and arouse maternal instincts in the female.

The first milk produced by the mother is called colostrum, is rich in proteins and minerals, and contains antibodies to help the newborn kitten fend off diseases.

SUMMARY

Cats are descendants of the same carnivorous, tree-climbing mammal *Miacis* from which dogs descended. Cats, like dogs, trace their lineage through *Cynodictis*. The skeleton of these early primitive carnivores was much like that of a civet or a weasel.

Modernized cat-like types, *Felis* and related genera, with "normal" canine teeth appeared early in the Pliocene epoch. Their origin has been much debated, for all the better-known older felids have much larger canines and hence were saber-tooths of one sort or another.

In Chapter 7, we learned that dogs were domesticated about 10,000 years ago. The cat, however, was not domesticated until much later. This was probably because of its independent, solitary nature. It is believed that domestication of the cat began around 4,000 years ago.

Cats are usually divided into three genus groups. One group is the large, roaring cats, *Panthera,* which includes the lion, leopard, tiger, and jaguar. Second are the cats that cannot roar, *Felis,* which includes the small cats and the domesticated cat. The third group, *Acinonyx,* includes only one member, the cheetah. The cheetah belongs to a special group because it has claws that do not fully retract.

The domestic cat breeds are commonly divided into two major groups: the shorthaired breeds and the longhaired breeds. There are thirteen longhaired breeds and twenty-six shorthaired breeds.

The body of the cat has been developed for speed and flexibility. The bones, joints, and muscles give the animal its size, shape, speed, and strength.

The skeleton of the cat contains 244 bones, which is about 40 more than a human; most of these extra bones are found in the spine and tail. Spread over this skeletal structure are approximately 500 separate muscles, compared with more than 650 in the human body. The largest muscles on the cat are found on the rear legs, which gives the animal its tremendous power for running and jumping.

Cats require almost twice as much protein in their diet as dogs. The best source is from animal products; 30 to 40 percent of the diet should be animal-type proteins.

A sleek, glossy hair coat is an indication of a cat's general health and its care. All cats benefit from grooming, although the longhaired breeds require more attention. A cat's fur usually sheds in spring and fall, although some shedding constantly occurs. Grooming removes old, dead hair and lessens the risk of hairballs. Regular grooming gives the opportunity to check for parasites, skin disorders, and eye and ear problems.

Common diseases in cats can be divided into six groups: infectious, noninfectious, internal parasites, external parasites, fungal, and poisonings. Common symptoms of illness that may occur are loss of appetite, sudden loss of weight, changes in behavior, apathy, neglect of grooming, dull fur, loss of hair, itching or scratching, swelling of the body, constant vomiting, increased thirst, and diarrhea. If any of these signs are observed, the animal should be taken to a veterinarian for diagnosis and treatment.

The age at which female cats first come into heat varies with breeds. The Siamese may come into heat as early as five months of age, whereas most longhaired breeds of the Persian type may not come into heat until ten months or more. Estrus in cats is induced by lengthening daylight and rising temperatures of spring. The heat period lasts for nine or ten days. If the cat is not bred during the heat period, she comes back in heat in fifteen to twenty-one days.

Pregnancy in cats averages about sixty-five days but may vary from fifty-one to sixty-eight days.

DISCUSSION QUESTIONS

1. What is the history of the cat?

2. What are the two major groups of domesticated cats?

3. List the different breeds of cats. How do they differ?

4. What is meant by the term *colorpoint*?

5. How should longhaired cats be groomed? Do shorthaired cats need grooming?

6. What is the most important thing to remember in feeding a cat?

7. What are the six groups of diseases that affect cats?

8. List the diseases in the six groups that affect cats. What are the symptoms of these diseases?

9. List the general health care practices for a pregnant cat before, during, and at birth.

10. What health care practices should be observed with the new litter?

SUGGESTED ACTIVITIES

1. Zoologists disagree on the evolution of the cat. Visit the library and research the evolution of the cat. Could you find a more detailed path of evolution than the one in this text? Did the modern cat evolve from the saber-tooth? Did the modern cat and saber-tooth evolve separately?

2. Bring the labels from various cat foods and compare the ingredients that are used. Compare these with labels obtained from dog foods.

3. With the approval of the instructor, invite the local veterinarian to talk about some of the common diseases of cats.

4. Visit a cat show. Watch how the show is conducted and how the cats are presented.

5. With the approval of the instructor, have local cat breeders bring their cats to the class and discuss characteristics of the breed.

ADDITIONAL RESOURCES

1. Cat Fanciers
www.fanciers.com

2. The Cat Fanciers Association (CFA)
www.cfainc.org

3. Purina
www.purina.com

4. The International Cat Association
www.tica.org

5. Feline Veterinary Medicine
www.fanciers.com/vetmed.html

6. The American Veterinary Medical
Association, Feline Health
www.avma.org/care4pets/animfelh.htm

7. Infectious Diseases of Cats and Dogs
http://microvet.arizona.edu/courses/MIC443/
notes/rand/cat_dog.htm

8. Wellsboro Veterinary Hospital & Reptile &
Bird Clinic
www.drkreger.com

CHAPTER 9

Rabbits

OBJECTIVES

After reading this chapter, one should be able to:

- discuss the history of the rabbit.
- discuss uses for rabbits.
- list the five weight categories of rabbits.
- list the breeds of rabbits.
- discuss the housing and equipment needs of rabbits.
- discuss the nutrient requirements of rabbits.
- discuss various diseases that affect rabbits.

TERMS TO KNOW

conjunctiva

coprophagy

dew-drop valve

dewlap

enteritis

fly-back fur

kindling

malocclusion

oocysts

roll-back fur

sore hocks

CLASSIFICATION

Kingdom — Animalia
Phylum — Chordata
Class — Mammalia
Order — Lagomorpha

Family — Leporidae
Genus — *Oryctolagus*
Species — *O. cunniculus*

HISTORY

Rabbits were once classified in the order Rodentia, but they are now placed in a separate order, Lagomorpha. The chief difference between rabbits and rodents is that rabbits have four upper incisor teeth arranged one pair behind the other, whereas rodents have two chisel-like incisor teeth in each jaw.[1]

Fossil remains of Lagomorpha have been found in Oligocene deposits, which date back 30 to 37 million years. These fossil remains indicate the animals have changed very little.

The order Lagomorpha includes the family Leporidae, the rabbits and hares; and the family Ochotonidae, which are the pika. Pika are small, furry animals that more closely resemble guinea pigs than rabbits or hares and are found in Asia, Europe, and western North America.

Among the family Leporidae are the American cottontail, genus *Sylviagus;* the true hares, genus *Lepus;* and the European wild rabbit, genus *Oryctolagus,* species *cunniculus.*

The European wild rabbit is the species from which all domestic rabbit breeds have been developed. Although many of our current domestic breeds have been developed in the United States, the American cottontail was not used in the development of our domestic rabbit.

Hares, genus *Lepus,* are usually larger than rabbits and have longer, black-tipped ears and are slightly different in appearance. Hares do not build nests, and their young are born fully furred with their eyes open; hares live above ground and do not dig tunnels.

Rabbits were fairly widespread and abundant in Europe during the late Tertiary period and early Pleistocene epoch. Abnormal climates during the Ice Age drove them to the southern parts of Europe.

First reports of rabbits were from Phoenician traders who visited the coast of Spain and islands along the coast during 1100 BC. The Phoenicians are probably responsible for transporting rabbits to other parts of the world.

Rabbits were of great economic importance. They were hunted for food, and their pelts were used to make clothing. Credit for domestication of the rabbit is given to French monks of the Middle Ages, who raised rabbits in walled cages kept in their monasteries. They served as an easily raised source of food, and the fur was used for clothing.

In the second half of the nineteenth century, the European wild rabbit was introduced into Australia and New Zealand, where it quickly spread and became a serious pest. Brought to Chile in the early twentieth century, it eventually spread over much of the South American continent. Some European wild rabbits were released on the San Juan Islands off Washington state in the United States at the beginning of the twentieth century and have flourished since, but the major rabbit of North America remains the cottontail.[2]

In wild populations, the European wild rabbit measures 13 to 18 inches long with a tail 1½ to 3 inches long, and weighs 3 to 5 pounds. The upper body is a mixture of black and light brown hairs; the ears are the same color except for the black-edged tip. The nape is buff colored, the collar is dark, the tail is white below and brownish-black above, and the underparts and inner surface of the legs are buffy-white. The hind legs are relatively long; the feet are well furred beneath and have large, straight claws. Many of the domestic forms bear little resemblance to the original wild stock.[3]

USES

Meat

Rabbits are unique among the animals covered in this book because they are not only used for pets but are also raised for meat, fur or wool, and laboratory use.

Those who encourage the use of rabbit meat stress the following advantages (see Tables 9–1 and 9–2):

- high in protein
- low in cholesterol
- low in fat
- low in sodium
- very palatable

Rabbit meat is primarily a white meat that is very fine in texture and has a very low fiber content. Because of the low fiber content, it is easily digested, which is desirable for individuals who may have difficulty chewing their food. Rabbit is a more popular meat in Europe than in the United States. Italy, Germany, France, and Spain are the world's largest producers and consumers of rabbits. The major reason for the low consumption of rabbit in the United States is probably because of the image we have of rabbits. We grow up with rabbits as friendly characters in cartoons and games. Only a small number of rabbits are produced for the meat, and the supply to the supermarkets is small. The two major breeds used for meat production are the New Zealand and the Californian.

Research

Almost 400,000 rabbits were used in research projects in the United States during 1991. They have been used to produce disease-fighting antibodies, to study reproduction, and to research several human diseases. Rabbit blood is one of the best mediums for growing the AIDS virus.

Rabbits have been used in controversial tests. The Draize Eye Test has been used by many companies to test cosmetics and other products. Rabbits do not have tear ducts and cannot shed tears to dilute chemicals or products put into their eyes. Many people have called for a ban on this type of test, and many cosmetic companies no longer use it.

Skin irritation tests have also been used with rabbits. A small patch of fur is removed along the back of the animal, and the product being tested is applied to the bare area. Irritations or reactions to the products are then noted.

Table 9–1 Composition of Some Common Foods				
Edible Portion Uncooked	Percent Protein	Percent Fat	Percent Moisture	Calories per 100 grams
Beef, good grade	16.3	28.0	55.0	317
choice and prime grade	13.7	39.0	47.0	406
Chicken fryers, total edible, including flesh, fat, skin, and giblets	20.0	11.0	67.6	179
Chicken fryers, flesh only	20.6	4.8	73.4	126
Lamb, intermediate	15.7	27.7	55.8	312
fat	13.0	39.8	46.2	410
Pork, medium	11.9	45.0	42.0	453
fat	9.8	55.0	35.0	534
Rabbit, domesticated, total edible, including flesh, fat, and giblets	20.8	10.2	67.9	175
Rabbits, domesticated, flesh only	20.8	10.2	67.9	175
Salmon, Atlantic	22.5	13.4	63.6	211
Salmon, Pacific (Chinook and King)	17.4	16.5	63.4	218
Tuna, blue fin	24.8	5.2	69.1	146
yellow fin	24.7	3.0	71.5	126
Turkey, medium birds, including flesh, skin, fat, and giblets	20.1	20.2	58.3	262
flesh only	24.0	24.0	68.6	156
light meat	24.5	4.6	69.2	139
dark meat	23.2	9.4	68.0	177
Veal, medium	19.1	12.0	68.0	184
fat	18.5	16.0	65.0	218
Whitefish	22.9	6.5	69.8	150

Information taken from United States Department of Agriculture Circular No. 549, *Proximate Composition of American Food Materials.*

Fur/Wool

Rabbit fur is divided into four types: normal, rex, angora, and satin. Each type of fur can be used in the manufacture of clothing, toys, coats, hats, and gloves. White is the preferred color because it can be used to match almost any type of animal fur.

Rex fur is short, with guard hairs being of the same length; the hair stands up straight at a right angle to the skin. This combination of properties gives the fur a very soft, plush characteristic. Pelts from Rex rabbits sell for up to $10 above ordinary fur pelts.

Rabbit wool comes from Angora breeds. Garments made from Angora are softer, finer, warmer, and much lighter in weight than those made from sheep's wool. Rabbit wool does not cause the irritation and is not as scratchy as sheep's wool either.

Table 9–2 Cholesterol Content of Foods	
Food	Cholesterol Content (milligrams per 100 grams)
Beef round (medium fat)	125
Beef round (lean)	95
Veal shank	140
Veal breast	100
Pork spare ribs	105
Rabbit	50
Chicken (light)	90
Chicken (dark)	60
Tuna	149–172
Salmon	60

Information taken from McLester, J. S., M.D., and W. J. Darby, M.D., Ph.D. 1953. *Nutrition and Diet in Health and Disease*, 6th Edition, Philadelphia: W.B. Saunders Company.

The Angora rabbit wool market in the United States is primarily with hand spinners. Angora rabbits are not bred in as great numbers as meat rabbits, and more time is necessary for their care.

The value of Angora wool depends on its thickness, quality, and length. The English Angora rabbits produce the finest-quality wool; these rabbits produce 8 to 17 ounces of wool per year and command the highest prices. French Angoras produce about the same amount of wool per year, but their wool is coarser and not as highly valued. The Giant Angora and German Angoras produce 18 to 36 ounces of wool per year, but their fur, too, is coarser and does not command high prices.

Satin fur is a mutation that first appeared in the 1930s. Satin fur has a smaller diameter and a transparent outer shell, which gives a more intense appearance to the color. The transparent shell also gives the fur its sheen, luster, and slick appearance.

Pets

Rabbits make very clean, gentle, and lovable pets that require little special care. They can easily be trained to use a litter box. Because of the diversity in sizes, they can fit into many different home situations. The small and dwarf size have been the most popular.

BREEDS

There are about seventy different breeds of domestic rabbits; however, only about forty-five breeds have been developed to the point that they are recognized by various rabbit organizations. These forty-five breeds are divided into five weight categories. These categories are listed below, along with the breeds.

Dwarf or Miniature	*Small*
Britannia Petite	Dutch
Netherland Dwarf	Tan
Himalayan	Florida White
Dwarf Hotot	Silver
Polish	Havana
Jersey Wooly	Mini Lop
Holland Lop	
American Fuzzy Lop	
Mini Rex	

Medium	*Large*
English Spots	Beveren
Standard Chinchilla	Californian
English Angora	Hotot
Lilac	Palomino
Silver Martin	Satin
Belgian Hare	Cinnamon
Rhinelander	Creme d'Argent
Harlequin	Champagne d'Argent
Sable	American
Satin Angora	American Chinchilla
French Angora	English Lops
Rex	New Zealand
	Silver Fox

Giant
Giant Angora
French Lops
Checkered Giant
Giant Chinchilla
Flemish Giant

The Britannia Petite. The Britannia Petite is an old English breed known in Britain as "Polish"; it is generally thought to have been developed from small common rabbits. The breed is currently recognized as the smallest of the standard breeds. This breed weighs less than 2½ pounds. The Britannia Petite is commonly recognized in the United States as being white with red eyes, although they are bred in many color varieties in Britain. See Figure 9–1.

Figure 9–1   Britannia Petite.

The Netherland Dwarf. The Netherland Dwarf was originally developed in Holland during the early part of the twentieth century from the Dutch rabbit and a wild rabbit. After many years of selective breeding, it was exported to England during the 1950s. British breeders improved the rabbit's stamina and standardized the colors. The modern Netherland Dwarf, as it was developed in England, first appeared in the United States in the mid-1960s. It has the distinction of being recognized in more different colors than any other breed. The breed is small, stocky, and compact. The head is round with a broad skull and a nose that is distinctly rounded. The eyes are round, bold, and bright; the ears are short and erect. Netherland Dwarfs usually weigh about 2½ pounds. See Figure 9–2.

Figure 9–2   Netherland Dwarf. Photo by Isabelle Francais.

The Himalayan. The Himalayan is one of the oldest established breeds, with a wider distribution than any other breed. It is believed to have originated in the Himalayan Mountain area. Some breeders believe it is an albino from the old Russian Silver rabbit. The young rabbits are white and slightly tinged with silver-gray. As the rabbit matures, the silver-gray disappears and its coat becomes snow white with the nose, ears, feet, and tail becoming a deep black. The Himalayan has ears that stand erect, a nose that is short and rounded, and red eyes. At maturity, Himalayans weigh 2½ to 4½ pounds. See Figure 9–3.

Figure 9–3   Himalayan. Photo by Isabelle Francais.

The Dwarf Hotot.   The Dwarf Hotot was developed in the 1970s by two different breeders in two different parts of Germany using two different approaches. A West German breeder tried to develop a Hotot marked Netherland Dwarf by crossing a Red-Eyed White and a Black Netherland Dwarf. An East German breeder crossed a Red-Eyed White Netherland Dwarf with a Blanc de Hotot. The two approaches merged as breeders exchanged stock, improving their size and vigor. Elizabeth Forstinger of Vista, California, imported the first Dwarf Hotots into the United States in 1980. See Figure 9–4.

The body of the Dwarf Hotot is short, compact, and well rounded. The entire body is white; the only markings being a very thin black eye band around each eye. This breed weighs a maximum of 3¼ pounds.

Figure 9–4   Dwarf Hotot. Photo by Isabelle Francais.

The Polish.   The exact history of the Polish breed is not known; it is believed to have been developed as an albino Dutch rabbit. The breed has a short, rounded body with short limbs and fine bones; the eyes are large. The Polish breed is commonly recognized in five colors: Black with brown eyes, Blue with blue-gray eyes, Chocolate with brown eyes, White with ruby-red eyes, and White with blue eyes. See Figure 9–5.

Figure 9–5   Polish. Photo by Isabelle Francais.

The Jersey Wooly.   The Jersey Wooly was developed in New Jersey in the late 1970s by Bonnie Seely in an effort to produce small animals to sell as pets. It is a cross between the Netherland Dwarf and Angoras. Jersey Woolies weigh less than 3½ pounds, and the ears are 2 to 3 inches long. The breed is shown in five color groups: Selfs, Shaded, Agouti, Tan, and Pointed. See Figure 9–6.

Figure 9–6   Jersey Wooly. Photo by Isabelle Francais.

The Holland Lops. Holland Lops were developed by Adrian DeCock of the Netherlands. They were developed by breeding Netherland Dwarfs to French Lops, then breeding their offspring to English Lops in order to enhance the ear carriage. They first appeared in the United States in 1976. The apple-shaped head of the Holland Lop is short and broad. The eyes are set wide, the muzzle is broad, and it has a short, rounded nose. The body of the Holland Lop resembles the French Lop but is smaller in scale. The breed presents a massive, thickset appearance with broad, deep hips. The Holland Lop weighs less than 4 pounds. It is recognized in solid or broken varieties. See Figure 9–7.

Figure 9–7   Holland Lop. Photo by Isabelle Francais.

The American Fuzzy Lops. The American Fuzzy Lops were developed during the 1980s. In an effort to get the broken pattern into the Holland Lop gene pool yet maintain the **roll-back fur** or **fly-back fur** (fur that gently returns to the original position) of the Holland Lop, breeders crossed the French Angora with Holland Lops. Although they produced the desired broken pattern with gentle, fly-back fur, they also produced occasional wooled rabbits. These wooled rabbits became the American Fuzzy Lops. They are a small lop-eared breed weighing less than 4 pounds. See Figure 9–8.

Figure 9–8   American Fuzzy Lop. Photo by Isabelle Francais.

The Mini Rex. The Mini Rex was developed by crossing standard Rex rabbits with the Netherland Dwarf in an effort to get the smaller animals. The Mini Rex weighs less than 4½ pounds. It has the same dense, plush fur as the Standard Rex and is shown in eleven varieties: Blue, Californian, Castor, Chinchilla, Lynx, Opal, Red, Seal, Tortoise, White, and Broken. See Figure 9–9.

Figure 9–9   Mini Rex. Photo by Isabelle Francais.

The Dutch. The modern Dutch rabbit was developed in England, although it originated in Holland; several rabbits were imported into England in 1864. The English refined their size, shape, and pattern. They were called "Dutch" rabbits and became very popular in England as well as in the United States. See Figure 9–10.

The breed has unique color markings. There are colored patches on each side of the head that encircle the eyes and ears. This leaves a white blaze on the forehead and down over the front part of the rabbit, including the front legs. The rear half of the animal is colored the same as the head patches. White also appears halfway up on the rear paws. The Dutch rabbit weighs 3½ to 5½ pounds. The animals are recognized in six color varieties: Black, Blue, Chocolate, Tortoise, Steel Gray, and Gray.

Figure 9–10   Dutch. Photo by Isabelle Francais.

The Tan. The Tan breed of rabbit was originally known as the Black and Tan breed. They were originally black with the underside tan. Today, they are also recognized in Blue, Chocolate, and Lilac with tan undersides. The tan color also appears in a narrow ring around the eyes and nostrils; the toes are also tan. These Tan rabbits originated in England and are a cross between a Dutch rabbit and a wild rabbit. The bucks of this breed weigh 4 to 5½ pounds, and the does weigh 4 to 6 pounds. See Figure 9–11.

Figure 9–11   Tan. Photo by Isabelle Francais.

The Florida White. The Florida White was developed in Florida. Dutch, Polish, and New Zealand White rabbits were crossed in an effort to develop a small meat rabbit and a small laboratory rabbit. The breed is short and compact with well-rounded hips and hindquarters; the head is small and the ears are short. See Figure 9–12.

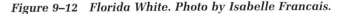

Figure 9–12   Florida White. Photo by Isabelle Francais.

The Modern Silver. The modern Silver breeds were developed in England, France, and Germany. They are believed to be descendants of silver-gray rabbits that originated in India. Portuguese sailors are credited with bringing the rabbits to Europe in the 1600s. Silvers are recognized in three varieties: Brown, Fawn, and Gray. The Brown variety was perfected in England with the introduction of Belgian Hare blood. The original Fawns came from France. The Gray-Silvers were also developed in England and were probably the first to be developed. The fur of the Silver has large numbers of pigment-free, silvery-white tips, giving the animal the silver color. This small breed of rabbit weighs from 4 to 7 pounds. See Figure 9–13.

Figure 9–13   Modern Silver. Photo by Isabelle Francais.

The Havana. The Havana rabbit is the result of a mutation that occurred in 1898. A chocolate-colored rabbit appeared in a litter from a common doe with markings resembling a black and white Dutch; the sire of the litter was unknown. The fur of the Havana is a rich, dark chocolate with intense luster and sheen. The eyes are dark brown but, under certain light conditions, show a ruby-red glow. Three color varieties are recognized: Black, Blue, and Chocolate. The Havana is a small, compact, meaty rabbit weighing between 4½ and 6½ pounds. See Figure 9–14.

Figure 9–14   Havana. Photo by Isabelle Francais.

The Mini Lop. The Mini Lop originated in Germany, where it is known as the Klein Widder. It was bred in two colors: Agouti and White. The breed is a cross between the German Big Lop and the Little Chinchilla rabbit. See Figure 9–15.

In 1972, Robert Herschbach sent back to the United States one pair of Agouti and one White doe, and he is given credit for developing the breed in the United States. The broken color variety was bred in the United States by mating a standard Chinchilla to a Broken French Lop. The Mini Lop is a scaled-down version of the French Lop. Mini Lops have a massive, thick-set body. They are recognized in Solid or Broken varieties and weigh 4½ to 6½ pounds.

Figure 9–15   Mini Lop. Photo by Isabelle Francais.

The English Spot. The English Spot has been bred extensively since 1880. It is a cross of the Flemish Giant, English Lop, Angora, Dutch, Silver, and Himalayan. They were introduced into the United States from England. The background color is white and is accepted in seven color varieties: Black, Blue, Chocolate, Gold, Gray, Lilac, and Tortoise. A unique chain of colored markings, about the size of peas, runs from the base of the ear to the rear flank, with an ideal chain consisting of two symmetrically arranged rows. The breed has a butterfly nose pattern, eye circles, cheek spots, and colored ears; other spots appear on the body. They weigh 5 to 8 pounds. See Figure 9–16.

Figure 9–16   English Spot. Photo by Isabelle Francais.

The Standard Chinchilla. The Standard Chinchilla was developed in 1913 by a French engineer, M. J. Dybowski. To get the original Chinchilla, he crossed a wild gray rabbit (the Garenne) with a Himalayan. He also crossed a wild gray rabbit with a blue. The two litters produced by these matings were interbred to produce the modern Chinchilla rabbit. The breed was introduced to England in 1917 and to the United States in 1919. Their popularity is probably due to the similarity of their fur to that of the Chinchilla of South America, genus *Chinchilla.* The undercolor is dark slate-blue at the base, the intermediate portion is pearl, and the top portion has a very narrow black band with a very narrow light band above that. This underfur is brightly ticked with black hairs in either a wavy or even pattern. The breed is of medium length, with a rather compact and chubby body. The bucks weigh 5½ to 7½ pounds, and does weigh 6 to 8 pounds. See Figure 9–17.

Figure 9–17   Standard Chinchilla. Photo by Isabelle Francais.

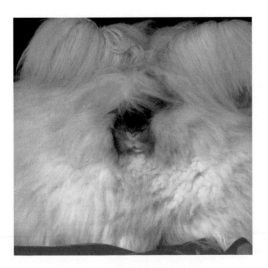

The Angora. The Angora rabbit originated in Ankara, Turkey, hundreds of years ago. There are four different breeds today: English, Satin, French, and Giant. The English Angora was developed from the French Angora and was primarily an exhibition animal; it is somewhat smaller than the French Angora.

The English Angora. The English Angora has a short, compact body and the head, ears, feet, and body are covered with wool. English Angoras weigh 5 to 7½ pounds. See Figure 9–18.

Figure 9–18   English Angora. Photo by Isabelle Francais.

The Satin Angora. The Satin Angora was created by L. P. Meyer of Canada. It is a cross between a copper Satin and a fawn French Angora. The wool of this breed has good length and density and a brilliant satin sheen. See Figure 9–19.

The body of the Satin Angora is of medium length, is firm, and resembles the French Angora. The wool is finer than normal Angora wool and has a soft, silky texture. Satin Angoras weigh about 8 pounds. They are recognized in six color groups: Agouti, Pointed White, Self, Shaded, Solid, and Ticked.

Figure 9–19   Satin Angora. Photo by Isabelle Francais.

The French Angora. The French Angora originated in the French departments of Vosges, Savoy, Magence, Normandy, and Jura. The French Angora is longer and narrower than the English and not covered as heavily by wool. The tail is covered by wool, but the feet and legs are furred only to the first joint, and the ears are not as heavily covered. The wool of the French Angora is coarse, and the ideal length is about 2¾ inches. Animals of this breed weigh more than 7½ pounds. See Figure 9–20.

Figure 9–20   French Angora. Photo by Isabelle Francais.

The Lilac. The Lilac is a diluted Havana Brown that sometimes appears in litters of Havana rabbits. Lilacs of various shades can also be obtained by crossing the Havana with blue rabbits and then interbreeding the young. The ideal color is a dove-gray with a slight pink sheen, which is generally referred to as pinky-dove. The body of the Lilac is the same as the Havana and weighs 6 to 7 pounds. See Figure 9–21.

Figure 9–21   Lilac. Photo by Isabelle Francais.

The Silver Martin. The first Silver Martin appeared about fifty years ago in the nest box of a Standard Chinchilla. As the rabbit matured, black became the predominant color. Others began to appear in litters of Chinchilla rabbits. When these were bred together, they bred true to color. As breeding continued, the eye circles of silver gradually became more distinct; the silver lining of the ears became more noticeable, and the black coat took on a new sheen. Today, the breed is recognized in the Black, Blue, Chocolate, and Sable varieties. The fur is about 1 inch in length and is interspersed with silver-tipped guard hairs along the sides, rumps, and hips. Bucks weigh 6 to 8½ pounds, and does weigh 7 to 9½ pounds. See Figure 9–22.

Figure 9–22   Silver Martin. Photo by Isabelle Francais.

The Belgian Hare. The Belgian Hare originated in Flanders in Belgium. It was introduced into England in 1874 and arrived in the United States in 1888. The English are credited with the selective breeding that produced the modern Belgian Hare. Contrary to its name, the Belgian Hare is a true rabbit; it has the normal rabbit gestation period of thirty to thirty-two days, and the young are born without fur and with eyes closed. The name is derived from the fact that the breed originated in Belgium and has always been bred by devoted fanciers to resemble the wild hare. See Figure 9–23.

The breed is known as the race horse or greyhound of the rabbit family. It has a long, fine body and a high arch to its back. The legs are long, straight, and slender. The head is rather long and set on a slender neck. The color varies from a fox to a mahogany red, with chestnut shading on the sides. Brilliant black ticking appears on the body and along the edges of the ears. The fur is close lying and of rather harsh texture. The Belgian Hare weighs 6 to 9 pounds.

Figure 9–23   Belgian Hare. Photo by Isabelle Francais.

The Rhinelander. The Rhinelander was developed in Germany. The ancestry is not known but is thought to be a cross of the English Butterfly, Japanese, and Flemish Giant. The breed has a butterfly marking on the nose, cheek spots, and eye rings. The rest of the body is white with a herringbone pattern or unbroken stripe running from the neck down the back to the tip of the tail; there are also color markings on the hip and flank. All colors are acceptable, and there is a Tricolored variety in white, black, and blue. The body type is rather unique in that it is rounded and almost cylindrical, avoiding any heaviness in either the shoulders or hindquarters. Bucks weigh 6½ to 9½ pounds, and does weigh 7 to 10 pounds. See Figure 9–24.

Figure 9–24   Rhinelander. Photo by Isabelle Francais.

The Harlequin.   The Harlequin rabbit originated in Northern France and Belgium in the 1880s. It is believed it was a cross mutation between Dutch and the common domestic rabbit. See Figure 9–25.

The markings of the Harlequin are its distinctive feature; each side of the "perfect" Harlequin has a different pattern. Each ear is a different color, and each side of the head has the opposite color. The colors should be distinctly separated with no mixing of the colors; no two Harlequins are marked the same. Harlequins are in two basic body colors: orange or white with black, chocolate, lilac, or blue markings. The orange color patterns are referred to as "Japanese," and the white color patterns are referred to as "Magpie." The eyes of both color patterns are dark brown. Harlequin bucks weigh 6½ to 9 pounds, and does weigh 7 to 9½ pounds.

Figure 9–25   Harlequin. Photo by Isabelle Francais.

The Sable.   The Sable rabbit originated in 1919, when rabbits with a deep sepia brown color and delicate shading began to appear in litters of English Chinchilla rabbits. Breeders decided to save them and developed the Sable breed. The Sable rabbit has a medium-length body with a slightly arched back. The head, ears, and feet have a purple or a shaded brown sheen. Bucks weigh 7 pounds or better, and does weigh 8 pounds or better. See Figure 9–26.

Figure 9–26   Sable.

The Shorthaired Rex.   The shorthaired Rex rabbit began to receive attention in 1919. The mutation had appeared frequently in litters of common gray rabbits over the years, but no importance was placed on them. In 1919, a farmer named Desire Caillon of Pringe Louche in the Sarthe department of France bred a couple of them and found they produced shorthaired young. A. M. Amedee Gillet, of Coulange, France, foresaw the possibility of introducing a new class of rabbit. See Figure 9–27.

The main characteristic of the Rex rabbit is the short hair coat, with guard hairs being of the same length as the underfur. The fur also stands at right angles to the skin, is about ⅜ inch long, and has a very soft, plush feel. Rex rabbits are recognized in fourteen different colors: Black, Blue, Californian, Castor, Chinchilla, Chocolate, Lilac, Lynx, Opal, Red, Sable, Seal, White, and Broken. The bucks weigh 7½ pounds, and does weigh 8 to 10½ pounds.

Figure 9–27   Shorthaired Rex. Photo by Isabelle Francais.

The Beveren. The Beveren rabbit originated late in the nineteenth century in the city of Beveren, Belgium. The Blue Beveren was imported into England in 1915 and became very popular. Although blue rabbits have been bred for centuries, none became as popular as the Blue Beveren. Blue-eyed White and Black Beveren were developed later. The White is the most popular in the United States, with its blue eyes distinguishing it from all other white breeds. The body is of medium length with a broad, meaty, slightly arched back. The head is full with an outward curvature between the eyes and nose that is characteristic of the breed. The bucks weigh 8 to 10 pounds, and does weigh 9 to 11 pounds. The Beveren can be tempermental at times and may not make the best family pet. See Figure 9–28.

Figure 9–28   Beveren. Photo by Isabelle Francais.

The Californian. The Californian was produced in 1928 by George West in an effort to produce a better commercial rabbit. He crossed the Himalayan and Chinchilla and produced a chin color buck that was crossed with several White New Zealand does. Since its development, it has become one of the most popular breeds. The rabbit is white with black-colored nose, ears, feet, and tail; the eyes are red. The bucks weigh 8 to 10 pounds, and does weigh 8½ to 10½ pounds. See Figure 9–29.

Figure 9–29   Californian. Photo by Isabelle Francais.

The Hotot. The Hotot originated in France, where it is known as the Blanc de Hotot (White of Hotot); the Baroness E. Bernhard originated the breed around 1912. It is believed that native French-spotted rabbits and the Giant Papillon Français (Checkered Giant) were used in developing the breed. It was imported into the United States in 1978. The color is a shiny white except for a thin black band surrounding the eyes. The bucks weigh 8 to 10 pounds, and does weigh 9 to 11 pounds. See Figure 9–30.

Figure 9–30   Hotot. Photo by Isabelle Francais.

Figure 9–31   Palomino. Photo by Isabelle Francais.

The Palomino.   The Palomino rabbit was developed by Mark Youngs of Washington. Several breeds were crossed to produce a breed with distinctive colors. Two colors are recognized: the Golden, which has light gold guard hairs and a white to creamy undercoat, and the Lynx, which has orange-colored guard hairs and a white undercoat. Bucks of this breed weigh 8 to 10 pounds, and does weigh 9 to 11 pounds. See Figure 9–31.

Figure 9–32   Satin. Photo by Isabelle Francais.

The Satin.   The Satin rabbit appeared as a mutation in a litter of Havana in 1930. Walter Huey, a Havana raiser in Indiana, noticed baby rabbits with especially shiny coats. Mr. Huey was interbreeding Chocolate Havanas in an effort to perfect the deep brown color and outstanding fur. The Satin mutation occurred sometime during the selection and eventually expressed itself. The Satin factor was later recognized as an entirely new coat mutation and was bred into a number of breeds with the result that there now is a "Satin" counterpart to many of the solid and normal-furred breeds. All of the Satins have been combined into a new breed having a uniform description of type and fur. The Satin breed is recognized in ten varieties: Black, Blue, Californian, Chinchilla, Chocolate, Copper, Red, Siamese, White, and Broken. Satins weigh 8½ to 10½ pounds, and does weigh 9 to 11 pounds. See Figure 9–32.

Figure 9–33   Cinnamon. Photo by Isabelle Francais.

The Cinnamon.   The Cinnamon rabbit was developed by Ellis Houseman of Montana. New Zealand White, Chinchilla, Checkered Giant, and Californian breeds were used to develop this breed. The Cinnamon is a medium-length rabbit with a well-filled, meaty body. The color is a rust or cinnamon ground color; the fur is ticked with smoke-gray across the back. The sides of the rabbit show a smoke-gray shading, and the underfur is orange; the fur around the head and ears shows a darker shading. Bucks of this breed weigh 8½ to 10½ pounds, and does weigh 9 to 11 pounds. See Figure 9–33.

The Creme d'Argent and the Champagne d'Argent. The Creme d'Argent and the Champagne d'Argent originated in France. The fur of the Creme d'Argent is a creme color with a bright orange under-fur; the guard hairs are also orange. The Champagne d'Argent has a body color of bluish-white. The underfur is a dark, slate blue, and the guard hairs are jet black. Both breeds made their appearance in the United States around 1934 and are becoming very popular because of their luxurious and richly colored fur. Creme d'Argent bucks weigh 8 to 10 pounds, and does weigh 8½ to 11 pounds. Champagne d'Argent bucks weigh 9 to 11 pounds, and does weigh 9½ to 12 pounds. See Figures 9–34 and 9–35.

Figure 9–34   Creme d'Argent. Photo by Isabelle Francais.

Figure 9–35   Champagne d'Argent. Photo by Isabelle Francais.

The American. The American rabbit breed was developed in the United States. It is recognized in Blue and White color varieties. The background of the American White variety is a mystery but is believed to have developed from throwbacks, called "sports" of blue rabbits. The ideal body type is described as "mandolin" because the arch of the back from the hindquarters to the shoulders has a man-dolin appearance to it. American rabbits were first shown in 1917 and became very popular during the 1920s and 1930s. Bucks weigh 9 to 11 pounds, and does weigh 10 to 12 pounds. See Figure 9–36.

Figure 9–36   American. Photo by Isabelle Francais.

The American Chinchilla. The American Chinchilla is a result of selective breeding for size from the Standard Chinchilla. It is larger than the Standard and has been referred to as the "Heavyweight" Chinchilla. The breed has the distinctive Chinchilla color pattern. Bucks weigh 9 to 11 pounds, and does weigh 10 to 12 pounds. See Figure 9–37.

Figure 9–37   American Chinchilla.

The English Lop. The English Lop is perhaps the oldest breed of domestic rabbit known to humans. Its exact origin is not known, but it first appeared in Algiers, North Africa, and from there it spread to England. The breed became very popular in England, and breeders competed to increase the ear length. Ears measuring 23 to 24 inches were common, and the first Lop to have an ear length of over 24 inches appeared in 1885; the record is 28½ inches. They are recognized in self and broken varieties. The body is of medium length with a well-arched back and the head carried very low. Bucks weigh 9 pounds and over, and does weigh 10 pounds and over. See Figure 9–38.

Figure 9–38   English Lop. Photo by Isabelle Francais.

The New Zealand. The New Zealand breed originated in California and Indiana almost simultaneously. The Red variety was the first to appear and is a cross between a Belgian Hare and a White rabbit. The White variety appeared later and was the result of crossing a number of breeds, including the Flemish, American White, Angora, and perhaps the Red variety. The Black variety came much later, and it too resulted from various crosses. The New Zealand, especially the White, has been the leading choice for the commercial rabbit producer in the United States. The White is preferred because when it is dressed, the small hairs that remain cannot be seen; thus, it is easier to clean and package. This breed is primarily raised for meat production. Mature bucks weigh 9 to 11 pounds, and does weigh 10 to 12 pounds. See Figure 9–39.

Figure 9–39   New Zealand. Photo by Isabelle Francais.

The Silver Fox. The Silver Fox rabbit was originated and developed during the 1920s by W. B. Garland of North Canton, Ohio. The various crosses used in the development of this breed are not known, but it is believed to have resulted from crosses of Checkered Giants and English Silvers. The breed was originally known as the American Silver Giant. Because the fur closely resembles the real Silver Fox, the name "American Silver Fox" was given to the breed in December 1929. The unique features of this breed are its evenly silvered fur resembling fox fur, fur length (1 to 1½ inches long), and the lack of fly-back fur (the fur returning to its original position) when stroked. Bucks weigh 9 to 11 pounds, and does weigh 10 to 12 pounds. See Figure 9–40.

Figure 9–40   Silver Fox.

The Giant Angora. The Giant Angora was originated by Louise Walsh of Taunton, Maine. The main purpose of developing the breed was to produce an animal that would have a maximum amount of wool production. See Figure 9–41.

The Giant Angora has the body structure of a commercial animal and possesses a unique coat structure. The wool comprises three distinct types: the underwool, which is medium-fine, soft, and delicately waved; the awn fluff, a soft, wavy wool with a guard hair tip found between the underwool; and the awn hair, which is strong, straight hair (guard hair) protruding above the awn fluff.

Giant Angora bucks usually weigh more than 8½ pounds, and the does weigh more than 9 pounds. There are two varieties: the Ruby-Eyed White and the Blue-Eyed White.

Figure 9–41   Giant Angora. Photo by Isabelle Francais.

The French Lop. The French Lop rabbit was first bred in France around 1850 by a Frenchman named Conderier. It is believed to be a cross of the English Lop and the Butterfly rabbit of France. The Butterfly rabbit is still bred in France and closely resembles the Flemish Giant. The ears of the French Lop are much shorter than those of the English Lop. The ears hang down in a horseshoe shape from the crown approximately 1½ inches or more below the muzzle. The body has a very massive, thick-set appearance. The breed is broad, deep, and heavily muscled. It is shown in any solid color and in any color in combination with white. The bucks weigh more than 9 pounds, and does weigh more than 10 pounds. See Figure 9–42.

Figure 9–42   French Lop. Photo by Isabelle Francais.

The Checkered Giant. The Checkered Giant originated in Germany in 1904. Otto Reinhardt of Rheinpfalz, Germany, took a Rhenish Checker (something like an English Spot) and crossed it with a Black or Gray Flemish Giant. This mating produced the foundation stock for modern-day Checkered Giants. See Figure 9–43.

In 1910, the first Checkered Giants came to America. Other shipments followed, and American breeders have developed a breed that is more racy than the larger-boned, heavier-shouldered European type. The American Checkered Giant has a long, well-arched body that stands well off the ground. The body color is white with either black or blue color markings. The colored butterfly nose, eye circles, cheek spots, spine markings, tail, and side markings are well formed and distinct. The bucks weigh 11 pounds and over, and does weigh 12 pounds and over.

Figure 9–43   Checkered Giant. Photo by Isabelle Francais.

The Giant Chinchilla. The Giant Chinchilla is a result of selective cross-breeding and the addition of Flemish Giant blood; the breed was developed by Ed Stahl of Missouri. This larger variety of Chinchilla rabbit still maintains the distinctive Chinchilla color pattern. Bucks weigh 12 to 15 pounds, and does weigh 13 to 16 pounds. See Figure 9–44.

Figure 9–44   Giant Chinchilla. Photo by Isabelle Francais.

The Flemish Giant. The early development of the Flemish Giant is not known, but it is known that the breed originated in Belgium. The Dutch probably deserve credit for domesticating and improving the breed, but the Germans deserve much of the credit for perfecting the breed and increasing its length. See Figure 9–45.

The breed was imported into the United States about sixty years ago, and breeders in the United States are given credit for producing the different colors. Flemish Giants are now recognized in seven color varieties: Steel Gray, Light Gray, Sandy, Black, Blue, White, and Fawn. This is the largest of all domestic rabbit breeds. The bucks weigh more than 12 pounds, and the does weigh over 13 pounds; 15- to 18-pound Flemish Giants are common.

Figure 9–45   Flemish Giant. Photo by Isabelle Francais.

ANATOMY

The anatomy and skeletal structure of the rabbit are identified in Figures 9–46 and 9–47.

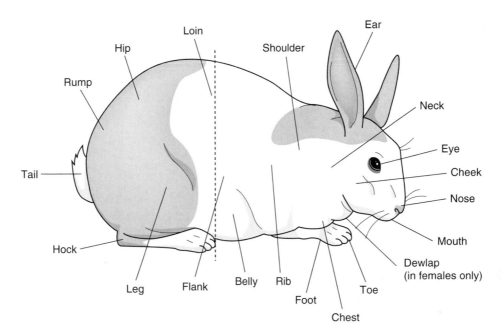

Figure 9–46 **The anatomy of the rabbit.**

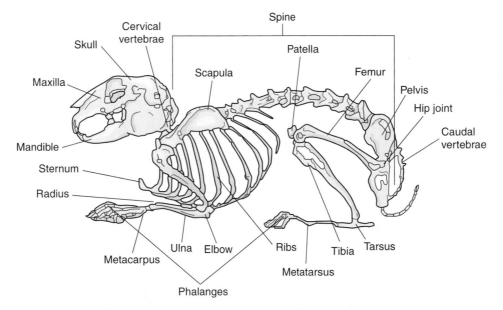

Figure 9–47 **The skeletal structure of the rabbit.**

HOUSING AND EQUIPMENT

Before purchasing a rabbit, some thought should go into the housing for the animal. Houses used to keep rabbits are referred to as "hutches." The size of the cage or hutch depends on the breed that will be housed in it. The miniature and small rabbits can be kept in wire cages that are 24 inches wide, 24 inches long, and 14 inches high; this size is common and can be purchased at most pet supply stores.

Cages for rabbits that weigh from 6 to 11 pounds should be 24 inches wide, 36 inches long, and 18 inches high. This size allows them plenty of space to move about and get exercise.

Rabbits that are 12 pounds and over need cages 24 inches wide, 48 inches long, and 18 inches high. Cages 30 inches wide, 48 inches long, and 24 inches high minimum should be provided for the Belgian Hare; cages 36 inches wide, 72 inches long, and 36 inches high are preferred for this active breed of rabbit.

Wire cages are recommended for most rabbits; however, the larger breeds over 12 pounds should be kept on solid wood floors to prevent sore hocks. Sore hocks are ulcerations to the foot pads caused by the animal's body weight pressing down on the foot pads, which becomes irritated on wire cages. See Figure 9–48.

Cages with wire floors have the distinct advantage of being easy to clean because the wire allows feces and urine to fall into trays below; these trays are easily emptied and cleaned. Wire cages are also easy to steam clean and disinfect.

The trays below the cages should be filled with wood shavings, kitty litter, or another type of absorbent material. Baking soda sprinkled in the corners helps cut down on the strong urine smell.

Wire used for the flooring of cages should have openings about ½ inch by 1 inch. The ½ inch side of the wire should be toward the animals' feet; this will cause less irritation.

Hutches with wood floors should be bedded with wood shavings, straw, or another type of absorbent material. It is important that this gets changed at least once a week. Rabbits bedded on wet bedding are subject to sore hocks, hutch burn, and other ailments.

Sanitation is of utmost importance, especially if large numbers of animals are involved.

The miniature and small rabbits are very clean and almost odorless. They can be kept in the house and need their cages cleaned only once or twice a week. Larger rabbits kept indoors can create an odor problem, and their cages may need to be cleaned daily.

Several conditions are important in housing rabbits: temperature, humidity, ventilation, proper lighting, and the absence of drafts. Rabbits can be raised outside year-round. They can stand extremely cold temperatures as long as they are out of the wind; however, they have trouble surviving extremely warm temperatures. When the humidity reaches 60 percent, ventilation must be provided. If temperatures reach 80°F, rabbits will be very uncomfortable. High temperature and high humidity can lead to heat prostration. Temperatures above 80°F over an extended time can affect the semen quality of a male rabbit. High temperatures with low humidity can be tolerated if the air is moving.

Rabbits keep cool by the breathing process and by the air movement about them, especially past their ears. Fans can be installed, and a bottle of frozen water can be put in the cage for the rabbit to lie against.

Ventilation is also important. Fresh air is a must if rabbits are to be kept healthy and productive. Stagnant and stale air must not be allowed to build up because this will lead to respiratory problems and diseases.

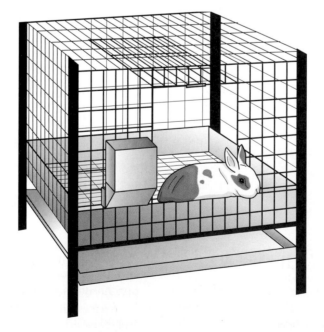

Figure 9–48 Cages made from wire are very nice for rabbits. Feces and urine fall through the wire and are easily cleaned from the slanted surfaces below the cages.

Figure 9–49   Galvanized-metal self-feeders are easily filled from outside the cage. Courtesy of FFA.

Lighting is important in priming a rabbit's fur and is important in the receptiveness of a breeding doe. A doe must have a minimum of 16 hours of daylight for proper breeding.

Cool, damp drafts can cause animals to become chilled. Again, this condition can lead to colds, pneumonia, and other ailments.

Galvanized-metal self-feeders are the easiest to use and prevent a lot of waste. Metal feeders usually clip onto the outside of the cage and fit through an opening into the cage. They can be filled quickly and easily from the outside. See Figure 9–49.

Other types of feeders can be used, such as metal tin cans, ceramic bowls, and plastic bowls; however, most of these take up floor space and can be tipped over.

Single rabbits can be watered with ceramic water bowls or with vacuum-type water bottles that hang on the outside of the cage. The disadvantage of the ceramic bowl is that it can be tipped over or become contaminated. Water bottles solve a lot of the problems associated with open bowl watering. Waterers can be constructed from plastic milk jugs and plastic soft drink bottles by using one of several water valves, such as the **dew-drop valve,** that are available on the market. All of these types of water systems are subject

to freezing during cold weather. Several types of heating wires and heating tapes are available for outside water systems to keep them from freezing.

Nest boxes should be provided for the does to use when they are giving birth (kindling). They can be purchased from pet supply stores, or they can be constructed from metal, wire, or wood. The size of the nest box will vary depending on the size of the doe, but it should be about 12 inches long, 7 inches wide, and 6 inches high for the small breeds; 18 inches long, 10 inches wide, and 9 inches high for the medium breeds; and 22 to 24 inches long, 14 inches wide, and 12 inches high for the giant breeds. Nest boxes need to be clean and sanitized before use, or the doe may refuse to use the box.

FEEDING

The easiest way to feed rabbits is to use one of the commercially prepared pellet-type feeds. See Figures 9–50 and 9–51. These feeds are especially formulated for rabbits and will supply all of the necessary nutrients needed. Several companies have made studies in preparing pelleted feed for various stages of rabbit growth. As Table 9–3 shows, protein and energy

BUNNETS*–17%
with Aqua Vite*
* TRADE MARKS
A Complete Cereal, Vitamin and Mineral
Ration in Pellet Form for Rabbits

INGREDIENTS:

Dehydrated and Sun Cured Alfalfa Meal, Pulverized Oats, Soybean Oil Meal, Feeding
Cane Molasses, Wheat Middlings, Wheat Bran, Vitamin A Acetate, D-Activated Animal
Sterol, delphine Trophopheryl Acetate, Choline Chloride, Folic Acid, Riboflavin, Niacin,
Vitamin B-12, Pyroxidine, Thiamin Hydrochloride, Ascorbic Acid, Carrot Oil Carotene,
Soybean Oil, Sodium Chloride, DiCalcium Phosphate, Magnesium Oxide, Magnesium
Sulfate, Iron Oxide from Sulfate, Copper Sulfate, Calcium Iodate, Cobalt Carbonate,
Zinc Oxide, BHT (preservative), Ligmin Sulfonate, Sodium Selente, Calcium Carbonate,
Verrets, Yucca, Extract, Sodium Benzonate, 6 Fermentation Product and Flavor.

GUARANTEED ANALYSIS:

Crude Protein, not less than	17.0%
Crude Fat, not less than	2.5%
Crude Fiber, not more than	25.0%

FEEDING DIRECTIONS:

The 17% protein ration is designed for rabbits in production, such
as for rapidly growing rabbits, gestating or lactating does, and working
bucks. It can be fed in either of two ways: free choice or limit feeding.
Free choice feeding, where the feeders are kept full of pellets all the
time, will result in faster gains of weaning rabbits, and is often used
for does with litters. We recommend limit feeding, where only as much
feed as will be cleaned up in a day is provided. This requires closer
observation by the feeder, but problems are often discovered sooner,
feed is kept fresher, and rabbits usually are healthier. Do not restrict
feed to the degree that gains are severely reduced.

Do not overfeed does at kindling. Bring does into full milk production
gradually by increasing feed as the litter grows and needs more milk.

Be sure to provide plenty of fresh clean cool water. Good quality hay
can also be offered.

Figure 9–50   Protein content is important in feeding rabbits. Always check the labels. Photo by Dean Warren.

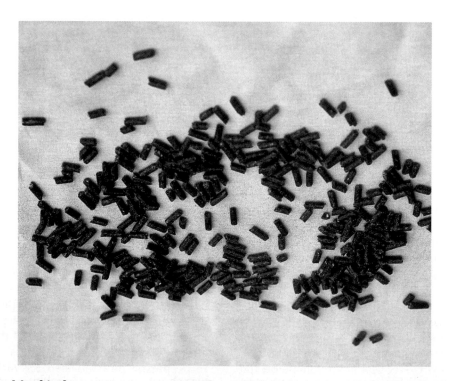

Figure 9–51   Pelleted feed is the easiest way to ensure that rabbits are receiving the proper diet. The main ingredient of these pellets is alfalfa meal. Photo by Dean Warren.

Table 9–3 Nutrient Requirements of Rabbits Fed Ad Libitum (Percentage or Amount per Kilogram of Diet)

Nutrients[a]	Growth	Maintenance	Gestation	Lactation
Energy and protein				
Digestible energy (kcal)	2500	2100	2500	2500
TDN (%)	65	55	58	70
Crude fiber (%)	10–12[b]	14[b]	10–12[b]	10–12[b]
Fat (%)	2[b]	2[b]	2[b]	2[b]
Crude protein (%)	16	12	15	17
Inorganic nutrients				
Calcium (%)	0.4	—[c]	0.45[b]	0.75[b]
Phosphorus	0.22	—[c]	0.37[b]	0.5
Magnesium (mg)	300–400	300–400	300–400	300–400
Potassium (%)	0.6	0.6	0.6	0.6
Sodium (%)	0.2[b, d]	0.2[b, d]	0.2[b, d]	0.2[b, d]
Chlorine (%)	0.3[b, d]	0.3[b, d]	0.3[b, d]	0.3[b, d]
Copper (mg)	3	3	3	3
Iodine	0.2[b]	0.2[b]	0.2[b]	0.2[b]
Iron	—[c]	—[c]	—[c]	—[c]
Manganese (mg)	8.5[c]	2.5[c]	2.5[c]	2.5[c]
Zinc	—[c]	—[c]	—[c]	—[c]
Vitamins				
Vitamin A (iu)	580	—[e]	>1160	—[e]
Vitamin A as carotene (mg)	0.83[b, c]	—[f]	0.83[b, c]	—[f]
Vitamin D	—[g]	—[g]	—[g]	—[i]
Vitamin E (mg)	40[h]	—[e]	40[h]	40[i]
Vitamin K (mg)	—[i]	—[i]	0.02[b]	—[i]
Niacin (mg)	180	—[j]	—[j]	—[j]
Pyridoxine (mg)	39	—[j]	—[j]	—[j]
Choline (g)	1.2[b]	—[j]	—[j]	—[j]
Amino acids (%)				
Lysine	0.65	—[g]	—[g]	—[g]
Methionine + cystine	0.6	—[g]	—[g]	—[g]
Arginine	0.6	—[g]	—[g]	—[g]
Histidine	0.3[b]	—[g]	—[g]	—[g]
Leucine	1.1[b]	—[g]	—[g]	—[g]
Isoleucine	0.6[b]	—[g]	—[g]	—[g]
Phenylalanine + tyrosine	1.1[b]	—[g]	—[g]	—[g]
Threonine	0.6[b]	—[g]	—[g]	—[g]
Tryptophan	0.2[b]	—[g]	—[g]	—[g]
Valine	0.7[b]	—[g]	—[g]	—[g]
Glycine	—[e]	—[g]	—[g]	—[g]

[a]Nutrients not listed indicate dietary need unknown or not demonstrated.
[b]May not be minimum but known to be adequate.
[c]Quantitative requirement not determined, but dietary need demonstrated.
[d]May be met with 0.5 percent Nacl.
[e]Converted from amount per rabbit per day using an air-dry feed intake of 60 g per day for a 1-kg rabbit.
[f]Quantitative requirement not determined.
[g]Probably required, amount unknown.
[h]Estimated.
[i]Intestinal synthesis probably adequate.
[j]Dietary need unknown.

Source: Reprinted with permission from *Nutrient Requirements of Rabbits,* 2d Edition. Copyright 1977 by the National Academy of Sciences. Courtesy of the National Academy Press, Washington, D.C.

requirements for maintenance are lower than the requirements needed for growth, gestation, and lactation.

Many owners like to supplement the diet with other types of foods. Several foodstuffs can be used as treats, but owners must be cautious because feeding too many fruits, vegetables, and green foods can cause severe gastrointestinal problems that may lead to fatal diarrhea. Any supplemented feeds should be kept to a minimum, especially if the animals are not used to receiving them.

Some types of foodstuffs that can be added to the rabbit's diet are high-quality grass hay, corn, oats, oatmeal, and wheat germ. Other treats may include pieces of carrot, carrot tops, slices of apple, green beans, and pieces of banana, pineapple, or papaya. Any green foods or vegetables should be washed to eliminate any chemical residue.

Green lettuce or other types of leafy green vegetables should not be fed to rabbits, especially young rabbits. Green, leafy vegetables have a high water content and cause diarrhea and dehydration because the rabbit ceases to drink water. Clean, fresh water should always be available for rabbits; this is especially true during warm, summer months.

Coprophagy is the term associated with the eating of fecal material; rabbits consume their own feces. The fecal material of rabbits is of two types: a hard, dry type that is normally seen in the cage or under the wire flooring; and the soft form that the rabbit consumes as it is being expelled. Because this type of feces is normally consumed at night, it is often referred to as "night feces," although the rabbit may consume this during the day as well.

HANDLING

The correct way to handle a rabbit depends on the size of the rabbit; however, under no circumstances should the animal be picked up by the ears because this may cause injury to small blood vessels or cartilage in the ear. Also, picking the animal up by the nape of the neck is not recommended, because this too can cause damage to tiny blood vessels in the skin and could cause damage to the pelt.

When approaching a rabbit cage, one should walk slowly and speak to the rabbit. If the rabbit knows someone is close, it won't get as frightened and bolt to the corner of the cage.

Next, one should reach in and gently stroke the animal from front to rear and rub and stroke the animal's head. Then, the rabbit can be gently moved into a position where it can be picked up by slipping one hand in under the chest and belly and placing the other hand behind the rabbit. The rabbit can now be lifted up and removed from the cage. It is important to remove it from the cage tail first to prevent the rabbit's legs from getting caught in the wire floor or wire door of the cage.

Small rabbits can be supported with one hand under the body, and large rabbits supported on one arm with the other arm cradling the animal securely. If the animal feels secure and comfortable, it won't struggle or try to escape. The head of the rabbit can be tucked in under the upper arm toward the elbow of the arm that is being used to support the rabbit. See Figure 9–52.

Small animals are easy to lift with the method just described; however, larger animals may pose a problem. This is especially true if they are not used to being handled. The handler should always wear long-sleeve clothing to prevent scratches.

When setting the rabbit down, one must do so gently and slowly, letting the animal see where it is going so that it will not be frightened. The handler must make sure the animal is set on a roughened sur-

Figure 9–52   Proper handling is important. Rabbits need to feel secure. Their powerful rear legs and sharp nails can inflict injury to the handler's arms.

face, never on slick table tops or floors. A rabbit's foot pads are covered with fur, and when placed on slick surfaces, they have no traction. Rabbits can easily dislocate a hip or their spine when they try to hop or push off on their rear legs.

The handler should be careful when returning the rabbit to its cage. The rabbit, recognizing a familiar area, may try to jump from the handler's arms and may inflict an injury to the handler with its strong, powerful rear legs and sharp toenails.

DISEASES AND AILMENTS

Control of temperature, humidity, ventilation, lighting, and drafts is extremely important in preventing diseases in rabbits; proper cleaning and disinfecting of equipment are also important.

Enteritis.   Enteritis is an inflammation of the intestinal tract and probably the most common cause of death in rabbits. There are four types of enteritis: mucoid enteritis, enterotoxemia, and Tyzzer's disease, which are caused by bacteria; and coccidiosis, which is caused by an internal parasite.

The exact cause of mucoid enteritis is unknown. The disease expresses itself as mucous discharges from the rectum and causes the animals to appear potbellied or to be filled with water. Diseased animals fail to eat but drink a lot of water. If water is supplied in a bowl, a diseased animal often puts its feet in the water and grinds its teeth. Death usually occurs four days to one week following the onset of symptoms; very little can be done to save affected animals. If there are several animals in the rabbitry, all animals should receive broad-spectrum antibiotics.

Mucoid enteritis may also be associated with stress, unsanitary conditions, and feeds with too high energy levels. Good management is important to reduce stress, and the animals should receive a ration containing less energy and higher levels of fiber.

Enterotoxemia is widely recognized as the leading cause of enteritis-type symptoms in young rabbits. The exact bacteria that cause the disease is not known, but it appears to be associated with the ingestion of high-energy feeds that are not completely digested in the stomach and small intestine. As this undigested material reaches the cecum, high amounts of carbohydrates allow for an increase in the growth of some types of bacteria. As the growth of bacteria increases, there is an increase in toxins produced, which causes dehydration and shock. Severe diarrhea, which is a common symptom of this disease, is caused by the increased amount of fluid taken into the cecum and intestines in an effort to dilute the toxins. Other symptoms include rapid weight loss, rough appearance, and sunken eyes; there is almost 100 percent mortality in affected animals.

To prevent the disease, sanitation is extremely important, as well as reduced stress and the feeding of rations containing at least 18 percent fiber. Broad-spectrum antibiotics may be effective in treating enterotoxemia.

Tyzzer's disease is caused by the bacterium *Bacillus piliformis.* This disease causes severe diarrhea and death among young rabbits six to seven weeks old. Other symptoms include rapid weight loss, dehydration, and death occurring within one to three days. There is no known treatment for Tyzzer's disease. Again, sanitation is especially important, and the control of rats and mice is important in preventing the organism from being carried into a rabbitry.

Coccidiosis is caused by several species of the protozoan parasite *Eimeria.* Hepatic coccidiosis is caused by the protozoan *Eimeria stieda,* and intestinal coccidiosis is caused by several species of *Eimeria: E. magna, E. irresiduca, E. media,* or *E. perforans.*

Adult coccidia live intracellularly in the linings of the bile duct (hepatic coccidia) or the intestinal tract (intestinal coccidia). These damage the linings of the bile duct and intestinal tract and produce eggs, oocysts, which pass in the feces.

Oocysts mature for varying periods of time, depending on physical conditions, until they are "infective" or ready to grow to adulthood if ingested. The mature oocyst is ingested by the rabbit while consuming contaminated feed or when licking its feet or contaminated cages. Once inside the digestive system, the oocyst matures to the infective stage and enters the cells of the intestinal lining, where it matures to adulthood; thus, the cycle is completed. The infective stages of the hepatic coccidia makes its way from the cells of the intestinal lining to the cells of the bile duct.

Symptoms are more pronounced in young animals; older animals may become immune to the organism. Common symptoms include diarrhea stained with blood and mucus, potbellies, and rough hair coat. The infected animals stand around with arched backs.

The development of coccidiosis is greatly influenced by unsanitary conditions and stress. Birds are carriers of coccidiosis and should not be allowed access into the rabbitry. A routine use of medicated water containing the coccidicide Sulfaquinoxaline at 0.04 percent is an effective treatment as well as a preventive measure.

Snuffles. Snuffles is a disease caused by the organism *Pasteurella multocida.* Symptoms of the disease include persistent sneezing with a white-colored nasal discharge; the fur on the front feet just above the paws will be matted from the rabbit constantly pawing at its nose. Abscesses caused by the *Pasteurella* organism may appear on the legs, neck, shoulders, and head of the animal. Affected animals are usually unthrifty and gain poorly. See Figure 9–53.

Many animals can carry the *Pasteurella multocida* organism in their upper respiratory tracts and

Figure 9–53 This rabbit has snuffles. Note the nasal discharge and wet nose. Reprinted from **Raising Rabbits,** *copyright 1977 by Ann Kanable. Permission granted by Rodale Press, Inc., Emmaus, PA 18098.*

not show any signs of the disease. The disease usually appears when infected animals are placed under stress conditions such as poor sanitation, poor ventilation, changing temperatures, shipping, or the stress of show conditions.

Mastitis. Mastitis is an inflammation of the mammary tissue. The mammary system becomes inflamed, swollen, hot, and dark red or blue-colored; the condition usually occurs just after kindling or weaning. Any type of bacteria that enters the teat opening is capable of causing mastitis; however, the most common type of bacteria is the *Staphylococcus* species. Bacteria of the *Pasteurella* species may be present in the blood of an animal and circulate to the mammary tissue and cause caked breast or trauma as a result of stress caused by producing milk.

Treatment for mastitis must begin as soon as signs of the disease are noticed. A veterinarian should be consulted. In treating caked breast, all concentrates should be withheld for 72 hours, and only roughages and water should be fed. If a large number of does in the rabbitry become infected with mastitis, cages and nesting boxes should be cleaned and disinfected.

Weepy Eye. Weepy eye is an inflammation of the conjunctiva, the mucous membrane lining the inner surface of the eyelids and covering the front part of the eyeball. This inflammation is usually caused by a *Pasteurella multocida* infection. Affected rabbits rub their eyes with their front feet; there is a discharge from the eye, and the fur at the corner and lower lid of the eye becomes matted. Common ophthalmic ointments containing sulfonamides or antibiotics are the recommended treatment. The eyes should be treated two or three times a day for three or four days. See Figure 9–54.

Syphilis. Vent disease or rabbit syphilis is a venereal disease of domestic rabbits. Symptoms first appear as bare areas around the genitals. Small blisters or ulcers usually form and eventually become covered with scabs; these scabs are usually confined to the genital areas but may appear on the lips and eyelids. Both sexes can become infected, and the disease is usually transmitted by mating infected animals. The specific organism responsible is the spirochete *Treponema cunniculi.* Infected animals should not be mated.

Wry Neck. Wry neck is a condition in which the head is tilted to the side; the tilt may be so severe that the animal has trouble maintaining its balance and falls over. The condition is caused by an inflammation of the inner ear caused by the organism *Pasteurella multocida.* See Figure 9–55.

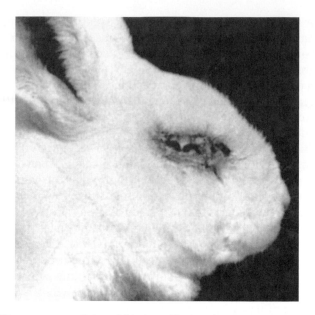

Figure 9–54   This rabbit is suffering from weepy eye, an inflammation of the conjunctiva of the eye. Note the discharge from the eye and the matting of the fur around the eyes. Reprinted from Raising Rabbits, *copyright 1977 by Ann Kanable. Permission granted by Rodale Press, Inc., Emmaus, PA 18098.*

Treatment of wry neck is difficult because it is hard to get antibiotics to the inflamed area. The incidence of wry neck is usually very high where there is a high incidence of snuffles. Treatment and prevention should be aimed at controlling the incidence of upper respiratory infections.

Figure 9–55   Wry neck is a condition in which the head is tilted to one side. Reprinted from Raising Rabbits, *copyright 1977 by Ann Kanable. Permission granted by Rodale Press, Inc., Emmaus, PA 18098.*

Infectious Myxomatosis. Infectious myxomatosis is a fatal disease of domestic rabbits that is primarily seen in the coastal areas of California and Oregon. The virus causing the disease is transmitted by mosquitos, biting flies, and direct contact. Incidence of the disease is greatest during the months of May to August, when mosquito populations are the greatest.

The first symptom is conjunctivitis with a milky discharge from the inflamed eye. The animal becomes listless, fails to eat, and may have a temperature as high as 108°F. Breathing usually becomes labored; upper respiratory infection occurs in later stages, followed by death. Advice and assistance from a local veterinarian is necessary. Dead animals should be burned and buried.

Papillomatosis. Papillomatosis are small warts or tumors that appear in or around the mouth and around the head, neck, and ears. Infected animals should be isolated and culled from the herd. See Figure 9–56.

Ringworm. Ringworm is an uncommon disease of domestic rabbits usually associated with poor management practices. It is caused by *Trichophyton mentagrophytes,* which also affects humans, guinea pigs,

Figure 9–56 Papillomatosis is characterized by warts or tumors. Reprinted from Raising Rabbits, *copyright 1977 by Ann Kanable. Permission granted by Rodale Press, Inc., Emmaus, PA 18098.*

mice, and rats. Affected areas are circular, raised, reddened, and capped with white, flaky material. Infected animals should be isolated and the infected area treated with daily applications of iodine. If there is a major problem in the rabbitry, equipment and cages should be disinfected, and powdered sulfur should be applied to nest boxes. The local veterinarian should be consulted.

The Ear Mite. The ear mite *Psoroptes cunniculi* is the most common external mite of rabbits. Common signs are a shaking of the head, flapping the ears, and scratching at the ears with the hind feet. The mites irritate the linings of the ears, causing a buildup of serum and blood that turns into brownish-colored scabs. Effective treatment consists of applying mineral oil in the ears for three days and repeating at ten-day intervals; this treatment should effectively suffocate the mites. Medicated ear drops are also available to control ear mites. Cages, nest boxes, and equipment should be cleaned and disinfected to effectively control mites.

Rabbits can be infected with the mange mites *Sarcoptes scabiei* or *Notoedres cati.* Infected animals will scratch themselves continuously, causing a loss of fur on the nose, head, base of the ears, and around the eyes. Prevention consists of cleaning and disinfecting cages, nest boxes, and equipment. Rodents, which can also carry these mites, should be kept out of the rabbitry.

Pinworms. Pinworms of the species *Passalurus ambiguus* may infect rabbits. If an animal is heavily infested, slow growth and poor coat condition will be a common sign. Sanitation and good management practices usually prevent infection from pinworms.

Rabbits are an intermediate host for the tapeworm species *Taenia pisiformis* and *Taenia serialis* in dogs, and *Taenia taeniaeformis* in cats. There are no visible signs of infection in rabbits, and evidence of infection is visible only upon necropsy or during processing. It is important to keep dogs and cats out of the rabbitry so that the feed cannot become contaminated.

Wet Dewlaps. Wet dewlaps is a condition caused by the rabbit dragging the **dewlap** into the water bowl; this causes the fur on the dewlap to become wet and matted. Infection of this area causes the skin to turn green and produce an odor.

The fur should be clipped from the infected area and an antibiotic ointment or cream applied. The use of water bottles or automatic waterers usually eliminates the problem. A rubber ball can be placed in the water bowl; this forces the rabbit to drink from the sides of the bowl and not drag the dewlap into the bowl.

Fur Chewing. Rabbits may pull fur from themselves or others in a pen and chew it. Fur chewing is thought to be caused by simple boredom, a low-fiber diet, or a nutrient deficiency. Increasing the fiber in the diet by feeding hay or straw may be effective in eliminating fur chewing. The addition of magnesium oxide has also been effective in some cases.

Hutch Burn. Hutch burn or urine burn is a chapped or burning condition around the external genital area of the doe; the condition is caused by the doe sitting for long periods of time in a wet, dirty area. The chemicals in urine-soaked bedding cause a burn to the tender skin of the genital area; these burned areas may develop secondary infections if conditions are not corrected.

Dirty, wet hutches should be cleaned and fresh bedding used. Antibiotic creams or ointments should be applied to the infected areas.

Sore Hocks. Sore hocks are an ulcerated area on the bottom of the foot pads. Although the condition can affect all breeds, the larger breeds with extremely long bodies are more commonly affected. The size of the foot pad is not large enough to support the weight of the large animal, and thus the foot pad becomes ulcerated from supporting the weight of the animal. Rough, wire cage floors, nervousness of the animal, and matting of the fur on the foot pad are conditions that increase the likelihood of sore hocks. An accumulation of wet bedding, hair, and feces in the corners of the cage may also be a contributing factor.

Treatment should consist of putting the affected animal on a solid surface and applying medication to the sore pads. Astringents, which contract body tissue and check secretions and bleeding, should provide effective treatment. Petroleum jelly, bag balm used on cow's udders, and ointments used to treat human hemorrhoids are common types of astringents.

Malocclusion. Malocclusion is a condition in which the upper and lower teeth grow to the extent that the animal cannot eat. The condition results because the teeth do not wear down properly, or it may be an inherited condition in which the teeth do not come together properly. Temporary treatment may sometimes be accomplished by clipping the teeth. Animals that have this condition should not be used for mating. A veterinarian can pull deformed incisors in case one wants to keep the animal.

REPRODUCTION

Rabbits become sexually mature at five to eight months of age depending on the breed. The miniature and small rabbits become mature at about five months. Males usually mature slower by thirty to sixty days.

Female rabbits do not have a regular cycle like most other animals. In nature, the female rabbit becomes sexually active based on the length of the day and the temperature, but rabbits kept inside under artificial light and warmth can be bred all year.

Even though the doe can be bred any time during the year, there may be times when the doe is more receptive. A breeder can check the external genital area or vulva of the doe. If the vulva is white, small, and dull, the doe will probably not be receptive; if it is mated, the chances of conception will be slight. If the vulva is pink, slightly swollen, and glistening in appearance, she will be more receptive and the chances of conception better; if the vulva is reddish-purple and swollen, she will probably be most receptive and chances for conception are highest.

To mate a doe, she should be placed in the male's cage; the male should *not* be placed in the female's cage. If the male is put in the female's cage, he will probably be too busy examining the new cage, and the female will be very protective and will probably fight. After the female is placed in the male's cage, the rabbits should be observed to see that no fighting takes place and that actual mating occurs.

The act of mating will be over very quickly, and the male will fall backward or to the side as the act is completed. The actual mating causes the ovaries of the female to release eggs 8 to 10 hours after mating. The sperm will remain viable for 30 to 36 hours in the female's reproductive tract but can remain viable for up to 120 hours or longer.

Another mating 8 to 10 hours after the first stimulates the ovaries to release more eggs and thus increases litter size. The fertilized eggs make their way to the uterus and attach themselves within four days.

An experienced handler can palpate a doe at twelve to fourteen days and determine whether the female is pregnant. Palpating is done by putting the doe on a flat surface and holding her gently with one hand and sliding the other hand under her belly. At ten to fourteen days, the fetuses feel like large marbles. The breeder should press slightly on both sides of the belly and back toward the groin area. If she is pregnant, birth will occur thirty to thirty-two days after mating.

Some breeders will put the doe back in with the buck fourteen days after the first mating. If the doe accepts the buck and mating takes place, the handler can assume she was not bred the first time; if she grunts and runs or wants to fight him, the handler can assume she is bred.

Kindling is the term used for the birth process in rabbits. About two days before the doe is expected to give birth, she should be provided with a nest box; some straw or shavings should be placed in the nest box. About a day before she gives birth, the doe will pull fur from her belly and use it for the nest. This does not hurt the doe and is part of the nesting instinct.

The expectant mother should be left alone at this point. Does usually have no difficulty kindling and no assistance is needed. Too much activity around the nest box may cause the doe to become excited and nervous. She may not use the nest box and scatter her newborn all over the cage, and she may become so nervous that she kills the young as they are born.

About 24 hours after the doe gives birth, the breeder should check the nest to see whether the young are doing well. It is important to approach the nest quietly and slowly, speaking to the doe so she does not get excited. Any dead fetuses should be removed; the female will not remove any dead babies from the nest, and these can become a source of contamination from bacteria and flies.

The litter should be checked daily. A doe nurses her young once or twice a day. There is no reason to get excited if the doe is not in the nest; nursing usually takes place at night. If the litter is large, young bunnies can be successfully fostered off onto another doe during the first five days of life. If the doe does not accept and nurse them, they can be fed successfully by hand. When feeding by hand, feed the young twice a day. A formula of 1 pint goat's milk or lactate-free cow's milk, plus 2 tablespoons of powdered milk, and 2 tablespoons of Karo syrup can be used. Care should be taken not to force the formula into the mouth because it can enter the lungs and is almost sure to cause death. It is also imperative that the young be stimulated to urinate and defecate every two or three hours by wiping their genital area with a soft, cotton swab moistened with warm water.

Young rabbits' eyes should open at twelve to fourteen days of age, and at three weeks, the young will start to venture from the nest, where they will start to nibble on straw and on feed pellets left for the doe.

The age at which the litter is to be weaned depends on the breed, the size of the litter, and the size of the individual bunnies. Usually, a litter can be weaned at four to eight weeks of age. After weaning and separating the young bunnies from their mother, they should be left together for a period of seven to fourteen days to help reduce stress. Young rabbits raised for fryers should reach a marketable weight of 4 to 5 pounds by eight to twelve weeks of age.

The doe can be rebred three days after kindling; however, this is not recommended. A better time is two weeks before weaning the litter; higher conception rates are usually achieved at this time.

SUMMARY

Rabbits were once classified in the order Rodentia but are now placed in a separate order, Lagomorpha. The chief difference between rabbits and rodents is that rabbits have four upper incisor teeth arranged one pair behind the other, whereas rodents have two chisel-like incisor teeth in each jaw.

Fossil remains of Lagomorpha have been found in Oligocene deposits, which date back 30 to 37 million years. These fossil remains indicate the animals have changed very little.

The European wild rabbit is the species from which all domestic rabbit breeds have been developed. Although many of our current domestic breeds have been developed in the United States, the American cottontail was not used in the development of our domestic rabbit.

Several conditions are important in housing rabbits: temperature, humidity, ventilation, proper lighting, and the absence of drafts.

The easiest way to feed rabbits is to use one of the commercially prepared pellet-type feeds. These feeds are especially formulated for rabbits and supply all the nutrients needed.

Rabbits become sexually mature at five to eight months of age depending on the breed; the miniature and small size become mature at about five months. Males usually mature slower by about thirty to sixty days.

Female rabbits do not have a regular cycle like most other animals. Rabbits kept inside under artificial light and warmth can be bred all year.

DISCUSSION QUESTIONS

1. What is the history of the rabbit?
2. What are the uses for rabbits?
3. What are the advantages of rabbit meat over other types of meat?
4. What are the five weight categories of rabbits?
5. List the breeds in each of the weight categories.
6. What are the houses for rabbits commonly called?
7. What are several important things to remember in housing rabbits?
8. Describe how a rabbit should be handled.
9. List and describe the common diseases and ailments of rabbits.

SUGGESTED ACTIVITIES

1. With the approval of the instructor, bring in breeds of rabbits or invite local breeders to bring in rabbits to the class and describe the characteristics of the breeds. The class can observe the animals. Correct methods of handling can be demonstrated.

2. Obtain tags from various rabbit feeds, bring them to class and explain the differences in the feed formulations.

3. With the approval of the instructor, invite a local breeder or veterinarian to demonstrate the correct methods of palpating a pregnant doe.

4. With the approval of the instructor, have a local veterinarian come to the classroom to discuss the common diseases and ailments of rabbits. The veterinarian can also discuss the prevention and treatment of rabbit diseases and ailments.

ADDITIONAL RESOURCES

1. Rabbits Only
   www.rabbits.com

2. The Rabbit Web
   www.rabbitweb.net

3. Island Gems Rabbitry
   http://littlegems.hypermart.net

4. Diseases of Laboratory Rabbits
   http://netvet.wustl.edu/species/rabbits/rabbits.txt

END NOTES

1. "Rabbit." 1979. *Encyclopedia International* 15:237. Copyright 1979 by Lexicon Publications, Inc. Reprinted by permission.

2. "Rabbit." 1984. *Encyclopedia Americana* 23:110. Copyright 1984 by Grolier Incorporated. Reprinted by permission.

3. Nowak, R. M. and J. L. Paradise. 1991. *Walker's Mammals of the World,* 5th Edition. Baltimore/London: Johns Hopkins University Press, 552. Reprinted by permission.

Hamsters

OBJECTIVES

After reading this chapter, one should be able to:

- recognize and describe the common types of hamsters.
- list common management practices for maintaining health.
- list common diseases and ailments of hamsters.

TERMS TO KNOW

demodectic mites	meningitis	solitary
demodicosis	nocturnal	wet tail
estivation	rectal prolapse	

CLASSIFICATION

Kingdom	— Animalia	Family	—	Muridae
Phylum	— Chordata	Subfamily	—	Cricetinae
Class	— Mammalia	Genus	—	*Mesocricetus*
Order	— Rodentia	Species	—	*M. auratus*
Suborder	— Myomorpha			

HISTORY

The next six chapters cover common rodents that are kept as pets. These common rodents are hamsters, gerbils, rats, mice, guinea pigs, and chinchillas. Rodents evolved during the Paleocene and Ecocene epoch 55 to 65 million years ago. *Paramys* is considered the ancestor of the rodents and was a large squirrel-like animal with clawed feet for grasping and climbing and a long tail for balancing.

Today, the order Rodentia contains 29 families, 380 genera, and 1,687 species and is further divided into three suborders based on the musculature of the jaw and other structures of the skull. These suborders are the Sciuromorpha (squirrel-like rodents), the Myomorpha (rat-like rodents), and the Hystricomorpha (porcupine-like rodents).

In studying small animals used for pets, two suborders are of primary importance to us: the suborder Hystricomorpha, because it contains the families Chinchillidae

Figure 10–1 (Left) *A longhaired Golden hamster.* (Right) *A shorthaired Golden hamster. Courtesy of Stacey Riggert.*

(chinchillas) and Caviidae (cavies or guinea pigs), and the suborder Myomorpha, because it contains the families Murinae (old world rats and mice), Cricetinae (hamsters), and Gerbillinae (gerbils).

MAJOR GROUPS

The Golden hamster, *Mesocricetus auratus,* is the most abundant of the hamsters. This species has been most often used in research and is the species most commonly found in pet stores.

The Golden hamster is native to the desert areas of Syria and is sometimes referred to as the Syrian Golden hamster. It was discovered in 1930 by Professor I. Aharoni of the Department of Zoology, Hebrew University, in Jerusalem. Professor Aharoni was exploring an animal tunnel near Aleppo, Syria, when he came across a mother and her litter. Of the litter of twelve, only a male and two females survived the trip back to the university. These three were later mated and are believed to be the foundation for all Golden hamsters in captivity today. See Figure 10–1.

In 1931, young hamsters were shipped to the United States Public Health Service Research Center at Carville, Louisiana, where they were used in medical research. While being used in research, it was discovered that the Golden hamster could be tamed and made into a pet.

An adult hamster is about 5 to 6 inches long with a short stump of a tail and weighs about 4 ounces.

A true Golden hamster has a rich mahogany or orange color on the back with a white or creamy-colored underside and legs; a black patch is usually present along the side of the cheeks. Hamsters have large black eyes and large erect ears. They have cheek pouches in which to store food.

Today, the Golden hamster can be found in several different varieties and many different colors and color combinations. Long-haired or "teddy bear" hamsters

with long, silky fur are popular pets. See Figure 10–2. Golden hamsters can be found in single or solid colors, two-colored or banded patterns, and the speckled or "piebald" pattern. Golden hamsters can be found with regular fur or satin fur. In addition to the regular black eye color, red or ruby eyes are also common. Hamsters can occasionally be found with one ruby eye and one red eye.

Male hamsters have prominent scent glands on their flanks. These appear as dark pigmented spots on the skin and become more obvious as the animal ages. Secretions from these glands are used to mark their territory. The female hamsters are very aggressive and usually dominate the males.

Another species of hamster that can be found in pet shops is the small desert or dwarf hamster, *Phodopus sungorus.* The dwarf hamster is light gray with

Figure 10–2 A Teddy Bear hamster. Courtesy of Stacey Riggert.

Figure 10–3 A dwarf hamster. Courtesy of Stacey Riggert.

a dark stripe down the back. It is smaller than the Golden hamster, being about 4 to 4½ inches long. These small hamsters are very active, make friendly pets, and are fun to watch. However, they are not recommended for children because their small size makes them harder to handle. See Figure 10–3.

ANATOMY

Figure 10–4 shows the anatomy of a rodent.

CHARACTERISTICS

Hamsters are **nocturnal** animals; they sleep during the day and are active at night. In the wild, hamsters travel long distances at night foraging for food, which they carry back to their burrows in cheek pouches. These cheek pouches are large and extend back to the front shoulders. The pouches can even be used by the female to carry her young. When food is plentiful, hamsters can carry and hoard large amounts in their burrows.

Hamsters are very protective of their territory and hoards of food. Even a tame hamster may bite when a finger is stuck into its cage and intrudes into its space. Once out of a cage, hamsters are usually more friendly and easy to handle.

Hamsters are **solitary** animals. Young can usually be kept together until they are about five weeks old, then they will begin to fight and may need to be separated. Cannibalism may also result if litters are left together for longer periods of time. Females usually tolerate the male only for breeding. In housing mature hamsters, it is best to keep a single hamster in a single cage.

The life span of hamsters is usually one to three years; with good care, proper feeding, and maintenance, they may live longer.

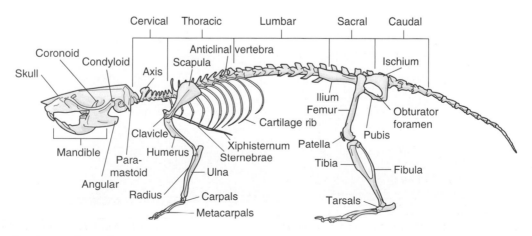

Figure 10–4 Anatomy of a rodent.

Hamsters prefer a temperature around 70°F. If the temperature gets much above 80°F, they go into a deep sleep-like hibernation. This is known as **estivation** and is a method by which the furry hamsters originally survived the hot, dry desert environment. In captivity, the hamster may go into hibernation and take considerable effort to wake. Sometimes, it appears that the hamster is dead and can be placed in the hand and gently bounced up and down before it awakens. To prevent estivation, room temperature should be maintained below 80°F.

If the temperature drops below 50°F, hamsters also go into hibernation. Hamsters catch colds easily if exposed to changing temperature or drafts; therefore, cages should always be kept out of drafts, and room temperature should remain constant.

HOUSING AND EQUIPMENT

Hamster cages are available on the market or can be constructed. The cage should be large enough to allow a hamster plenty of room for exercise. It is generally recommended that a single hamster should have a cage 10 inches by 16 inches by 10 inches tall. See Figure 10–5.

A cage should be constructed of gnaw-proof materials or constructed such that the hamster cannot gnaw its way out. Most of the commercial types of cages are constructed with stainless steel side and top bars and have plastic floors. The plastic floors are usually constructed in such a way that the hamster cannot get its teeth into the plastic and start gnawing. These cages are usually easy to clean, allow the hamster to see out, and provide exercise because the hamster can climb up the sides and across the top bars.

Water bottles can be hung on the outside of the cage with the drinking tube extending into the cage through the side bars. By hanging the bottle on the outside, the hamster is prevented from gnawing the bottom of the water bottle.

Aquariums make nice cages; they are easy to clean and allow the hamster to see out. A water bottle can be hung from the top of one of the sides but may provide a means of escape as the hamster can climb up the water bottle and over the sides; wire mesh tops solve this problem (see Figure 10–6). Water bottles hung on the inside of an aquarium must have a protective metal hanger to prevent the hamster from gnawing holes in the plastic bottle; a fruit juice can cut in half can be slipped over both ends of the water bottle.

When constructing a cage, consideration should be given to the ease of cleaning and in making sure that it is escape-proof. Cages with wood bottoms are harder to clean, and the wood will soak up urine and smell. If wood is used, the wood should be coated with linseed oil. Wood bottoms should be thick enough to prevent the hamster from gnawing through and escaping.

Cages can be constructed easily using galvanized metal for the sides and a wood floor. A small hole can be drilled in one side and a water bottle hung on the outside with the drinking tube extended into the cage through the hole. Sheet metal screws are used to hold the floor in place and allow the floor to be easily removed and replaced.

If the cage is large enough, an exercise wheel can be added; exercise wheels are available in plastic or metal. Hamsters will gnaw on the plastic ones and may

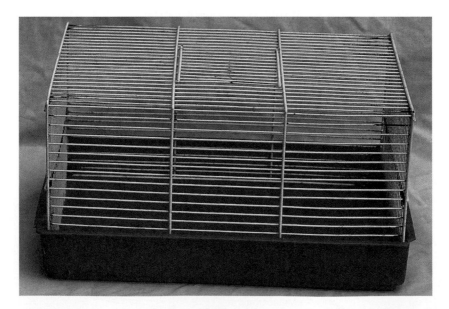

Figure 10–5 Wire cages with plastic bottoms are available in most pet and discount stores. Photo by Dean Warren.

Figure 10–6   Aquariums make nice cages and the wire mesh top prevents the hamster from escaping. Courtesy of Carloyn Miller.

soon destroy them; plastic exercise wheels do have the advantage of making very little noise. Remember the hamster is nocturnal and will run an exercise wheel all night long. A squeaky, metal exercise wheel may quickly become a nuisance. See Figure 10–7.

The hamster cage should have plenty of clean, fresh bedding. Bedding can be paper confetti, paper strips, wood chips, wood shavings, hay, straw, or pieces of cotton. Wood shavings usually work the best. They absorb most of the urine, and the hamster chews on the shavings, which helps keep its teeth worn down. Coarse sawdust can be used, but sawdust that has a lot of fine, dusty material in it should be avoided. This material can cause irritation to the eyes and get in the nose and lungs of the animal. See Figure 10–8.

Figure 10–7   Metal exercise wheels are available in most pet stores. Courtesy of PhotoDisc.

Figure 10–8 *Wood shavings made from pine make ideal bedding for all small animals. Large bags like this 50-pound bag are available at feed stores and grain elevators at economical prices. Photo by Dean Warren.*

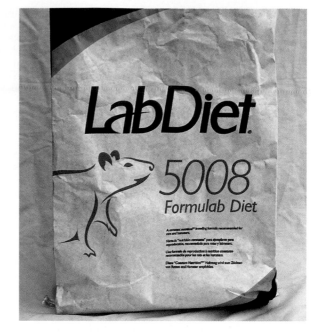

Figure 10–9 *Purina Mills, Inc., markets Formulab chow for small animals. Photo by Dean Warren.*

FEEDING

The easiest way to feed hamsters is to purchase commercially prepared pellets specifically designed for small animals. These pellets are especially formulated and contain all of the vitamins and minerals necessary. The pellets are hard, and in order for the animal to consume them, they must gnaw on the pellet. This also aids in wearing down the continuously growing front incisors of these rodents. See Figures 10–9 and 10–10.

In mixing a ration, it is important that the animal receive a wide range of foods to ensure a nutritionally balanced ration. The diet should contain seeds such as corn, millet, wheat, oats, sorghum, rape, and sunflowers. Lettuce, carrots, pieces of potato, fresh clover, or alfalfa hay should be supplemented into the diet as well as dried peas, beans, and nuts. Dry dog pellets or biscuits can be fed also. These contain trace minerals and vitamins and are also helpful in grinding the teeth down. See Figure 10–11.

It is important that the diet be consistent. One must not change diets suddenly or feed greens and fruits in large amounts because this may cause diarrhea. Any soft-type food not eaten by the hamster should be removed from the cage before it becomes spoiled. Consumption of spoiled food can throw the animal off its feed and cause diarrhea or illness. Soft foods may also get lodged in the pouch of a hamster

Figure 10–10 *Purina Formulab pellets are oval shaped. They are ½ to 1 inch long, ½ inch wide, and ⅜ inch thick. Photo by Dean Warren.*

MICE — Adult mice will eat 4 to 5 grams of pelleted ration daily. Some of the larger strains may eat as much as 8 grams per day per animal. Feed should be available.

Crude protein not less than	23.0%
Crude fat not less than	6.5%
Crude fiber not more than	4.0%
Ash not more than	8.0%
Added minerals not more than	2.5%

Ingredients: Ground yellow corn, soybean meal, ground wheat, fish meal, wheat middlings, animal fat preserved with

Figure 10–11 *The back of a Formulab bag shows the analysis and ingredients, plus feeding instructions for all small animals.*

and become spoiled, causing infection and illness. For special treats, hamsters love sunflower seeds, bee moths, crickets, and grasshoppers.

HANDLING

As mentioned earlier, hamsters are nocturnal animals and should not be disturbed while sleeping. If a sleeping hamster is poked or touched, it may awaken quickly and its reaction is to bite in self-defense. Occasionally, a hamster may need to be awakened during the day to clean its cage. To awaken a sleeping hamster, one should approach it carefully, tap lightly on the cage, and speak to it. Hamsters learn to recognize their handler's voice, and speaking to it will calm it. The handler should attempt to stroke or pet it and see how it reacts. It may allow itself to be picked up; however, if it retreats to a corner or stands on its rear legs with its paws upraised, one must be very careful because this is the hamster's defensive posture. The handler should continue to talk to the animal and, if it leaves the corner, try to pet or stroke it. When the hamster allows it, it can be picked up with the thumb and forefinger placed right behind the front quarters. Hamsters are quite strong and very dexterous in their attempts to escape. They can twist their bodies and use their rear feet to free themselves. See Figure 10–12.

The more a hamster is handled, the tamer it will become. The hamster will soon learn the voices of its

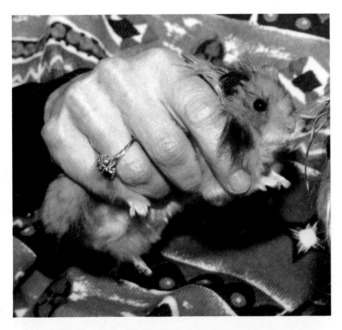

Figure 10–12 A hamster should be picked up firmly between the thumb and forefinger. Holding it in this manner usually prevents it from wriggling free or from turning and biting. Courtesy of Stacey Riggert.

handlers and come readily to the edge of the cage to be picked up.

A way to get a hamster familiar with members of the family is to place it on a table with the family members seated around the table, letting the hamster roam the table freely, talking to it, and attempting to pet or stroke it. The hamster will soon become familiar with the voices of the family members, and within a short time, the hamster will become a gentle, friendly pet.

Young hamsters should not be handled until their eyes open; this will be at about fourteen days of age. They quickly get used to human smell and will allow themselves to be picked up without any fear. Care must be taken when handling young hamsters because they can move so quickly and are hard to hold without hurting them. They may jump from the handler's grasp in an effort to escape and fall to the floor, which could result in serious injury. Children need to be watched carefully when they handle small hamsters because they tend to want to hold them too tightly.

Because the hamster's eyesight is not especially good is another reason for approaching a hamster carefully. Hearing and sense of smell are its most acute senses.

If they evade the handler's hands, squeal, or make a chattering noise with their teeth, they want to be left alone. Remember to move slowly and be patient. Most people get bitten by hamsters because they poke them when the hamster is asleep. Hamsters may also bite the end of a finger because they think they are being offered a treat.

DISEASES AND AILMENTS

Hamsters are generally healthy animals, and there are only a few diseases that may cause problems. The most important disease is hamster enteritis, commonly called **wet tail**. The exact cause is not known, but several different causes such as bacteria, viruses, diet, and poor sanitation may be involved. Many times, the disease is associated with general neglect and poor care.

Symptoms of wet tail include wetness around the tail and rear area of the animal, caused by a runny diarrhea. Affected animals go off their feed, have rough hair coats, become weak, and are feeble. **Rectal prolapse,** seen as a red protrusion from the anus, may occur. High death rates often occur within one to two days after the clinical signs.

An infected hamster should be taken to a veterinarian. Treatment results are often poor. The bedding of the infected animal should be changed, and the cage, food bowls, and water container should be disinfected. Infected animals should be kept separate, and all animals should be closely observed.

Bacillus piliformis and *Salmonella* species may also affect hamsters. Affected animals may show signs of diarrhea, go off their feed, lose weight rapidly, or become weak and lethargic. Death usually results in a large number of cases, and treatment is often ineffective. *Salmonella* species can also be transmitted to humans, so caution should be taken.

Common diarrhea, caused by incorrect diet, should not be confused with wet tail or other serious ailments. Feeding an overabundance of green leafy materials, vegetables, or fruits may cause diarrhea. The soft foods should be removed and the dry grains and seeds increased in the hamster's diet; this should solve the problem.

Baldness or loss of hair in old hamsters is normal; loss of fur usually begins around the rump. This loss occurs at ten to twelve months of age; if it occurs to younger hamsters, other causes should be suspected.

Fleas and lice are common among hamsters, especially if cats, dogs, rats, and mice have access to them. If the presence of fleas or lice is suspected, the cage must be cleaned regularly and flea powder sprinkled in the bedding; a flea powder recommended for cats should be used. Flea powder used for dogs is too strong; hamsters and cats lick their fur and will get an overdose of the medication if dog powder is used.

Notoedres notoedres species of mites are common in hamsters; they spread by direct contact from one animal to another. This family of mites burrow within the skin, and the entire life cycle is spent on the host animal.

Demodex criceti and *Demodex aurati* are two species of **demodectic mites** that may occasionally be found on hamsters. These mites are usually considered normal inhabitants on the skin. In large numbers, they can cause the skin disease called **demodicosis** or demodectic mange.

Lymphocytic choriomeningitis (LCM) is a **meningitis** that normally affects young animals. Common symptoms are a loss of weight, drowsiness, ruffled coat, and lethargic movement. This type of meningitis usually is not fatal to the animal but causes slower growth than normal. It is usually spread by mice, and the disease can be spread to humans, causing flu-like symptoms. Pregnant women must take care to avoid exposure because the disease can cause premature birth and malformations. Pregnant women should not clean cages or handle hamsters.

REPRODUCTION

Young litters of hamsters should be separated at five or six weeks of age, not only to prevent fighting, but

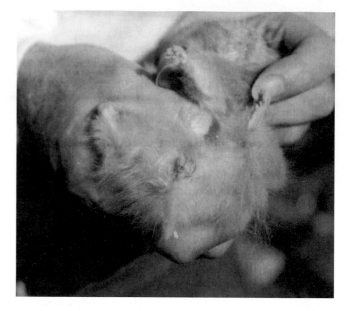

Figure 10–13   The genital area of a female hamster. Photo by Isabelle Francais.

also to prevent unwanted matings. Hamsters may become sexually mature around six weeks of age. Like other animals, the females should not be mated at too early an age because small, weak litters may result and the life span of the female may be shortened. See Figures 10–13 and 10–14.

Hamsters can reproduce rapidly; females come in heat every four days and, if bred, produce a litter in sixteen days. Females should be ten to twelve weeks old when mated and of good size (approximately 3½ ounces); small females should not be

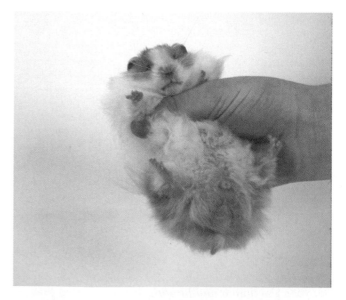

Figure 10–14   The genital area of a male hamster. Photo by Isabelle Francais.

mated until later. Increased activity is usually a sign of heat in the female hamster. The heat period usually occurs in the evening hours between 5 and 11 o'clock. During the hot summer days, they may come in heat earlier in the day.

When mating hamsters, the female should always be placed in the male's cage; this usually results in less fighting. They should be observed carefully. If the female is not in heat and not receptive to the male, fighting will soon occur, and the female should be removed immediately. If the female is in heat and receptive, she usually accepts the male within a couple of minutes. The male will sniff the female and lick her fur, checking to see whether she is receptive. If the female is receptive, she will go very rigid and arch her back with her tail raised. Actual mating occurs very quickly.

After mating, the male will rest for a few moments, clean himself, and then pursue the female again. They may mate several times in a 15- to 20-minute period.

When the female is satisfied, she will turn on the male and bite him. At this time, the female should be removed from the male's cage and placed back in her cage. If the female is not removed from the male's cage, she will attack the male and fierce battles will ensue. The female may even kill the male.

One must be very careful in attempting to separate a pair of fighting hamsters; in the fury of the fight, they may bite. One should use a brush, a piece of cardboard, or anything available to slip between the two animals and separate them.

When the female is introduced into the male's cage and is observed rubbing her rear on the floor of the cage, this is a sign that she is not receptive and should be removed immediately. When she rubs her rear on the floor, she leaves an offensive-smelling fluid that is intended to distract the male. If the female is not removed, fighting will soon occur.

The gestation period is sixteen days and seldom varies. Two or three days before the female is due, her cage should be cleaned and fresh bedding provided. The mother and her young should not be handled until the young are about fourteen days old. Litters consist of two to sixteen young, with six to eight being an average. The young are born without fur, are blind, are totally helpless, and weigh about ¼ to ⅛ ounce. Attempts to handle the newborn or disturb the female and her litter may result in the mother biting and killing the babies because of nervousness and fright.

The young grow rapidly; at about ten days, they begin to move about the cage. The mother will often be seen picking them up and carrying them back to the nest. The eyes of the young hamsters open at fourteen days. Their appetite will increase, and they will begin consuming large amounts of food.

The young should be weaned at about twenty-four to twenty-eight days. If left longer, the mother may bite them in an attempt to drive them away. When separating the young, the males should be placed in one cage and the females in another; this will prevent inbreeding and unwanted matings. Good homes should be found for the young hamsters because they will soon start fighting if left together. Fighting is common when too many hamsters are kept in too small a cage; torn ears, torn faces, punctured eyes, severe injuries, death, and cannibalism may result.

Hamster breeders who do not have the time to carefully observe their animals may want to try leaving a single male and a single female together in a cage. The cage should be fairly large and have a house of some type so that the male has room to escape the female and hide. If fighting has been intense and either hamster has received injuries, the pair should be separated and their injuries allowed to heal. If fighting does not appear too intense and neither animal has received any major injuries, the pair can be left together for about ten days. During this period, the female should have come in heat twice, if not bred the first time.

Colony mating is another method that can be used with hamsters. In this method, three or four females are placed in a large cage with a single male. The females are usually left with the male for about ten days and then removed to separate cages. Fighting is almost sure to occur, and small houses should be provided so that the smaller, less aggressive animals can hide.

After the females are removed, the male should be allowed to rest for a week and then new females added to the cage. If large numbers of hamsters are wanted, ten or twelve females can be introduced with three or four males. Close observation is necessary to prevent severe injury to any of the animals.

Females produce only five or six litters a year and then no longer produce offspring. Males may become infertile at about one year of age; although some will remain fertile longer.

SUMMARY

The Golden hamster, *Mesocricetus auratus,* is the most important of the hamsters. This species has been most often used in research and is the species most commonly found in pet stores.

The Golden hamster is native to the desert areas of Syria and is sometimes referred to as the Syrian Golden hamster. It was discovered in 1930 by Professor I. Aharoni of the Department of Zoology, Hebrew University, in Jerusalem.

Another species of hamster that can be found in pet shops is the small desert or dwarf hamster, *Phodopus sungorus.*

Hamsters are nocturnal animals; they sleep during the day and are active at night. In the wild, hamsters travel long distances at night foraging for food, which they carry back to their burrows in cheek pouches.

Hamsters are solitary animals. The young can usually be kept together until they are about five weeks old; after that they will begin to fight and may need to be separated.

Hamster cages are available on the market or can be constructed. Cages need to be large enough to allow plenty of room for exercise and should be gnaw-proof.

The easiest way to feed hamsters is to purchase commercially prepared pellets specifically designed for small animals. These pellets are especially formulated and contain all of the vitamins and minerals necessary. The pellets are hard, and in order for the animal to consume them, they must gnaw on the pellet. This also aids in wearing down the continuously growing front incisors of these rodents.

Hamsters can reproduce rapidly; females come in heat every four days and, if bred, produce a litter in sixteen days. This litter usually consists of two to sixteen young with six to eight being an average. Young litters should be separated at five or six weeks of age, not only to prevent fighting, but also to prevent unwanted matings. Hamsters may become sexually mature around six weeks of age.

Hamsters are generally healthy animals, and there are only a few diseases that may cause problems. The most important disease is hamster enteritis, commonly called wet tail. The exact cause is not known, but several different causes such as bacteria, viruses, diet, and poor sanitation may be involved. Many times, the disease is associated with general neglect and poor care.

DISCUSSION QUESTIONS

1. What is the history of the common hamster?

2. What are common types of hamsters found in pet shops?

3. What foods can be fed to hamsters?

4. How should a hamster be handled?

5. List the common diseases and ailments of hamsters. What are the symptoms? How can we control each of them?

6. What are common mating methods used with hamsters?

SUGGESTED ACTIVITIES

1. With the approval of the instructor, bring hamsters to the classroom. The class can observe the animals. Correct methods of handling can be demonstrated to the class.

2. With the approval of the instructor, bring a hamster or pair of hamsters to the classroom. Leave the cage and animals for a few days. Class members can then become familiar with all aspects of hamster care and management.

3. With the approval of the instructor, have a local veterinarian come to the classroom to discuss common diseases and ailments of hamsters. The veterinarian can also discuss treatments and prevention of hamster diseases and ailments.

ADDITIONAL RESOURCES

1. Diseases of Hamsters
   www.afip.org/vetpath/POLA/
   HAMSTER.BUNTE.94

2. Syrian or Golden Hamsters
   www.webcom.com/~lstead/rodents/
   hamsters.html

3. Humane Society
   http://humanesociety.dane.wi.us/
   SAHMSTR.HTM

4. American Animal Hospital Association
   (AAHA)
   www.healthypet.com/Library/index.html

5. Animal Health Center
   www.caringtogether.com/index.html

6. MicroVet–University of Arizona
   http://microvet.arizona.edu/Courses/MIC443/
   notes/disease.htm

Gerbils

OBJECTIVES

After reading this chapter, one should be able to:

- describe the characteristics of gerbils.
- list practices for general care of gerbils.
- describe the common diseases and ailments of gerbils.

TERMS TO KNOW

agouti

camouflage

monogamous

mutations

obesity

red nose

selective breeding

Tyzzer's disease

CLASSIFICATION

Kingdom — Animalia

Phylum — Chordata

Class — Mammalia

Order — Rodentia

Suborder — Myomorpha

Family — Muridae

Subfamily — Gerbillinae

Genus — *Meriones*

Subgenus — *Pallasiomys*

Species — *M. unguiculatus*

HISTORY

There are four subgenera and fourteen species of common gerbils. The best known of these is the Mongolian desert mouse or Mongolian gerbil, *Meriones unguiculatus.*

Japanese scientists were the first to breed Mongolian gerbils in captivity; a Tokyo research institute received twenty pairs in 1935. In 1954, eleven pairs were sent to Dr. Victor Schwentker in the United States. They were found to be easy to work with, gentle, and active during the day; had no special food or housing requirements; drank little water; were virtually odorless; and seldom would bite. Because of these qualities, their popularity grew rapidly, and today gerbils are common household pets.

Gerbils, also called sand rats, include a large number of species of burrowing rodents forming the subfamily Gerbillinae, of the family Muridae. There are about 100 different species of these rat-like rodents, which are found in the dry, desert regions of Africa and Asia.

Figure 11–1 A normal-colored gerbil. Photo by Isabelle Francais.

The Mongolian gerbil is found in parts of China, in the former Soviet Union, and throughout most of Mongolia. In the wild, they live in colonies and dig burrows in sandy soils for shelter. Mongolian gerbils are active both day and night. Their diet consists of seeds, leaves, roots, stems, and occasionally insects.

GROUPS AND BREEDS

The Mongolian gerbil reaches a size of 6 to 8 inches long from nose to tail; the body is about 3 to 4 inches long. A mature gerbil weighs 3 to 4 ounces.

The body is short and thick and has a hunched appearance when squatting on its hind legs. The forelegs are very short, with the forepaws being very similar to hands and used to hold food. The hind legs and feet look very similar to those of a kangaroo and are large and furry with long toes and claws. The large feet enable the gerbil to stand firmly in the sand of its desert environment. The gerbil walks on all fours but, when stopped, will stand up on its hind legs to observe the immediate area. In the wild, gerbils use their strong hind legs to leap and jump, but those raised in captivity rarely jump, except in an effort to escape from the cage. See Figure 11–1.

The tail is covered with fur, has a bushy tip, and is used for support when standing and acts as a rudder and stabilizer when jumping or leaping. The skin covering the tail is loose and, if handled roughly, can become torn and pulled from the tailbone.

The head is broad and wide, and the eyes are large, usually jet black, and protrude slightly. The ears are small, rounded, and covered with soft fur; its sense of hearing is very acute because of the enlarged cavities in the skull that amplify the slightest sounds.

The natural color of the Mongolian gerbil is a reddish brown to dark brown. The fur is composed of dark guard hairs that give the overall coat a darker appearance. The underside fur is usually lighter brown, cream, or even white. This natural color pattern is referred to as an **agouti** color and serves as a **camouflage** against the sand and rock of its environment. The light underside fur serves to reflect heat from the desert sand.

Through **mutations** (a sudden change in an inherited trait) and **selective breeding,** gerbils are found today in several different color patterns. Black, white, black and white, white-spotted agouti, charcoal and white, and orange and white are common color patterns. The red-eyed white gerbil is not considered a true albino because some of the fur on the back and tail turns light brown as the animal matures. See Figures 11–2 and 11–3.

Figure 11–2 A black gerbil. Courtesy of Stacey Riggert.

Figure 11–3 A white gerbil. Courtesy of Stacey Riggert.

CHARACTERISTICS

The male gerbil has a scent gland on his stomach that looks like a bald patch or strip; this patch is more visible on mature males. The adult male leaves his scent by sliding his stomach along an object or area.

Gerbils are very quiet animals. A shrill squeak is sometimes made as an alarm and during mating by the female. Young gerbils can be heard making squeaking sounds, but these sounds diminish as they get older.

Another method of communication is a drumming sound made by standing upright and pounding with the rear feet; this drumming sound is also used as an alarm and by the male during mating.

Gerbils are short-lived animals. The life expectancy is usually two years but may be up to four years.

HOUSING AND EQUIPMENT

Most of the housing and equipment available for hamsters is also suited for gerbils. Being more active, gerbils should have more cage space than hamsters. A breeding pair should have approximately 150 square inches of floor space. Several gerbils kept in a cage together should have approximately 36 square inches per animal. If cage space is limited, cannibalism may occur; in the case of a breeding pair, they may not mate.

Like hamsters, gerbils are very adept at finding ways to escape. Gerbils can jump, so it is important that the cage be covered. A wire mesh cover can be purchased or constructed to prevent animals from escaping.

Several types of cages available on the market are suitable for gerbils; they all have advantages and disadvantages. Plastic cages are lightweight, inexpensive, and easy to clean. They are not, however, chew-proof, and a gerbil will gnaw at areas of the plastic that are not protected.

Metal cages are chew-proof but generally are more expensive, heavier, and harder to handle. Metal that is not protected with paint will soon rust. Gerbil urine is very concentrated and acidic and will quickly eat away paint and cause bare metal to rust.

If wood cages are used, they need to be made of heavy materials to prevent the gerbil from gnawing its way out. Wood will soak up urine and is harder to clean, especially if cleaning is delayed for long periods.

The common fish aquarium makes an excellent cage for gerbils; its greatest advantage is visibility. Gerbils are very curious and will watch people and come up to the glass when humans get close. Aquariums also have the advantage of being easy to clean, although children may find them heavy and hard to handle.

Several different types of bedding materials are available. Wood chips or shavings are the most desirable. They soak up urine, and the gerbil will spend endless amounts of time gnawing and chewing on the larger pieces.

Sawdust can be used, but sawdust that contains fine, dusty materials should be avoided. Gerbils dig and kick in their bedding, and the fine materials can get in their eyes, causing irritation, and can also get in the nose and lungs.

If a large cage is available, an area with sand would be a nice addition; the gerbils will spend time digging and playing in the sand. An attractive-looking cage can be set up using sand and rocks that will simulate a gerbil's natural environment.

Cotton and wool should be avoided as bedding and nesting materials because these can be ingested by the animal and cause blockage of the digestive system.

Gerbils, being very active, like to be provided with toys to keep them busy. Remember that gerbils will gnaw on anything soft, so toys should be made of a solid material.

Cardboard tubes from rolls of toilet paper and paper towels make excellent tunnels for gerbils but will soon be destroyed as the gerbils gnaw and chew on them.

Care should be taken in providing gerbils with an exercise wheel; the tail can easily get caught in the spokes. The skin on the tail is loose, and as the gerbil fights to free itself, the skin may break loose from the tail. Metal exercise wheels should especially be avoided. Some plastic wheels are made solid without spokes; these are safe for gerbils but will probably be gnawed on.

Glass or ceramic cups and toys used in fish aquariums, toys used for small birds, and tin cans provide gerbils with hours of play and exercise. Be careful to avoid any items with sharp or pointed edges that could cause injury.

FEEDING

The easiest way to feed gerbils is using commercial pellets made especially for small animals. These will be complete balanced rations, with all of the required vitamins and minerals added.

If preferred, rations can be prepared by mixing different types of foodstuffs. Common grains used are corn, oats, wheat, and barley. Mixing these grains can be a good base for a balanced diet. Gerbils eat only the heart out of the corn kernel, and just the kernel from oat seeds. Waste can be prevented by feeding breakfast foods made from corn flakes and rolled oats.

Sunflower seeds, linseed, millet, canary seed, rape, and hemp seeds found in mixtures for parakeets and other birds add variety to the diet. Although gerbils love sunflower seeds, they should be fed carefully because they contain a high percentage of fat and overconsumption may lead to obesity.

Green foods such as lettuce, cabbage, carrots, turnips, and beets also add variety to the diet and are usually rich in minerals. Green foods should be washed to remove any dirt, chemical residues, and insects. These materials should be fed sparingly because they can cause diarrhea and upset stomachs.

To be safe, vitamin supplements can be added; vitamin supplements are available that can be added to both food and water.

Gerbils consume very little water; however, a supply should always be available. The easiest way to provide water is by way of the glass drip-feed or automatic drinking bottle; several types are available. These should be hung on the outside of the cage with the drinking spout extending into the cage. Plastic bottles and the rubber stoppers on glass bottles will soon be chewed if hung on the inside of the cage. If water bottles have to be hung on the inside, one should provide a metal protective covering for the bottle. See Figures 11–4 and 11–5.

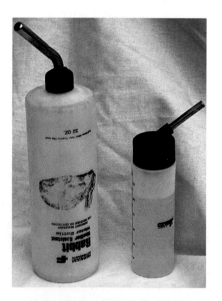

Figure 11–4   Vacuum-type water bottles are commonly available in 8-, 16-, and 32-ounce sizes. The 8-ounce size is ideal for hamsters, gerbils, rats, and mice. Pictured here are the 8- and 32-ounce sizes. Photo by Dean Warren.

Figure 11–5   Feed bowls are available in a wide range of sizes and made from several different materials. Ceramic bowls like the ones pictured here are heavy, so they won't tip over, and durable so small animals can't chew them. Photo by Dean Warren.

HANDLING

Gerbils are gentle and curious animals, and little skill is needed to tame one. Seldom will a gerbil bite, and then only if it is mistreated or provoked. Moving slowly and gently usually gets a gerbil to feel comfortable so that it can be picked up. Many times, they will readily climb into an owner's hand.

The easiest way to pick up a gerbil is to grasp the tail of the gerbil near the base, with the thumb and forefinger. After a gerbil has become accustomed to being handled, it will normally climb right into the owner's hand. The end of the tail should never be grabbed because the hair may come out or the skin can tear and pull loose, leaving the tailbone exposed. See Figure 11–6. This is extremely painful to the animal and may cause death. While holding the gerbil, one should stroke the head and back; the animal should be calm before any attempt is made to pick it up. Gerbils can also be picked up by grasping loosely with a hand around its back and the thumb and forefinger around the neck. The handler should be careful not to squeeze the animal too tightly or it will struggle to escape. Many times, a gerbil will nibble on a handler's finger, not to inflict a bite, but to let the handler know that it has been held long enough and wants down.

Young gerbils are especially hard to handle. They are small, nervous, and very quick. Be careful in handling them, because they may jump from the hand in an attempt to escape and fall to the floor; this can result in injury.

Tame gerbils readily learn to accept food from the handler's fingers. Most will readily come forward to

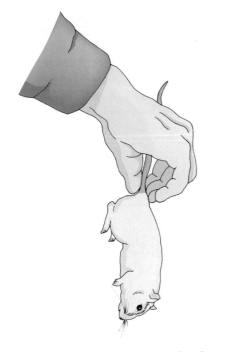

Figure 11–6 Gerbils can be picked up by the tail, but the tail must be grasped up close to the body. The skin can break and pull off the tailbones.

take sunflower seeds from the fingers. If they hesitate, be patient and they will soon get over any fear.

DISEASES AND AILMENTS

Gerbils are very hardy and, if well cared for, will seldom be affected by disease.

Colds are the most common ailment of gerbils; common symptoms are loss of appetite, constant sneezing, and runny eyes and nose. The animal may also appear lethargic or inactive. Colds are usually caused by temperature changes, overcrowded conditions, or drafty or damp conditions. The easiest way to treat colds is to remove the cause.

Red nose is a common condition that causes hair loss and red, swollen areas of the skin around the nose and muzzle. *Staphylococcus* bacteria are usually the cause, although other bacterial organisms or trauma may be involved. Gerbils usually recover without any medication, although local or systemic antibiotics may be needed.

The bacteria *Bacillus piliformis,* the cause of Tyzzer's disease, also affects gerbils and may cause high death losses. Affected animals have rough hair coats, lose weight, and appear listless and inactive; treatment is usually ineffective.

Gerbils generally are not troubled by external parasites, but they can become infected with fleas, lice,

and *Demodex* species of mange mites, which are transferred from other animals. General signs of these pests are dull fur, fur falling out, and unusual or continuous scratching by the animal. Animals suspected of having these parasites should receive a dusting from a flea powder formulated for cats or small animals. The animal should be dusted and the powder brushed; into the fur. 20 to 30 minutes should pass and then the powder brushed out of the fur. The dusting should be repeated in 24 hours. It is important that the cage be cleaned, disinfected, and dusted also.

Gerbils are occasionally seen with what appears to be fits or seizures; these attacks are usually observed during handling. In some cases, the animal appears to go into a trance with the body remaining motionless or slightly trembling; these attacks last only a few seconds. Treatment is not needed because the animal will return to normal activity. These attacks may be observed when cleaning the animal's cage and during handling. No cause is known at this time. Animals that show signs of fits or seizures should not be used for mating because this condition appears to be genetic.

REPRODUCTION

A pair of gerbils should be introduced to each other as early as possible; older animals may not accept each other, and fights may occur when they are introduced. Putting the female in the male's cage usually results in less fighting, but the pair should be observed closely. If introduced at about nine to twelve weeks of age, they will readily accept each other and mate shortly. However, it may be five or six months before a litter is produced, and many times, first-litter parents may neglect their young. Gerbils are primarily monogamous, and the pair bond may last throughout the life of the animals.

Young gerbils become sexually mature between nine and twelve weeks of age. The female comes into heat every four to six days, and mating usually takes place during the evening hours. Actual mating is very brief but occurs several times. During events leading up to actual mating, the male will sniff the female and make drumming noises with his rear feet. High-pitched squeaks may be heard from the pair. The gestation period varies between twenty-four and twenty-six days. See Figures 11–7 and 11–8.

Newborn gerbils are born naked, blind, and completely helpless. Litters consist of one to ten young, with three or four being the average. Their ears are folded shut, they do not have any teeth, and they rely entirely on mother's milk for nourishment. Newborn gerbils are about 1 inch long. During the first few days after birth, high-pitched squeaks can be heard coming

Figure 11–7 *The genital area of a female gerbil. Photo by Isabelle Francais.*

Figure 11–8 *The genital area of a male gerbil. Photo by Isabelle Francais.*

from the nest; these squeaking noises diminish as the young get older. At about five or six days of age, the body will become covered with a light covering of fur.

Young gerbils are very tough. They take a lot of bumping about from the mother and often will be kicked from the nest. Later, the mother will gather them up and return them to the nest. The male may also help gather the young up and many times will hover over them, keeping them warm and protected.

Gerbils do not seem to mind early handling of their young, and the handler can assist in returning lost young to their nest.

Young gerbils grow very rapidly and begin to crawl out of the nest even before their eyes are open. At about five days of age, the ears unfold; at about fourteen days of age, the eyes open. Their incisors have developed by this time, and they will start gnawing on solid food and drinking water. They are very active and will jump and run about the cage and engage in fights with their littermates.

SUMMARY

There are four subgenera and fourteen species of common gerbils. The best known of these is the Mongolian desert mouse or Mongolian gerbil, *Mericones unguiculatus.*

The Mongolian gerbil is found in parts of China, the former Soviet Union, and throughout most of Mongolia. In the wild, they live in colonies and dig burrows in sandy soils for shelter. Mongolian gerbils are active both day and night. Their diet consists of seeds, leaves, roots, stems, and occasional insects.

Most of the housing and equipment available for hamsters is also suited for gerbils; however, being more active, gerbils should have more cage space than hamsters. Like hamsters, gerbils are very adept at finding ways to escape. Gerbils can jump, and it is important that the cage be covered.

The easiest way to feed gerbils is using commercial pellets made especially for small animals. They are complete balanced rations, with all of the required vitamins and minerals added.

Young gerbils become sexually mature at between nine and twelve weeks of age. The female comes into heat every four to six days; mating usually takes place during the evening hours. The gestation period varies between twenty-four and twenty-six days.

Gerbils are very hardy and, if well cared for, will seldom be affected by disease.

DISCUSSION QUESTIONS

1. What are the general characteristics of gerbils?

2. How do hamsters and gerbils differ?

3. What types of food are usually fed to gerbils?

4. How should one pick up and handle a gerbil?

5. What are the common diseases and ailments of gerbils? What are their symptoms and how should they be treated?

SUGGESTED ACTIVITIES

1. With the approval of the instructor, bring gerbils to the classroom. The class can observe the animals. Correct methods of handling can be demonstrated.

2. With the approval of the instructor, bring a gerbil or pair of gerbils to the classroom. Leave the cage and animals for a few days. Your classmates can then become familiar with all aspects of gerbil care and management.

3. With the approval of the instructor, have a local veterinarian come to the classroom to discuss the common diseases and ailments of gerbils. The veterinarian can also discuss the prevention and treatment of gerbil diseases and ailments.

ADDITIONAL RESOURCES

1. Gerbils
   www.gerbils.org

2. The National Gerbil Society
   www.rodent.demon.co.uk/gerbils

3. Gerbils–Your guide To These Lovable Pets
   http://home.clara.net/david.hinsley/
   About_Gerbils.html

4. Mongolian Gerbils as Pets
   www.webcom.com/lstead/rodents/gerbils.html

5. Gerbil Care
   www.aspca.org/learn/fs_petcare

6. Pet Care Library
   www.healthypet.com/Library/index.html

7. Gerbils–Origin
   http://users.bart.nl/~fredveen/origin1.htm

CHAPTER 12

Rats

OBJECTIVES

After reading this chapter, one should be able to:

■ detail the history of the brown and black rat.

■ describe the characteristics of common rats.

■ list the common methods of feeding and caring for rats.

■ describe common diseases and ailments of rats.

TERMS TO KNOW

albino   carriers   inbreeding
Black Plague   gregarious   rodents
caped

CLASSIFICATION

Kingdom — Animalia   Family   — Muridae
Phylum — Chordata   Subfamily — Murinae
Class   — Mammalia   Genus   — *Rattus*
Order   — Rodentia   Species   — *R. norvegicus*
Suborder — Myomorpha   — *R. rattus*

HISTORY

Rats belong to the order of gnawing mammals called **rodents** (Rodentia). Together with mice, they compose the subfamily Murinae of the family Muridae. True rats form the genus *Rattus*. Rats are larger than mice and have more rows of scales on their tail.

The genus *Rattus* contains from 78 to 570 species, depending on which classification is used. Only two species, the black rat *(Rattus rattus)* and the brown rat *(Rattus norvegicus),* have been domesticated and used for research and as pets.

The black rat is sometimes referred to as the roof rat, climbing rat, or gray rat. It is believed to have originally come from Southern Asia. This species was well established in Europe during the 1200s and was the major reservoir for the **Black Plague** that killed more than one-fourth of the population during the 1300s.

The black rat reached North America during the 1500s and today is found along both coasts and throughout the southeastern states.

The brown rat is sometimes referred to as the Norway rat, barn rat, sewer rat, or wharf rat. It originated in Eastern Asia and Northern China; it found its way to Europe and North America during the 1500s. Today, the brown rat is well established throughout the United States.

Rats have spread to all parts of the world. Their success is attributed to their ability to adapt to many different habitats, environments, and food sources, as well as their ability to reproduce rapidly.

Rats in the wild cause tremendous damage in the United States every year. They destroy eggs, fruits, stored grain, and vegetables, and they attack other animals. Rats cause damage to buildings and cause fires by chewing through insulation of electrical wires. Rats are **carriers** of disease and have attacked human beings, especially babies and young children.

With all their destructiveness and bad characteristics, the black and brown rats, in the form of the white **albino** laboratory rats, have been of major importance in medical, biological, and psychological research. Laboratory rats have been used extensively in developing drugs, studying diseases, learning about nutrition, studying aging, and many other studies. Because they are intelligent and have shown an ability to learn, they have also been used in many behavioral studies.

MAJOR COLOR GROUPS

The Black Rat.   The black rat has a head and body length of 7 to 8 inches and weighs 4 to 12 ounces. The tail is longer than the head and body, and the ears are about half as long as the head. The color is usually black or dark gray with a brown or gray-white under-

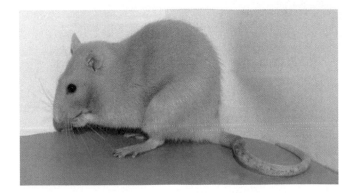

Figure 12–2   A brown rat (Rattus norvegicus). *Courtesy of Stacey Riggert.*

side. The black rat is an excellent climber and jumper. See Figure 12–1.

The Brown Rat.   The brown rat is larger than the black rat; its body and head are 7 to 10 inches long, and the tail is always shorter than the head and body. The brown rat is thicker and more robust than the black rat and weighs 7 to 17 ounces. Its ears are shorter and more rounded, and the fur is a dark to gray-brown on the back, with lighter colors of brown and gray on the underside. The brown rat is not a climber and prefers to live in tunnels, burrows, sewers, and basements. See Figure 12–2.

Varieties.   Because of mutations and selective breeding, both the black and brown rat can be found in several colors. The white laboratory rat is a descendant of albino strains. Creams, fawns, and light gray varieties are common today; another common variety is the hooded rat. It has a colored head and dorsal stripe on a white body; the colors are brown, black, fawn, and cream. The **caped** variety has a colored head but lacks the colored dorsal stripe. See Figures 12–3 through 12–8.

Figure 12–1   A black rat (Rattus rattus). *Courtesy of Stacey Riggert.*

Figure 12–3   A rat with hooded markings. Courtesy of Stacey Riggert.

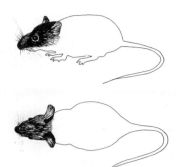

Figure 12–4   Caped markings.

Figure 12–5   Variegated markings.

Figure 12–6   Dalmatian markings.

Figure 12–7   Masked markings.

Figure 12–8   A rat with Berkshire markings. Courtesy of Brian Yacur.

CHARACTERISTICS

Rats are agile climbers, especially the black rat. They use their scaly tail to help maintain balance and to help them hold onto objects. Rats can jump up to 2 feet. They are also excellent swimmers and can stay submerged under water for 2 to 3 minutes.

Rats are very curious. They will climb a cage or stand on their rear legs in an attempt to get a better view of their surroundings, and they will soon learn their owner's voice and will come running when food is offered.

Rats are surprisingly clean and odor-free; they will groom and clean themselves and each other. If the bedding is changed regularly, very little odor is noticed.

Domesticated rats seldom bite; young rats bite only if they are being handled roughly or being abused. Rats, like other rodents, are very protective of their territory; adult rats will bite if surprised or scared. Handlers should go slowly and be patient when handling or feeding adults, especially if the hand is put into the cage. Females with young are very protective and will bite their handler if they feel the litter is being threatened.

Rats can become very upset and angry; when upset, they stiffen their body and arch their back. Every hair on the body will stand on end, giving it an imposing appearance. The tail will wag back and forth, and the animal will shake. Mature males will usually fight when put together, and fierce battles can result. Extreme care must be taken when handling rats that are upset or that are fighting.

Rats are **gregarious.** They do best when kept with other rats. Males that have been raised together usually get along. Care should be taken, however, when introducing a strange male into an occupied cage because the stranger will not be accepted and fighting will occur.

Rats are primarily nocturnal. Domestic rats that are kept as pets are active all day and do not mind being bothered; however, care should be taken when arousing a sleeping rat.

Rats in the wild will rarely live more than a year because of the large number of predators; domestic rats may live up to three years.

HOUSING AND EQUIPMENT

Cages available for housing hamsters and gerbils are acceptable for rats. Cages that are 10 inches by 16 inches by 10 inches allow sufficient room for a pair and their litter. A cage 10 inches by 20 inches by 10 inches works well for two females and a male; a cage 12 inches by 24 inches by 10 inches is more desirable. A cage that is 12 inches by 30 inches by 10 inches works well for three or four females and a male.

Remember that rats are rodents and will gnaw on soft materials. Wire mesh cages with solid bottoms or pull-out trays work well and are easy to clean. These cages are available with doors at the end or with tops that hinge open.

Cages can be constructed from wood and wire mesh; materials should be thick enough to withstand gnawing of the rats. Wood will soak up urine and will be harder to clean, however.

Glass aquariums work very well for rats. They are easy to clean and with a lid are escape-proof; a weight may need to be applied to the lid to keep the rats from escaping.

Water bottles should be hung on the outside of the cage with the spout extending into the cage. If the water bottle has to be hung on the inside, one must make sure it is protected with a metal shield. Feed bowls should be made of glass or ceramic materials; these are easy to clean, will not rust, and are gnaw-proof.

Rats are very active, and if a large cage is available, it should be equipped with ladders and ropes for the rats to get exercise. Small rats also enjoy playing on exercise wheels.

Several materials can be provided for bedding, but wood shavings or chips work best. Other materials can be provided for the female to build a nest, such as paper towels, paper napkins, or toilet tissue. The female will shred these into a nest.

FEEDING

The easiest way to feed rats is to use commercially prepared pellets formulated for small animals; these contain all of the nutrients that are necessary. Dry dog food can also be fed to rats; however, some fruits and vegetables may need to be added. Crackers, cereals, and breads also make ideal foods. These too should be supplemented with occasional fruits and vegetables.

If left too long in the cage, fresh foods will spoil and can cause upset stomach and diarrhea.

Bananas, apples, and strawberries, along with lettuce, carrots, cucumbers, and tomatoes, can be added to the diet; they should be added to dry rations. Over-

feeding a ration of fruits and vegetables may cause diarrhea.

Vitamins and mineral supplements are also available at most pet stores. These supplements are available in liquid form and can be added to the water or given directly to the rat.

Rats will consume an ounce of water or more per day. Water should not be given in an open dish because rats will generally make a mess.

HANDLING

Young rats can easily be picked up by grasping them around the body just behind the front legs. See Figure 12–9. They may also be picked up by the tail; the tail should be grasped at the base close to the body and the rat lifted. See Figure 12–10.

When handling older rats, be careful and talk so that they will relax. How the handler approaches the animal depends a great deal on how tame it is. If the animal is not tame, one must go slowly, get its attention, grasp it by the base of the tail, and lift. If further restraint is needed, the handler should take the other hand and grasp the animal around the body just behind the front legs and restrict the movement of its head with the thumb and forefinger.

A rat should never be grasped by the tip of the tail. In an effort to free itself, it will twist and squirm. Its tail or the skin on the tail can break loose and be pulled off, leaving the tailbone exposed. The rat is also very dexterous and can turn and climb up its tail. Leather gloves should be used if unsure of the situation.

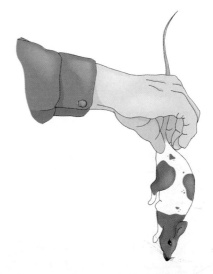

Figure 12–10 *An adult rat that may not be completely tamed can be picked up by the tail if it is grasped up close to the body. Rats that are tame can be picked up by grasping them around the body.*

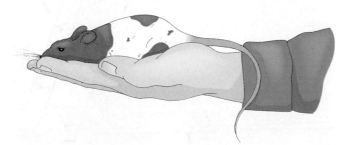

Figure 12–9 *Young rats are easily tamed and can be picked up and handled very easily.*

DISEASES AND AILMENTS

Rats in the wild have been recognized as carriers of many types of diseases and parasites; however, rats that are kept in captivity are fairly free of diseases and ailments.

Respiratory disease caused by the organism *Microplasma pulmonis* is one of the more common diseases. Signs include nasal discharge, snuffling, rattled breathing, rubbing the eyes and nose, tilted head, incoordination, and circling. Acute outbreaks can be controlled with antibiotics added to the drinking water.

Secondary invaders such as *Streptococcus pneumoniae, Pasteurella, Bordetella, Mycoplasma,* and *Pseudomonas* may cause pneumonia-like symptoms in animals. Increased respiratory difficulties and sudden death may also occur.

Treatment of respiratory diseases with antibiotics is often helpful, but complete isolation or elimination of affected animals may be the only way to eliminate the problem.

Good sanitation and ventilation are extremely important in controlling respiratory disease. The buildup of ammonia in unsanitary cages exacerbates the symptoms.

Polyplax spinulosa is a type of louse that may infest rats and causes loss of hair and intense itching. *Dermanyssus, Notoedres muris, Radfordia ensifera,* and *Ornithonyssus bacoti* are species of mites that may affect rats; these mites may cause hair loss and skin irritation. In more serious infestations, small fluid-filled lesions, swellings, and inflammation of the skin may be evident.

External parasites such as lice and mites may be controlled by dusting with flea powder. One should dust the area that appears to be infected, leave the powder on for 20 or 30 minutes, and then brush the powder from the animal. The treatment should be repeated in three or four days.

REPRODUCTION

Rats may become sexually mature at six weeks of age. Litters should be weaned and separated at four weeks to prevent **inbreeding** and unwanted litters. Young females bred at too young an age usually have small litters, and the babies are small and weak. Breeding at an early age may also shorten the productive life of the female. See Figures 12–11 and 12–12.

Breeding should be delayed until the rats are sixteen weeks old. At this age, the female is strong enough and mature enough to raise a litter.

The heat cycle occurs every four days and lasts about 12 hours. When the female is in heat, the male

Figure 12–11   Genital area of a male rat. Courtesy of Dean Warren.

will sniff the female and attempt to mount her. The actual act takes place very quickly and will be repeated several times.

Several different methods may be used in breeding rats; the easiest way is to put one male in with one to three females. This depends on the size of the cage and how many offspring are wanted.

The gestation period is twenty-one to twenty-four days. As the delivery day approaches, the female will busy herself shoving bedding up in a corner to make a nest.

The young are born pink, naked, blind, deaf, and completely helpless. There will be up to sixteen in a litter. The young can be heard making bird-like peeps. In a couple of days, soft fur will become visible; if the young rat is going to be two-colored, the darker color

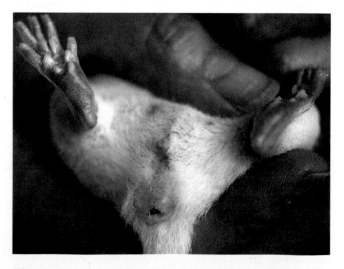

Figure 12–12   Genital area of a female rat. Courtesy of Dean Warren.

will be more visible. At about four days, the ears open. Young rats will start to explore the cage at about ten days of age. Teeth will begin to develop at ten days, and they will begin to nibble on hard foods. The water bottle should be hung low enough so that they can start drinking water. Their eyes will open at about fourteen days.

The male can be left in the cage or removed after mating. He will play no part in the care of the young but will not harm them either. The male is usually very tolerant of the young, allowing them to crawl or lie all over him. The female comes in heat very soon after giving birth, and the male should be removed to prevent another litter. Both parents will be very protective of the young, and care must be taken when feeding and caring for rats that have litters.

The cage should be cleaned a couple of days before the litter is expected. The mother and litter should not be disturbed until the young start exploring from the nest. If the cage gets especially messy, the litter may need to be moved so that the cage can be cleaned.

The young will be fairly self-sufficient at three weeks of age and should be weaned and separated at four weeks of age. At four weeks of age, young rats are very curious, friendly, and easy to handle. They can be tamed at this time and will soon learn to eat from the handler's fingers.

If two females occupy a cage and both have litters, they will share responsibilities for nursing and caring for the young. If the litter is large, the female may divide the litter into two or three groups and then spread her time among the groups. Nursing a large litter can be very taxing on the female.

SUMMARY

Rats belong to the order of gnawing mammals called rodents (Rodentia). Together with mice, they compose the subfamily Murinae of the family Muridae; true rats form the genus *Rattus*. Rats are larger than mice and have more rows of scales on their tail.

The genus *Rattus* contains 78 to 570 species, depending on which classification is used. Only two species, the black rat *(Rattus rattus)* and the brown rat *(Rattus norvegicus)*, have been domesticated and used for research and as pets.

Because of mutations and selective breeding, both the black and brown rat can be found in several colors. The white laboratory rat is a descendant of albino strains of the brown and black rat. Creams, fawns, and light gray varieties are common today. Another common variety is the hooded rat; it has a colored head and dorsal stripe on a white body. The caped variety has a colored head but lacks the colored dorsal stripe; colors are brown, black, fawn, and cream.

Rats are surprisingly clean and odor-free. They will groom and clean themselves and each other. If the bedding is changed regularly, very little odor is noticed.

Domesticated rats seldom bite; young rats bite only if they are being handled roughly or abused.

Any of the cages available for housing hamsters and gerbils are fine for rats. Cages 10 inches by 16 inches by 10 inches allow sufficient room for a pair and their litter.

Rats can be fed commercially prepared pellets formulated for small animals; these pellets contain all of the nutrients that are necessary.

Young rats may become sexually mature as early as six weeks of age. The heat cycle of the female rat occurs every four days and lasts about 12 hours; the gestation period will be twenty-one to twenty-four days. The young are born blind, naked, deaf, and completely helpless; there can be up to sixteen in a litter.

Rats in the wild have been recognized as carriers of many types of diseases and parasites; however, rats that are kept in captivity are fairly free of diseases and ailments.

DISCUSSION QUESTIONS

1. What are the two species of rat that have been domesticated?

2. Describe the common types of rats used for research and as pets.

3. How should one handle pet rats?

4. What are the common foods for pet rats?

5. What are the common diseases and ailments of pet rats?

SUGGESTED ACTIVITIES

1. With the approval of the instructor, bring rats to the classroom. The class can observe the animals. Correct methods of handling can be demonstrated.

2. With the approval of the instructor, bring a rat or pair of rats to the classroom. Leave the cage and animals for a few days. Class members can then become familiar with all aspects of rat care and management.

3. With the approval of the instructor, have a local veterinarian come to the classroom to discuss the common diseases and ailments of rats. The veterinarian can also discuss the prevention and treatment of rat diseases and ailments.

4. With the approval of the instructor, bring rats and cages to the classroom. Conduct feeding trials using different types of foods (e.g., candy bars, cakes, and cookies).

5. If available and with the approval of the instructor, invite research people in to speak on the research they do using rats.

6. Write to research companies for information on research using rats. If Internet access is available, one may find information on companies doing research with rats.

7. With the approval of the instructor, construct a maze for rats. Set up different situations and use different treats to see if the rats react differently and learn some mazes more quickly than others.

ADDITIONAL RESOURCES

1. Fancy Rats
   www.altpet.net/rodents/rats/rf32.html

2. The National Fancy Rat Society
   www.cableol.net/nfrs

3. Rats
   www.webcom.com/lstead/rodents/rats.html

Mice

OBJECTIVES

After reading this chapter, one should be able to:

- detail the history of the common house mouse.
- describe the clan or colony structure of the house mouse.
- list practices for the general care of pet mice.
- list common feeds for mice.
- describe the common methods of handling mice.
- describe common diseases and ailments of pet mice.

TERMS TO KNOW

clan
colony structure

condo
piebald

subordinate

CLASSIFICATION

Kingdom — Animalia
Phylum — Chordata
Class — Mammalia
Order — Rodentia
Suborder — Myomorpha

Family — Muridae
Subfamily — Murinae
Genus — *Mus*
Species — *M. musculus*

HISTORY

The common mouse is the most familiar and most widely distributed rodent in the world; there are four subgenera and thirty-six species. The best known species is the common house mouse *(M. musculus)*; it dwells wherever humans live.

The word *mouse* comes from an old Sanskrit word meaning "thief"; Sanskrit is an ancient language of Asia, where scientists believe the house mouse originated.[1] The house mouse was originally a wild field animal. As humans turned from hunter to farmer, the house mouse learned that it was easier to live on the stored grain of farmers and that the presence of people meant safety from enemies.

From Asia, the house mouse spread throughout the world. The common house mouse is one of the world's most successful mammals; they multiply extremely fast and are able to adapt to almost any condition. Their reputation has been as a dirty animal associated with carrying diseases and as a very destructive animal capable of causing tremendous economic losses of stored grains and other property.

Subsequently, mice have gained a place in history. The ancient Egyptians and Romans had a phrase that described the mouse's reproductive ability; they used to say, "it's raining mice," or, "mice are made of raindrops." The Greeks and the people of India believed that mice were lightning bolts born of thunderstorms. In some places, mice were kept in temples and worshiped as sacred animals. They were also viewed as an instrument of punishment. People would present sacrificial offerings to stop or prevent a mouse plague.

More than 4,000 years ago, the Cretans built a temple in Tenedos, Pontus, where they fed and worshiped mice. In the Cretan victory over Pontus, according to legend, mice helped by chewing through the leather straps of the shields of the Pontic soldiers so that they were unable to defend themselves.

Paintings on ancient bowls and other clay artifacts tell us that Egyptians also kept mice about 4,000 years ago. The first white and spotted mice appeared about 4,000 years ago in Greece, Egypt, and China. White mice in particular were considered sacred; people used them to predict the future, as lucky charms, or to keep wild mice away. Waltzing mice originated in China 2,000 to 3,000 years ago. The Japanese bred white and colored mice systematically 3,000 years ago. From there, these mice were introduced into Europe and North America about 150 years ago. See Figure 13–1.

In ancient Rome and during the Middle Ages, mice were used as medicine against all kinds of diseases. They were dried, pulverized, sliced, marinated in oils, and then used as a compress or taken internally. Their blood was a favored ingredient of drugs and tonics. Mice were supposedly good for flesh wounds, snake bites, warts, bladder irregularities, diabetes, enlarged thyroid glands, diseases of the eye, and loss of hair.

Even though mice were used for medicinal purposes, they were also targets for extermination. All kinds of methods were recommended for getting rid of them, except for the use of cats. During the Middle Ages, cats were not recognized as a natural predator of the mouse nor were they considered pets. Cats were hated as accomplices of sorcerers and witches and suffered extreme cruelty.

In more modern times, the house mouse, especially the albino strains of *M. musculus,* has been

Figure 13–1 *A white mouse. Courtesy of Stacey Riggert.*

widely used for medical and biological research, particularly in the study of heredity.

By careful breeding over the years, today's mice are more gentle and less timid than their ancestors. Pet mice are relatively free of disease, and if handled frequently, they show little tendency to bite or escape. Today, these curious and interesting mammals are available in many different colors and color combinations.

CHARACTERISTICS

Mice have a pointed nose and a slit upper lip. The tail is usually hairless with visible scale rings. Mice have four toes on their forefeet and five toes on their hindfeet.

The house mouse is approximately 2½ to 3½ inches long excluding the tail; the tail is generally the same length as the body. This small mammal will weigh ½ to 1 ounce. The house mouse has a small head and a long, narrow snout. Several long, thin whiskers are used by the animal to help find its way about in dark, tight areas. The fur of the house mouse is grayish-brown on the animal's back and lighter colored to white on the underside. See Figure 13–2.

The eyes of the house mouse are fairly large, round, and black; they cannot see very well. Because of the large, spherical shape of the eye, they have an almost perfect visual field, but because of the position of the eyes on the side of the head and the shape of the eye, they are incapable of detailed vision. Mice can detect movement but may be unable to determine what the movement is. Cats, in their pursuit of a mouse, move ever so slowly and may remain motionless for long periods.

Figure 13–2 A brown house mouse. Courtesy of PhotoDisc.

Mice have fairly large ears and possess a highly developed sense of hearing. They perceive sounds in very high frequency ranges, up to 100,000 Hz.

Smell is the most highly developed sense that mice possess. Mice use smell for locating food, identifying family and colony members, finding their way around in darkness, and identifying enemies.

Mice are primarily nocturnal in habit and usually seek hiding during the day, but pet mice, feeling safe around humans, may come out during the day. Pet mice can recognize by smell those who feed and handle them and will feel at ease in their presence.

Mice are gregarious and prefer the company of other mice. The house mouse lives in a **clan** or **colony structure.** This colony structure gives the colony more protection against enemies and makes it easier for the colony members to find and store large quantities of food. Females in the colony share duties of raising the young, providing warmth for the young, and nursing and feeding the young.

Members of the colony leave urine markings as they explore new areas and search for food. Mice have glands in the soles of their feet that leave secretions that other members of the colony can follow. These scent markings also mark the territory of a colony and serve as a warning to members of other colonies. Members of colonies, upon coming in contact with markings or scent from another colony, will not venture into the territory.

The mouse colony is ruled by a lead or head buck; all other males in the colony are **subordinate** to him. The lead buck is the only male in the colony allowed to mate with the females. The subordinate males are kept from mating the females by the lead buck, and they also are rejected by the females.

Subordinate males may challenge the lead buck for the position by attacking him; subordinate males attempt to bite and fight. During the fight, the males attack each other with their front feet and continuously try to bite each other. If one animal in the fight wants to stop, he will make a squeaking noise and standing on his rear legs, he may press his forefeet against his body, which signifies defeat and submission.

Should the lead buck be defeated in a fight, all males in the colony will bite him, and he will move to the bottom of the social structure. Many times, the lead buck will be an older male and will be defeated by a younger, stronger male. The defeated male may also be bitten by all the females in the colony. The defeated male may essentially be an outcast and go off alone, refuse to eat, and eventually die.

Today, mice can be found in a wide variety of body colors, combinations of colors, and with red or black eyes. There are approximately fifty different varieties of fancy mice. These are divided into five basic groups:

1. Self colors or mice of one color

2. Tans, mice of any recognized color with tan belly color (See Figure 13–3.)

3. **Piebald** or pied marked mice—mice with spots, patches, or broken patterns (See Figures 13–4 and 13–5.)

Figure 13–3 A mouse with a tan or fox pattern.

Figure 13–4 A mouse with a broken color pattern.

Figure 13–5   A mouse with a variegated pattern.

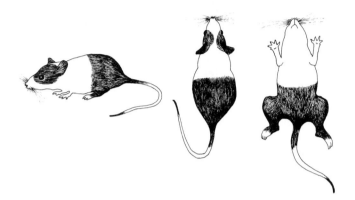

Figure 13–6   A mouse with a Dutch color pattern.

4. Satins of any color or markings, but with satinized coat

5. Any other variety that does not fit into any other group (See Figure 13–6.)

Males tend to produce more odor than females and leave more scent markers; however, males may learn tricks faster and usually are easier to tame and handle.

HOUSING AND EQUIPMENT

Homes for pet mice can be purchased from pet shops, or they can be constructed from wood and wire screen. A cage for a pair of mice should have at least 72 square inches of floor space and a height of 8 inches. Larger size cages would allow for small nest boxes, exercise wheels, and toys to be added to the cage. Also, in the case of a breeding pair, a larger cage would be needed for growing litters. Mice will stay healthier and be more active if they have plenty of space to move around.

Many of the cages and the equipment used for other types of small animals can be used for mice. Cages that have plastic bottoms and metal bars will work for mice, but the bars need to be close together to prevent mice from escaping. The bars should have about ⁵⁄₁₆ inch between them. Cages of this type are easy to clean and disinfect because the plastic bottom can be easily removed.

Aquariums and terrariums also make excellent cages for mice. Small aquariums with dimensions of 6 inches by 12 inches by 8 inches will work for a pair of mice; however, larger aquariums will allow the mice to have more space for exercising. Exercise wheels and nest boxes can be easily added to larger aquariums.

Cages can be constructed from wood and wire screen. Wood cages, however, are harder to clean. The urine will soak into the wood and make cleaning difficult. Mice are fairly clean and odor free, but mouse urine is very strong. Treating the wood with linseed oil or other nontoxic preservative will make wooden cages a little easier to clean and make them last longer.

All cages for mice must have some type of cover or lid because mice are excellent climbers. Most of the cages with metal bars have hinged covers. Covers or lids for aquariums can be purchased at most pet shops or can easily be constructed with wood and wire screen.

Another interesting type of "home" for mice is the open concept or mouse **condo**. As mentioned earlier, mice do not have very good vision. They cannot see detailed objects at ranges of more than 20 inches. Because of this poor vision, mice will not jump from tables or objects that are more than 20 inches from the floor. A small table can be used as a base for the condo. The top of the table should extend beyond the legs so that the mice cannot climb down the legs. Two or more smaller platforms can be constructed above the base. These platforms can be connected with ropes, slides, or ladders. Exercise wheels, teeter-totters, swings, and other toys can be placed on the various platforms to provide the mice plenty of exercise. The top layers may be furnished with a nest box or two so the mice can hide or escape. Small strips of wood should be placed around the edges of the platforms to keep food, nesting materials, and toys from falling off the edges. Mice will enjoy climbing and playing on a mouse condo, and watching them can provide hours of entertainment. See Figure 13–7.

In a classroom situation, mice may run for hiding in the nest boxes because they are wary of all the noise and activity, but as soon as the activity dies down in the late afternoon, they will venture out to feed and play. Occasionally, a mouse may fall off the edges of the condo as it leans out over the edges, but it will never jump from the edges.

Wood shavings make the ideal bedding for mice. Mice will chew on the shavings and shred them for making nests. Sawdust and wood shavings with a lot of dust in them should be avoided. The dust can cause breathing and respiratory problems for the mice.

Bedding should be changed at least once a week or more often if the cage begins to smell. Care should be taken when cleaning the cages containing female

Figure 13–7 A mouse "condo" can be constructed and set up in a classroom. Students can observe the activities of the mice. Photo by Dean Warren.

mice with newly born litters so that the female does not get overly excited or nervous.

Newspaper, paper towels, and toilet paper also make suitable bedding material. Mice will also chew and shred these materials to make comfortable nests.

Mice usually urinate in the corners of their cages. Placing cat litter or baking soda in the corners of the cage will help minimize the mousy odor. Although this minimizes the smell, it does not eliminate the need for cleaning and disinfecting the cages.

Mice do not drink very much water, but water should be available at all times. The common vacuum-type water bottles that are available at most pet shops are ideal. Water should not be placed in small bowls or plates because they will soon be contaminated with urine and feces.

FEEDING

Mice in the wild will eat almost anything, but their main diet consists of grain and seeds. Either a formulated commercial feed or a mixture of grain and seeds can be used.

Commercial feeds come in a hard pellet form and contain all necessary nutrients. The hard pellet is also beneficial in keeping the continuously growing front incisors worn down.

A ration can be prepared using corn, oats, and wheat, and then adding small amounts of millet, barley, and buckwheat. Mice love sunflower seeds, which should be added sparingly because they are high in fat content. Small amounts of oily seeds are good for adding shine to the hair coat. Oatmeal, crackers, and bread also are good for young mice.

In feeding a ration, various greens and vegetables can be added. Fresh grass, lettuce, dandelions, carrots, apples, dates, and raisins can be added. Clean, fresh hay can supplement the diet and will also serve as a nesting material. It is important that these foodstuffs be free of any chemicals, contaminants, or molds that can affect the animal's health.

Mice will not overeat. Care should be taken so that waste is minimized and soft foods are not allowed to spoil. Too much soft foods may cause upset stomachs and may lead to diarrhea.

HANDLING

Mice can provide hours of entertainment as they climb ropes, climb ladders, and run on exercise wheels. When purchased at a young age, mice will tame very quickly; they will take food from fingers and eventually climb onto the owner's hand.

Small children need to realize that mice are not toys; they are delicate animals that can easily be crushed by excited children.

A mouse should not be picked up by the end of the tail or nape of the neck because this causes discomfort and pain.

Until a mouse becomes tame enough, it can be scooped up in a cup. A mouse will usually go into a cup because of its curious nature. See Figure 13–8.

A mouse brought home should be left alone for a day or two. After this period of adjustment, taming can begin. Treats can be offered from the fingertips; the mouse should not receive any other food. Later,

Figure 13–8   One way of picking up a mouse is to use a cup and scoop the mouse up. Hold the cup down and the mouse will usually go into the cup.

Figure 13–9   Mice can also be picked up by lifting the base of the tail. This is probably the easiest way.

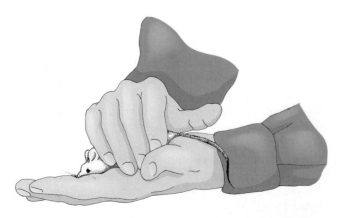

Figure 13–10   Mice that are tame can be picked up by "cupping" in the hands. Mice that are familiar with their handler and not afraid will usually climb up in the hands.

food can be put in the palm of the hand; soon, the mouse will come onto the hand and eat the food. An attempt can then be made to pet and stroke the mouse; a mouse likes to have its head and ears stroked. See Figures 13–9 and 13–10.

DISEASES AND AILMENTS

If mice are given proper diet, housing, and sufficient exercise, they will usually be free of disease. Problems that usually arise are nutrient deficiencies or obesity because of improper diets. Mice that do not receive sufficient exercise may become lethargic and lazy; their overall life span will also be shortened.

Several organisms can cause respiratory diseases in mice: *Mycoplasma pulmonis, Sendai virus, Pasteurella pneumotropica, Pseudomonas aeruginosa, Klebsiella pneumoniae,* and *Bordetella bronchiseptica.*

Many of the respiratory diseases are a result of changes in temperature, drafts, and high humidity. Symptoms may appear as squeaking or rattling breath, running nose, and watering eyes. The fur loses condition and the animals sit around in a huddled position with their backs arched; they are also listless and inactive. Prevention and control of causative factors are the most effective controls.

Salmonella and related bacteria are responsible for many of the serious infectious diseases. Unclean cage conditions, contaminated feed or water, sour milk, wet greens, and flies and parasites transmit disease agents.

Mice are commonly infected with a variety of mites *(Myobia musculi, Mycocoptes musculinus, Myocoptes rombutsi, Radfordia affinis, Ornithonyssus bacoti)* that can cause severe irritation and inflammation due to self-trauma and hair loss. Repeated use of insecticides is generally effective in controlling these parasites.

Mice can be infested with pinworm *(Syphacia ob-velata, Aspiculturis tetraptera)* and tapeworm *(Hymenolepis nana* and *Hymenolepis diminuta).* These parasites rarely cause clinical diseases.

The protozoans *Spironucleus muris* and *Giardia muris* are capable of causing high death rates in mice between three and five weeks of age. Affected animals become lethargic, have hunched posture, do not gain weight, develop sticky stools, and have gas- and fluid-filled intestines because of the presence of large numbers of parasites.

REPRODUCTION

Female mice become sexually mature and have their first heat period as early as four weeks of age; males become mature as early as five weeks of age. A non-pregnant female comes in heat every three to six days. The heat period lasts approximately 12 hours, depending on the age. Young females usually have short cycles, and older females have longer cycles.

When a female is in heat, the male mouse continuously follows her around in an attempt to mate. When the female is ready, she remains still and moves her tail to one side. The actual mating takes only a few seconds and is repeated several times. See Figures 13–11 and 13–12.

The gestation period is normally twenty-one days but may vary by one to three days.

As gestation progresses, the female seeks out a nesting area that will provide security, darkness, and warmth. Newspaper, paper towels, straw, or other materials should be provided for a nest.

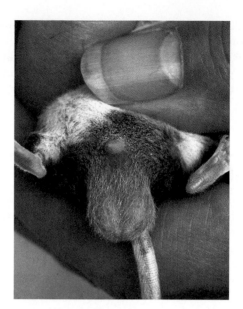

Figure 13–12   The genital area of a male mouse. Courtesy of Dean Warren.

As with other animals, the female will tear open the sack, bite off the umbilical cord, and clean each newborn. The female will eat the placenta and afterbirth and prepare for the next delivery. The actual birth of the litter will take about 15 to 30 minutes, depending on the size of the litter.

A mouse giving birth should never be disturbed; she may become nervous and panic. In her panic, she may kill and eat the newborn. The new litter should not be touched for at least a week.

A litter normally consists of five to ten but may vary from one to twenty. The young are extremely small, weighing about 1 to 2 grams and measuring about ½ inch long. The newborn will be completely helpless, hairless, and blind; they will depend on their mother's milk for the first two weeks. Hair will start to develop at two or three days, and they will have all their hair at eight to ten days. They will start to move around the nest at about five days, but they will not venture out until about ten days of age.

During the first eight to ten days, the mother mouse will lick away the urine and soft feces of the newborn. This licking and massaging is important to stimulate the digestive and excretory systems of the newborns.

The incisors of the newborn become visible at about ten days of age. Their eyes open at about fourteen days, at which time they become more active. They begin to explore their surroundings and start to nibble on solid foods.

The owner may want to start handling the young when they reach fourteen days of age; the young soon become tame through continued handling. Care needs

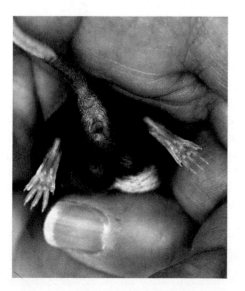

Figure 13–11   The genital area of a female mouse. Courtesy of Dean Warren.

to be taken because they may become frightened and leap from the hands in an attempt to escape.

The young should be separated by sex at about three weeks of age to prevent unwanted matings. Mice can reproduce rapidly, and if the owner is not keeping a close watch, the numbers can get out of control quickly.

SUMMARY

The common mouse is the most familiar and widely distributed rodent in the world; there are four sub-genus groups and thirty-six species. The best known species is the common house mouse *(M. musculus)*.

By careful breeding over the years, today's pet mice are more gentle and less timid than their ancestors. These mice are relatively free of disease, and if handled frequently, they show little tendency to bite or escape. These curious and interesting mammals are available in many different colors and color combinations.

Mice are gregarious and prefer the company of other mice. The house mouse lives in a clan or colony structure. This gives mice more protection against enemies, and it is easier for colony members to find and store large quantities of food. Females in the colony share duties of raising young, providing warmth, nursing, and feeding the young.

Suitable homes for pet mice can be purchased from pet shops or can be constructed from wood and wire screen. A cage for a pair of mice should have at least 72 square inches of floor space and a height of 8 inches. Larger size cages would allow for small nest boxes, exercise wheels, and toys to be added to the cage; also, in the case of a breeding pair, a larger cage would be needed for growing litters. Mice will stay healthier and be more active if they have plenty of space to move around.

Mice in the wild eat almost anything, but their main diet consists of grain and seeds. When feeding a pet mouse, either a formulated commercial feed or a mixture of grains and seeds can be used.

Mice can make enjoyable pets and provide hours of entertainment as they climb ropes, climb ladders, and run on exercise wheels. When purchased at a young age, they usually tame very quickly. As they become familiar with their owner, they will take bits of food from their owner's fingers and will eventually climb onto the owner's hand.

If mice are given proper diet, adequate housing, and sufficient exercise, they will usually be free of disease. Problems that usually arise are nutrient deficiencies or obesity because of improper diets. Mice that do not receive sufficient exercise will become lethargic and lazy, and their overall life span will be shortened.

Female mice become sexually mature as early as four weeks of age. The heat cycle occurs every three to six days and lasts approximately 12 hours. The gestation period is normally twenty-one days but may vary by one to three days. A litter consists of five to ten but can vary from one to twenty.

DISCUSSION QUESTIONS

1. What is the history of the domesticated mouse?

2. Mice were used in ancient Rome for medicinal purposes. What were they used to treat?

3. What are the characteristics of the common house mouse?

4. What is a clan or colony structure? Describe the advantages of such a structure.

5. What is the main diet of mice in the wild?

6. How should a pet mouse be handled?

7. What are the common foods for pet mice?

8. What are the common diseases and ailments of pet mice? What are the symptoms and treatment for them?

SUGGESTED ACTIVITIES

1. With the approval of the instructor, bring a pair of mice to the classroom. Classmates can observe the habits of these small animals and the class can discuss their observations.

2. With the approval of the instructor and if materials are available, build a mouse "condo" for the classroom. Classmates can observe the mice in an open, cage-free environment.

3. With the approval of the instructor, bring in mice and cages or an aquarium to house them. Class members can conduct feeding trials with mice. Use various types of food. Class members can keep records and observe the results.

4. With the approval of the instructor and if materials are available, build a maze for mice. Class members can conduct various tests to see how quickly the mice learn the maze.

ADDITIONAL RESOURCES

1. American Fancy Rat & Mouse Association www.afrma.org

2. Northeast Rat & Mouse Club International
   www.geocities.com/Heartland/Ranch/3220/
   mice.html

3. The Rat & Mouse Club of America
   www.rmca.org

END NOTE

1. Adapted from *The World Book Encyclopedia.*
   © 1993 World Book, Inc. By permission of the
   publisher.

CHAPTER 14

Guinea Pigs

OBJECTIVES

After reading this chapter, one should be able to:

- detail the history of the guinea pig.
- list and describe the seven varieties of guinea pigs.
- list practices for maintaining the good health of guinea pigs.
- describe how to pick up and handle a guinea pig.
- describe the common diseases and ailments of the guinea pig.

TERMS TO KNOW

abscess

cavy

crest

kinked

malocclusion

mane

oral mucosa

peripheral vision

rosettes

ticking

vivariums

CLASSIFICATION

Kingdom — Animalia
Phylum — Chordata
Class — Mammalia
Order — Rodentia

Suborder — Hystricomorpha
Family — Caviidae
Genus — *Cavia*
Species — *C. porcellus*

HISTORY

The guinea pig is more accurately called a **cavy;** there are seven species of cavy. The exact history of the domesticated cavy *C. porcellus* is not known, but it is believed to be derived from *C. aperea, C. tschudii,* or *C. fulgida.* The domesticated guinea pig has been bred for meat production in South America for at least 3,000 years. During the period of the Inca Empire, from about 1200 to 1532 AD, highly selective breeding produced a variety of strains, differentiated by color pattern and meat flavor. The range of domestication extended from Northwestern Venezuela to Central Chile. Subsequent to the Spanish conquest, there was a disruption of breeding and a contraction of range. The guinea pig is still widely kept as a source of food by the native people of Ecuador, Peru, and Bolivia. The guinea pig was brought to Europe in the sixteenth century and, since the mid-1800s, has been used by laboratories around the world for research on pathology, nutrition, genetics, toxicology, and development of serums. It also makes an ideal pet.[1]

The guinea pig is neither a pig nor does it come from Guinea; the actual reason for the name is not known, but there are several explanations. The species name *porcellus (C. porcellus)* for the domestic guinea pig means "little pig" in Latin. The only similarities to a pig are the low grunts, the squeals they make when they are hungry, and their fat little bellies.

After obtaining some of the animals, Dutch and English traders stopped off in Guinea on their way back to Europe. The first guinea pigs were sold in England for one English coin called a *guinea.* Another explanation could be that guinea comes from the word *guine* in Portuguese, which means "far away and unknown lands."

Guinea pigs have been instrumental in research on nutrition, tuberculosis, and other infectious diseases, as well as in the discovery and production of serum against diphtheria. They were also used by Louis Pasteur in his research on rabies. Guinea pigs' contribution to medical science has saved countless human lives.

MAJOR VARIETIES

There are seven common varieties of guinea pigs. They are the Abyssinian, American, Peruvian, Satin, Silkie, Teddy, and the White Crested.

The Abyssinian.   The Abyssinian has a rough, wiry hair coat. The hair is made up of swirls or cowlicks called **rosettes.** The more rosettes on the animal, the more desirable. The Abyssinian is found in all colors and color combinations. See Figure 14–1.

Figure 14–1   Abyssinian guinea pig. Courtesy of Isabelle Francais.

The American.   The American is the most common of the guinea pig varieties. The hair of the American is short, very glossy, and fine in texture. This shorthaired guinea pig can be found in a wide variety of colors and color combinations; its short hair makes it easy to care for. See Figure 14–2.

Figure 14–2   American guinea pig. Courtesy of Isabelle Francais.

The Peruvian.   The Peruvian is the longhaired variety. Length, evenness, and balance of the hair coat are the deciding features when judging this variety. The Peruvian can be found in many of the same colors and color combinations as the shorthaired, American variety. The hair coat takes a great deal of time and effort to keep clean and may reach a length of 20 inches. Because the guinea pig does not have a tail, the Peruvian has the appearance of an animated mop, and it may be difficult to tell which end is which. See Figure 14–3.

Figure 14–3   Peruvian guinea pig. Courtesy of Dean Warren.

The Satin. The luxurious, shiny fur of the Satin is its major characteristic. The coat is fine, dense, and soft and has a sheen. Satins are found in the same colors and color combinations as the other guinea pig varieties. See Figure 14–4.

Figure 14–4   Satin guinea pig. Courtesy of Isabelle Francais.

The Silkie. The Silkie, sometimes referred to as the Sheltie, is another longhaired variety, but unlike the Peruvian, there is no long frontal sweep over the head. The Silkie has a **mane** that sweeps back from the head, between the ears, and back over the back and down the sides. The Silkie is also found in the same colors and color combinations as the other varieties. See Figure 14–5.

Figure 14–5   Silkie or Sheltie guinea pig. Courtesy of Isabelle Francais.

The Teddy. The Teddy variety has short, kinky hair. The fur is short, resilient, and thick and lies close to the body; the whiskers are also **kinked.** The Teddy can be found in the same colors and color combinations as the other guinea pig varieties. See Figure 14–6.

Figure 14–6   Teddy guinea pig. Courtesy of Isabelle Francais.

The White Crested. The White Crested guinea pig has short hair and resembles the American shorthair except for the **crest,** which is a rosette and radiates evenly with a clearly defined center from the forehead. The White Crested is primarily found in the self, solid, and agouti colors; the crest is, of course, white. See Figure 14–7.

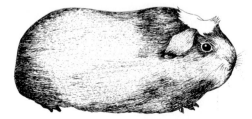

Figure 14–7   White Crested guinea pig.

CHARACTERISTICS

Adult guinea pigs reach a length of 8 to 14 inches and weigh 1 to 4 pounds. The males are larger and have larger, more rounded heads. The body of the guinea pig is short, stocky in build with short legs; there are three toes on the forefeet and four toes on the hindfeet. They have short, sharp claws.

The nose of the guinea pig is short, blunt, and rounded. The mouth is small, and the upper lip, like that of all members of the rodent family, is split. They have long whiskers on the sides of the upper lip; these whiskers serve the same function as the whiskers on a cat, allowing the animal to make its way in dim light. The ears are short and have very little fur on them.

A guinea pig has very sensitive hearing; it can detect frequencies beyond the range of the human ear. The upper levels of the human ear is around 15,000 to 20,000 Hz. The guinea pig, however, can hear frequencies up to 30,000 Hz. Guinea pigs communicate with a series of various sounds; many are high-pitched squeaks or whistles. When threatened, the fur on their neck stands up, and they chatter their teeth in an attempt to scare off an enemy.

Their sense of smell is also highly developed; they learn to recognize human scent and can identify individual humans. They recognize their owner and whistle when the owner comes into the room.

Guinea pigs also have very keen sight and excellent peripheral vision that allows them to spot approaching enemies from all directions.

Guinea pigs in the wild live in colonies or clans; there usually is one dominant male and five to ten females in a colony. Two adult males cannot exist peacefully in a colony; younger and subordinate males usually exist on the edges of the colony. As younger males become independent of the colony, they try to lure females away from existing colonies and start their own.

Colors and color combinations of guinea pigs are agouti, selfs, solid, and marked. Agouti is described as having a ground or undercolor, determining the base color, with dark ticking over the entire animal except the belly, which should be devoid of ticking. Selfs or self-colored animals have the same color hair over the entire body. Solid or solid-colored animals have color uniformly over the entire animal. This color uniformity may be obtained by the intermingling of different-colored hair shafts (Brindles, Roans) or Agouti-colored (one base color, one tip color) hairs over the entire body. The solids should be devoid of markings and shadings. Animals with marked coloring have a basic color broken up by orderly placement on a white background.

HOUSING AND EQUIPMENT

Cages are available at most pet shops. The typical cage has a plastic bottom that is 3 or 4 inches deep and has vertical wire bars for the upper part; the plastic bottom can be easily removed for cleaning. The plastic bottom trays should be at least 3 or 4 inches deep, with deeper trays preferred to prevent the animals from scattering bedding or shavings out of the cage. Cages for a single guinea pig should be at least 12 inches by 24 inches; smaller cages will not allow the animal to have sufficient room to exercise.

Cages do not need covers or tops; guinea pigs usually will not climb out of cages. Larger aquariums or vivariums will also work but are somewhat heavy, although easy to clean. Cages with vertical wire bars and glass aquariums allow the guinea pig to see out; they like to see what is going on in their surroundings.

Square or rectangular cages can be made of wood for very little cost. Wood cages are not as suitable because the wood will soak up urine and is hard to clean; eventually, these cages will become a permanent source of odors.

Plastic tubs normally used to store engine or machinery parts make ideal houses and can usually be purchased from hardware stores. These tubs should be at least 6 to 8 inches deep. See Figure 14–8.

If more than one animal is going to be housed, each adult should have a floor space of at least 180 square inches.

Guinea pigs can be kept outside during the summer once the temperature moves above 50°F. Small houses similar to a dog house can be constructed to protect the animals from the elements, and a run can be constructed from boards and chicken wire. Each animal should have 3 square feet of run. Portable-type houses and runs can be moved as the grass gets eaten and tracked down. The runs should have covers for protection from dogs, cats, and wild animals. Runs also need to be anchored down to prevent preadators from digging under the wire and attacking the guinea pigs and to prevent the guinea pigs from escaping.

Food bowls should be made of heavy glass or earthenware to prevent the bowl from being knocked over. Guinea pigs drink a lot of water, which should be provided in a vacuum-type water bottle. Guinea pigs should not be provided water in a bowl, which can be tipped over or contaminated with feces and urine.

Figure 14–8   Plastic tubs that are designed for machine parts can be used to house guinea pigs. This black plastic tub was purchased at a local hardware store. Courtesy of Dean Warren.

FEEDING

Guinea pigs are vegetarians. Their large front incisor teeth are very sharp and are used to bite food. They have eight molars in both the upper and lower jaws that are used to grind food. All of their teeth grow continuously; therefore, they must receive solid foods to keep their teeth worn down.

Guinea pig pellets are available that contain all of the nutrients needed. These pellets are hard and serve to wear the teeth down.

Guinea pigs, unlike many other animals, cannot synthesize vitamin C in their body and must be supplied with it in their diet. A lack of vitamin C causes scurvy, which results in dehydration, poor appetite,

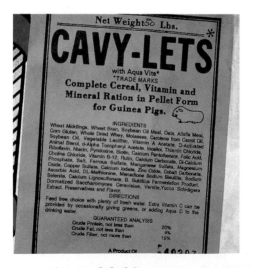

Figure 14–9   This is a label from a bag of guinea pig pellets. Note that the feed contains ascorbic acid (vitamin C). Courtesy of Dean Warren.

diarrhea, rough hair coats, lethargy, and weight loss. Small pinpoint hemorrhages on the gums and joint swellings with lameness also occur. The animals have severely reduced resistance to infectious agents causing respiratory disease. Most commercially prepared pellets have vitamin C added; however, it breaks down rapidly, and opened bags of pellets should be used within thirty days. See Figure 14–9.

Fresh greens, fruits, herbs, green vegetables, root vegetables, potatoes, clover, and dandelion greens are all readily consumed by guinea pigs. It is important that these foodstuffs be rinsed to remove any residue of chemical pesticides. Lettuce is a favorite food, as well as carrots, apples, pears, turnips, beets, and cucumbers.

Dry alfalfa or clover hay should be a main staple in the diet; the dry hay helps wear the animal's teeth down. It is important that hay not be moldy, musty, or dusty; moldy hay can be fatal. Dry grass hay and bean hay can also be used.

Wheat, corn, and oats are high-protein foods that should be added. Peanuts and sunflower seeds are readily consumed but should be used only as treats because of their high fat content.

Feeding should be done twice a day. Leaving too much feed in the bowls results in obese guinea pigs that are not as productive. Fresh foods left in the cage can mold and spoil; consuming moldy or spoiled foods can result in upset stomachs, diarrhea, or even death.

Fresh water should always be available for guinea pigs. Vacuum-type bottles work very well. Guinea pigs will drink when their mouth is full of food, and this food can get pushed up inside the tube. The bottle and tube need to be cleaned out frequently to prevent growth of bacteria.

CARE AND HANDLING

A guinea pig should be brushed daily to remove old, dead hair. A guinea pig with rosettes and long hair will need to have the hair combed out in addition to brushing. Longhaired animals may also need the hair clipped around the rear to prevent it from becoming matted with feces. Longhaired animals exhibited in shows have the rear hair tied up in bows and wrapped with cloth. Well-groomed animals not only look nice, but also will be more healthy. Longhaired guinea pigs may need their long hair trimmed occasionally to keep it from dragging in the bedding and becoming dirty and matted.

Guinea pigs that are exhibited in shows may also need to be given an occasional bath. Guinea pigs are not especially fond of water. When bathing guinea pigs, a mild shampoo similar to what would be used on cats should be used; baby shampoo also works. Care must be taken not to get the guinea pig chilled; they are especially prone to colds and need to be dried and kept warm during and after the bath.

Guinea pigs are usually friendly and easy to handle; however, they may be timid and hide when approached. A guinea pig brought home should be allowed to explore its new home for several hours before handling. If the animal seems unusually frightened, leave it alone for a day or two.

If the handler offers treats of fresh greens, carrots, or apple, the guinea pig will get over this shyness and venture forth. If the animal remains shy, the treat should be dropped in the cage. By continuing to offer treats, the animal should soon take treats from the hand. Sometimes picking the animal up and petting it will help it get over its shyness.

A guinea pig should be grasped firmly around the front shoulders with one hand and have its rear supported with the other hand. This will prevent the

Figure 14–10   When picking up a guinea pig, place one hand in front of the animal and with the other hand behind the animal, scoop it up. In this manner, the animal's rear is supported. Courtesy of Dean Warren.

Figure 14–11   When holding a guinea pig, make sure its rear is supported. A guinea pig can injure its back if it is not held properly. Courtesy of Dean Warren.

guinea pig from struggling. Guinea pigs can injure themselves if allowed to struggle while being picked up with one hand. Once picked up, it should be cradled in the palm and forearm and held close to the body. In this position, the animal will feel safe and secure. The handler must not allow the animal to fall; guinea pigs are not agile and can injure themselves if allowed to fall on a hard surface. See Figures 14–10 and 14–11.

DISEASES AND AILMENTS

Guinea pigs are usually very healthy animals, especially if they are well cared for. Care must be taken to eliminate any major changes in temperature, drafts, and humidity. All conditions that might cause stress to the animals must be eliminated.

Common cold and respiratory diseases are the major problems of guinea pigs. *Bordetella bronchiseptica, Streptococcus pneumoniae, Pseudomonas aeriginosa, Klebsiella pneumoniae,* and *Streptococcus pyogenes* are types of bacteria responsible for respiratory problems.

Animals with respiratory diseases are usually lethargic and listless, have discharges from the nose, and sneeze. Increased doses of vitamin C and the use of broad-spectrum antibiotics are needed. Guinea pigs do not appear to have a great deal of strength or resolve to fight off these types of colds. Animals showing signs of respiratory diseases should be taken to a veterinarian and treatment begun immediately, or death may occur.

Guinea pigs may occasionally have large swellings on the neck. These swellings are a result of enlarged abscessed lymph nodes caused by the organisms *Streptococcus zooepidemicus* and *Streptococcus moniliformis.*

The organism usually enters through cuts or abrasions of the oral mucosa or upper respiratory tract.

Affected animals should be isolated because the drainage from abscesses can infect other animals in the colony. They then should be taken to a veterinarian. If this cannot be done immediately, with the appropriate guidance, a slit with a sterile razor on the abscess will allow it to drain thoroughly. The abscess should be cleansed with hydrogen peroxide. After cleansing, Neopocin, Mycitracin Ointment, or Panalog should be applied. Warm compresses can be applied to the area several times during the day to aid in healing.

Cuts and bites should be cleaned with hydrogen peroxide and then Neopocin, Mycitracin Ointment, or Panalog applied to the wound.

Occasionally, a baby guinea pig will be born with or develop a cloudy eye shortly after birth. This condition can be treated by cutting the end off a vitamin A capsule and squeezing some of the contents into the baby's mouth. Any remaining contents of the capsule can be put into the eye. This treatment should continue twice a day until the eye is clear.

Pregnant females may develop toxemia in late pregnancy. Affected females will be lethargic, go off their feed, refuse to eat, and have difficulty breathing. Females that are fat and overweight appear to be most affected. A high-quality diet is essential during the late stages of pregnancy; adding one-half teaspoon of sugar in the water bottle may help avoid toxemia.

The mites *Chirodiscoides caviae* and *Trixacarus caviae,* and the biting lice *Gyropus ovalis* and *Gliricola porcelli* are the most common external parasites. *Chirodiscoides* mites usually cause severe lesions, and hair loss may occur over the entire body. Insecticide preparations used on cats are generally effective.

If guinea pigs do not receive a diet containing solid foods, their teeth can grow to the point that they cannot bite or even close their mouth. If their teeth overgrow to this point, they will need to be clipped or the animal will starve. A veterinarian or experienced breeder can do this. Malocclusion is a condition where the upper and lower teeth do not come together properly; the condition is probably an inherited trait, and animals showing this condition should not be used in breeding. The condition can possibly be corrected by a trained veterinarian, depending on the severity.

Male guinea pigs should be checked occasionally to make sure the penis is free of shavings and hair, which will occasionally mat around the penis and render the male unable to mate.

Guinea pigs may develop a toxic reaction to several antibiotics, such as penicillin, chlortetracycline, oxytetracycline, lincomycin, bacitracin, erythromycin, streptomycin, and tylosin. Only broad-spectrum antibiotics, those effective against both gram-positive and gram-negative bacteria, should be used.

REPRODUCTION

Guinea pigs breed throughout the year, with increased births in late spring and summer. Females (sows) can produce up to five litters in a year, with an average being two to four, although litters of six are not uncommon.

The heat cycle in females averages about sixteen days. Females come into heat within 24 hours after giving birth and can subsequently be bred if a male is present.

The heat period normally lasts 24 hours; the male (boar) chases the female around the cage until she is ready to mate. She then flattens herself and raises her hindquarters. The actual mating lasts only a few seconds but may be repeated several times.

About four to five weeks into the pregnancy, the female becomes large and looks similar to a water balloon. The young weigh about half the weight of the female. As the delivery date approaches, the young can be seen moving about in the womb. The female guinea pig does not attempt to make a nest.

The gestation period ranges from fifty-six to seventy-four days, with an average of sixty-eight days. The young will weigh 1½ to 4½ ounces. Depending on the size of the litter, the actual birth will take 10 to 30 minutes. During the delivery, the mother is usually in a seated position and receives the newborn under her. She will immediately tear off the enveloping membrane, eat it, and then lick the newborn's nose, eyes, and mouth to remove any fluid.

The mother is usually very clean and will eat and clean up the placenta and afterbirth. If the birth takes place at night or early morning, the mother will have all traces of birth cleaned up by daylight; except for some soiled bedding, it would be hard to recognize that a birth took place.

The young are fully developed at birth and will be running around and consuming solid food within 24 hours. Their eyes are open, and they are fully furred. It is hard for the novice owner to believe that they are newborn.

In the wild, guinea pigs have very little defense against their enemies except to be able to run and hide. The young, being fully developed at birth, are able to run and hide shortly after birth.

The young normally nurse for three weeks. The female guinea pig has only two teats located far back on her belly. While nursing, the mother sits upright and the young sit on each side of her. If there are more than two in the litter, the others patiently wait their turn. See Figure 14–12.

The young can be separated from the mother at four weeks of age. Sexual maturity in females occurs at two months of age and in males at three months, al-

Figure 14–12   Occasionally guinea pigs and other small animals may lose their mothers and need to be hand fed. A substitute for mother's milk can be prepared using goat's milk, Karo syrup, and powdered milk. Photo by Dean Warren.

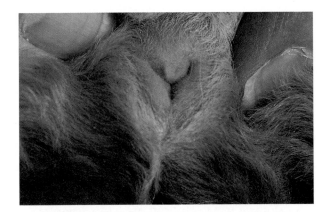

Figure 14–13   The genital area of a female guinea pig. Courtesy of Dean Warren.

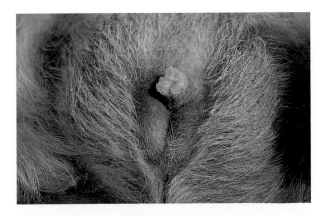

Figure 14–14   The genital area of a male guinea pig. Courtesy of Dean Warren.

though it is not uncommon for them to become sexually mature earlier. To prevent unwanted matings, the litter should be separated by sex at four weeks of age.

Breeding of guinea pigs is usually done by either pair breeding or in colonies. In colony breeding, one male is put with two to twenty females. Remember that each adult animal should have approximately 180 square inches of floor space; an adult female with a litter should have about 2 square feet of space. See Figures 14–13 and 14–14.

SUMMARY

The guinea pig is more accurately called a cavy; there are seven species of cavy. The exact history of the domesticated cavy *C. porcellus* is not known, but it is believed to be derived from *C. aperea, C. tschudii,* or *C. fulgida.*

There are seven common varieties of guinea pig. They are the Abyssinian, America, Peruvian, Satin, Silkie, Teddy, and White Crested.

Cages are available at most pet shops. The typical cage has a plastic bottom that is 3 or 4 inches deep and has vertical wire bars for the upper part.

Guinea pigs are vegetarians. Their large front incisor teeth are very sharp and used to bite food. Guinea pig pellets are also available that contain all of the nutrients the animal needs; these pellets are hard and serve to wear the teeth down. Guinea pigs, unlike many other animals, cannot synthesize vitamin C in their body and must be supplied with it in their diet.

When picking up a guinea pig, grasp it firmly around the front shoulders with one hand and support its rear with the other hand; this will prevent the guinea pig from struggling. A guinea pig can injure itself if allowed to struggle.

Guinea pigs are usually very healthy animals, especially if well cared for. Care must be taken to eliminate any major changes in temperature, drafts, and humidity.

Guinea pigs breed throughout the year, with increased births in late spring and summer. Females produce up to five litters in a year, with an average being two to four, although litters of six are not uncommon.

The heat cycle in females averages about sixteen days and lasts for 24 hours. The gestation period ranges from fifty-six to seventy-four days with an average of sixty-eight days. The young are fully developed at birth and will be running around and consuming solid food within 24 hours. The young normally nurse for three weeks.

DISCUSSION QUESTIONS

1. Where were guinea pigs originally found?

2. Where does the name *guinea pig* come from?

3. What is the more correct name for the guinea pig?

4. What are the seven varieties? What are the characteristics of each type?

5. How should a guinea pig be picked up?

6. What is the importance of vitamin C in their diet?

7. What are the common foodstuffs?

8. What are the common diseases and ailments of guinea pigs? What are the causes, symptoms, treatment, and control?

SUGGESTED ACTIVITIES

1. With the approval of the instructor, bring guinea pigs of various varieties to class and talk about the characteristics of the varieties. Class members can observe the guinea pigs.

2. With the approval of the instructor, invite a local veterinarian into the classroom to talk about the diseases and ailments of guinea pigs.

ADDITIONAL RESOURCES

1. Guinea Pig Hutch
www.halcyon.com/integra/pigs.html

2. Guinea Pig Care
www.aspca.org/learn/fs_petcare

3. Guinea Pigs: Dr. Sue's Pocket Pets
www.virtual-markets.net/vme/DrSue/guinea.html

4. Lovable, Cuddable, Small, Gross Rodents
http://tqjunior.advanced.org/3882/index.htm

5. Diseases of Guinea Pigs
http://netvet.wustl.edu/species/guinea/gpigs.txt

6. Allerpet
www.allerpet.com/products/care/care3.htm#top

END NOTE

1. Nowak, R. M., and J. L. Paradise. 1991. *Walker's Mammals of the World,* 5th Edition. Baltimore/London: Johns Hopkins University Press, p. 808. Reprinted by permission.

CHAPTER 15

Chinchillas

OBJECTIVES

After reading this chapter, one should be able to:

■ detail the history of the chinchilla.

■ describe the two species of chinchilla.

■ list practices for care of chinchillas.

■ describe the methods of handling a chinchilla.

■ describe the common diseases and ailments of the chinchilla.

TERMS TO KNOW

enteritis	pathogenic organisms	progeny
grotzen	polygamous	trophozoites
impaction	prime fur	veil
otitis	priming line	

CLASSIFICATION

Kingdom — Animalia
Phylum — Chordata
Class — Mammalia
Order — Rodentia
Suborder — Hystricomorpha

Family — Chinchillidae
Genus — *Chinchilla*
Species — *C. laniger*
C. brevicaudata

HISTORY

Chinchillas belong to the order Rodentia and are descended from Paramys of the Paleocene and Eocene epoch, although the actual connection is obscure. In *Modern Chinchilla Fur Farming* by Willis D. Parker, reference is made to Megamys, a chinchilla-like animal that existed millions of years ago. Megamys is described as being the size of an ox or larger. Because of the abundant vegetation that existed millions of years ago, it is believed the predecessor of the modern, small-sized chinchilla was probably a much larger animal.[1]

A thousand years ago, the Incas used chinchillas as a source of furs. They used the skins for garmets and blankets and even made thread from the fur. The softness and distinctiveness of the fur was valued highly and provided them with a badge of distinction. Subsequently, chinchillas were used by a tribe of Indians known as Chinchas; the animal was named for them and means "Little Chincha." Chinchilla fur reached Europe during the 1500s but did not become popular until the 1700s. Not until the 1800s did a real volume of pelts begin satisfying a market hungry for this prized fur.

South American pelts were shipped to Richard Gloek of Leipzig, the "Chinchilla King." He handled 300,000 pelts in 1901, making Leipzig the fur capital of the world and chinchilla the toast of European royalty. About the same time, South American pelts reached the New York fur markets.

Chinchilla nearly became extinct in the wild state but were revived as a ranch-raised animal. Greedy trappers invaded regions of South America and killed the little animals indiscriminately to feed the fur markets.

South American governments realized that destruction of this resource must be stopped, so they placed an export duty on all skins leaving the country. Unfortunately, this only encouraged smuggling. Later, the governments established regulations to prohibit killing the animals and encouraged trapping for the purpose of raising chinchillas in captivity. This was the beginning of fur ranches supervised by the governments.

A few private ranches were set up, and the industry as we know it today was started by Mathias F. Chapman, a mining engineer with Anaconda Copper Company. Chapman applied to the Chilean government for permission to capture several animals and transport them to the United States. After three years, he succeeded in trapping only eleven quality animals.

Because the chinchilla's native habitat was approximately 12,000 feet above sea level, Chapman gave the animals time to acclimate themselves before boarding a ship for California. His chinchilla cages were lowered down the mountains gradually during a twelve-month period, cooled with blocks of ice when necessary, and shaded from the direct sunlight. His careful handling paid off, and all eleven chins survived the trip down the mountains.

The animals required special conditions and care aboard the ship, too.[2] A constant supply of ice was needed to keep the specially constructed cages and passengers cool.

Over the years, Chapman experimented with both housing and diet. As the herd adjusted and thrived, livestock was finally offered for sale. Most people purchasing chins at that time did so on a group basis, with members buying shares in a pair of animals hoping to sell lots of livestock in the future.[3]

The first organized offering of chinchilla skins, approximately 10,000, took place at the New York Auction Company on June 21, 1954.[4] Today, it is estimated that there is a world production of 150,000 skins per year.[5] Chinchillas were quite expensive, especially in the early years. As the numbers continued to increase, some animals were offered as pets.[6]

MAJOR COLOR GROUPS

The Standard. The most common color of chinchillas is the "standard" or blue-gray color; however, they are available in other colors. Common mutant colors include white, beige, and black. See Figure 15–1.

Many breeders have been reluctant to breed the mutant colors. It takes 120 to 150 pelts to make a full-length coat and would take considerable time before the mutant strains could be produced in large enough quantities to supply a market. However, with the increase in the pet industry, the demand for mutant colors continues to increase.

The White. The first important mutant was a white male born in 1955. It was not a true albino because it had black eyes. The male was used in further crossings, and the offspring are known as Wilson Whites. Mating whites with standards usually produces 50 percent whites and 50 percent standards in the first generation. Further matings may produce silver-white and platinum-white animals. See Figure 15–2.

Figure 15–1   A "standard" or gray-colored chinchilla. Courtesy of Brian Yacur and Guilderland Animal Hospital.

Figure 15–2   A white chinchilla. Courtesy of Michael Gilroy.

Figure 15–4   A black-colored chinchilla. Courtesy of Stacey Riggert.

Figure 15–3   A beige-colored chinchilla. Courtesy of Brian Yacur and Guilderland Animal Hospital.

The Beige.   Another mutant was a beige female born in 1955 on an Oregon ranch. Matings between beige bucks and standard females give light beige (pearl-colored) and medium to dark beige shades (pastel-colored). Mating beige with white produces cream-colored **progeny** that are called Rose or Apricot. See Figure 15–3.

The Black.   Black mutations called Black Velvet were developed in 1956 by an American breeder. The fur on these animals has a black undercoat, a very narrow gray-white band, jet-black **veil** (tips of the fur), and a very high density. Mated with clear-colored standards, it produces the most beautiful mutation ever seen in chinchillas, the Blue Black Velvet.

Matings of black males with females from other mutations have produced Sapphire Velvets, Pastel Velvets, and Brown Velvets.[7] See Figure 15–4.

The Sullivan Violet.   One of the newest mutations to be raised in the United States is called the Sullivan Violet, an animal with a clear white belly and a lavender (pinkish-violet) **grotzen** (the center back strip of a fur pelt). The first violets were born on a ranch in Rhodesia (now Zimbabwe) Africa in the mid-1960s. The animals were put up for sale and are now housed on the Loyd Sullivan ranch in California.[8]

CHARACTERISTICS

The native habitat of the chinchilla is the barren areas of the Andes Mountains at elevations up to 20,000 feet. Chinchillas shelter in crevices and holes among the rocks; they are primarily nocturnal. To eat, they sit upright and hold food in their forepaws; the diet consists of any available vegetation.

Chinchillas are classified into two species: *C. brevicaudata* and *C. laniger*. *C. brevicaudata* is native to Peru, Bolivia, and Northwestern Argentina. The native range of *C. laniger* is Northern Chile.

C. brevicaudata has a thick neck and shoulders and is heavily furred with light gray coarse hair that is often tinged with a yellowish cast. Its ears are shorter than those of *C. laniger,* and its nose is flatter, giving the animal a stocky appearance. *C. laniger* has a narrow neck and shoulders; the fur is very silky and medium to dark gray, with a bright bluish cast.[9]

Today, most of the chinchillas raised domestically are of the *C. laniger* species or a close relative designated the "Costina" type. The *C. brevicaudata* is a larger animal, but its fur is not as desirable as the crisp blue fur of the *C. laniger*.

Some attempts were made to cross-breed the two species; however, the male offspring of the crossings were found to be sterile.

Chinchillas resemble small rabbits with short ears and a short, bushy tail much like that of a squirrel. Chinchillas range from 9 to 15 inches long, have a tail of 3 to 10 inches, and weigh from 1 to 2 pounds.

Their thick, shiny fur is about 1 inch long; the softness of the fur is due to fewer number of guard hairs as compared with other fur bearers. Chinchillas usually shed about every three months. New growth starts at the neck and progresses toward the rear of the animal. Where the new hair is coming up through the old hair a darker, distinct line known as the priming line can be seen. When the priming line reaches the tail, it is time to take the pelt. The hair coat remains prime fur for several weeks, and then the process begins again.

The head of the chinchilla is broad, with large ears and eyes; they have very long, stiff whiskers. Both the short forefeet and the narrow hindfeet have four digits, with stiff bristles surrounding soft, weak claws; the pads of the feet are large and rubbery. Females have one pair of mammary glands on the lower abdomen and two pair higher up near the chest.

Chinchillas make very little noise but do make a barking sound similar to a squeak toy. When annoyed, they can be heard making this sound.

Chinchillas are very clean, almost odor-free animals. The feces are normally dry and produce very little odor; the urine does have an unpleasant odor if allowed to accumulate.

HOUSING AND EQUIPMENT

Chinchillas are nocturnal; when planning housing for them, one should consider a location where it will be fairly quiet during the day. This is important to the successful raising of chinchillas. The area should be fairly dry, have adequate ventilation, and not get too hot. As temperatures reach 80°F, chinchillas become uncomfortable; at 90°F, they are subject to heat prostration.

Cages used for other small animals can be used to house chinchillas. A cage of approximately 14 inches by 24 inches by 12 inches is adequate for a single chinchilla; however, cages of approximately 24 inches by 24 inches by 14 inches are more desirable.

Cages should be made of metal and wire; any plastic used will soon be chewed through. Cages can be constructed of wood and wire mesh screen; however, the wood must be on the outside of the wire mesh or it will be destroyed.

The size of openings in the wire mesh is very important, especially if the owner plans to breed the

chinchillas and produce young ones. Small chinchillas can crawl through openings about an inch square. They are very active shortly after birth, and the owner needs to take precautions not to lose any young.

Chinchillas must be kept in individual cages or they will fight and may inflict serious injuries on one another. Chinchillas raised as littermates can sometimes be kept together, but trying to keep two animals raised in a single cage could be disastrous.

Polygamous breeding cages are available for owners who want to breed more than one pair of animals. See Figure 15–5. This type of cage usually has three separate cages connected together with holes in the back of the cages that lead to a tunnel which connects all three cages. Females are fitted with a plastic collar that prevents them from leaving their cage and entering the tunnel. See Figure 15–6. Two females and one male are placed in the cages, and the male can use the tunnel to go from one cage to the other. If one of the females allows him to stay in her cage, a third female

Figure 15–5 Polygamous breeding cages are developed so that one male can breed with three or more females. A tunnel at the upper back of the cage connects to three or more cages. The male can move through the tunnel and enter any of the female's cages. Courtesy of Michael Gilroy.

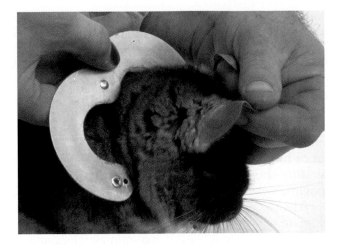

Figure 15–6 Females are fitted with a plastic or aluminum collar that prevents them from entering the tunnel. Courtesy of Isabelle Francais.

can be added to the cage where the male was put first. Additional cages can be added to house up to twenty females. Allowing the male to run the tunnel keeps him in good condition.

Cages usually have wire mesh or solid bottoms. With the wire bottoms, newspaper can be placed below the cage to catch feces and litter. Solid-bottom cages usually have a pan that can be removed for cleaning. This pan is usually filled with some type of absorbent material; wood shavings are probably best to use.

Wood shavings make good, inexpensive bedding material. Ground corn cobs are also inexpensive and absorbent but may carry parasites. Hay and straw can

be used, but these materials can become moldy and may stain the animal's fur.

An owner may want to put a wooden box or a large can in the cage to give the animal a place to hide and sleep. If a tin can is used, the ends should be cut out and the can flattened on one side so it cannot roll around.

A small wooden block should be placed in the cage to give the animal something to chew on and to play with. Remember that the chinchilla is a rodent with continuously growing teeth, and it will need something to chew on to wear the teeth down.

Chinchillas need regular baths to rid their fur of excess moisture and oil; they do not bathe in water but in finely ground powder. In their natural setting, chinchillas bathe in volcanic ash of the Andes Mountains. A very similar powder can be purchased from pet stores for this purpose. A small rectangular pan with about 2 or 3 inches of powder can be used; the pan needs to be large enough for the animal to roll around in. The powder may be used several times. See Figure 15–7.

The animal should receive a bath about twice a week. During hot, humid days, daily baths may be needed. The pan with the powder can be placed in the cage and left for about 5 minutes; that should be sufficient time for the animal to finish a bath. Chinchillas should not be given a bath in water.

Cages should be fitted with metal, creep-type feeders; these are the easiest for the animal to use. Heavy ceramic feeders can also be used. These will not tip over, and the animal cannot destroy them.

Cages should be fitted with vacuum-type water bottles. Chinchillas do not drink very much water, but a clean, fresh supply should always be available.

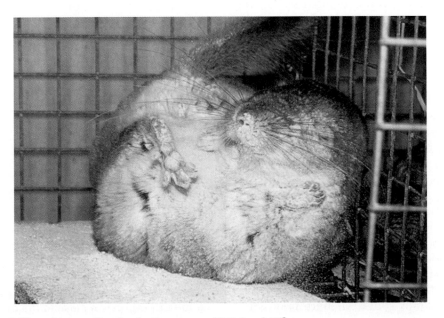

Figure 15–7 A chinchilla taking a dust bath. Courtesy of Michael Gilroy.

FEEDING

A variety of pelleted foods are available today for chinchillas. These pellets consist primarily of ground alfalfa hay, corn gluten, corn tailings, wheat germ, and bran. These foods usually have all of the vitamins and minerals necessary. If pelleted chinchilla feed is not available, pelleted guinea pig feed will be sufficient. The pelleted foods may be supplemented with fresh, well-dried alfalfa or timothy hay. The key is to make sure it is not moldy or dusty or contains weeds.

There appears to be some disagreement as to whether the pelleted feeds are adequate; continued nutritional research needs to be conducted. Some believe that pelleted feeds may lead to developmental abnormalities, inadequate milk supply in lactating females, and fur biting because of nutritional deficiencies in the animal's diet. They believe the diet should also include green foods.

If an owner so desires, the diet can be supplemented with grass, lettuce leaves, carrots, and celery. Pears and apple slices are readily consumed by chinchillas. Care must be taken when feeding greens and other supplements because these can easily upset the digestive process and lead to serious problems. As special treats, chinchillas love raisins. The animals will jump and run around their cages upon hearing a bag of raisins being opened.

Owners should avoid wet grass and hays that may spoil and cause digestive problems and diarrhea. Although chinchillas readily consume nuts and sunflower seeds, these should not be fed because they are fattening.

Clean, fresh water should be available to chinchillas on a pelleted diet, or they will become dehydrated. If the diet is being supplemented by feeding lettuce leaves or similar green foods, the chinchilla may not need water.

HANDLING

When a chinchilla is first brought home, it should be allowed to get used to its new surroundings. When approaching the cage, one should go quietly and speak to the animal softly. Chinchillas dislike loud noises and quick, sudden movements; they may become frightened and keep to the corner of the cage. Being nocturnal animals, they may not appreciate being approached during the day. Even a very tame chinchilla may raise up on its rear legs and bark; females are able to shoot urine at an intruder.

Several methods are used to handle chinchillas. Ranchers will grasp the animal's tail close to the body and pick the animal up; some ranchers may grasp hold of both of the animal's ears and pick it up this way. Ranchers are very careful not to grab the animal in any way that may cause it to lose some of its fur. An animal with a patch of fur missing would mean a pelt that is worthless; it would take several weeks or months for the fur to grow back.

Chinchillas are able to release their fur as a defense mechanism. If the animal is grabbed by a predatory animal, it will leave the predator with a mouthful of fur. The release of fur in chinchillas is called *fur slip.* If the chinchilla owner handles the animal roughly, the same thing can happen.

The easiest and gentlest method of picking up a chinchilla is to cup both hands and place one hand in front and one hand behind the animal. The handler should then cup the animal in the hands. This needs to be done fairly quickly, or the animal will jump over the handler's hands and head back to a safe corner. The animal should be held close to the handler's body so that it feels safe and secure. For a chinchilla that is somewhat frightened, the handler may want to hold on to the tail also. Young animals that have been handled frequently can be trained to ride around on their handler's shoulders. See Figures 15–8 and 15–9.

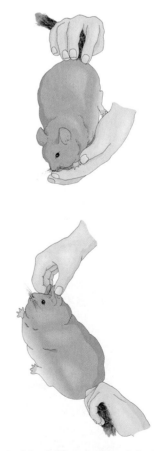

Figure 15–8 *A chinchilla can be picked up by its tail if the tail is grasped close to the body.*

Figure 15–9   Chinchillas that are tame can be picked up by cupping the hands around the animal. Place one hand behind the animal and one hand in front and cup it in the hands.

DISEASES AND AILMENTS

Chinchillas are very healthy if they receive proper nutrition; have a clean, dry, draft-free environment; and are free of any stressful conditions. However, there are diseases and ailments that can affect these animals.

Pseudomonas aeruginosa is an organism that causes infections in wounds, inflammation of the eyes and ears, pneumonia, inflammation of the intestines, inflammation of the uterus in females, and poisoning to the circulatory system.

Swelling and redness around the eyes, sensitivity to light, watering, and pus formation are common symptoms of eye inflammation (conjunctivitis). The eyes should be washed gently and frequently with a warm boric acid solution; several ophthalmic medications are available. Advanced cases should receive the attention of a veterinarian. Cages and feed containers should be washed and disinfected; infected animals should be isolated.

Twisting and lowering of the head, hanging the head to one side, and running around in a circle are common signs of an inner ear infection (otitis). The ear should be cleaned with a warm boric acid solution; several ear drop medications are available. The animal should be kept warm and free of drafts.

Listlessness, failure to eat, difficulty breathing, and distended or swollen abdomen are common signs of pneumonia. Pneumonia usually results from unsanitary conditions or cold, drafty, or high-humidity environments. Treatment with various antibiotics (aureomycin and penicillin) should begin as soon as possible.

Watery diarrhea, mushy droppings, constipation, listlessness, loss of appetite, rough hair coat, and swollen abdomen are common symptoms of intestinal infections (enteritis). Poor-quality food, moldy hay, or unsanitary feeding and watering equipment may be involved, as well as infectious agents. The animals should be supplied with clean water and fresh food. Animals should be vaccinated and given yearly booster shots.

Uterine infection can cause disturbances and irregular breeding cycles. Vaginal discharges in the form of watery, white pus, increased urination, loss of weight, and fur chewing at the flank and abdomen are additional symptoms. Infections to the uterus can spread from other parts of the body, can be caused by retained placentas, and can be spread by males in a polygamous breeding system. Infected animals should be isolated, and injections of antibiotics (streptomycin) should begin as soon as possible.

Pathogenic organisms and their toxins can be absorbed into and distributed throughout the body by the circulatory system. Animals appear listless, go off their feed, and lose weight.

Listeria monocytogenes is a pathogenic organism that infects chinchillas. Affected animals go off their feed, lose weight, develop paralysis in the rear quarters, and go into a coma. Many animals may die without any clinical signs, and mortality is high.

Giardia species is a common intestinal parasite found among chinchillas. Cysts and trophozoites can be seen in intestinal scrapings from chinchillas that have died after a period of intermittent watery diarrhea or impaction. Mortality may be high in such outbreaks, and often, no other cause of death can be found.

Trichophyton mentagrophytes is a common cause of ringworm in chinchillas. Affected animals usually have loss of hair and reddening around the nose, eyes, and base of tail. Affected animals should be isolated and treated.

Impaction is a condition in which the small intestines, cecum, and large intestines become tightly packed with food material or feces; many times, impaction follows diarrhea. The condition is usually caused by moldy or spoiled feed, poor nutrition, stress, shock, insufficient water intake, pressure on the intestines from pregnancy, and infectious agents. Common symptoms include the reduction or absence of feces, lack of appetite, rough hair coat, and an animal humped up with its chin between the front feet and resting on the cage floor. The animal may show no desire to move about and may appear glassy-eyed. Several remedies may be successful in relieving impaction; a dropper full of mineral oil can be administered daily, or two or three dropperfuls of grapefruit juice can be given. If the condition is a result of an infectious agent, a veterinarian may need to be consulted.

Fur chewing is a common problem of chinchillas; the animals will chew the fur on the flanks, leaving patches of short fur. Poor nutrition, small cages, drafty conditions, boredom, and high humidity have been suggested as causes.

Malocclusion, which is a problem among rodent-type animals, is a dental deformity where the molars and incisor teeth are overgrown or do not meet properly. Affected animals fail to eat, lose weight, and may salivate or slobber; death may result from starvation.

Two types of convulsions can occur in chinchillas. One is caused by a lack of calcium, and the other is caused by a lack of vitamin B. Both conditions cause muscle spasms and body trembling. A veterinarian may prescribe intramuscular injections of calcium gluconate as the immediate solution to lack of calcium. Calcium gluconate, calcium lactate, dicalcium phosphate, and bone meal are common supplements that supply adequate calcium to the diet. Thiamine and B-complex injections provide rapid relief from convulsions caused by vitamin B deficiency; convulsions from vitamin B deficiency may lead to paralysis and death.

Mastitis may appear in nursing mothers; swelling, redness, and hardness of the breast are common symptoms. Pain may occur while the young are trying to nurse, and the mother will often scold them and not allow nursing. The condition is caused by infection following biting of the nipples by the young, weaning the young too early, unsanitary conditions, or incomplete nursing. The owner must try to determine the cause and eliminate it; the young may need their teeth clipped. The cages and nest box should be cleaned and disinfected thoroughly. The amount of hay in the ration should be increased and pellets decreased. The owner will have to supplement milk to the young and may want to consult with a veterinarian.

REPRODUCTION

The estrus cycle in female chinchillas varies from twenty-eight to thirty-five days. A young female chinchilla first comes in heat at about five months of age but should not be bred until about eight months. Females that have not mated and those that have given birth but are not in heat form a "plug" in the vagina that closes it off. When the heat cycle occurs, this plug is ejected and a successful mating can occur.

A female in heat shows general signs of restlessness, frequently urinates, and scratches the bedding. Males become aroused when around females in heat. See Figures 15–10 and 15–11.

In pair mating of chinchillas, the male should be placed near the female's cage for a while so that they can become familiar; the male is then placed in the female's cage. A box should be placed in the cage so that the male can find an area of safety in case the fe-

Figure 15–10   The genital area of a female chinchilla. Courtesy of Michael Gilroy.

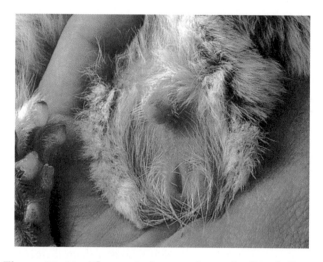

Figure 15–11   The genital area of a male chinchilla. Courtesy of Michael Gilroy.

male is unwilling to mate and prefers to fight. The pair should be watched closely; if the female continues to fight, the male should be removed from the cage. If the female rejects the male, she will usually stand on her rear legs, aim, and shoot urine at the male. Fighting almost always occurs, but the pair should settle down after a while. Mating usually takes place in the evenings.

The male and female can be left together after mating. The male may even remain in the cage during birth, but the female comes into heat again within 24 hours. Unless a second mating is desired, the male should be removed; he may be returned to the cage after a couple of days.

In polygamous matings, the male is left in the tunnel and can visit the cages of several females at any time. The opening to the tunnel can be closed off to prevent a female from being rebred right after giving birth.

Whichever mating method is used, the owner should always observe the chinchillas and be aware of overly aggressive animals; serious fighting may ensue and death can result.

The gestation period for chinchillas is 111 days. Any stress can cause premature births, abortion, or resorption of the fetuses. Pregnant females should be handled very carefully, if at all. At about two months, she will almost stop eating and may lose weight. After a short period, she will regain her appetite and continue to gain weight until birth of the litter.

Chinchillas usually do not need any assistance at birth; birth usually occurs during the early morning hours. A heated nest box or heating pad must be available to keep the newborns warm. If the newborns are delivered slowly, the mother usually has time to dry each one off. If they come fairly fast, she may not have time to dry them. The additional heat is critical, or the young become chilled and catch cold.

When delivery is near, the female stretches, rears up, and emits low, moaning sounds. During a normal birth, the female pulls out the young, removes the placental sac, cleans fluid from around the young's nose and mouth, and dries it off.

Litters vary from one to eight, but the average is two per litter. The young are born fully furred, eyes open, with a full set of teeth, and have lots of energy. Most parents do not mind the young being handled, but handling should be put off until the young are dry and warm.

If a polygamous breeding system is used, the young should not be allowed into the cages of other females because they will probably be killed.

The owner should observe the young carefully during the first few days to see if they are getting enough to eat. If the litter consists of more than two, the mother may not be able to provide enough milk. Babies that do not get enough to eat fight and fail to grow; young chinchillas can be successfully fed using a dropper or a small bottle. The young grow fast and are ready for weaning at about six weeks of age.

SUMMARY

Chinchillas belong to the order Rodentia and are descended from the Paramys of the Paleocene and Eocene epoch, although the actual connection is obscure. A thousand years ago, chinchillas were used by the Incas.

In the 1800s, chinchilla fur became highly prized in Europe. This demand became so great that the animals almost became extinct in the wild; they eventually recovered by being raised in ranches. M. F. Chapman is responsible for the ranch-raised chinchilla industry.

Chinchillas are classified into two species: *C. brevicaudata* and *C. laniger*. Because of its more desirable blue fur, the *C. laniger* species was more commonly raised and bred.

Chinchillas are nocturnal; when planning housing for them, one should consider a location where it will be fairly quiet during the day.

Chinchillas are able to release their fur as a defense mechanism. If the animal is grabbed by a predatory animal, it will leave the predator with a mouthful of fur; if the chinchilla owner handles the animal roughly, the same thing can happen.

A variety of pelleted foods are available on the market for chinchillas. These pellets consist primarily of ground alfalfa hay, corn gluten, corn tailings, wheat germ, or bran. If pelleted chinchilla feed is not available, pelleted guinea pig feed is sufficient.

The estrus cycle in females varies from twenty-eight to thirty-five days; the gestation period is 111 days. Litters vary from one to eight; the average is two. The newborn are fully furred, eyes open, with a full set of teeth, and have lots of energy.

Chinchillas are very healthy if they receive proper nutrition; have a clean, dry, draft-free environment; and are free of any stressful conditions.

DISCUSSION QUESTIONS

1. Where did chinchillas originate?

2. What are the two species of chinchilla?

3. Who is responsible for the chinchilla industry in the United States?

4. What are the mutant colors of chinchilla?

5. What are common foods for chinchillas?

6. Describe a polygamous mating system.

7. List the common diseases and ailments of chinchillas. What are the causes, symptoms, and treatment?

SUGGESTED ACTIVITIES

1. With the approval of the instructor, invite local breeders or pet shop owners to bring chinchillas into the classroom to talk about the animals.

2. With the approval of the instructor, invite a local veterinarian into the classroom to discuss the diseases and ailments of chinchillas.

ADDITIONAL RESOURCES

1. Etc. Chinchilla Page
   www.etc-etc.com/chin.htm

2. Underhill Chinchillas
   http://members.aol.com/UHChins/
   UnderhillChinchillas/main.htm

3. Chinchillas: Dr. Sue's Pocket Pets
   www.virtual-markets.net/vme/DrSue/
   chinchil.html

END NOTES

1. Parker, W. D. 1975. *Modern Chinchilla Fur Farming.* Alhambra, CA: Borden Publishing Company.

2. Zeinert, K. 1986. *All About Chinchillas.* Neptune City, NJ: T.F.H. Publications, Inc., pp. 12–14.

3. Zeinert, K. *All About Chinchillas.*

4. Bowen, E. G., and R. W. Jenkins. 1969. *Chinchilla, History, Husbandry, Marketing.* Hackensack, NJ: Adler Printing Company, pp. 1–10.

5. Bickel, E. 1987. *Chinchilla Handbook.* Neptune City, NJ: T.F.H. Publications, Inc.

6. Zeinert, K. *All About Chinchillas.*

7. Mosslacher, E. 1968. *Feeding and Caring for Chinchillas.* Neptune City, NJ: T.F.H. Publications, Inc., p. 116.

8. Zeinert, K. *All About Chinchillas.*

9. Zeinert, K. *All About Chinchillas.*

CHAPTER 16
Ferrets

OBJECTIVES

After reading this chapter, one should be able to:

■ detail the history of the domestic ferret.

■ list and describe the various color patterns of ferrets.

■ list the general care practices for ferrets.

■ list the common feeds for ferrets.

■ describe the correct method for handling a ferret.

■ describe the common diseases and ailments of ferrets.

TERMS TO KNOW

aplastic anemia	ferreting	kits
bib	hobs	mitt
estrogen	jills	ovulation

CLASSIFICATION

Kingdom	— Animalia	Genus	—	*Mustela*
Phylum	— Chordata	Subgenus	—	*Mustela*
Class	— Mammalia	Species	—	*M. putorius*
Order	— Carnivora			(European polecat)
Family	— Mustelidae	Subspecies	—	*M. putorius furo*
				(Domestic ferret)

HISTORY

The domestic ferret belongs to the family Mustelidae and the genus *Mustela,* which also includes weasels, ermines, mink, and polecats. There are four subgenus of *Mustela;* all members of this genus have long, limber, slender bodies. The tail is long and is usually about half the length of the head and body. They have short legs and small, rounded ears. They have rather long, oval-shaped heads and a pointed snout; the eyes are bright and clear.

The European polecat *(M. pulorius)* is a wild animal that is common to the open forest areas and meadows of Europe. It makes its den in hollow logs, crevices, and burrows made by other animals. It is nocturnal and capable of climbing. The European polecat has been tamed for centuries and used in **ferreting** or hunting rabbits. The polecat will readily go into a burrow and chase the rabbits out. Polecats have also been used to drive rats from their burrows and tunnels. The domestic ferret *(M. pulorius furo)* is believed to be a descendant of the European polecat *(M. pulorius).*

Ferrets have been in the United States for more than 300 years; they were used in the 1800s for rodent control. The "ferret meister" would come with his ferrets to a farm or granary and release his ferrets. These working ferrets ran into holes and hiding places of rodents and the rodents ran out. The people outside would wait with shovels and terrier-type dogs and kill as many rodents as they could. In the 1800s, ferrets were often kept on small farms and at feed mills for rodent control. The normal range of a ferret is about 200 yards from the place it considers home. Ferrets traveling through rat tunnels leave trace odors that trigger fear in rats and mice, causing them to flee.

Ferrets have also been used successfully to help wire airplanes in hard-to-reach places. Because ferrets love to enter anything that looks like a tunnel, it is a simple matter to attach a wire and have them run down a tunnel and return. Ferrets have also been used in scientific research since they catch the same colds as humans.[1]

People have recently found that ferrets make wonderful pets. Although ferrets have been in the United States for more than 300 years, no wild colonies of the domestic ferrets exist. After many generations of domestication, they are unable to survive in the wild.

The only wild ferret in the United States is the black-footed ferret *(M. nigripes),* which is found in small numbers in Arizona. The black-footed ferret is very similar in size and build to the domestic ferret but is lighter in color and has very dark feet and legs. It has a broad black stripe across the eyes, and the tail is black-tipped. The ears are also larger, more rounded, and higher on the head. The black-footed ferret is on the endangered species list.

MAJOR COLOR GROUPS

Several colors are now being bred: sable (the most common), red-eyed white, silver mitt, sterling silver, white-footed sable, butterscotch, white-footed butterscotch, and the rare cinnamon. See Figure 16–1.

Figure 16–1   A Common Sable ferret. Note the mask across the face that is characteristic of the sable ferret. Photo by Isabelle Francais.

The Common Sable.   The sable ferret ranges from light to dark, depending on the shade of both the underfur and guard hairs; the underfur ranges from white to beige. The guard hairs are longer and are usually black. A well-marked sable ferret should have a definite mask or hood pattern over the face. The preferred nose color is black.

The White.   The red-eyed white is often referred to as an albino and has the characteristic red eyes of an albino. The fur of the albino ranges from white to yellowish-white; the nose of the albino is pink. See Figure 16–2.

Black-eyed white ferrets are white or yellowish-white with black eyes; only a few exist.

The Silver Mitt.   A silver mitt ferret has underfur of white or off-white and guard hairs of black and white. This combination of white and black gives the ferret a silvery appearance. The feet of the silver **mitt** (feet)

Figure 16–2   A White ferret. This is a red-eyed white ferret. There are black-eyed white ferrets, but they are not as readily available. Photo by Isabelle Francais.

Figure 16–3   A Silver Mitt ferret. The underfur of the silver mitt ferret is white. The guard hairs are black and white. Photo by Isabelle Francais.

Figure 16–5   A Butterscotch ferret. Butterscotch ferrets will have white to beige-colored underfur and butter- scotch guard hairs. Photo by Isabelle Francais.

are white. The silver mitt has a white **bib** (the patch of white below the chin and on the throat) and eyes that appear black but are often a deep burgundy. See Figure 16–3.

The Sterling Silver.   A sterling silver ferret is very similar to a silver mitt but has more white guard hairs that give the animal a lighter appearance. See Figure 16–4.

The White-Footed Sable.   White-footed sables are marked like the regular sable ferret but have four white feet and a white bib.

The Butterscotch.   The underfur of the butterscotch ferret is the same as the sable, but the guard hairs, legs, mask, and hood coloring are butterscotch in- stead of black. See Figure 16–5.

The white-footed butterscotch ferrets have the same markings and color combinations, except for having four white feet and a white bib.

The Cinnamon.   The cinnamon ferret has underfur that is white or off-white and guard hairs that are a rich red-brown or cinnamon color. Some believe the cinnamon ferret is more docile and more easily managed.

Figure 16–4   A Sterling Silver ferret. The sterling silver ferret is similar to the silver mitt, but there are more black guard hairs, giving the ferret a darker shading. Photo by Isabelle Francais.

CHARACTERISTICS

The domestic ferret, like the other members of the genus *Mustela,* has a body that is elongated, lean, slender, and muscular. The legs are short, and the feet have five toes with claws. The fur of ferrets is dense and very soft; they change coats completely twice a year.

They have oval-shaped heads and pointed snouts. The head of the male usually is broader and less pointed than that of the female.

Male ferrets **(hobs)** are 16 to 20 inches long and weigh 3 to 5 pounds. Females **(jills)** are a little smaller, usually 12 to 14 inches long, and weigh 1½ to 3 pounds.

There is little difference between a male and fe- male ferret as a pet. Males are usually about twice the size of the females. Both develop a strong musky odor as they reach maturity; the odor is strongest in the males. The odor originates from anal glands that can be surgically removed. In males, however, this procedure alone does not eliminate the odor; castration is also necessary. With occasional baths, the male ferret that has been descented and castrated has very little odor. If a pet owner is not interested in breeding, a spayed and descented female would make the best pet.

Ferrets that have been descented, spayed, or neutered may not have the same definite color pattern around the eyes. These patterns usually come back in a paler color following shedding.

Male ferrets that have not been neutered and descented also have the nasty habit of leaving small drops of urine in areas they have traveled. These urine spots leave strong, musky scent marks and are almost intolerable. The male ferret that had been kept in the house before will now probably be kept outside.

Ferrets do not have a very well-developed sense of sight; being primarily nocturnal in nature, they do not see well in bright light. They do, however, have a highly developed sense of hearing, smell, and touch. When put in new surroundings, they use their sense of smell to check out the area thoroughly. Ferrets learn their owner's voice and usually come to the door of a cage when the owner approaches.

Ferrets that are well cared for live from eight to eleven years. Normal body temperature is 101.8°F; the heart rate is between 300 and 400 beats per minute, and respiration is between 30 and 40 per minute.

Ferrets have a total of forty teeth. On each side, there are three incisors on top and three on the bottom, two canine teeth on the top and bottom, four premolars on top and three on the bottom, and one molar on top and two on the bottom.

HOUSING AND EQUIPMENT

The type of housing depends on the owner's situation. A single ferret should have a cage at least 12 inches wide, 24 inches long, and 10 inches high. This amount of space will sufficiently house a ferret and provide space for a litter, feeding, and sleeping area; however, a ferret is very active and, if kept in this size cage, should be allowed out frequently for exercise.

Cages that are 24 inches wide, 24 inches long, and 14 inches high are more desirable. A cage approximately 2 feet wide and 4 feet long is appropriate for a pair of ferrets; this allows plenty of room for a sleeping or nesting box, a litter box, and space for exercising. Cages of this size can easily be constructed from wood and wire screen. See Figure 16–6.

Glass aquariums can be used to house ferrets but are not desirable because they do not allow air to circulate and can become overly warm in the summer. Aquariums must be fitted with a lid to prevent escapes.

Whatever is used to house a ferret, two things should be kept in mind:

1. If there is a way to escape from a cage, a ferret will find it. All materials should be of good quality and latches on the doors should lock securely.

2. Ferrets enjoy watching activity around them. Cages should be constructed to allow ferrets to see out and observe what's going on; they want to be part of the action.

Ferrets can be kept outside all year long. Hutches similar to the type used for rabbits and mink can eas-

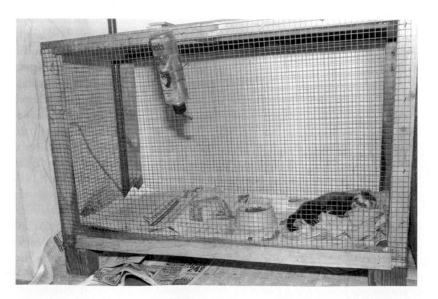

Figure 16–6   This cage is constructed from plywood and wire screen. Courtesy of USDA.

ily be constructed. The hutches should be constructed in such a way to prevent the animal from getting wet during inclement weather. The hutch should also be free from drafts and should provide shade during the hot summer months. Ferrets cannot tolerate temperatures above 95°F and must be provided with shade and plenty of water during hot weather.

Food bowls should be made of earthenware. These are heavy and almost impossible for the ferret to upset. A constant supply of fresh water should be available; the larger vacuum-type bottles used for rabbits and guinea pigs are ideal.

Ferrets are very playful, and almost anything makes a suitable toy. Rubber balls and squeak toys provide ferrets with hours of fun. Plastic pipe that they can crawl through and plastic milk jugs with holes cut in them provide ferrets with a place to explore and play.

FEEDING

Ferrets are easy to feed; the major part of their diet should consist of a high-quality commercial dry cat or kitten food. It is important that the cat food contain approximately 35 percent animal protein.

Dry cat foods help keep the ferret's teeth and gums in good condition, cost less than moist foods, and are easier to feed. Foods formulated for kittens may be of higher protein than foods formulated for mature cats. See Figure 16–7.

Figure 16–8   This ferret is being kept in a wire cage typical of the type that can be purchased at pet shops and discount stores. The cage is equipped with a vacuum-type water bottle so that the ferret has a supply of water at all times. Photo by Isabelle Francais.

Figure 16–7   Ferrets can be fed dry cat food. A large, heavy bowl prevents the ferret from tipping the bowl over. Courtesy of Carolyn Miller.

When feeding dry foods, it is important to remember to make sure the animal has plenty of water available because no water is contained in the dry product. See Figure 16–8.

Cow's milk should never be given to ferrets because it almost always leads to diarrhea. An exception to this would be in the case of a pregnant and nursing female to supplement the feeding of young **kits**. Goat's milk can be given to young animals to supplement mother's milk.

Young ferrets reach 90 percent of their adult size in about fourteen weeks; large amounts of food are consumed during this time. Young, growing ferrets should be fed all they will consume, twice a day.

If an animal is receiving a well-balanced diet, it should not be necessary to provide additional vitamin or mineral supplements.

Older ferrets may need vegetable oil added to their diet to aid in digestion and bowel movements and help to maintain a healthy coat; about ½ teaspoon per day will be sufficient.

Meat scraps, cracklings, fruits, red licorice, and ice cream can be used for treats. Fed in limited amounts, they will not harm the animal.

Bones should never be used as treats; slivers may be broken off and eaten by the animal, causing internal injury.

HANDLING AND TRAINING

A ferret should be allowed to get used to its new home for a couple of days before the handler begins training. Young ferrets, although they may be used to humans, need some time to get used to new surroundings. When approaching a ferret, one should speak to it in a soft, gentle voice before attempting to pick it up.

When ferrets are young, their mother will pick them up by the skin on the nape of the neck; handlers should pick up ferrets in a similar manner. One should grasp the ferret firmly around the body at the front shoulders. The ferret will go limp and allow itself to be picked up. Placing the other hand under the rear of the animal supports its body. See Figures 16–9 and 16–10.

Ferrets have very tough skin and, while playing with their littermates, will play rough and bite one another. This playful bite may be painful to a human, and the animal should be disciplined. A firm "no" each time it bites should be sufficient for the young ferret to realize this is not approved behavior. If the "no" is not sufficient, the owner can thump the ferret on the end of the nose with a flip of the forefinger, accompanied by a firm "no." If the ferret bites while one holds it, the ferret should not be put down because it will learn to use this behavior when it wants down.

Figure 16–10 After picking the ferret up, continue to hold the animal with one hand under the front legs and support the rear of the animal with the other hand. Photo by Isabelle Francais.

Two items necessary in training ferrets are Linatone and Bitterapple; these items are available at pet shops. Ferrets love Linatone, and it should be used as a treat and reward for good behavior or in association with training. Linatone is also an excellent supplement to help keep the animal's coat in good condition. Linatone should be limited to three to five drops per day because it contains vitamin A, and too much is not good for the animal. See Figure 16–11.

Figure 16–9 Ferrets are easy to pick up and handle. Grasp the animal around the body just behind the front legs. The animal will usually go limp and allow itself to be picked up. Photo by Isabelle Francais.

Figure 16–11 Two items necessary in training ferrets are Linatone and Bitterapple. Courtesy of Carolyn Miller.

Figure 16–12 Brushing a ferret's fur will help to remove loose hair and stimulate the skin. Photo by Isabelle Francais.

Bitterapple works in the opposite manner. Ferrets dislike it and will not touch anything to which Bitterapple has been applied. It should be used on electrical and telephone cords until the animal learns not to chew them.

Ferrets enjoy being stroked and talked to. The owner's voice and hands are the best training tools for ferrets. See Figure 16–12.

Ferrets prefer to relieve themselves in the same place every time and are easy to litter train; a litter box

Figure 16–13 Ferrets allowed to run free should be fitted with a collar and a tiny bell. Photo by Isabelle Francais.

should be placed in the cage and some of the ferrets feces placed in the litter. The ferret will soon learn to use the litter box. If the ferret is allowed to run free in the room, it will usually return to the litter box to relieve itself. Ferrets usually relieve themselves shortly after waking up, so some time should be allowed for this before letting it out of the cage.

Ferrets are considered very intelligent animals and can learn tricks. They can learn to sit and beg for treats, to ride around on the owner's shoulder, or to ride in a coat pocket or hood of a sweater.

Ferrets allowed to run free should be fitted with a collar and a tiny bell. When they are outside, they will be easier to locate, and the collar will signal to others that this is a pet and not a wild animal, see Figure 16–13.

DISEASES AND AILMENTS

Ferrets are highly susceptible to canine distemper. This highly contagious viral disease is usually fatal to ferrets; vaccination is very important. Young kits should receive vaccinations at twelve weeks of age, and a booster should then be given annually. Discharge from the eyes and nose, breathing difficulty, and diarrhea are early signs of the disease. Once an animal has canine distemper, there is very little hope of saving it.

Ferrets also are susceptible to feline distemper, although not all sources agree. To be safe, vaccination for feline distemper would be inexpensive insurance for a pet ferret.

Rabies among ferrets is rare because the animals are usually confined to cages; however, there is a risk of a ferret escaping to the outdoors and becoming exposed to rabid animals. An owner may choose to have the pet vaccinated as a precaution.

Hemorrhagic enteritis, or bloody diarrhea, is a common disease in some areas. In acute cases, animals have bloody diarrhea, go off their feed, and die within three or four days. In chronic cases, diarrhea may come and go; the animal loses weight, becomes dehydrated, and dies within a month. Antibiotics and sulfur preparations may be given. Quick recognition and treatment are important in saving the animal; the cause of bloody diarrhea is not known.

Pasteurella multocida is a bacteria that may cause sudden death in ferrets without any clinical sign. Postmortem examination will find abscesses on the lungs, liver, and spleen. Affected animals may go off feed, develop breathing problems, become lethargic, and have locomotor problems. Antibiotics are helpful if treatment is started early.

Ferrets are very susceptible to botulism caused by the toxin produced by the bacterium *Clostridium botulinum.* Clinical signs include breathing difficulty and

Figure 16–14 The ears of a ferret can be cleaned with cotton tipped applicators and peroxide. Dip the cotton tip into the peroxide and squeeze out any excess. Carefully clean out any wax or dirt from the outer ear. Make sure the ear is dry. Photo by Isabelle Francais.

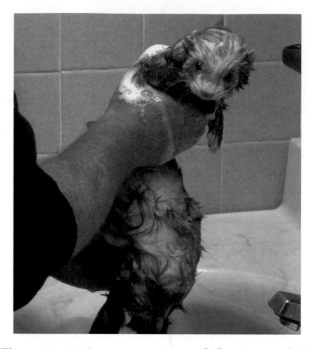

Figure 16–15 Proper grooming and cleaning can help keep a ferret healthy. Ferrets may need to be bathed occasionally. Wet the ferret thoroughly, apply a pet shampoo or baby shampoo, rinse the ferret to make sure no soap remains in the fur, and then dry the ferret completely and place it in a warm area. Photo by Isabelle Francais.

paralysis; however, death may occur without clinical signs. The disease is usually seen in large colony operations and is caused by contaminated food.

Ear mites of the species *Otodectes cynotis* are a common problem and cause discomfort, scratching, head shaking, and a buildup of debris in the external ear canal. Ear drops used for dogs and cats can be used on ferrets. See Figure 16–14.

The mite *Sarcoptes scabei* may infect ferrets, causing hair loss and skin lesions. This is fairly rare but, when it occurs, can be easily taken care of with cat-type flea dips.

Fleas may infect ferrets; again, flea dips for cats will eliminate the problem. Remember to also treat the cage and bedding of the ferret. See Figure 16–15.

Physical injuries are common with ferrets. Doors, rocking chairs, and refrigerator and freezer motors may be hazardous to these small, agile animals. Houses where ferrets are allowed to run loose should be "ferret-proofed" so that physical injuries can be eliminated. Their nails should also be kept trimmed to avoid injury. See Figure 16–16.

REPRODUCTION

Female ferrets come into heat for the first time at about ten months of age. Under natural conditions, ferrets are seasonal breeders; the female comes in heat the spring following birth. The males are also sexually in-

Figure 16–16 The nails of a ferret should be kept trimmed. Clippers used to cut the nails of cats can be used, or regular clippers used by humans will work. Be careful not to cut into the blood vein that runs down the inside of the nail. Photo by Isabelle Francais.

active during the winter months because of a decline in the production of testosterone. This yearly cycle of seasonal activity is regulated by the length of the day. As spring days lengthen, sexual activity starts to increase. If mating does take place before the male is seasonally ready, the sperm that is produced will usually be sterile, and pregnancy will not occur. Ferrets kept indoors under artificial light may come in season at anytime during the year. See Figures 16–17 and 16–18.

When the female comes in heat, the vulva will show signs of swelling; this swelling continues until the vulva is about fifty times normal size. A thin, watery fluid is secreted to cause wetting to the area of the female; there is an obvious increase in odor, even if the female has been descented. A female ferret in heat usually shows very little change in behavior.

As spring approaches, there is an increase in the level of testosterone production. This increase in testosterone production causes the male to become sexually active. As the level of testosterone increases further, a production of fertile sperm occurs.

The female ferret should be put with the male for breeding when she comes in heat. When the female is first put into the male's cage, she may seem disinterested and go about exploring; this is normal behavior. The male will become very aggressive and grab the female by the nape of the neck with his teeth. It is not uncommon for the male to throw the female back and forth and drag her around the cage. Fur may be pulled from the neck of the female, and

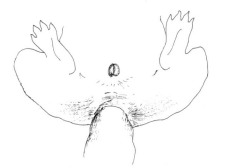

Figure 16–17   The genital area of a female ferret.

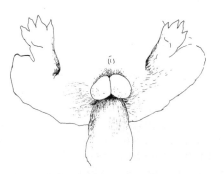

Figure 16–18   The genital area of a male ferret.

the skin may be broken causing bleeding. When he has a firm hold, the female will become limp and passive and mating can then occur.

After successful mating, the two may remain locked together for up to three hours. After they separate, the female may be removed from the cage, although it is not necessary because the two will get along together. If they are left together, additional mating will probably occur during the next few hours. After mating, ovulation occurs. If the male was fertile, pregnancy will result.

The pregnant female can go about normal activity up to a couple weeks before the young are due. The average gestation period will be approximately six weeks. The female should be allowed to eat as much as she wants; however, she should not be allowed to put on excess fat. Her diet should be supplemented with warm milk and powdered calcium to guard against calcium deficiency. Her activity may decline somewhat, and she will probably sleep more often.

If the male was left in with the female, he should be taken from the cage at about two weeks before the arrival of the litter. A bed or nest box should be provided for the female to have her young in. A clean towel, an old shirt, or a piece of old sheet can be added for bedding.

Female ferrets usually have no problems giving birth and probably will not need any assistance. Sometimes, first-time mothers may need some assistance as they may be nervous and not know exactly what is happening.

Most mothers are very protective of their young, and they should not be handled unless absolutely necessary. As each young kit is born, the sac that surrounds the newborn is broken open by the mother and cleaned away from the nose of the young kit so that it can breathe; if this is not done, the animal will suffocate. Some assistance may be necessary at this point. A clean, damp washcloth can be used to gently wipe the membrane from the young kit's face.

Occasionally, a first-time mother will have nothing to do with the young; she may even eat each new kit as it is born. Unless another female is available to foster the kits, the litter will be lost. Second litters will usually be well cared for, however.

If everything is going well, the newborns will arrive at intervals of 5 to 10 minutes; nervous or upset females may take longer to deliver.

The average litter has six to eight kits but may vary from one to fourteen. The young are pink and covered with an almost invisible, thin, white fuzz. They are born with their eyes closed and ears sealed shut. Each kit weighs about ½ to ¾ of an ounce and is about 1½ inches long, with a tail about ½ inch long.

Figure 16–19   A female with her litter. Photo by Isabelle Francais.

The young will cry and be very noisy at first but will quiet down after a few days.

After delivery, they should be left alone so that the female remains calm. The progress of the litter can be checked every three or four days during the next two weeks. Always remember to approach the female and her litter slowly and with caution.

The young litter will grow very fast, and with the approval of the female, the owner may start to handle the young kits at two weeks of age. Handling at an early age is important so that the young kits become familiar with humans, even though their eyes are not open yet. At four weeks of age, the young start to crawl around the nest and may even venture out. Their eyes should start to open at this time. The young can start to receive bread or dry cat food soaked with water or milk. At six weeks, the young kits can be weaned from their mother. See Figure 16–19.

When a female ferret comes in heat, she remains in heat if she is not bred. The breeding cycle is controlled by induced ovulation, meaning the eggs are not released into the womb for fertilization until mating has taken place. The continued production of estrogen during the heat period can lead to aplastic anemia. A female that is allowed to remain in heat for a long period of time is very susceptible to death. The seriousness of these ailments is increased because during the continued heat period, the female ferret is losing weight, and her general condition deteriorates. Symptoms of these ailments include listlessness, depression, and apathy along with pale white gums, dehydration, and fever. The fever is usually followed by a rapid decrease in temperature that is shortly followed by death.

One of three things should be done with a female ferret in heat: she should be bred, spayed, or brought out of heat with hormones.

Chorionic gonadatropin is one type of hormone used to promote ovulation and terminate the heat period. Ovaban is a cat birth control pill, and Ovarid is another heat suppressant that may be effective. It is essential that a veterinarian be consulted in obtaining and prescribing these hormones.

Ovulation induced by hormones may result in a false pregnancy. During a false pregnancy, the abdomen and mammary glands of the female may enlarge, thus showing the signs of a true pregnancy. At three weeks, a veterinarian can determine a false or true pregnancy by palpating for the growing fetuses.

SUMMARY

The domestic ferret belongs to the family Mustelidae and the genus *Mustela,* which also includes weasels, ermines, mink, and polecats. All members of this genus have long, limber, slender bodies. The tail is usually about half the length of the head and body. They have short legs and small, rounded ears. They have long, oval-shaped heads and pointed snouts; the eyes are bright and clear.

The domestic ferret *(M. pulorius furo)* is believed to be a descendant of the European polecat *(M. pulorius);* the two are very similar in appearance.

Several colors are now being bred: sable (the most common), red-eyed white, silver mitt, sterling silver, white-footed sable, butterscotch, white-footed butterscotch, and the rare cinnamon.

The type of housing used depends on the owner's situation. Two things should be kept in mind: (1) If there is a way to escape a cage, a ferret will find it. All materials should be of good quality, and all latches on the doors should lock securely. (2) Ferrets enjoy watching the activity around them. Cages should be constructed to allow ferrets to see out; they want to be part of the action.

Ferrets are easy to feed; the major part of their diet consists of a high-quality commercial dry cat or kitten food. It is important that the cat food contain approximately 35 percent animal protein.

Ferrets are highly susceptible to canine distemper. This highly contagious viral disease is usually fatal to ferrets. Ferrets are also susceptible to feline distemper. Ferrets should receive regular vaccinations for both.

Female ferrets come into heat for the first time at about ten months of age, usually the spring following birth. The gestation period is approximately six weeks. The average litter has six to eight kits but may vary from one to fourteen.

When a female ferret comes in heat, she remains in heat if not bred. The breeding cycle is controlled by induced ovulation, meaning the eggs are not released into the womb for fertilization until mating has taken place. A female that is allowed to remain in heat for a long period is very susceptible to death from uterine and reproductive tract infection and from aplastic anemia.

DISCUSSION QUESTIONS

1. What is the ancestor of the domestic ferret?
2. What are the characteristics of the ferret?
3. What are the eight color variations of ferrets?
4. How should ferrets be handled?
5. What type of housing and equipment are necessary for ferrets?
6. What are ferrets commonly fed?
7. What happens if a female ferret is not bred when she comes into heat? Why does this happen?
8. List the common diseases and ailments of ferrets. What are the causes, symptoms, and treatment?

SUGGESTED ACTIVITIES

1. With the approval of the instructor, invite local breeders, pet shop owners, or class members who have ferrets to bring the animals to the classroom and talk about the characteristics of the ferrets. Class members can observe the animals and their habits.
2. With the approval of the instructor, invite a local veterinarian into the classroom to talk about the diseases and ailments of ferrets.
3. If one has access to the Internet, research and report to the class on the black-footed ferret *(M. nigripes)*.

ADDITIONAL RESOURCES

1. Ferret Net
   www.ferret.net
2. Grin and Ferret
   http://homearts.com/depts/pastime/37ferrf1.htm
3. Electronic Zoo/NetVet-Ferret Page
   http://netvet.wustl.edu/ferrets.htm
4. The American Ferret Association, Inc.
   www.ferret.org
5. Safari Animal Care Centers - Ferrets
   www.safarivet.com/ferrets.htm

END NOTE

1. Morton, C., and F. Morton. 1985. *Ferrets, A Complete Pet Owner's Manual.* New York: Barron's Educational Series, Inc.

Amphibians

OBJECTIVES

After reading this chapter, one should be able to:

- describe the characteristics of amphibians.
- compare the three orders of amphibians.
- describe and construct an aquatic, semiaquatic, and land habitat aquarium.
- describe the methods used in handling and feeding amphibians.

TERMS TO KNOW

amphibians	metamorphosis	osmosis
amphiumas	newts	sirens
cloaca	olm	spermatophore

HISTORY

Evidence indicates that the lobe-finned fish were ancestors of amphibians. Lobe-finned fish were abundant during the Devonian period, 345 to 405 million years ago. This period of time was believed to be a period of changing weather conditions during which the weather varied from periods of rain and flood conditions to dry conditions. When ponds and flooded areas dried up, lobe-finned fish were forced to move to another area with water. Their strong lobe-fins allowed them to do this. These fish also possessed a type of lung developed as an outgrowth of the pharynx. Over time, the lobe-fins developed into legs, and these fish were able to utilize atmospheric oxygen. The skull and tooth structure of lobe-finned fish are very similar to the earliest known amphibians of the late Devonian period.

CHARACTERISTICS

Amphibians live the larval part of their lives in water and their adult lives partially or completely on land. They are cold-blooded and do not have scales.

Approximately 4,000 species of amphibians exist, belonging in three different orders: Gymnophiona, which is made up of worm-like amphibians called caecilians; Caudata, which is made up of newts and salamanders; and Salientia, which is the frogs and toads.

Amphibians have thin, moist skin, which serves a variety of purposes. They can breathe through their skin by the process of osmosis. During osmosis, oxygen is absorbed through the skin from the surrounding air into the bloodstream. To keep their thin skin from drying out, most amphibians must live in moist environments. Amphibians also absorb water through their skin and lose water from the skin by evaporation. Amphibians that live in drier areas have thicker skin, which enables them to conserve moisture.

The skin of amphibians is covered with fluid-secreting glands that produce a slimy mucus. This mucus serves several purposes. It helps conserve moisture, prevents too much water from being absorbed into the body when the animal is in the water, and makes the animal slippery, which aids in its defense. Some amphibians have toxin-producing glands. The toxic substance produced by these glands serves as a means of defense against predators; some species of frogs produce a toxin that can be deadly to predators and humans.

The tongues of amphibians may vary considerably. Some do not have tongues, whereas others have very long tongues with sticky tips that they can stick out very quickly to capture small insects. Amphibians do not have teeth; they crush their prey with their jaws and swallow it whole. Amphibians are capable of consuming large numbers of insects and prey; they feed primarily on spiders, insects, and insect larvae. Larger species can also feed on mice and small rodents.

CLASSIFICATION— NEWTS AND SALAMANDERS

Kingdom — Animalia
Phylum — Chordata
Class — Amphibia
Order — Caudata (Newts and Salamanders)
Family — nine different families
Genus — sixty genus groups
Species — 358 species

MAJOR FAMILIES— NEWTS AND SALAMANDERS

The Family Hynobiidae

The family Hynobiidae is referred to as the Asiatic salamanders because they are found in Northern and Eastern Asia; this family is made up of nine genus groups consisting of thirty-three different species. The characteristics of this family vary considerably; they range in size from about 4 to 8 inches.

The Family Cryptobranchidae

The family Cryptobranchidae is referred to as giant salamanders. The family is made up of two genus groups and consists of three different species.

The Hellbender. The Hellbender, *Cryptobranchus alleganiensis,* which is native to the Central and Eastern United States, may reach a length of 1½ feet. The Hellbender lives in fast-moving mountain streams. The Japanese giant salamander, *Megalobatrachus japonicus,* may reach a length of 5 feet and weigh more than 200 pounds. See Figure 17–1.

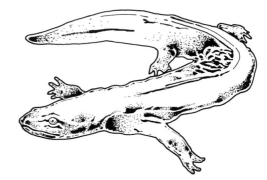

Figure 17–1   Hellbender.

The Family Ambystomatidae

The family Ambystomatidae, or mole salamander, is made up of four genus groups and consists of thirty-five different species. Mole salamanders burrow in decaying, damp soil, where they live most of their lives. They are small, although some species may reach a length of 10 inches. They are more commonly seen during the breeding season, when they move to water to mate and lay eggs. Mole salamanders have thick bodies and smooth skin with bright, colorful markings; they are found in North America.

The Spotted Salamander. The spotted salamander, *Ambystoma maculatum,* is a member of the family Ambystomatidae, which are commonly called mole salamanders. The spotted salamander is blue-black with yellow spots. See Figure 17–2.

The Tiger Salamander. The tiger salamander, *Ambystoma tigrinum,* is another member of the family Ambystomatidae. It is an olive-green color with yellow spots. See Figure 17–3.

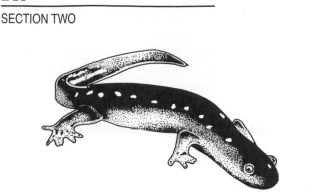

Figure 17–2 Spotted salamander.

Figure 17–3 Tiger salamander.

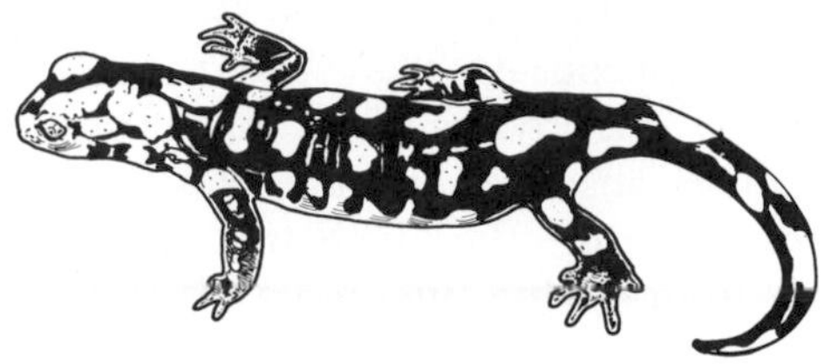

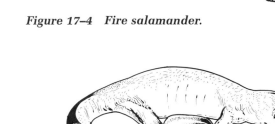

Figure 17–4 Fire salamander.

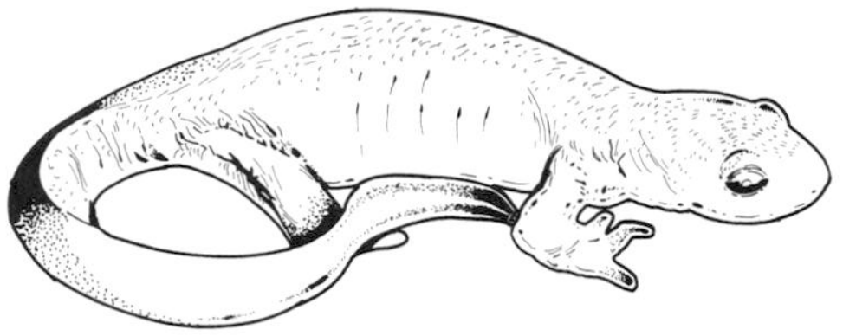

Figure 17–5 California newt.

Figure 17–6 Eastern newt.

The Family Dicamptodontidae

The family Dicamptodontidae is more commonly referred to as Pacific mole salamanders. This family is very similar to mole salamanders but is found along the Pacific coast of North America.

The Family Salamandridae

The family Salamandridae is made up of the more commonly known fire salamanders, brook salamanders, and **newts.** The family is made up of fourteen genus groups and consists of fifty-three different species. Fire salamanders are brightly colored and are found in Europe. They have black and yellow or black and orange coloration. They have glands in their skin that produce an irritating fluid. This fluid is especially irritating to the mucous membranes of the mouth; many predators avoid this animal. Fire salamanders return to the water only to breed. Three species of brook salamanders are found in isolated, highland areas of Europe.

The Fire Salamander. The fire salamander, *Salamandra salamandra,* belongs to the family Salamandridae. It is black with yellow spots. See Figure 17–4.

The California Newt and The Eastern Newt. Two species of newts are found in the United States: The Californian newt, *Taricha torosa,* and the Eastern newt, *Notophthalmus viridescens.*

Newts spend a good part of their time in water; they need the moisture to keep their thin, soft skin from drying out. They are able to replace lost or damaged parts of their bodies by regeneration and periodically shed their skin. See Figures 17–5 and 17–6.

The Family Amphiumidae

The family Amphiumidae is made up of one genus group and consists of three different species of **amphiumas.**

The Congo Eel. Commonly called Congo eels, amphiumas have long, round bodies with tapered heads. They have two pairs of tiny, weak limbs with one, two, or three toes on each limb. They range from 18 to 46 inches in length; all three species are found in muddy waters and swamps of the Southern United States. Amphiumas usually spend the day submerged in the mud or among stands of water weed and hunt for food at night. See Figure 17–7.

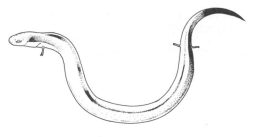

Figure 17–7   Congo eel.

The Family Proteidae

The family Proteidae is made up of water dogs, mud puppies, and the **olm**. There are two genus groups and six different species.

The Mud Puppy.   Water dogs and mud puppies are very similar in appearance; they may reach a length of up to 18 inches. They have rather broad, flattened heads, with bodies that taper down to a short, flattened tail. They breathe by means of external, bushy gills found on both sides of the head. The mud puppy, *Necturus maculosus,* is fairly common in the Eastern and Central United States. Four species of water dog are found in the Eastern and Southern United States. The olm, *Proteus anguinus,* is a dull, white animal with red feathery, external gills. It is about a foot long and is found in underground lakes, rivers, and streams in Austria, Italy, and Yugoslavia. The olm is basically blind. See Figure 17–8.

Figure 17–8   Mud puppy or water dog.

The Family Plethodontidae

The family Plethodontidae is commonly referred to as lungless salamanders. There are twenty-four genus groups consisting of 209 species; this is the largest group of salamanders. They are mostly found in caves, under rocks, and in other damp habitats throughout the United States.

Figure 17–9   Red salamander.

The Red Salamander.   This is a lungless salamander. Lungless salamanders breathe through their skin and throats; the oxygen is absorbed directly into blood vessels in the skin and throat. See Figure 17–9.

The Family Sirenidae

The family Sirenidae is made up of two genus groups and three species of **sirens.**

The Siren.   Sirens are eel-like amphibians that live in freshwater areas of the Southern and Central United States. They have feathery, external gill structures. Sirens have small, weak front legs and no rear legs. The greater siren, *Siren lacertina,* may grow to a length of 29 inches. The lesser siren, *Siren intermedia,* grows anywhere from 7 to 27 inches in length. The third member of the family, the dwarf siren, *Pseudobranchus striatus,* reaches a length of only 4 to 8 inches. See Figure 17–10.

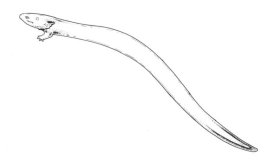

Figure 17–10   Siren, Siren lacertina.

CLASSIFICATION—FROGS AND TOADS

Kingdom — Animalia
Phylum — Chordata
Class — Amphibia
Order — Salientia (Frogs and Toads)
Family — Twenty different families
Genus — 234 genus groups
Species — 3494 different species

MAJOR FAMILIES—FROGS AND TOADS

The Family Leiopelmatidae

The family Leiopelmatidae is made up of two genus groups and three species. Two species are commonly referred to as New Zealand frogs because they are found in New Zealand. The third species is the 2-inch-long, tailed frog *Ascaphus truei* that is found in the Western United States. These species of frogs do not have tails but have retained two tail-moving muscles from their tadpole days. The muscles extend for about ¼ inch. This extension allows the male to fertilize the female internally; this is the only species to use this method of fertilization.

The Family Discoglossidae

The family Discoglossidae is made up of five genus groups and fourteen different species. They are commonly referred to as disk-tongued toads. They are found in Europe, Western Asia, the Philippines, Borneo, and other areas of the Far East. These toads do not extend their tongues to catch their prey but catch them with their mouths.

The Family Pipidae

The family Pipidae is made up of four genus groups and twenty-six different species. In this family are the clawed and the Surinam toads that are found in Africa and parts of Central and South America. These toads do not have tongues but use their front toes to dig for and pick up food. Some references refer to the clawed and Surinam toads as "frogs" because they are aquatic animals. This species has fairly flattened bodies, with small eyes on the top of their heads that aid them in looking up from ponds or streams. They also have large, webbed rear feet. The Surinam toad of Mexico and Central America is unique in that the female carries her eggs on her back; they are placed there by the male during fertilization.

The African Clawed Frog. The African clawed frog, *Xenopus laevis,* is an aquatic frog that seldom leaves the water. Its forelimbs are very weak and do not support the frog's body while on land. See Figure 17–11.

The Family Rhinophrynidae

The family Rhinophrynidae is made up of only one genus group and consists of one species. The burrowing toad, *Rhinophrynus dorsalis,* is found in Central America. The species has spade-like structures on the first toe of its hind feet that enable it to dig tunnels. It is about 3 to 4 inches long and feeds on termites and other soil insects.

The Family Pelobatidae

The family Pelobatidae is made up of ten genus groups and consists of eighty-eight different species. In this family are the spade-foot frogs, horned toads,

Figure 17–11   African clawed frog. Photo by Isabelle Francais.

and parsley frogs that are found in North America, Europe, Indonesia, the Philippines, and parts of Asia.

The Spade-Foot Frog. The spade-foot frogs have developed spade-shaped structures on their hind feet that enable them to burrow into the soil. They are nocturnal and usually spend the day in their damp tunnels. The horned toads get their name from the horn structures that are found on their head. Parsley frogs closely resemble other frogs in their general shape. Their name is derived from the spotted, green patterns on their backs. They live among rocks or dig tunnels. Their hind feet are not as well developed for digging as the spade-foot or horned toads. See Figure 17–12.

Figure 17–12   Spade-Foot frog. Photo by Isabelle Francais.

Figure 17–13   Malaysian horned frog. Photo by Isabelle Francais.

The Malaysian Horned Frog. The Malaysian horned frog, *Megophrys monticola,* is native to Southeast Asia, Indonesia, and the Philippines. This is a ground-dwelling frog that lives among the dead leaves on the forest floor. See Figure 17–13.

The Family Ranidae

The family Ranidae are the true frogs. There are forty genus groups and 611 different species. True frogs can be found in almost all parts of the world and range in size from ½ inch up to 14 inches in length. Frogs have slender bodies and long, muscular legs that enable them to be excellent jumpers and swimmers; a few species are adapted to tree life.

Figure 17–14   Green frog. Photo by Isabelle Francais.

The Green Frog. Included in the family Ranidae is the American bullfrog, *Rana catesbeiana,* and the green frog, *R. clamitans,* both of which are common in the United States. See Figure 17–14.

Also belonging to this family is the goliath frog, *Gigantorana goliath,* the largest member. It may reach a length of 16 inches from snout to rear, and with its legs stretched out, it may exceed 30 inches in length.

The Family Dendrobatidae

The family Dendrobatidae is made up of four genus groups and consists of 116 different species. These frogs are commonly referred to as poison-arrow frogs. This family is found in the tropical areas of South and Central America. Poison-arrow frogs are brightly colored and are noted for the extremely potent poisons that are secreted from their skins. The bright colors serve as a warning to would-be predators. The poison from one species, the Kokoi Poison-Arrow Frog, *Phyllobates bicolor,* found in Colombia is used by the local Choco Indians to tip the points of their arrows. The Kokoi Poison-Arrow Frog is bright red with black spots. They are very small in size. The male poison-arrow frog usually carries the fertilized eggs on his back until they hatch; then he carries the tadpoles to the water, where they continue to develop.

The Family Rhacophoridae

The family Rhacophoridae is made up of ten genus groups and 184 different species. This family is commonly referred to as Old World tree frogs, which have adapted themselves to life in the trees. They have flattened bodies and disks on their feet that aid them in clinging to tree limbs. About one-third of the species are referred to as flying frogs because they are able to stretch their bodies and limbs and glide from tree to tree. The female tree frog lays her eggs in moist, foamy masses on branches that overhang ponds and streams. The male mounts the female from the rear and fertilizes the eggs as the female lays them. The foamy mass then hardens. After the tadpoles have developed, the foamy mass breaks open and the liquid within the foamy mass runs down to the water below, carrying the tadpoles with it. The Old World tree frogs are found in tropical areas of Africa and Asia.

The Family Hyperoliidae

The family Hyperoliidae is made up of twenty-three genus groups and consists of 292 different species. This family is referred to as sedge frogs and bush frogs. They are found in habitats throughout Africa. They resemble tree frogs in that they have sucker disks on the toes of their feet. Most species, however, are found around swamps and ponds. The female deposits eggs in a jelly mass on reeds or limbs above the water. The developed larval stage drops to the water, where they continue to grow. These frogs can change their color as a result of temperature changes, humidity changes, or nervous stimuli.

The Family Sooglossidae

The family Sooglossidae is made up of two genus groups and three different species. These species are found on the islands of Mahe and Silhouette, which are part of the Seychelles Islands. The small, 1-inch-long Seychelles frog, *Sooglossus seychellensis,* is the best known of this group. Because there are no still, breeding ponds available, the female lays her eggs in gelatinous masses on the ground. When the eggs hatch, the larval stage of the tadpoles make their way to the back of the male, where they are attached by means of a sticky substance secreted by the male; here they develop until they are able to fend for themselves.

The Family Microhylidae

The family Microhylidae is made up of sixty-one genus groups and consists of 281 species; they are referred to as narrow-mouthed frogs. These frogs have small, tapered heads; narrow mouths; and pointed snouts. They dig and burrow into the soil. They are found in Indonesia, Northern Australia, South America, Africa, southern parts of North America, and Asia.

The Family Pseudidae

The family Pseudidae is made up of two genus groups and consists of four different species. This family is commonly referred to as shrinking frogs because they grow smaller as they mature. They live in shallow lakes and seldom leave the water. One species lives in shallow ponds that evaporate during the summer. They bury themselves in the mud and lie dormant until the next rain.

The Family Bufonidae

The family Bufonidae is made up of twenty-five genus groups and 339 different species. This family represents the true toads, which are found in most parts of the world. Toads have short, thick bodies and

Figure 17–15   American toad. Photo by Isabelle Francais.

short legs. Although toads can jump, their legs are too short to allow them to jump the distances that frogs are capable of jumping. The toad's skin is dry and has a rough appearance; true toads have skin glands that produce a poison that can be irritating to the mucous membranes of predatory animals. Toads eat a wide variety of foods, but insects make up the main part of their diet. During the winter months, toads hibernate under rocks, under logs, or in underground holes. They emerge in the spring and head for lakes and ponds to breed. Toads do not make noises like frogs and use their sense of smell to locate the opposite sex for mating.

The American Toad.   The American toad, *Bufo terrestria,* is a common toad of the eastern parts of the United States. See Figure 17–15.

Figure 17–16   Marine or cane toad. Photo by Isabelle Francais.

The Marine Toad. The marine or cane toad, *Bufo marinus,* is native to extreme Southern Texas and to Northern South America. It is a large toad that grows to 9 inches. The marine toad is poisonous to dogs, cats, and other possible predators. See Figure 17–16.

The Family Brachycephalidae

The family Brachycephalidae is made up of only one genus group and consists of two different species. These are commonly called gold frogs and are found in the southern parts of Brazil. The frogs are about 1 inch long and are brilliant orange in color. A feature of the frog is a bony, cross-shaped structure on the back of the animal.

The Family Hylidae

The family Hylidae is made up of thirty-seven genus groups and 637 different species. These are the true tree frogs and resemble the Old World tree frogs in the flattened shape of their bodies and the adhesive disks on their toes. These tree frogs live in South and Central America; species can also be found in North America, Central Asia, and Europe. See Figure 17–17.

The Red-Eyed Tree Frog. The red-eyed tree frog, *Agalychnis callidryas,* is an arboreal frog of Central America. Its eyes are red with vertical pupils. Its primary color is green. The color markings on its sides are yellow and blue, and the toes are red. It also has

Figure 17–18   Red-eyed tree frog. Photo by Isabelle Francais.

opposable thumbs, which allow it to take a strong grip on leaves and branches. See Figure 17–18.

The White's Tree Frog. The White's tree frog, *Litoria caerulea,* is native to Australia. This small, bloated-appearing, light green frog is commonly found around houses and breeds in wet grassland areas. See Figure 17–19.

The Green Tree Frog. The green tree frog, *Hyla cinerea,* is found in the Mississippi Valley and South-eastern United States. It is yellow-green with a yellow stripe running from the lower jaw back along its side. See Figure 17–20.

The Family Leptodactylidae

The family Leptodactylidae is made up of fifty-one genus groups and consists of 722 different species. This is the largest family in the order Anura. Because

Figure 17–17   Tree frogs have long, slender limbs and adhesive disks on the ends of their toes. Photo by Isabelle Francais.

Figure 17–19   White's tree frog. Photo by Isabelle Francais.

Figure 17–20   Green tree frog. Photo by Isabelle Francais.

Figure 17–22   Colombian horned frog. Photo by Isabelle Francais.

of the large number of species, there is a large variety of colors, characteristics, and behavior within this family. Most of these frogs are found in tropical areas of South and Central America. One of the most common members of this family is the South American bullfrog, *Leptodactylus pentadactylus.*

The Ornate Horned Frog.   The ornate horned frog, *Ceratophrys ornata,* belongs to the family Leptodactylidae. It is a carnivorous frog that lives among the leaves and debris on the rainforest floors of South America. Its primary color is a dark green with brown markings above and yellow below. The legs are yellow with green and brown markings. See Figure 17–21.

The Colombian Horned Frog.   The Colombian horned frog, *Ceratophrys calcarata,* also belongs to

Figure 17–21   Ornate horned frog. Photo by Isabelle Francais.

the family Leptodactylidae. It is primarily light brown with dark brown markings. See Figure 17–22.

The Family Rhinodermatidae

The family Rhinodermatidae is made up of one genus group and consists of two different species. This group is commonly referred to as mouth-brooders. They are unique because of their parental rearing habits. The female deposits her eggs on the ground; when the larvae start to move within the egg, the male gathers them up in his mouth and holds them there as they develop. The male retains the eggs in his mouth for about three weeks; during that time, the tadpoles hatch. When they are able to survive on their own, the male spits them out.

The Family Heleophrynidae

The family Heleophrynidae is made up of one genus group and consists of four different species. This family of frogs is commonly called ghost frogs. These small frogs live in fast-flowing mountain streams in Africa. The tadpoles have a large suction disk around the mouth that is used to help them cling to rocks. The skin on the belly of the adults is white and so thin that their internal organs and muscles can be seen.

The Family Myobatrachidae

The family Myobatrachidae is made up of twenty genus groups and 100 different species. This family of frogs is found in Australia, New Zealand, and Tasmania. The family varies a great deal in appearance and habits, but their feet are usually less than half webbed, and their toes may have small disks. Included in this family is the pseudophryne frogs, which are among the smallest frogs.

The Family Centrolenidae

The family Centrolenidae is made up of two genus groups and consists of sixty-four different species. These are small, bright green frogs that have transparent skin on their bellies; their internal organs and muscles are visible. Because of their visible internal structure, these tree-dwelling frogs of Central America, Northern South America, and Southern Brazil are commonly referred to as glass frogs.

CLASSIFICATION—CAECILIANS

Kingdom — Animalia
Phylum — Chordata
Class   — Amphibia
Order   — Gymnophiona (Caecilians)
Family — Caecilliidae
Genus   — sixteen genus groups
Species — fifty different species

MAJOR FAMILY—CAECILIANS

The family Caecilliidae is made up of sixteen genus groups and consists of about fifty different species. The caecilians are seldom seen, and not very much is known about their habits. References differ as to the actual number of families, genus groups, and species that make up the order. Caecilians are worm-like in appearance; they do not have limbs and are usually without sight. They burrow into the decayed material on top of the soil and into the soil. They vary in length from 3 to 59 inches and have tentacle-like structures on their heads that enable them to find

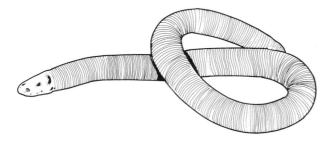

Figure 17–23 The family Caecilliidae consists of about fifty species of worm-like amphibians called caecilians.

their way in their tunnels. Caecilians are found in tropical areas of Southern Asia, Africa, and South Africa. See Figure 17–23.

HOUSING AND EQUIPMENT

In discussing housing and equipment, one must realize the vast numbers of different species and the large variation in environments. The information in this text is intended as a general guide only. One should duplicate as closely as possible the animal's natural environment.

The best container for most types of amphibians is the common rectangular aquarium. Consideration for the size of the animal is very important when deciding what size aquarium to use.

Three specific-type habitats need to be considered for amphibians: (1) aquatic habitat for tadpoles and completely aquatic species, (2) semiaquatic habitat for species that live partially in the water and partially on land, and (3) woodland habitat for species that live almost entirely on land.

When preparing an aquarium for tadpoles or aquatic species, it can be set up like a true aquarium with filtration and aeration systems; no land area is needed. One should prepare the aquarium by placing a 1-inch layer of rock or gravel in the bottom; then a 3-inch layer of sand can be added. The sand should be sloped gradually so that one end is more shallow. Water, to a depth of 1 to 2 inches on the shallow end, should be added; tadpoles will use the shallow end to rest, and this is where they are fed. Other species may prefer a different habitat, so it is important to research the species that is being put into the aquarium. Depending on the species, a source of heat may be needed.

Unchlorinated well water should be used for the aquarium if it is available; if chlorinated water is used, it should be allowed to sit in open containers for two days to allow the chlorine to evaporate.

Plants, rocks, and tree branches can be added if desired; they provide hiding places. As tadpoles develop legs, they will crawl up onto the rocks and branches.

If the aquarium does not have a filtration and aeration system, the water will need to be changed every two or three days; a buildup of excrement and the loss of oxygen will kill aquatic animals.

Semiaquatic aquariums are set up for species that spend part of their lives in water and part on land. As tadpoles develop into toads and frogs, they require semiaquatic habitats.

To set up this type of habitat, an aquarium needs to be divided in half. A piece of glass or Plexiglas can be cemented to the bottom to divide it into two watertight halves; the divider should be about one-third the height of the aquarium. This again depends on the species.

To set up the land habitat half, a 1-inch layer of small rocks or gravel should be placed in the bottom of the aquarium. A 1-inch layer of charcoal is placed over this; then, a 2- to 3-inch layer of sand or sandy loam is placed over the charcoal. On top of this, a layer of topsoil is added so that it is level with the divider. Finally, leaves, branches, bark, and some plants are added to give the aquarium a natural look. The water in the semiaquatic habitat should be from an unchlorinated source if possible. Water is added to the aquatic side of the aquarium until it is level with the land habitat. This allows the animals to move freely between the water habitat and the land habitat. The water should be changed every two or three days if a filtration and aeration system is not used. Plants and soil should be sprinkled daily with unchlorinated water so that the habitat does not dry out, and a desirable, damp environment is maintained.

The aquarium needs to have a cover to keep the animals from escaping; a piece of window screen or hardware cloth can be used. Temperature of the aquarium should be maintained at 60 to 80°F, depending on the species. A small light bulb can be added if heat is needed. Salamanders prefer temperatures around 60 to 70°F, whereas frogs and toads should be kept in an environment around 70 to 80°F.

In preparing a woodland aquarium, one should proceed just as when preparing the land half of the semiaquatic aquarium. A small bowl of water should be provided; amphibians absorb the water through their skin. Make sure the bowl is level with the soil so that the animals can crawl in and out freely. Remember that plants and soil should also be kept moist. A small light bulb can be added if the temperature drops too low.

During the winter months, most amphibians from northern climates hibernate; the aquarium should be placed in a cool place. When temperatures begin to get cool, the amphibians burrow into the soil. At about 40°F, they stop breathing and absorb the oxygen they need through their skin. One must make sure the aquarium is not in an area where it will receive a hard freeze; basements and cellars are ideal locations.

FEEDING

Most amphibians are insect eaters. Depending on the size and species, they also consume earthworms, small birds, and small rodents. Tadpoles and aquatic species feed on vegetation, aquatic plant life, and algae; they also feed on dead animals.

Figure 17–24 *Frogs and toads can easily be picked up by carefully grasping them around the body. Photo by Isabelle Francais.*

In an aquarium habitat, tadpoles can be fed bits of rabbit, dog, or cat food pellets, as well as baby food. As the tadpoles change to adults, they need to be fed insects. Bait and pet stores are sources of crickets, mealworms, bee moths, earthworms, and night crawlers for tadpoles. Flies, crickets, grasshoppers, and other insects can be collected with a butterfly net along roadways and in open fields.

Salamanders prefer to be fed earthworms or night crawlers but can learn to eat moist dog food, cat food, and even hamburger.

HANDLING

Frogs and toads are best picked up by grasping them around the body just in front of the rear legs. They should be held firmly enough to prevent escape but gently enough to prevent injury. They should not be held by the rear legs because they can injure themselves as they struggle to escape; small frogs and toads can be cupped in the hand. See Figure 17–24.

When amphibians are transported, they should be placed in a moist cloth bag to prevent excess moisture loss.

Salamanders have soft, tender skin, and before picking them up, a person should run water over the hands; this prevents injury to the skin of the salamander. One should gently grasp the salamander

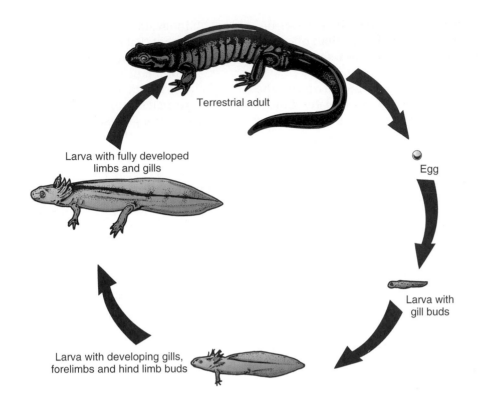

Figure 17–25   The life cycle of an amphibious salamander consists of stages referred to as larval stages, where the small salamanders have external gills. As they develop, some species lose the external gills, whereas other species retain them and remain completely aquatic. Terrestrial salamanders are usually miniature adults and do not have external gills.

around the body; its tail should not be grabbed because the tail may be released in one's hand. Some salamanders can release their tail as a defense mechanism; the tail usually grows back.

The skin glands on some frogs and toads produce toxic substances. After handling frogs and toads, one should wash his or her hands and be careful not to touch the face and eyes before washing. Again, it is important to know the species one is handling.

DISEASES AND AILMENTS

Red Leg.   Red leg is a disease of frogs caused by the bacterium *Aeromonas hydrophila,* which is found in standing water. It is primarily a disease of frogs kept in laboratory or aquarium situations. Normal, healthy frogs living in a natural environment are usually not susceptible. The added stress of capture and captivity in aquariums seems to be a contributing factor in causing the disease. Red blotches appear on the skin, especially on the inside of the thighs. Antibiotics given by stomach tube may be effective.

Fungus Infection.   Fungus infection may affect newts and salamanders. Their skin can develop abrasions from handling or crawling over rocks; these abrasions may be attacked by fungus. Some of the fungus treatments used for fish are effective.

REPRODUCTION

Male frogs and toads call females to breeding sites with distinctive calls that each species makes. Most species mate in the water; the male grasps the female from the back, and as she releases the eggs, he fertilizes them. The eggs then usually attach to vegetation in the water.

Eggs hatch into the larval form of tadpoles. The newly hatched tadpoles lack legs, eyelids, lungs, and true jaws; they have a long tail with a high fin. Tadpoles breathe through external gills and feed on vegetation.

Tadpoles go through a series of physical changes called **metamorphosis.** During these changes, the tadpole develops rear legs, then front legs; eyelids and jaws also develop. The tail is absorbed, and internal lungs develop. Tadpoles gradually develop into the adult form over a period of a few weeks to three years, depending on the species. See Figure 17–25.

Most salamanders and newts breed during the spring; they do not make any noise, so males and females must find each other by smell. Breeding in most cases occurs in the water; in some cases, the male deposits sperm on the eggs as the female releases eggs into the water. In some species, the male deposits a packet of sperm on the bottom of the pond; this packet of sperm is called a spermatophore. The female rubs her cloaca (the cavity where the intestinal and genitourinary tract empties) on the packet and picks up the sperm; fertilization takes place internally. The eggs are then released into the water. The larval forms that hatch resemble their parents in appearance and have bushy external gills. Those species that move onto land lose their external gills.

SUMMARY

Amphibians live the larval part of their lives in water and their adult lives partially or completely on land.

Approximately 4,000 species of amphibians exist, belonging in three different orders: Gymnophiona, which is made up of worm-like amphibians called caecilians; Caudata, which is made up of newts and salamanders; and Salientia, which is the frogs and toads.

There is great diversity among the species; however, most of them have thin, moist skin and live in moist environments. Dryness and excessive heat are enemies of most amphibians.

The skin of amphibians is covered with fluid-secreting glands that produce a slimy mucus. This mucus serves to conserve moisture and makes the animal slippery, which aids in its defense. Some amphibians also have skin glands that produce toxic substances.

Most amphibians feed on spiders, insects, insect larvae, and other small prey. Larger species can also feed on mice and other small rodents.

There is a wide variation in the habits and environments of amphibians. Three specific types of habitat need to be considered: (1) water or aquatic habitat for tadpoles and completely aquatic species, (2) a semiaquatic habitat for species that live partially in water and partially on land, and (3) a woodland habitat for species that live almost entirely on land.

If interested in keeping amphibians in captivity, it is important that one duplicates as closely as possible the animal's natural environment.

Amphibians return to water to reproduce; most species lay eggs that cling to vegetation in the water. The eggs hatch into the larval form of tadpole; the tadpoles go through a series of changes called metamorphosis. During this metamorphosis, the tadpoles develop into the adult form.

DISCUSSION QUESTIONS

1. What is an amphibian?

2. What are the three different orders of amphibians?

3. What are caecilians?

4. What are the characteristics of a newt?

5. What are the characteristics of a salamander?

6. What are the differences between a frog and a toad?

7. Describe how to set up an aquatic, semi-aquatic, and land habitat in an aquarium.

SUGGESTED ACTIVITIES

1. Visit the local pet shop and make a list of the different species of amphibians that are available. Using the local library, compile as much information as you can about the species.

2. Set up aquatic, semiaquatic, and woodland habitats in aquariums. Obtain some amphibians from pet shops or capture them in local ponds to place in the aquariums; observe them and their habits.

ADDITIONAL RESOURCES

1. Caudata: The Salamanders
   http://biodec.wustl.edu/~larsontl/caudata.html

2. NetVet: Amphibians
   http://netvet.wustl.edu/amphib.htm

3. Amphibian Species Identification Guide
   www.npwrc.usgs.gov/narcam/idguide/specieid.htm

4. Living Amphibians
   http://phylogeny.arizona.edu/tree/eukaryotes/
   animals/chordata/living_amphibians.html

5. Checklist of Florida Amphibians and Reptiles
   www.flmnh.ufl.edu/natsci/herpetology/
   fl-guide/flaherps.htm

6. The Complete Tree Frog Page
   www.megsinet.net/~treefrog

7. Introduction to Amphibians
   www.cmnh.org/research/vertzoo/frogs/
   amphibians.html

8. The EMBL Reptile Database
   www.embl-heidelberg.de/~uetz/
   LivingReptiles.html

18 Reptiles

OBJECTIVES

After reading this chapter, one should be able to:

■ describe the characteristics of reptiles.

■ compare the four orders of reptiles.

■ describe the methods used in handling and feeding reptiles.

■ describe the housing and equipment needs of reptiles kept in captivity.

TERMS TO KNOW

arboreal
brille
brood
carapace
casque
crepuscular
dimorphism

ectotherms
hemipenes
lamellae
oviparous
ovoviviparous
plastron

scutes
terrapins
terrarium
tympanum
vivarium
viviparous

CHARACTERISTICS

Reptiles are cold-blooded vertebrates that possess lungs and breathe air. They also have a bony skeleton, scales, or horny plates covering the body, and a heart that has two auricles and in most species one ventricle.

There are approximately 6,500 species of reptiles belonging in four different orders: Testudines (called Chelonia in some references), which are made up of turtles, tortoises, and terrapins; Serpentes, which includes snakes, pythons, and boas; Squamata, which includes of iguanas and lizards; and Crocodilia, which includes crocodiles, alligators, caimans, and gharials.

Fossil remains of reptiles date back to the Carboniferous period 340 million years ago. Other fossil remains dating back 315 million years are of small agile, lizard-like animals. The evolution of reptiles continued through the Permian and into the Triassic periods. During the Triassic period, reptiles became the dominant group of terrestrial vertebrates inhabiting the earth; they continued their dominance for about 160 million years. The reptiles reached their greatest number during the Cretaceous period; then many became extinct. Today, only four orders of the fifteen that existed during the Cretaceous period have survived.

Present-day reptiles have three body types. The first have long bodies and clearly defined tails like the crocodiles and lizards; the seacond have long bodies that taper into tails, like the snakes; and the third have short, thick bodies encased in shells like the turtles and tortoises.

The limbs of crocodiles and lizards are usually paired and attached to the body at right angles. This enables the animal to lift its body up off the ground when moving. The limbs of crocodiles and alligators are strong and powerful; the limbs of most lizards are weak, with some having very short, stump-like limbs.

Snake-like reptiles do not have limbs, and their movement is a result of undulating movements of the body. The scales on the underside of the body project outward as the muscles are contracted and relaxed; these scales exert pressure on the surface and move the animal forward.

A few turtles and tortoises have limbs that enable them to walk on land. This group of reptiles has modified limbs that enables it to swim as well. Some turtles and tortoises spend most of their time in water, with their limbs primarily adapted for swimming, and move about on land with creeping and crawling motions.

The skin of reptiles has a horny surface layer. In lizards and snakes, this layer forms a hard, continuous covering of scales. These scales lie beneath the outer layer of the epidermis so the body can grow; this outer layer is shed, allowing for further growth. Crocodiles, alligators, and some lizards have bony, dermal scales covered by a horny epidermal layer that must also be shed to allow for body growth. Turtles and tortoises do not molt their thick epidermal skin; each year, a new

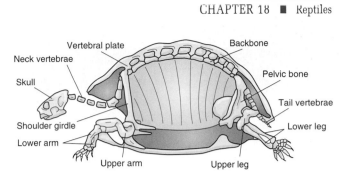

Figure 18–1 ***The parts of a turtle's skeleton.***

epidermal scale is formed beneath the old one. This allows for growth of the animal. These epidermal scales, or **scutes,** form rings that can be counted, and from the rings, an estimate of the animal's age can be obtained. The scutes are divided into three types: the marginal scutes are located around the outer edge of the shell, the central scutes are located down the center, and the coastal scutes are located between the marginal and central scutes. The hard, tough layer of scales prevents moisture from dissipating from the body. In turtles and tortoises, this horny epidermal layer forms an exoskeleton that has two parts: the upper part is called the **carapace** and forms a part of the vertebrate and ribs of the animal; the lower portion is called the **plastron.** See Figures 18–1 and 18–2.

Snakes cannot close their eyes; instead of eyelids, they have a transparent layer, or **brille,** that permanently covers the eye. During the shedding process, this layer over the eyes is shed and replaced with a new covering. Nocturnal species have vertical pupils that open very wide in dim light and close to

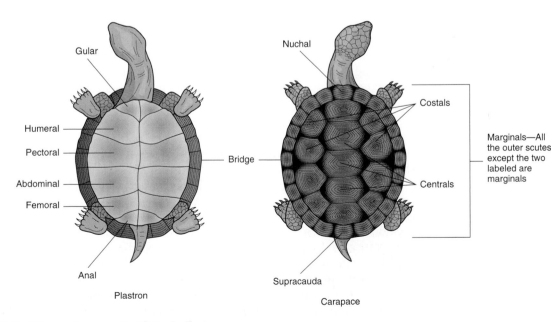

Figure 18–2 ***The scutes on a turtle's shell.***

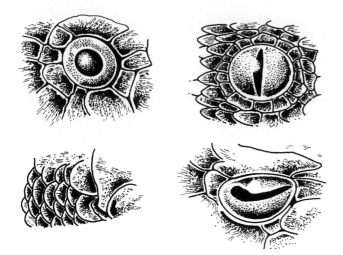

Figure 18–3   The eyes of snakes.

a small slit or a pinhole in bright light; diurnal species have round pupils. Burrowing species have an opaque scale covering the eye for protection and are almost blind. Some arboreal (see Glossary) snakes have horizontal pupils, large protruding eyes, and tapered snouts, which give them a binocular vision. Some lizard species also do not have eyelids. See Figure 18–3.

Lizards and snakes have teeth that are fused to the jaw bones; some snakes also have teeth fused to the palate bones. In some snakes and in two species of lizards, the teeth are connected to poison glands. Crocodile teeth are set in sockets and are similar to the teeth of mammals. Turtles and tortoises do not have teeth; their jaws form sharp crushing plates. The front part of the jaws form a horny beak.

The tongues of reptiles vary considerably. Some have short, fleshy tongues that have little movement, whereas others have tongues that are long, slender, and forked like some lizards and all snakes.

The lungs of most lizards are the same size, but in some lizards, the left lung is reduced in size, with the right lung being very long and lying between internal organs. All reptiles have a three-chambered heart, which has two auricles and one ventricle. The combination of reduced lung size and three-chambered heart allows the animal to have a greatly reduced metabolic rate. This low metabolic rate means that not much heat is produced within the body; therefore, they must function at the same temperature as their surroundings. Most reptiles are found in the tropical areas where they can function both day and night. Outside the tropical areas, they must spend a great amount of their time basking in the sun to obtain the body warmth they need. When temperatures get cool, they seek places to hibernate to survive the cold weather.

Tortoises, crocodiles, and many lizards and snakes are **oviparous,** meaning they lay eggs that hatch after leaving the body of the female. Some species are **ovoviviparous,** meaning the eggs hatch within the body of the female, with the live young emerging from the female's body.

Hibernation

Reptiles that live in temperate zones (area between the tropical zones and polar circles) must hibernate during the winter; winter temperatures are too cold for them to remain active. The length of the hibernation period varies depending on how long the cold weather lasts. In the wild, reptiles seek out dens or tunnels; burrow under rocks, logs, and leaves; or burrow into the soil to find a suitable place to hibernate. If they are to survive, they must be below the frost line. During hibernation, the reptile's body temperature drops to a point where the body's systems barely function. Respiration and heartbeat are greatly reduced, and oxygen is absorbed through the skin. Hibernation is associated with bringing the reptile into breeding condition; temperate-zone reptiles that are not allowed to hibernate do not breed.

Plans should be made for temperate-zone reptiles in captivity to go through a period of hibernation. As the time nears, one should reduce the amount of food and gradually reduce the temperature in the vivarium. Temperature for the hibernating reptile should be between 39 and 50°F, depending on the species. The length of the hibernation period varies from four to twelve weeks, again depending on the species and its original environment. The vivarium can be put in the basement where the temperature is lower. To bring the animal out of hibernation, one can simply reverse the procedure, gradually increasing the temperature until normal temperatures are reached. Most reptiles coming out of hibernation do not eat for a few days.

CLASSIFICATION—TURTLES, TORTOISES, AND TERRAPINS

Kingdom — Animalia
Phylum — Chordata
Class — Reptilia
Order — Testudines or Chelonia (turtles, tortoises, and terrapins)
Suborder — (1) Pleurodira (Side-necked turtles)
Suborder — (2) Cryptodira (Hidden-necked turtles and tortoises)
Family — twelve families

MAJOR FAMILIES—TURTLES, TORTOISES, AND TERRAPINS

The Family Chelidae

The family Chelidae consists of the long, side-necked turtles of South America, Australia, and New Guinea. These turtles have very long necks, and only a few can withdraw them into their shells; those that can withdraw their necks do it in a sideways manner so that the head and neck are on the side of the shell when retracted. There are nine genus groups consisting of thirty-seven species in this family. See Figure 18–4.

Figure 18–4   A member of the family Chelidae, which consists of long-necked turtles. Photo by Isabelle Francais.

The Family Pelomedusidae

The family Pelomedusidae consists of side-necked turtles found in tropical South America and Africa. There are five genus groups and twenty-four species in this family. See Figure 18–5.

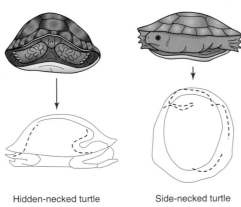

Hidden-necked turtle
(a)

Side-necked turtle
(b)

Figure 18–5   (a) Most common turtles are hidden-necked turtles and withdraw their head and legs back into the shell. (b) Side-necked turtles bring their heads backward and to the side as pictured.

The Family Kinosternidae

The family Kinosternidae is made up of the mud and musk turtles. These are small-sized turtles that are found in North America, Mexico, and Central America. They are aquatic turtles that live in rivers, swamps, and ponds, where they feed on insects, worms, small fish, and plants. See Figures 18–6 and 18–7.

Figure 18–6   Common mud turtle. Photo by Isabelle Francais.

The Mud and Musk Turtles.   Mud turtles have flattened shells and are able to completely retract their head, tail, and webbed feet into their shells. Musk turtles get their name from the strong, musky smelling liquid they produce from glands between their upper and lower shells. Their lower shell does not hinge like that of the mud turtles.

Figure 18–7   Common musk turtle of the genus Sternotherus. *Photo by Isabelle Francais.*

The Family Chelydridae

The family Chelydridae consists of two genus groups and two species. *Chelydridae serpentina* is the common snapping turtle. *Macroclemys temminckii* is the alligator snapping turtle.

The Common Snapping Turtle. The common snapping turtle, *Chelydridae serpentina,* is the most widespread turtle in North America. It can reach a length of over 12 inches and weigh up to 60 pounds. Snapping turtles feed on fish, frogs, small ducks, geese, and small mammals. They usually live in ponds and swamps but will venture out to find food. They are very aggressive and, if threatened, will open their mouth and snap their powerful jaws. Snapping turtles are eaten and are becoming scarce. See Figure 18–8.

Figure 18–8   Common snapping turtle. Photo by Isabelle Francais.

The Alligator Snapping Turtle. The alligator snapping turtle, *Macroclemys temminckii,* is found in parts of the Southwestern United States. This is the largest of all freshwater turtles, reaching a total length of up to 5 feet and weighing 200 pounds. It usually lies motionless on the bottom of ponds and slow-moving streams, where it waits for its prey to swim by. Attached to its tongue is a red, worm-like growth that lures small fish in close so they are captured in the turtle's powerful jaws.

The Family Platysternidae

The family Platysternidae consists of one genus group and one species.

The Big-Headed Turtle. The big-headed turtle, *Platysternon megacephalum,* is native to streams in China, Thailand, and Burma; its head is huge and out of proportion with the rest of its body. It feeds primarily on shellfish, which it crushes with its powerful jaws, but will also feed on fish and small animals. This species of turtle can climb trees and will emerge to sun itself. This turtle has a long tail that is almost as long as its shell. See Figure 18–9.

Figure 18–9   Big-headed turtle, Platysternon megacephalum. *Photo by Isabelle Francais.*

The Family Emydidae

The family Emydidae, with approximately thirty different genus groups and eighty-five different species, is the largest family within the order Testudine/Chelonia. Commonly called pond and river turtles, they are found in parts of Europe, North Africa, Asia, South America, Central America, and North America. These turtles are primarily found in freshwater habitats but live in brackish habitats or on land. Those that live in freshwater spend warm, sunny days basking on logs and rocks.

Turtles of the Emydidae family vary from about 4 inches up to 2 feet in length. The bog turtle, *Clemmys muhlenbergi,* at 4 inches in size, may be the smallest of all turtles and is found in shallow bogs along the East Coast of the United States. The largest Emydidae turtle is the batagur, *Batagur baska,* which reaches a length of 2 feet and weighs 50 to 60 pounds. The batagur is found in Sumatra, Burma, and Thailand.

The Emydidae turtles are noted for their bright colors and distinct patterns. This group of turtles makes up most of the turtles that are traded as pets and used for research.

The Red-Eared Turtle.   The red-eared turtle, *Pseudemys scripta elegans,* is one of the most popular turtles kept as pets. It is found from the Southwestern United States eastward to Kentucky and Indiana. Red-eared turtles have a characteristic red streak behind the eye; they may reach an adult length of up to 12 inches. See Figure 18–10.

Closely related to the red-eared turtle is the yellow-bellied turtle, *Pseudemys scripta scripta.* This turtle is found in the Southeastern United States. They are usually dark-green with a yellow plastron and a broad yellow crescent-shaped marking behind the eyes. They are 9 to 12 inches long.

Figure 18–10   Red-eared turtle. Photo by Isabelle Francais.

The Red-Bellied Turtle.   Red-bellied turtles, *Pseudemys nelsoni* and *Pseudemys rubriventris,* have reddish-colored plastrons. *Pseudemys nelsoni* are found in Florida, and *Pseudemys rubriventris* are found in areas along the East Coast of the United States. Adult red-bellied turtles are usually 9 to 15 inches long. See Figure 18–11.

Figure 18–11   Red-bellied turtle. Photo by Isabelle Francais.

The Coastal Plain Turtle.   *Pseudemys floridana floridana,* or Coastal Plain turtle, is found along the Southeastern Coast of the United States. It is brown with yellow stripes and a yellow plastron; it reaches a length of 15 inches. *Pseudemys floridana peninsularis,* or peninsula turtle, is a common turtle found throughout the peninsula of Florida; this turtle has a greenish tint to its yellow plastron.

The Missouri Slider.   The Missouri slider, *Pseudemys floridana hoyi,* is found from Alabama north to Southern Illinois and in parts of Kansas, Oklahoma, and Texas. It is somewhat smaller than its southeastern relatives. Its plastron is yellow or orange. There are no markings on the plastron of the *floridana* species of turtles.

The Concinna Species.   The *concinna* species of turtles are similar to the *floridana* except for the dark markings along the plastral seams; these markings appear as loops and whorls. Adults of this species range from 9 to 16 inches in length. Among the *concinna* species are the river cooter, *Pseudemys concinna concinna,* found in Southeast Virginia, North and South Carolina, and the eastern edge of Alabama; *Pseudemys concinna hieroglyphica* or the hieroglyphic turtle, found in Texas, Oklahoma, Kansas, and southern Illinois; *Pseudemys concinna mobiliensis,* or mobile turtle, found along the Gulf Coast from Florida to Texas; and *Pseudemys concinna texana,* found in southern Texas and northern areas of Mexico.

Blanding's Turtle.   Blanding's turtle, *Emydoidae blandingi,* is found in the Great Lakes area of the United States. This semiaquatic turtle is found in ponds, in small streams, and around lakes. The color of the Blanding's turtle varies from black, brown, or olive with yellow spots and markings. The neck is speckled, and the lower jaw, chin, and throat are a bright yellow. This turtle is usually 7 or 8 inches long. It is carnivorous, feeding on worms, small fish, crustaceans, and insects. See Figure 18–12.

Figure 18–12   Blanding's turtle. Photo by Isabelle Francais.

The European Pond Turtle. The European pond turtle, *Emys orbicularis,* is a semiaquatic turtle found in North Africa and most of Europe. This is a small turtle; adults usually measure 4 to 5 inches long. The color of the European pond turtle varies, but the carapace (upper shell) is usually black with yellow spots, specks, or short lines. The head, neck, and limbs are also covered with yellow markings. This small turtle is carnivorous, feeding on worms, frogs, mollusks, and small fish.

Painted Turtles. The painted turtles of the genus *Chrysemys* are another group of brightly colored turtles. They are different from the genus *Pseudemys* because they have smooth shells, lack the serrations to the rear shell margins, and are somewhat smaller. There is one species of painted turtle, *Chrysemys picta,* and four subspecies.

The Southern Painted Turtle. The Southern painted turtle, *Chrysemys picta dorsalis,* is probably the most familiar of the painted turtles. This turtle is dark in color with yellow stripes along its head and an orange-yellow stripe running down the center of its carapace. The Southern painted turtle is found in Louisiana and surrounding states.

The Eastern Painted Turtle. The Eastern painted turtle, *Chrysemys picta picta,* is found along the Eastern coast of the United States and as far north as Nova Scotia. It has a dark olive to dark brown upper shell and a yellow plastron. Two characteristic yellow marks are found behind the eyes, and there are yellow stripes along the chin and neck; these stripes turn into orange as they run down the neck.

The Midland Painted Turtle. The Midland painted turtle, *Chrysemys picta marginata,* is found in the area around the Great Lakes, from Canada south to northern Alabama. It is very similar in appearance to the Eastern painted turtle but lacks the yellow margins on the carapace scutes, and there is a dark marking along the midline of the plastron.

The Western Painted Turtle. The Western painted turtle, *Chrysemys picta belli,* is the largest of the painted turtles; it may reach a length of 10 inches. The carapace is green with yellow lines; the head and limbs have yellow stripes. The plastron is pink or red with a center marking that branches out along the seams. This turtle is found from Kansas and Missouri westward and from Southern Canada to Northern Mexico. See Figure 18–13.

Figure 18–13   Western painted turtle. Photo by Isabelle Francais.

Box Turtles

Box turtles belong to the genus group *Terrapene.* They are easily recognized by their domed shells and a hinged plastron. They can withdraw their head, tail, and limbs, and the hinged plastron closes, completely sealing the animal from outside predators.

The Common Box Turtle. The common, or Eastern, box turtle, *Terrapene carolina carolina,* is usually found from Illinois and

Mississippi eastward. They are brown to black with yellow or orange markings; the plastron is dark brown to yellow. Males have bright red eyes. The common box turtle is usually found in open woodland areas. They may be found near water but rarely enter water over a few inches deep. Adults are usually 4 to 5 inches long. Box turtles are long-lived and may easily live forty to fifty years.

The Florida Box Turtle.   The Florida box turtle, *Terrapene carolina bauri,* is found throughout the peninsula of Florida. The carapace of the Florida box turtle is dark brown to black with a yellow marking down the center. The scutes of the carapace have yellow streaks radiating outward; the plastron usually lacks any color. See Figure 18–14.

Figure 18–14   Florida box turtle. Photo by Isabelle Francais.

The Gulf Coast Box Turtle.   The Gulf Coast box turtle, *Terrapene carolina major,* is found along the Gulf Coast from the Florida panhandle to Texas. This is the largest box turtle, reaching a length of 7 to 8 inches. It is usually a dark color and, in many cases, lacks the markings common to other subspecies. See Figure 18–15.

Figure 18–15   Gulf Coast box turtle. Photo by Isabelle Francais.

Map Turtles

Map and sawback turtles belong to the genus group *Graptemys;* these are deepwater, diving turtles. Adults are wary and hard to capture. They are recognized by their striped heads and limbs, serrated rear shell margins, and center scutes that have raised or pointed projections. Females may be twice as large as males, and develop huge heads with large jaws. See Figure 18–16.

Figure 18–16   Common map turtle, Graptemys geographica. *Photo by Isabelle Francais.*

Barbour's Map Turtle.   Barbour's map turtle, *Graptemys barbouri,* is found in the Chipola, Apalachicola, and Flint rivers of the Florida panhandle and in southwestern Georgia. Females usually reach 10 inches in length and have large, toad-like heads; males reach only 5 inches in length and have small, narrow heads. The center scutes have raised projections. The carapace is usually olive-green with yellow, wavy lines on the coastal scutes and inverted L-shaped markings on the marginal scutes. There are yellow-green stripes on the head, neck, limbs, and tail.

The Alabama Map Turtle. The Alabama map turtle, *Graptemys pulchra* is found in most of the rivers that empty into the Gulf of Mexico. The females reach lengths of up to 11 inches, and the males reach about 4 inches in length. The carapace is olive with distinctive yellow, curved markings on each of the marginal scutes; the plastron is yellow. The head marking is very distinctive and has a single pale green or white marking between the eyes that tapers toward the nose; there are also large markings behind the eyes. There are several yellow lines on the neck and limbs.

The Black-Knobbed Sawback Turtle. The black-knobbed sawback turtle, *Graptemys nigrinoda,* is found in several rivers of Alabama and the Tombigbee River of Mississippi. The central scutes have raised projections that are blunt, rounded, and knob-like. The carapace is gray-brown, and there are dark-edged circles on the coastal scutes. Females of this species reach 8 inches in length, and males reach about 4 inches in length.

The Yellow-Blotched Sawback Turtle. The yellow-blotched sawback turtle, *Graptemys flavimaculata,* is a smaller sawback turtle found in the Pascagoula River system of Mississippi. The females reach a length of only 7 inches, with the males being much smaller. There are large yellow blotches on the coastal scutes; the rear margin scutes have large serrations.

The Ringed Sawback Turtle. The ringed sawback turtle, *Graptemys oculifera,* is very similar to the yellow-blotched sawback except the markings form circles on the coastal scutes and incomplete circles on the marginal scutes. The ringed sawback turtle is found in the Pearl River of Louisiana and Mississippi.

The Mississippi Map Turtle. The Mississippi map turtle, *Graptemys kohni,* is a common species found in pet stores. This species is found in most of the lower Mississippi River drainage system. It can be recognized by its distinctive eye; the iris appears as a white ring. There is also a distinctive crescent-shaped marking behind the eyes.

The Spottle Turtle. The spotted turtle and the wood turtles belong to the genus *Clemmys*. The spotted turtle, *Clemmys guttata,* is common in the Eastern United States. This is a small turtle that reaches about 4 inches in length. These turtles are easily recognized by the yellow spots found on the black carapace; young turtles do not have spots, they develop as the turtle matures. Yellow spots are also found on the head, neck, and limbs.

The Wood Turtle. The wood turtle, *Clemmys insculpta,* is found in the Eastern United States, southern Ontario, and northern Wisconsin. It is easily recognized by its rough carapace and orange throat and limbs. It is primarily terrestrial and usually reaches 6 to 7 inches in length.

Terrapins

The genus *Malaclemys* is commonly referred to as terrapins. The term *terrapin* is applied to turtles found in fresh and brackish water that are considered excellent eating; these turtles were very popular as a gourmet food before World War I.

The northern terrapin, *Malaclemys terrapin terrapin,* is found from the Cape Cod area to Cape Hatteras. The southern terrapin, *Malaclemys terrapin centrata,* is found from the Cape Hatteras area south to Florida. The Florida East Coast terrapin, *Malaclemys terrapin tequesta,* is found along the East Coast of Florida.

Other species include the diamond terrapin, *Malaclemys terrapin rhizophorarum,* which is found in the lower Gulf Coast area of Florida and the Florida Keys. The ornate diamondback, *Malaclemys terrapin macrospilota,* is found along the Gulf Coast of Florida. The Mississippi diamondback, *Malaclemys terrapin pileata,* is found along the Gulf Coast from Florida to Louisiana. The Texas diamondback, *Malaclemys terrapin littoralis,* is found from Mississippi to Texas.

Females of terrapin species are larger than the males and range in size from 6 to 7 inches; the males are usually 4 to 5 inches long. The Mississippi diamond back is the largest species.

The Family Trionychidae

The family Trionychidae consists of the soft-shelled turtles. They are commonly found in freshwater habitats of North America, Africa, Asia, Indonesia, and the Philippines. There are six genus groups consisting of twenty-two species. They have flat, flexible shells that lack scutes; soft-shelled turtles have three claws on each foot. Three species are found in the United States; the largest is *Trionyx ferox,* the Florida soft-shelled turtle. Females are approximately 12 to 15 inches long, and the males are somewhat smaller. They are found throughout most of Florida. The spiny soft-shelled turtle, *Trionyx spinifer,* is the most widespread species. This species has cone-shaped projections along the front margin of the carapace. There are six subspecies of spiny soft-shelled turtles. The smooth or spineless soft-shelled turtle, *Trionyx muticus,* is common in the central United States; two subspecies are recognized. See Figure 18–17.

Figure 18–17   Soft-shelled turtle. Photo by Isabelle Francais.

Soft-shelled turtles settle on the bottom of streams, ponds, and lakes. Most have snorkel-shaped snouts that enable them to breathe while they remain beneath the water's surface. Most species are carnivorous and feed on crustaceans, mollusks, insects, worms, frogs, and fish.

Soft-shelled turtles are able to exchange oxygen through their shell and skin. This allows them to remain submerged longer than other turtles. Soft-shelled turtles can often be seen sunning themselves on the banks of streams, ponds, and lakes.

The Family Testudinidae

The tortoises make up the family Testudinidae. There are ten genus groups consisting of forty-one different species. Tortoises are terrestrial and primarily herbivorous; they vary in size from 6 inches to more than 50 inches. The Egyptian tortoise, *Testudo kleinmanni,* is the smallest species, with the Aldabra tortoise, *Aldabrachelys elephantina,* being the largest. The Aldabra tortoise is found on the Seychelles, Mascarene, and Aldabra islands. The most famous tortoises are the large Island tortoise, *Chelonoidis elephantopus,* found on the Galapagos Islands. See Figure 18–18.

Figure 18–18   Large Island tortoise, **Chelonoidis elephantopus.** *Photo by Isabelle Francais.*

The Gopher Tortoise. Gopher tortoises of the genus group *Gopherus* are found in North America. The desert tortoise, *Gopherus agassizi,* is found in the Southwestern United States. Adults are about 11 inches long. The desert tortoise has a uniform brown carapace and wanders over desert areas during the day and retires to burrows at night. It digs short burrows at the base of bushes to escape from the cool night temperatures.

The Texas Tortoise. The Texas tortoise, or Berlandier's tortoise, *Gopherus berlandieri,* is found in southern Texas. This tortoise is about 6 to 7 inches long and dark brown. The color usually becomes lighter as the tortoise ages. The Texas tortoise's front limbs are not as well developed for digging as the desert tortoise. They may dig shallow trenches to protect themselves from the night cold. The Texas tortoise feeds on grass, cactus pods, flowers, and fruits.

The Florida Gopher Tortoise. The Southeastern United States is home for the Florida gopher tortoise, *Gopherus polyphemus.* This tortoise is 9 to 10 inches long and has a uniform brown color with some light mottling; the young are bright yellow-orange. This species is well adapted to digging and may make burrows up to 30 feet long and 12 feet deep. The Florida gopher tortoise feeds on grass and leaves. Although no laws prevent the keeping of these tortoises, it is illegal to buy and sell them in Florida. See Figure 18–19.

Figure 18–19 Florida gopher tortoise and plated lizard. Photo by Isabelle Francais.

The Family Dermochelyidae

The family Dermochelyidae consists of one genus group and one species, the leatherback sea turtle or leathery turtle, *Dermochelys coriacea.* The leatherback sea turtle is the largest species of turtle in the world. The average carapace length is about 61 inches, and leatherbacks average about 800 pounds. Some leatherbacks have been reported at 8 or 9 feet in length and weighing more than 1,000 pounds. The leatherback is primarily black with numerous white spots. The carapace of the leatherback is without scutes and is somewhat flexible.

The Family Cheloniidae

The family Cheloniidae consists of four genus groups and six species of hard-shelled sea turtles. The hawksbill turtle, *Eretonochelys imbricata;* the loggerhead turtle, *Caretta caretta;* and the green sea turtle, *Chelonia mydas,* are found in the Atlantic, Indian, and Pacific Oceans. The flatback turtle, *Chelonia depressa,* has its nesting area only in the Northern Coast of Australia. There are two species of Ridley sea turtle. The Pacific (or olive) Ridley sea turtle, *Lepidochelys olivacea,* is found in the tropical areas of the Pacific, Indian, and Atlantic Oceans. The Kemp's (or Atlantic) Ridley sea turtle, *Lepidochelys kempi,* is found in the Gulf of Mexico and the North Atlantic.

Both the leatherback and the hard-shelled sea turtles have been hunted for their meat and eggs; today, most species are protected.

The Family Dermatemyidae

The family Dermatemyidae consists of one genus group containing one species, the Central American river turtle, *Dermatemys mawi.* This species of turtle is found in Belize, Guatemala, and Southern

Mexico. The carapace of both the male and female of the species reaches a length of about 18 inches. The shell scutes of this turtle are very thin. The turtle spends most of its time in the water. It floats in the water and does not climb on logs or river banks to bask in the sun.

The meat of the Central American river turtle is considered excellent; therefore, the turtle has been hunted extensively for human consumption.

The Family Carettochelyidae

The family Carettochelyidae consists of one genus group and one species, the New Guinea plateless river turtle, *Carettochelys insculpta.* This species is found in the remote areas of New Guinea and Northern Australia. Its carapace reaches a length of about 18 inches. The upper surface of the turtle is gray, and the underside is white. Its carapace does not have scutes but consists of a layer of skin. The tail is covered with a series of crescent-shaped scales.

CLASSIFICATION—SNAKES, PYTHONS, AND BOAS

Kingdom — Animalia
Phylum — Chordata
Class — Reptilia
Order — Serpentes (Snakes, Pythons, and Boas)
Family — Eleven families, 2,400 species

MAJOR FAMILIES—SNAKES, PYTHONS, AND BOAS

The Family Leptotyphlopidae

The family Leptotyphlopidae is commonly referred to as thread snakes. They are among the smallest in the world and range from 3 to 16 inches in length. There are seventy-eight species of thread snakes found in arid areas of Africa, Arabia, North America, Central America, and South America. The western blind snake, *Leptotyphlops humilis,* and the Texas blind snake, *Leptotyphlops dulcis,* are found in the Southwestern United States. Their eyes are very small, and they spend most of their time in tunnels. They feed on insects, primarily termites and ants. They are light brown to pink.

The Family Typhlopidae

The family Typhlopidae is made up of 180 different species commonly referred to as blind snakes. They inhabit moist, tropical areas where they burrow under leaves, around roots, and in decaying plant material. These snakes have small eyes, and their scales are smooth, which aids in their movement through tunnels. They differ from thread snakes in that they have teeth on the upper jaw only, whereas thread snakes have teeth on the lower jaw only.

The Family Anomalepidae

The family Anomalepidae consists of about twenty species, commonly referred to as dawn blind snakes. They range in size from 5 to 6 inches in length. They also have small eyes and burrow among the leaves and mats of tropical Central and South America. They differ from the first two families because they show no evidence of pelvic limb bones, which the families Leptotyphlopidae and Typhlopidae possess.

The Family Acrochordidae

The family Acrochordidae consists of two species, the Javan wart snake (or elephant's trunk snake), *Acrochordis javanicus,* and the Indian wart snake (or Asian file snake), *Chersydrus granulatus.*

The Wart Snake. Wart snakes are aquatic and found in streams, canals, and rivers in Asia and Australia; they feed primarily on fish. Wart snakes can reach a length of 6 feet.

The Family Aniliidae

The family Aniliidae consists of three genus groups and ten species. They are commonly referred to as pipe snakes. All species except one is found in Asia and Indonesia; the one exception is found in South America.

Pipe snakes get their name from the shape of their bodies, which is almost the same circumference from head to tail.

The Coral Pipe Snake. The coral pipe snake (or fake coral snake), *Anilius scytale,* of South America is ringed with brilliant scarlet and black that resembles the poisonous coral snakes.

The Family Uropeltidae

The family Uropeltidae is made up of eight genus groups containing forty-four different species. They are commonly referred to as shield-tail snakes. These snakes have an enlarged, thick tail with spines on the upper surface. This adaptation may serve to help the snake move through its tunnel as well as serving as a plug for the tunnel. They have brightly-colored red, orange, or yellow markings, and they feed primarily on earthworms.

The Family Xenopeltidae

The family Xenopeltidae is made up of one genus group containing only one species, the sunbeam snake (or sunbeam python), *Xenopeltis unicolor.*

The Sunbeam Snake. The sunbeam snake grows to about 3 feet in length. It burrows under leaves and decaying matter and is primarily active at night. It is brown, and its smooth scales shimmer in the sunlight, giving rise to its name.

The Family Boidae

The family Boidae contains the pythons and boas; there are twenty-seven genus groups containing about eighty-eight species. Included in this family are the largest species of snakes, the reticulated python, *Python reticulatus,* and the anaconda, *Eunectes murinus.* Both may reach a length of up to 33 feet and weigh more than 400 pounds. The family Boidae is divided into several subfamilies.

The subfamily Pythoninae includes about twenty-four species found in Africa, India, Southeast Asia, and Australia. Members of this subfamily are egg layers. The female may build a nest; after the eggs are laid she coils around the eggs and protects them until they hatch.

The Reticulated Python. The reticulated python is found in Southeast Asia and the Philippines. It has a network pattern of brown, buff, yellow, and reddish-brown. These large pythons feed on mammals, including deer, pigs, and poultry. The reticulated python, *Python reticulatus,* may reach a length of more than 33 feet. See Figure 18–20.

Figure 18–20   Reticulated python. Photo by Isabelle Francais.

The India Python. The India python, *Python molurus,* looks very similar to the reticulated python but is smaller; the colors are similar. There is a distinctive arrow-shaped marking on the head of the India python. This python is found in India, Ceylon, and the East Indies. India pythons that are light in color are said to tame more readily than the darker-colored ones. See Figure 18–21.

Figure 18–21   India python. Photo by Isabelle Francais.

The Burmese Python. The Burmese python, *P. m. bivittatus,* is a subspecies and is darker in appearance. Also included in this subspecies is the albino Burmese python. See Figures 18–22 and 18–23.

Figure 18–22   Burmese python. Photo by Isabelle Francais.

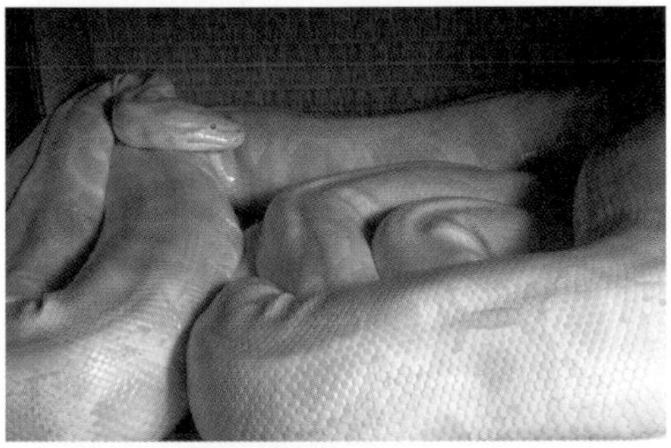

Figure 18–23   Albino Burmese python. Photo by Isabelle Francais.

The Ball Python. The Ball python, *Python regius,* is a small species found in West Africa; it reaches a length of 6 to 7 feet. This python is dark brown with light brown markings. The color helps it blend into the dry forest and bush lands where it is found. It protects itself from predators by coiling up into a tight ball; the ball python is sometimes referred to as the royal python. See Figure 18–24.

Figure 18–24   Ball python. Photo by Isabelle Francais.

The Rock Python. The rock python, *Python sebae,* is a large python found in Africa that reaches a length of more than 20 feet. It is dark brown with light brown to yellow markings; it has a dark arrow-shaped marking on its head. This python feeds on antelope, gazelle, and other mammals found on the grassy plains.

The Green Tree Python. The green tree python, *Chondropython viridis,* is a tree-dwelling python of Northeast Australia, New Guinea, and the Solomon Islands. It is bright green with small blue, red, and white markings; its underside is yellow or white. This python has a prehensile tail that allows it to cling to tree limbs. The eggs of the tree python are laid on the ground, where they hatch into yellow or red young; the young soon take to the trees. These pythons feed on roosting birds that they seize with extremely sharp teeth. They usually rest during the day, curled up on the limbs of trees. See Figure 18–25.

Figure 18–25   Green tree python. Photo by Isabelle Francais.

The Carpet Python. The carpet python, *Morelia spilotes variegata,* and the diamond python, *Morelia spilotes spilotes,* are found in Australia, New Guinea, and other Pacific Islands. These pythons reach 6 to 10 feet in length. The diamond python is strictly an **arboreal** (tree-dwelling) species, but the carpet python feeds both in trees and on the ground; they feed on birds, mammals, and lizards. The diamond python is light brown with light brown to yellow diamond-shaped markings; the carpet python is similarly marked. See Figure 18–26.

Figure 18–26   Carpet python. Photo by Isabelle Francais.

Boelen's Python. Boelen's python, *Liasis boeleni,* is found in the high mountain jungles of several Pacific Islands. It is black with yellow markings.

The Amethystine Python. The amethystine python, *Liasis amethystinus,* is a long, slender python reaching lengths of 25 feet. It is found in Australia and New Guinea. See Figure 18–27.

Figure 18–27   Amethystine python. Photo by Isabelle Francais.

The Black-Headed Python. The black-headed python, *Aspidites melanocephalus,* is different from other snakes because it feeds on other snakes, although it also eats other warm-blooded animals. Its head, neck, and throat are black in contrast to the rest of its body, which is light brown.

The Blood Python. The blood python, *Python curtus,* is found in Asia. This small python reaches a length of only 4 to 5 feet. They vary in color from yellow, orange, or blood-red. See Figure 18–28.

Figure 18–28   Blood python. Photo by Isabelle Francais.

The subfamily Boinae is made up of about thirty-nine species. Boas are very similar to pythons except that they bear live young instead of laying eggs. Most boas live in Central and South America, although two species are found in North America.

The Anaconda.   The anaconda, *Eunectes murinus,* is a semiaquatic boa found in the river basins of the Amazon and Orinoco rivers of South America. This large boa may reach a length of up to 33 feet. It is grayish-brown with large black markings. Anacondas are excellent swimmers; their nostrils are located on the top of their snout, which allows them to breathe easily while in the water. They feed in trees, on land, and in the water. Included in their diet are birds, peccaries, caiman, tapirs, capybara, fish, and turtles. Anacondas do not adapt well to captivity.

The Emerald Tree Boa.   The emerald (or green) tree boa, *Corallus caninus,* is similar to the green tree python in appearance. It lives in trees and feeds primarily on birds. It grows 6 to 7 feet in length and is bright green above with a yellow underside. The young vary in color from rusty red to blue-green. See Figure 18–29.

Figure 18–29   Emerald (or green) tree boa. Photo by Isabelle Francais.

The Boa Constrictor.   The boa constrictor, *Boa constrictor,* is one of the most popular of the common boas. They are popular in the pet trade because they tame quickly and breed well in captivity. Their color varies from a light brown, yellow, or orange background with dark bars and markings. They grow up to 18 feet in length but average 12 to 14 feet. They can be found in trees and on the ground and are excellent swimmers. Boa constrictors are found throughout much of South America and north into Central America and Mexico. They feed on all types of small animals, birds, eggs, and fish.

The Red-Tailed Boa.   The Mexican red-tailed boa and the Peruvian red-tailed boa are two subspecies of boa constrictor that are commonly found in the pet trade. The red-tailed boa is found in Mexico and Central and South America. See Figure 18–30.

Figure 18–30   Red-tailed boa. Photo by Isabelle Francais.

The Yellow Anaconda.   The yellow anaconda, *Eunectes notaeus,* is a smaller anaconda that grows up to 15 feet in length. The young are bright yellow with black markings; the colors fade somewhat as the animal ages.

The Rainbow Boa.   The rainbow boa, *Epicrates cenchris,* is one of the most attractive species because of the blue-green iridescence of its scales. Its color ranges from light brown to bright reddish-orange. They grow to about 8 feet in length and feed at night on rodents and birds. This boa is found throughout the Caribbean region, Central America, and Northern South America.

The Central American Dwarf Boa.   The Central American dwarf boa, *Ungaliophis panamensis,* is a small boa that reaches a length of only 30 inches. It is found in the rainforests of Central and South

America. It is light brown with dark oblong spots, and it feeds on lizards, frogs, small birds, and rodents.

The Rubber Boa.   Two boas are found in North America. The rubber boa, *Charina bottae,* is found in the Western United States; its olive-colored body scales have a rubbery feel. This small boa reaches a length of 6 feet, although it usually averages about 30 inches. It is harmless and usually rolls up into a ball when threatened. It feeds on lizards, rodents, and small birds. See Figure 18–31.

Figure 18–31   Rubber boa, Charina bottae. *Photo by Isabelle Francais.*

The Rosy Boa.   The rosy boa, *Licanura trivirgata,* is found in the drier areas of the Southwestern United States. Its background color varies from pink to gray, and it has three black stripes that run the length of its body. Both boas are nocturnal and spend the day hiding among rocks and logs; the rubber boa may also burrow in search of food. Both boas have blunt tails that may be used to lure predators to attack the tail instead of their head. See Figure 18–32.

Figure 18–32   Rosy boa. Photo by Isabelle Francais.

The Family Colubridae

The family Colubridae is the largest family of snakes; there are approximately 300 genus groups consisting of 1,562 species. They are commonly referred to as harmless snakes. They can be found on all five continents and as far north as the Arctic Circle and south to the tip of South America. Only the more common genus groups are covered here. There are twenty subfamilies; eight subfamilies, although considered harmless snakes, do have grooved fangs at the rear of their jaws that can inject venom into their prey. These snakes are referred to as rear-fanged snakes. Most of these snakes are small and cannot inflict injury to humans. Two species, however, can grow to over 5 feet. The boomslang, *Dispholidus typus,* and the twig snake, *Thelotornis kirtlandii,* both found in Africa, have been known to inflict serious injury and even death to humans. Species of rear-fanged snakes are found in Southeast Asia, New Guinea, and Northern Australia. Because rear-fanged snakes may be dangerous to handle, they are not discussed further here.

Common Water Snakes.   The subfamily Natricinae consists of the common water snakes and garter snakes found in North America. This group ranges in length from 24 inches to more than 6 feet; they all give birth to live young. Water snakes are semiaquatic and are seldom found far from water; they spend a lot of their time basking in the sun. Most water snakes, although harmless, are very aggressive. If provoked or cornered, they will strike. To keep water snakes in captivity, the snakes must have an aquatic environment; cleaning and maintaining this aquatic environment can be difficult and time consuming. Water snakes feed on fish, frogs, salamanders, and toads. These snakes are gray-brown with darker cross-bands. One common species is the northern water snake, *Natrix sipedon.* There are about eighteen different subspecies of northern water snakes. Three other common species in North America include the

Figure 18–33   Brown water snake. Photo by Isabelle Francais.

red-bellied water snake, *Natrix erythrogaster;* the green water snake, *Natrix cyclopion;* and the brown water snake, *Natrix taxispilota.* See Figure 18–33.

The Common Garter Snake. The common garter snake, *Thamnophis sirtalis,* and about fifteen subspecies of garter snakes are the most widespread snakes in North America. Almost all youngsters have encountered garter snakes in the outdoors, where the snakes can be found among leaves, under logs, and under rocks. Young garter snakes feed on earthworms, mealworms, and other insects; larger garter snakes feed on rodents and frogs. Garter snakes can be tamed and adjust readily to captivity. Common species are the eastern garter snake, *Thamnophis s. sirtalis;* the checkered garter snake, *Thamnophis marcianus;* and the black-necked garter snake, *Thamnophis cyrtopsis.* Most species of garter snakes average 2 to 3 feet in length. The smallest species is the Butler's garter snake, *Thamnophis butleri,* which reaches a length of only 18 inches at maturity. The giant garter snake, *Thamnophis couchi gigas,* may grow to over 4 feet. See Figure 18–34.

Most species of garter snakes are easily recognized by their vivid stripes running the length of their body on a dark background. Some species have a checkerboard pattern (checkered garter snake), whereas others may be solid black.

Figure 18–34   Common garter snake. Courtesy of PhotoDisc.

Ribbon Snakes. Ribbon snakes are closely related to garter snakes; they are semiaquatic and more slender. The eastern ribbon snake, *Thamnophis sauritus sauritis,* is black or dark brown with yellow stripes running the length of the body. They can usually be found near water and can be seen basking in the sun on branches and logs. They feed on small fish, frogs, and salamanders.

Racers. The genus group *Coluber* is made up of the racers; they are common in North America and are found from coast to coast. They are among the fastest moving snakes, reaching speeds of 3.5 miles per hour. Species found in the Western United States are usually brown or olive with a lighter underside. Those found in the East are usually black or blue-black. Racers may be difficult to keep in captivity; they are aggressive and have been known to chase those who encounter them. Many times when encountered, they will stand their ground and move their head back and forth while flicking their tongue. In captivity, angry racers kept in glass aquariums may strike at the glass as people approach it. If provided with a hiding place, they may stay in hiding and not venture out. Racers do not like to be handled; when attempting to capture a racer, great care should be exercised to limit the stress to the snake. Racers feed on earthworms, insects, frogs, salamanders, and rodents; racers in the wild swallow their food live. The young are hatched from eggs.

Whipsnakes. Whipsnakes are closely related to racers. The coachwhips, *Masticophis flagellum,* may reach a length of 8 feet; whipsnakes need dry environments.

Rat Snakes. Rat snakes make up the genus *Elaphe.* There are more than forty species of rat snakes. Several species are brightly colored and boldly marked. Rat snakes are constrictors; they feed almost entirely on mice and rats.

The Black Rat Snake. The black rat snake, *Elaphe obsoleta obsoleta,* is black with white on the chin and throat. It may reach lengths of 4 to 8 feet and is usually found in wooded and rocky hillsides. See Figure 18–35.

Figure 18–35   Black rat snake. Photo by Isabelle Francais.

The Corn Snake. The corn snake, *Elaphe guttata,* is found from New Jersey to Florida and across to Utah. It is yellow or gray with black edged, red blotches. This snake, also referred to as a red rat snake, may reach a length of 6 feet. See Figure 18–36.

Figure 18–36   Corn snake. Photo by Isabelle Francais.

The Fox Snake. The fox snake, *Elaphe vulpina,* is found in the central states from Wisconsin to Missouri. This snake receives its name from the fox-like odor it releases when captured. It is yellowish to pale brown above with dark blotches; its underside is yellow with black, checkered markings. The fox snake reaches a length of about 4 feet and feeds on rodents and birds.

The Yellow Rat Snake. The yellow rat snake, *Elaphe obsoleta quadrivittata,* is a bright yellow snake with four black stripes running the length of its body. It is found in the southern coastal states of North and South Carolina, Georgia, and Florida. See Figure 18–37.

Figure 18–37   Yellow rat snake. Photo by Isabelle Francais.

The Everglades Rat Snake. A subspecies, the Everglades rat snake, *Elaphe obsoleta rossalleni,* is found in the Everglades. It is orange or pinkish with black stripes. See Figure 18–38.

Figure 18–38   Everglades rat snake. Photo by Isabelle Francais.

The Gray Rat Snake. Another rat snake found in the Southern Mississippi drainage area is the gray rat snake, *E. o. spiloides;* it is gray with darker blotches. See Figure 18–39.

Figure 18–39   Gray rat snake. Photo by Isabelle Francais.

The genus *Lampropeltis* consists of the king snakes and milk snakes. They are constrictors and feed primarily on small animals and birds; however, they have been known to feed on other snakes. King snakes are immune to the venom of venomous snakes.

The Common King Snake. The common king snake, *Lampropeltis getulus,* is found over the southern third of the United States. It is also referred to as a chain snake because its black body has cream-colored markings that form a chain-like pattern. See Figure 18–40.

Figure 18–40   Common king snake. Photo by Isabelle Francais.

The Scarlet King Snake. There are several subspecies of common king snake. The scarlet king snake, *L. doliata,* has a scarlet red background with black and yellow bands down the length of its body; it is often confused with the venomous coral snake. See Figure 18–41.

Figure 18–41   Scarlet king snake. Photo by Isabelle Francais.

The Sonora Mountain King Snake. One of the most beautiful snakes is the Sonora Mountain king snake, *L. pyromelano.* This snake is black with white markings down the length of its body. The speckled king snake, *L. getulus holbrooki,* is black with yellow specks all over the body. Most king snakes adapt to captivity and can be fed mice and small rats.

The Eastern Milk Snake. The eastern milk snake, *L. triangulum,* is found in the Northeast United States and south to the Gulf Coast and west to Texas. In the East, they are usually a dark brown with black-edged, gray chain bands the length of the body. In southern areas, they are usually found with yellow and red colorings that can be mistaken for the venomous coral snake. Milk snakes are fairly small, reaching lengths of only 30 inches. Milk snakes usually do not do well in captivity and must be given some branches or bark to hide under. See Figure 18–42.

Figure 18–42   Southern milk snake. Photo by Isabelle Francais.

The Ring-Neck Snake. The ring-neck snakes belong to the genus group *Diadophis.* Ring-necked snakes are small snakes that inhabit forested areas; they are usually less than 2 feet in length. They hide under leaves and logs, where they feed on insects and earthworms. The common ring-neck snake, *Diadophis punctatus,* is black with a red or orange underside. A narrow ring is located behind the head.

The Pine Snake. The genus *Pituophis* consists of pine snakes and bull snakes. The pine snake, *Pituophis melanoleucus melanoleucus,* is a fairly large snake reaching lengths of about 5 feet. It is white with black markings and is found in pine forested areas of the East Coast, where they feed primarily on rodents. See Figure 18–43.

Figure 18–43   Pine snake. Photo by Isabelle Francais.

The Bull Snake.   The bull snake, *P. m. sayi,* is an even larger snake, reaching lengths of over 8 feet; it is yellow-brown. They feed on mice and adapt readily to captivity. See Figure 18–44.

Figure 18–44   Bull snake. Photo by Isabelle Francais.

The Rough and Smooth Green Snake.   The genus group *Opheodrys* consists of the rough green snake, *O. aestivus,* and the smooth green snake, *O. vernalis.* The smooth green snake is small, reaching a length of only 30 inches. The rough green snake is somewhat larger, reaching a length of up to 3 feet. They are both bright green and feed primarily on insects.

The final two families of snakes are the Elapidae and Viperinae. These are not discussed here because they are venomous snakes and should be kept only by those with considerable experience.

CLASSIFICATION— IGUANAS AND LIZARDS

Kingdom — Animalia
Phylum   — Chordata
Class   — Reptilia
Order   — Squamata (Iguanas and Lizards)
Family   — Sixteen families, 3,750 species

MAJOR FAMILIES— IGUANAS AND LIZARDS

The Family Gekkonidae

The family Gekkonidae are commonly referred to as geckos. The family includes eighty-five genus groups consisting of approximately 800 species. Geckos are found in tropical and semitropical regions throughout the world. They have four limbs with five toes on each limb; the toes of most geckos have adhesive pads called lamellae. These pads give the gecko the ability to climb up almost any surface, including the glass walls of an aquarium. Most geckos are nocturnal and have vocal cords, allowing them to communicate in the darkness.

The geckos that are nocturnal have large eyes; the pupil changes depending on the light intensity. The pupil closes to a narrow, vertical slit or to four small pinhole openings when the light intensity increases. Geckos do not have eyelids but have a transparent covering over the eye. They clean this covering with their large, flat, fleshy tongues; the covering is replaced each time the animal sheds its skin. See Figures 18–45 and 18–46.

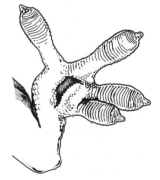

Figure 18–45   The toes of most geckos have adhesive pads called lamellae. These pads enable geckos to climb almost any surface.

Figure 18–46   Most geckos are nocturnal and have large eyes with vertical pupils that close to a small slit or to four small pinhole openings when the light intensity increases.

Most geckos are insect eaters. Others feed on the nectar of plants, small mammals, and birds. Some geckos eat their skin after shedding.

The Tokay Gecko.   The tokay gecko, *Gekko gecko,* is one of the largest geckos; it reaches a length of about 12 inches. The tokay gecko is found in the rainforests of Southern Asia, India, and New Guinea. The body is a grayish-blue with small orange spots. They have large yellow eyes with slit pupils and feed on insects and small rodents. In captivity, they feed on insects and baby mice and are especially fond of bee moths. They obtain water by licking it from plants and logs. Tokay geckos get their name from the "to-kay," "to-kay" sound they make. Tokay geckos have powerful jaws and will bite.

The Leopard Gecko.   The leopard gecko, *Eublepharus macularus,* is a medium-sized gecko that reaches a length of about 8 inches. It is found in dry areas from Iraq to Northern India. This gecko has claws instead of the adhesive lamellae and has a movable eyelid. Its body is a gray-buff color with dark spots and blotches. The skin has wart-like projections all over its body. It feeds on insects and small rodents; water is licked from plants, logs, and rocks. See Figure 18–47.

Figure 18–47   Leopard gecko. Photo by Isabelle Francais.

The House Gecko.   The house gecko, *Hemidactylus frenatus,* is commonly referred to as a chitchat gecko. It is found around human dwellings in Southeast Asia and on many islands in the Pacific Ocean. It reaches a length of about 6 inches and has slit pupils and adhesive lamellae; the lamellae do not reach the tip of the toes. House geckos feed on insects, spiders, and small rodents. Water is licked from plants, logs, and rocks. They are sandy-brown with darker spots and blotches.

The Gold Dust Madagascar Day Gecko.   The gold dust Madagascar day gecko, *Phelsuma laticauda,* is found on the island of Madagascar and several surrounding islands. It is diurnal. It reaches a length of about 5 inches, and its eyes are large, round, and black and do not have slits like the nocturnal geckos. The day gecko feeds on insects, spiders, and fruit nectar; water is usually licked from plants. It is light yellow-green with a dusting of yellow spots. It has small blue blotches above the eyes and two orange stripes; one running across the nose and one between the eyes. See Figure 18–48.

Figure 18–48   Gold dust Madagascar day gecko. Photo by Isabelle Francais.

The Large Madagascar Gecko. The large Madagascar gecko, *Phelsuma madagascariensis,* grows to 10 inches in length. It is found on the island of Madagascar and neighboring islands. This gecko is light blue-green below with the upper body a darker green. Numerous pinkish-orange blotches and spots appear on its back and on the head. This gecko is diurnal and has large, round, black eyes. It is found in wooded areas, where it feeds on spiders, insects, and fruit nectar. Water is obtained from the leaves of plants. See Figure 18–49.

Figure 18–49   Large Madagascar gecko. Photo by Isabelle Francais.

Kuhl's Flying Gecko. Kuhl's flying gecko, *Ptychozoon kuhli,* has large, webbed feet and a flap of skin along each side of the body. These two adaptations allow the gecko to "glide" from branch to branch. This gecko is native to Southeastern Asia. It is brownish-gray with darker blotches and markings, and it grows to a length of about 6 inches. It is nocturnal, has slit pupils, and feeds on insects; water is obtained from leaves.

The Banded Gecko. The banded gecko, *Coleonyx variegatus,* is a small, thin-skinned gecko found in the arid areas of the Southwestern United States. This gecko reaches a length of only about 3 inches. It is light yellow with dark bands across its back and tail, and feeds on small insects.

The Mediterranean Gecko. One of the gecko species that has been introduced into the United States is the Mediterranean gecko, *Hemidactylus turcicus.* It is about 3 inches long and can now be found in the Southern states, usually around homes; it feeds on insects. This gecko is also referred to as the Turkish gecko.

The Family Pygopodidae

The family Pygopodidae is made up of eight genus groups consisting of thirty-one species, and is commonly referred to as snake lizards; most species are found in Australia. They are limbless lizards that move like snakes. Most species feed on insects and rodents. Little is known about their habits, and very few are kept in captivity.

The Family Xantusiidae

The family Xantusiidae is made up of two genus groups consisting of four species; this family is commonly referred to as night lizards. They are nocturnal and are found in arid, rocky areas.

The Yucca Night Lizard. The yucca night lizard (or desert night lizard), *Xantusia vigilis,* is a small lizard found in the Southwestern United States and Mexico. It feeds on insects and spiders.

The Island Night Lizard. The island night lizard, *Klauberina riversiana,* is found on the islands off the southern coast of California; these lizards grow to about 8 inches in length. This species is primarily nocturnal but may be seen during the day. They feed on leaves, blossoms, and seeds of plants. Night lizards give birth to live young, usually one to eight after a gestation period of about ninety days. Night lizards are fairly secretive, and if kept in captivity, they should be provided with areas to hide.

The Family Dibamidae

The family Dibamidae is made up of one genus group consisting of four species; the family is commonly referred to as blind lizards. Little is known about these lizards. Three species are found in Asia and a fourth, the Mexican blind lizard, *Anelytropsis papillosus,* is found in Eastern Mexico.

The Mexican Blind Lizard. The Mexican blind lizard is flesh-colored to purplish-brown. It lacks limbs and burrows under leaf litter, in the soil, and around rotten logs. The eyes are very small.

The Family Iguanidae

The family Iguanidae is made up of fifty-five genus groups consisting of more than 650 different species; most species are found in the Americas, with a range from Southern Canada to the tip of South America. A few are found on the islands of Fiji and Tonga in the Pacific Ocean, the wetlands of Madagascar, and the Galapagos Islands. Members are active during the day and most lay eggs, although a few give birth to live young. There is a great variation in body size and body shape among the members of this family.

The genus *Anolis* includes the anoles and related species that makes up about 50 percent of the Iguanidae family. They are primarily found in South America and the West Indies Island group.

The Green Anole. This species, however, is found in the Southern United States. The green anole or American anole, *Anolis carolinensis,* is found in trees, in shrubs, and in and around houses; it is also commonly found in pet shops. They have adhesive lamellae. The green anole are able to change color from various grays to browns, depending on its surroundings. The male is usually a little larger than the female and displays or extends a bright pink to red dewlap during sexual activity or to scare predators. The green anole reaches a size of about 6 to 8 inches in length. It feeds on insects and obtains water from vegetation.

The Cuban Brown Anole. Another species found in southern Florida is the Cuban brown anole, *Anolis sagrei sagrei.* This anole is brown with light brown to gray markings. The male displays a reddish-orange dewlap during sexual activity or when threatened. It is 6 to 7 inches long and is found on and around trees and shrubs. See Figure 18–50.

Figure 18–50   Cuban brown anole. Photo by Isabelle Francais.

The Knight Anole. The Knight anole, *Anolis equestric,* is the largest of the anoles. This species, found in Cuba, reaches a length of about 18 inches. It is green with white markings; the eyes have a blue circle around them. The males extend a pink-colored dewlap during sexual activity or when threatened. It feeds on insects but is large enough to feed on mice and small birds. Water is usually obtained from vegetation. See Figure 18–51.

Figure 18–51   Knight anole. Photo by Isabelle Francais.

The Common Green Iguana. The genus *Iguana* consists of three similar species. The common green iguana, *Iguana iguana,* is common in the pet trade. Found in Central and South America, this arboreal lizard reaches a length of about 6½ feet. Young iguanas are about 8 inches long when they hatch from eggs. The young feed on insects and earthworms; when they get older, they feed on rodents, birds, and fish. Mature iguanas eat fruits, leaves, blossoms, and buds. The common iguana is green and has a crest running from the neck to the tail. Two similar species are the *Iguana iguana delicatissima* and the *Iguana iguana rhinolopha.* The *Iguana iguana delicatissima* lacks the large, round scale under the ear and has large scales along the side of the lower jaw. *Iguana iguana rhinolopha* has the large scale below the ear but differs because it has several erect spines on its snout. See Figure 18–52.

Figure 18–52   Common (or green) iguana. Photo by Isabelle Francais.

The Galapagos Islands Marine Iguana. The genus *Amblyrhynchus* consists of one species, the Galapagos Islands marine iguana, *Amblyrhynchus cristatus.* This marine species lives near seawater and feeds only on certain types of seaweed. Because of its special diet, it is found only in a few zoos. It is dark green to black and grows to about 4½ feet in length.

The Galapagos Land Lizard. The genus *Conolophus* contains one species, the Galapagos land lizard, *Conolophus subcristatus.* This land-dwelling iguana of the Galapagos Islands feeds on bark, flowers, fruits, and insects. This species grows to about 4 feet.

The Rhinocerios Iguana. The genus *Cyclura* consists of about five species. The rhinocerios iguana, *Cyclura cornuta,* is found in Haiti and the Dominican Republic. It has three blunt horns on its snout. It is dark brown to dull gray and grows to more than 5 feet in length. It is primarily a vegetarian but also eats rodents, birds, and other small mammals.

Turks Island and Bahama Iguanas. Two other species are the Turks Island iguana, *Cyclura carinata,* and the Bahama iguana, *Cyclura baeolopha,* found on the islands of Bahama and Jamaica. They are smaller but similar in appearance to the rhinocerios iguana.

The Spiny-Tailed Iguanas. The genus *Ctenosaura* contains species commonly referred to as spiny-tailed iguanas. The common spiny-tail iguana, *Ctenosaura hemilopha,* is found from Central Mexico to the border of the United States. It is primarily terrestrial, is gray, and has a short, spiny tail. Authorities disagree on the number of different species in this genus group.

The Desert Iguana. The genus group *Dipsosaurus* contains three species. The desert iguana, crested iguana, or northern crested iguana, *Dipsosaurus dorsalis,* is found in the Southwestern United States and Northern Mexico. It grows to about 18 inches in length. It is light brown with shadings of gray and brown; black lines and white spots may also appear on the body. It feeds on insects and the flowers and buds of plants.

Spiny Lizards

The genus *Sceloporus* consists of the largest number of species. These lizards are referred to as spiny lizards, fence lizards, and swifts. They are found from Canada to Southern South America and the Caribbean; most species reach a length of about 10 inches. They have overlapping scales with a ridge that ends in a sharp point or spine. They are primarily terrestrial but can be found on tree stumps and among rocks.

The Eastern Fence Lizard. The eastern fence lizard, *Sceloporus undulatus,* is found along the East Coast of the United States from New Jersey and across the Southern United States. It is commonly referred to as a fence lizard because it can be found on fences basking in the sun or as a pine woods spiny lizard because it can sometimes be found in pine woods habitats. Fence lizards have sharp claws that enable them to climb almost any rough surface. They can move very quickly. Males have bright blue-green markings on their stomachs and throats. Fence lizards feed primarily on insects.

The Rock Crevice Spiny Lizard. The rock crevice spiny lizard, *Sceloporus poinsetti,* lives among the rocks and boulder habitats of New Mexico and Texas. This lizard has a distinguishing black collar around its neck and black bands around its tail; insects are its primary food source. It gives birth to live young.

The Clark's Spiny Lizard. The Clark's spiny lizard, *Sceloporus clarki,* is found in the Western United States; it is one of the larger spiny lizards. It is light brown to gray-brown with small, darker markings along the back and head.

The Emerald Swift. The emerald swift, *Sceloporus malachitus,* is found in Mexico and Northern Central America. This is a bright green lizard; the male has patches of blue-green on its stomach and throat. The emerald spiny lizard feeds on insects and gives birth to live young.

The Curly-Tailed Lizard. The genus group *Leiocephalus* is represented by the curly-tailed lizard, *Leiocephalus carinatus.* This lizard is found on several Caribbean Islands and has found its way to southern Florida. It is a dark brown lizard with lighter stripes and markings on its body; the tail has bands and ends in a curl.

The Helmeted Iguana. The genus *Corythophanes* is represented by the helmeted iguana, *Corythophanes cristatus,* found in Central America. Not much information exists about these iguanas. They have a large crest or "helmet" on the neck and are arboreal.

The Keel-Tailed Lizard. The genus *Tropidurus* is made up of keel-tailed lizards. Two species from South America are *Tropidurus semitaeniatus* and *Tropidurus torquatus.* They have a raised, spiny keel that runs from the lower back down the top of the tail. They primarily feed on insects and spiders but also eat small rodents and plant material.

Basilisks. The genus *Basiliscus* consists of lizards commonly referred to as basilisks. They are found from Southern Mexico, throughout Central America, and into Northern South America. These tree-dwelling lizards can reach lengths of up to 3 feet, with the major portion of their length being a long tail. Males have a large crest or casque along their backs and tails that look like sails or fins. Basilisks are primarily carnivorous, feeding on insects, small mammals, and birds. These lizards have the ability to run short distances very quickly on two legs. Some species have long toes with webbing between them that allows them to run for short distances across the surface of water. Most basilisks live near water and are powerful swimmers. Species that may be found in pet stores are the common basilisk, *Basiliscus basiliscus;* the double-crested basilisk, *Basiliscus plumifrons;* and the banded basilisk, *Basiliscus vittatus.* The common basilisk is brown with darker stripes on the back. It has a white throat and lower jaw, with a white stripe running from each eye back across the head and down the back. Another white strip runs from the white throat patch back across the front shoulders. The double-crested basilisk is green with orange eyes.

The Family Agamidae

The family Agamidae is made up of over thirty genus groups consisting of approximately 300 species; this family is referred to as chisel-teeth lizards. Their teeth are fused at the base on the ridge of the upper and lower jaws, and the front teeth are long and chisel-like. These lizards are found in Africa, the Middle East, South and Central Asia, and Australia. Very few are available in the pet trade; several species live in special habitats that would be hard to duplicate.

The genus group *Physignathus* is commonly referred to as water lizards or water dragons; they are similar in appearance to iguanas.

Water Dragons. The Asian Water Dragon, Chinese Water dragon, or Oriental water dragon, *Physignathus cocincinus,* is commonly found in pet stores. They are native to tropical forest areas of Southeast Asia. They reach a length of about 30 inches, two-thirds of it being the tail. These lizards are green with light bands around the body and tail. There may be light blue-green patches on the throat. They feed on insects, earthworms, fish, and small rodents. Their diet is more carnivorous than the common iguana.

Figure 18–53   Young water dragon. Photo by Isabelle Francais.

The water lizard, *Physignathus lesueurii,* of Australia and New Guinea grows to more than 3 feet in length. It is olive-brown with darker bands around the tail. A dark stripe extends from the eye back toward the shoulders; the stomach is red. It is carnivorous and feeds on insects, small fish, small birds, and rodents. See Figures 18–53 and 18–54.

Figure 18–54   Asian water dragon. Photo by Isabelle Francais.

The Bearded Dragon. The genus *Amphibolurus* contains the bearded lizard, *Amphibolurus barbatus;* it is found in Australia and New Guinea. This lizard inhabits the grassland area and is usually found around tree stumps and among rocks. The scales on its neck are elongated into spines that can be raised when the lizard is threatened. Its primary diet is insects, but it will eat vegetative material. This lizard is brown with lighter spots along the sides. Light bands encircle the tail, the throat and stomach area is light brown, and a light brown stripe extends from the eye backward. The bearded lizard reaches a length of 18 to 22 inches. See Figure 18–55.

Figure 18–55   Bearded dragon. Photo by Isabelle Francais.

The Sail-Tailed Water Lizard. The sail-tailed water lizard (or soasoa), *Hydrosaurus amboinensis,* is found on the Indonesian island of Sulawesi. This is the largest member of the Agamidae family, reaching a length of more than 3 feet. This lizard has a large crest that runs from head to tail; the crest on the tail is especially large. The sail-tailed water lizard is brown and is a tree-dwelling species that takes to the water when alarmed; it is an excellent swimmer. Its diet consists of a large variety of items, including insects, earthworms, fish, birds, small rodents, and fruit.

Spiny-Tailed Water Lizard. The genus *Uromastyx* is made up of the spiny-tailed lizards, which have short, thick, spiny tails. They are found in hot, dry habitats; in captivity, this habitat must be duplicated if they are to do well. Representatives of this group are the black spiny-tailed lizard, *Uromastyx acanthinurus;* the Egyptian spiny-tailed lizard, *Uromastyx aegyptius;* and the Hardwick's spiny-tailed lizard, *Uromastyx hardwicki.* Spiny-tailed lizards change skin color during the day; in the morning, they are usually darker to help absorb the sun's rays; they become lighter as temperatures rise.

The Common House Agama. The common house agama (or common agama), *Agama agama,* is found throughout Africa. Most of them are very colorful, and several species change color during the day. In the morning, they have dark brown coloration, and as they absorb the sun's rays, they become lighter. The male of the Namibian rock agama, *Agama planiceps,* is dark blue to purple with a red or orange head and tail; the females are usually brown or gray.

Flying Lizards. Flying lizards of the genus *Draco* have folds of skin that they extend and use to glide among the trees. Several species are known; most are native to Southeast Asia and the East Indies, where they feed on ants. Most are small and delicate and do not live long in captivity.

The Family Cordylidae

The family Cordylidae consists of sungazers, girdle-tails, and plated lizards. This family is made up of ten genus groups and consists of about fifty species. They are inhabitants of the African deserts, where they are usually found among rocks; they feed primarily on insects and spiders.

Sungazers. The sungazers get their name because they lie with the front portion of the body raised and appear to be staring into the sun. Their scales are sharply pointed. One or two large young are born in February to March.

The Giant Girdle-Tailed Lizard. The giant girdle-tailed lizard, *Cordylus giganteus,* is the largest of the family, growing to about 18 inches. They give birth to live young.

Plated Lizards. Plated lizards also belong to the Cordylidae family. The giant plated lizard, *Gerrhosaurus validus,* is the largest member, reaching a length of more than 2 feet. It is found among the rocks and boulders of the African savanna. Its diet includes insects, spiders, rodents, and plant material. The desert plated lizard, *Angolosaurus skoogii,* is found in the sand dunes of the Namib desert. It feeds on insects and, when disturbed, it buries itself in the sand. Plated lizards are egg-laying species. See Figure 18–56.

Figure 18–56   Giant plated lizard. Photo by Isabelle Francais.

The Yellow-Throated Plated Lizard. Another species is the yellow-throated plated lizard, *Gerrhosaurus flavigularis.* It is found in South Africa and Madagascar. Plated lizards have a smaller lower jaw, and the throat area is covered with large scales. In some references, the plated lizards are placed in a separate family, Gerrhosauridae. See Figure 18–57.

Figure 18–57   Yellow-throated plated lizard. Photo by Isabelle Francais.

The Family Teiidae

The family Teiidae are commonly referred to as whiptails, race runners, and tegus. There are thirty-nine genus groups consisting of 227 species. They are found from the northern parts of the United States down through Central America to Argentina and Chile.

The Six- and Seven-Lined Race Runner. The six-lined race runner, *Cnemidiphorus sexlineatus,* and the seven-lined race runner, *Cnemidiphorus inornatus,* are found in the Southern United States. The seven-lined race runner is brown with seven lighter lines running from the head down the back. The underside, throat area, legs, and tail are a light blue. Race runners live among the rocks in dry desert habitats. They feed on insects and spiders.

The Common Tegu. The common tegu, *Tupinambis teguixin,* grows to 4 feet in length. Tegus are heavy-bodied lizards that are native to South America, where they are found from the tropical rainforest to fairly dry, arid habitats. Tegus are terrestrial and dig tunnels; they are diurnal and return to their tunnels at night. They are dark brown to black with numerous white patches and markings over the entire body, limbs, and tail. They feed on snails, mice, small birds, fish, and eggs. See Figure 18–58.

Figure 18–58 Common tegu. Photo by Isabelle Francais.

The Golden Tegu. The golden tegu, *T. t. nigropunctatus,* is black with yellow markings. Tegus may be difficult to keep in captivity because of their size, difficulty in getting them to eat, and difficulty in maintaining proper temperature and humidity.

The Jungle Runners. The jungle runners (sometimes referred to as rainbow runners), *Ameiva ameiva,* are very colorful lizards found from Central America to Northern South America; they have also been introduced into Florida. The color of this lizard varies considerably, but it usually has a brown background with green vertical stripes on the sides of the body. Blue stripes may appear on the stomach and underside of the tail as well. They may grow to 20 inches in length. They are diurnal and return to their tunnels at night. They feed on insects, small birds, and rodents. A smaller but similar lizard is the rainbow runner, *Cnemidiphorus lemniscatus.*

The Family Lacertidae

The family Lacertidae is referred to as the true lizards. Other common names are sand lizards and collared lizards. Most species have a horizontal fold between the throat and chest scales. There are twenty-five genus groups consisting of about 200 species.

The Green Lizard. The green lizard, *Lacerta viridis,* is a bright green lizard found in Central and Southern Europe. It is a diurnal species that prefers the open woodland, fence rows, and hedge rows, and it feeds primarily on insects. A closely related species is the eyed lizard or ocellated green lizard, *Lacerta lepida,* found in Southern Europe. Both lizards are considered delicate and may be difficult to raise in captivity.

The Wall Lizard. The wall lizard, *Podarcis nuralis,* is a small lizard growing to about 7 inches. It may be found in a wide range of colors, from brown to green with specks and patterns of blues and yellows. Native to Central and Southern Europe, wall lizards are found on the walls of buildings, around ruins, and on rocky walls; a related species is the ruin lizard, *P. sicula.* Like the green lizards, wall lizards are considered delicate and not easily kept in captivity.

The Canary Island Lizard. The Canary Island lizard, *Gallotia galloti,* grows to about 18 inches. It is greenish-brown, and males may have blue spots on the shoulders and sides. This species is diurnal and is found around rocks. Their diet consists of insects, earthworms, snails, small rodents, fruits, and seeds.

The Pityuses Lizard. The Pityuses lizard, *Podarcis pityusensis,* is found on the islands of the Pityuses group in the Mediterranean. It is found among rocks, stone fences, stone piles, and buildings. It is diurnal and feeds on insects and spiders.

The Family Scincidae

The family Scincidae is made up of eighty-five genus groups consisting of more than 1,275 species; they are found in tropical areas throughout the world. Skinks are secretive, and very little is known about them. Most skinks are small, but the largest members may reach 2 feet in length. Skinks are usually terrestrial and burrow under leaves and debris on the forest floor. They are almost round in cross section, and they have indistinct necks and short limbs. The limbs of some species may be so short that they are unable to lift their bodies off the ground; some species may not have limbs.

The Five-Lined Skink. The five-lined skink, *Eumeces fasiatus,* is found in the Eastern United States. They are black when young, with white or yellow stripes running from the head to the tail, which is blue. As they mature, the body turns reddish-brown. Males may have a reddish tinge to their head. The five-lined skink is diurnal and feeds on insects and earthworms.

The Australian Blue-Tongued Skink. The Australian blue-tongued skink, *Tiliqua scincoides,* is native to Eastern and Northern Australia. Its habitat ranges from the rainforests to dry forest and scrubland areas. It has a large, broad, blue tongue that it extends when threatened. They have been bred in captivity and can produce up to twenty-five live young. Skinks become very tame in captivity and are considered excellent pets. They feed on snails, earthworms, and insects that they find among the ground cover and rotten vegetation. In captivity, they feed on baby mice, lean raw meat, and high-quality dog and cat food. See Figure 18–59.

Figure 18–59 Astralian blue-tongued skink. Photo by Isabelle Francais.

The Solomons Giant Skink. Other related species are *T. occipitalis, T. nigrolutea, T. gigas,* and *T. gerrardii.* The Solomons giant skink, *Cornucia zebrata,* is an arboreal skink that has a prehensile tail. It too has been bred in captivity and is commonly found in pet stores. They are **crepuscular** (active at twilight or before sunrise) to nocturnal and grow to 24 inches in length. This giant skink is native to the forests of San Cristobal Island on the edge of the Solomon Island chain. They feed primarily on plant material; in captivity, they feed on chopped fruit and vegetables. They give birth to live young. See Figure 18–60.

Figure 18–60 Solomons giant skink. Photo by Isabelle Francais.

The Family Anguidae

The family Anguidae is made up of glass lizards, slow worms, and alligator lizards. There are eight genus groups consisting of seventy-five species. The limbs of glass lizards are usually reduced in size and are basically useless, whereas the limbs of slow worms are absent.

The Eurasian Glass Lizard. The Eurasian glass lizard, *Ophisaurus apodus,* is the largest of this family, reaching a length of up to 5 feet. It is tan to bronze-brown with a yellowish head and underside. They are found in open woodland and rocky areas, where they hide among rocks and burrows that they have made in the loose soil. They feed on snails, earthworms, small rodents, and eggs. Eurasian glass lizards are native to Eastern Europe and parts of Asia.

The Slender Glass Lizard. The slender glass lizard, *O. attenuatus,* is native to North America. It is found in tallgrass prairie habitats, where it hides among grass and vegetation; it may also occupy the burrows of other animals. The slender glass lizard is primarily active during the day, but if the weather gets too hot, it becomes active at twilight and during the night.

The Northern and Southern Alligator Lizards. The northern alligator lizard, *Elgaria coerulea,* and the southern alligator lizard, *E. multicarinata,* are found in the Western United States from southern Washington to Baja, California; both are about 12 inches in length. The southern alligator lizard lays eggs in burrows, whereas the northern alligator lizard gives birth to live young. The northern species prefers cooler, moister habitats. Both are primarily terrestrial but can climb trees and branches aided by a partially prehensile tail.

Arizona and Texas Alligator Lizards. Two other species are the Arizona alligator lizard, *E. kingi,* and the Texas alligator lizard, *E. liocephalus.* The Arizona species has black and white spots on its upper jaw. The Texas species is the largest of the alligator lizards, reaching a length of about 16 inches. Alligator lizards have the ability to drop off their tails as a decoy for predators as the lizard makes an escape. See Figure 18–61.

Figure 18–61 One of the Alligator lizards, genus Elgaria. *The exact species is unknown. Photo by Isabelle Francais.*

The California Legless Lizard. The California legless lizard, *Aniella pulchra,* is found in California and Mexico. This limbless lizard is about 6 to 10 inches long. It burrows in the soil along stream banks, beaches, and sand dunes. It feeds on insects and earthworms. In appearance, it looks very much like a snake, but has a moveable eyelid.

The Family Xenosauridae

The family Xenosauridae contains two genus groups with only four species. They are commonly referred to as crocodile lizards. These lizards have large scales along the back. They are primarily terrestrial and live in damp habitats. Their main diet consists of fish, tadpoles, and amphibians. They give birth to live young. Three species of the genus *Xenosaurus* are found in the rainforests of Mexico and Guatamala. The fourth species, belonging to the genus *Shinisaurus,* is found in Southern China.

The Family Varanidae

The family Varanidae is made up of one genus group consisting of thirty-one species. They are commonly referred to as monitor

lizards; most are found in Australia and Southeast Asia. Monitor lizards vary greatly in size from the 8-inch short-tailed pygmy monitor, *Varanus breicauda,* to the 10-foot giant Komodo dragon, *Varanus komodonsis.* All monitors have long necks, powerful limbs, strong claws, and powerful tails, and all are egg layers. Most monitors are large animals and can be dangerous; therefore, they are not recommended as pets.

Figure 18–62   Sand (or Gould's) monitor. Photo by Isabelle Francais.

The lace monitor, *V. varius;* the yellow monitor, *V. flavescens;* and the sand monitor, (see Figure 18–62) *V. gouldii,* are species that may occasionally be found in pet shops. The savannah monitor, *V. exanthematicus,* is another monitor that is sometimes found in the pet trade, but it can reach an adult length of more than 6 feet and become hard to handle. The savannah monitor feeds on mice, rats, small birds, eggs, and raw meat. In captivity, it feeds on dog and cat food. See Figure 18–63.

Figure 18–63   Savannah monitor. Photo by Isabelle Francais.

The Family Chamaeleonidae

The family Chamaeleonidae consists of the true chameleons. The family is made up of four genus groups with 855 species. Most are found in Africa and Madagascar, with a few found in Europe and Asia. Chameleons have the ability to change color quickly. This is believed to be controlled by their nervous system. Their color may also be controlled by their surroundings, the temperature, and the animal's mood. Chameleons have tongues as long as 5½ inches, and the tip is covered with mucus. They can shoot their tongue out in a flash to catch insects.

Chameleons also have excellent eyesight. Their eyes can move independently of each other and can focus on different objects at the same time if necessary. Most are arboreal with opposing toes that allow them to grasp branches and limbs; most have prehensile tails. True chameleons are delicate and are hard to raise in captivity; most do not live long in captivity.

The Common Chameleon.   The common chameleon, *Chamaeleo chamaeleon,* grows to about 10 inches in length, and is found along the coasts of North Africa and Southern Europe. It feeds on mealworms, flies, and spiders.

Meller's Chameleon.   Meller's chameleon, *Chamaeleo melleri,* is one of the largest species, growing up to 2 feet in length.

Jackson's Chameleon.   The males of some species, such as the Jackson's chameleon, *Chamaeleo jacksonii,* have projections or horns on their head. Several species, including the South American dwarf chameleon, give birth to live young. See Figures 18–64 and 18–65.

Figure 18–64   Female Jackson's chameleon. Photo by Isabelle Francais.

Figure 18–65   Male Jackson's chameleon. Photo by Isabelle Francais.

The Family Helodermatidae

The family Helodermatidae is made up of one genus group with two species; they are referred to as beaded lizards and are the only venomous lizards in the world. Beaded lizards are heavy-bodied, terrestrial lizards that inhabit dry areas. Their scales are small and bead-like and do not overlap. The venom glands are on the lower jaw, and when the lizard bites, the venom flows by capillary action upward into the wound; they do not inject the venom like a snake. The venom of beaded lizards is usually not fatal to humans but can cause swelling around the bite wound, severe pain, vomiting, and fainting. The beaded lizard has powerful jaws and can hold onto its prey for long periods.

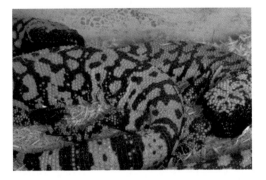

The Gila Monster and Mexican Beaded Lizard.   The Gila monster, *Heloderma suspectum,* of the Southwestern United States and Mexico, and the Mexican beaded lizard, *H. horridum* of Mexico and Guatemala are the two species in the family Helodermatidae. Both grow to about 24 inches in length. The Gila monster is black with yellow or white markings, and the Mexican beaded lizard is black with pinkish markings. They have small eyes, and their vision is not very good. They rely on their sense of hearing and smell to catch their prey. They feed on small mammals, birds, and eggs; beaded lizards are oviparous. See Figure 18–66.

Figure 18–66   Gila monster. Photo by Isabelle Francais.

The Family Lanthanotidae

The family Lanthanotidae contains only one genus group and one species—the Borean earless lizard, *Lanthanotus borneensis.* This nocturnal lizard lives in the lowland areas of Northern Borneo. They are semiaquatic and feed on earthworms and small prey they find around rivers and streams. The Bornean earless lizard lacks a tympanum, or external ear, but is capable of detecting sounds.

CLASSIFICATION—CROCODILES, ALLIGATORS, AND GHARIALS

Kingdom — Animalia
Phylum — Chordata
Class — Reptilia
Order — Crocodilia (Crocodiles, Alligators, and Gharials)
Family — Three families exist

MAJOR FAMILIES—CROCODILES, ALLIGATORS, AND GHARIALS

The order Crocodilia consists of the largest reptiles. Today's living crocodilians are survivors of the Mesozoic era, when giant land and marine reptiles ruled the world. Crocodilians look very similar to their relatives of that era, 230 to 70 million years ago.

Crocodilians are easily recognized, and all are semiaquatic. They can walk on land but can move more easily in the water, propelled by their powerful tails. General characteristics are the large head and long jaws that are lined with teeth. They are almost all brown, olive-green, or black, which helps them blend into their habitat. Their bodies are covered with heavy, rough skin that is made up of scales called scutes. Bony plates under the scutes help give them added protection. Crocodilians are cold-blooded and spend long periods basking in the sun.

The Family Alligatoridae

The family Alligatoridae is made up of four genus groups and consists of seven species. Alligators and caimans differ from crocodiles in the way their teeth fit into the jaws. In alligators and caimans, the fourth tooth of the lower jaw fits into sockets in the upper jaw, and when the mouth is closed, the tooth is not visible. In crocodiles, the fourth tooth is visible when the mouth is closed.

There are two species of alligators: the American alligator, *Alligator mississippiensis,* and the Chinese alligator, *A. sinensis.*

The American Alligator. The American alligator is found along the East Coast of the United States from North Carolina south and across the Gulf of Mexico to Texas. They are black and reach a length of about 14 feet. The American alligator was once on the endangered species list. They were hunted for their meat, for their skins, and as pets. Several years of protection have brought them back to where the public is allowed to hunt them. Small alligators feed on insects, snails, and small aquatic life. Adults feed on larger fish, small mammals, turtles, and birds. See Figure 18–67.

Figure 18–67 American alligator. Photo by Isabelle Francais.

Caimans. Caimans are found from Central America southward to the central parts of South America; they commonly find their way into the pet trade. Caimans are heavily protected by an armored underside made up of overlapping, bony plates. The largest member of the caiman family is the black caiman, *Melanosuchus niger,* which can reach lengths of 15 feet. The common caiman, *Caiman crocodilus,* and the large-snouted caiman, *C. latirostris,* are found in tropical marsh areas of South America. The common caiman has been widely hunted for its skin. Dwarf caimans make up the genus *Paleosuchus;* two species are the dwarf caiman, *P. palpebrosus,* and the smooth-fronted caiman, *P. trigonatus.* Caimans adapt readily to captive situations. They are tropical animals and must have temperatures of 78 to 85°F. They are carnivorous and, in the wild, feed on earthworms, snails, minnows, frogs, and small animals. In captivity, they feed on small pieces of meat sprinkled with vitamins and bone meal to provide adequate nutrition. See Figure 18–68.

Figure 18–68 Small caiman of the genus Paleosuchus. *Photo by Isabelle Francais.*

The Family Crocodylidae

The family Crocodylidae is made up of three genus groups consisting of fourteen species. They range in size from 4 feet to the giant crocodiles, which may reach 25 feet in length and weigh more than 2,000 pounds. One species, the American crocodile, *Crocodylus acutus,* is found in extreme southern Florida, where it survives in brackish waters. Crocodiles are generally not considered part of the pet trade. See Figure 18–69.

Figure 18–69   American (or Florida) crocodile. Photo by Isabelle Francais.

The Family Gavialidae

The family Gavialidae contains two genus groups and two species: the false gharial, *Tomistoma schlegelii,* and the true gharial, *Gavialis gangeticus.* Gharials have very long, narrow snouts that are used for catching fish, which is their main diet. Gharials are seldom exported and are not considered part of the pet trade. See Figure 18–70.

Figure 18–70   False gharial. Photo by Isabelle Francais.

HOUSING AND EQUIPMENT

It would be impossible in the scope of this text to cover all of the different housing and equipment requirements to keep reptiles in captivity. In the following material, some general guidelines are given as a starting place in providing housing and equipment for reptiles. These guidelines can be altered to meet the requirements of specific species.

Habitats for reptiles generally fall into one of four types:

1. Terrestrial habitats for reptiles that live on land

2. Semiaquatic habitats for reptiles that live on land or in trees but spend part of their time in the water

3. Aquatic habitats for reptiles that live primarily in the water

4. Arboreal habitats for reptiles that live in trees

Reptiles can be kept in a terrarium, a vivarium, or an aquarium. A terrarium is a cage used for keeping land or terrestrial animals. A vivarium is a cage in which the environment is duplicated as close as possible to the original environment of the species. An aquarium is a water environment for aquatic animals. A combination of two of these would be needed for animals that are semiaquatic. *Vivarium* is the term that is used in this discussion. See Figure 18–71.

It is difficult to give general guidelines for the size of a vivarium. It would depend on the size of the species, the number of animals, how active they are, and how large the species will be when they mature. Some of the larger species of reptiles are less active than the smaller species; therefore, they may not need a proportionally larger vivarium.

Following are some general guidelines. A land turtle needs a vivarium that is about five times, in length and width, the length of the turtle's shell. If the turtle's shell is 6 inches long, the vivarium should be approximately 30 inches long by 30 inches wide. If rocks, logs, or decorations are to be added, the vivarium should be larger. If additional turtles are going to be added, figure one-third more area needed for each additional turtle. For a single turtle with a 6-inch shell, a vivarium 30 inches by 30 inches, or approximately 900 square inches, is needed. Each additional turtle would require about one-third that area, or

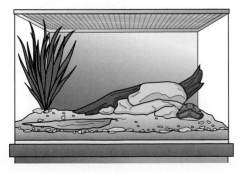

Figure 18–71   A very nice, well-decorated vivarium for land turtles. A vivarium does not need to be this elaborate.

about 300 square inches; therefore, a vivarium approximately 35 inches long and 35 inches wide would be needed. For aquatic turtles, an area five times as long and three times as wide as the turtle's shell should be provided. Platforms made from wood or rocks should be provided so that the turtles can climb out of the water and bask in the rays of a heat lamp. The size of the basking area will influence the overall size of the vivarium. The depth of the water in the vivarium should be about 1 foot. The height of the sides should be sufficient to prevent the turtle from climbing out and escaping.

The substratum in the bottom of a land turtle's vivarium can be made by first putting in a 1-inch layer of small stones or gravel, placing a 1-inch layer of charcoal over the stone or gravel, and then adding about 3 inches of coarse sand. Depending on the species, a layer of soil, leaves, and branches can be added for a wetland or forest habitat. Large rocks and logs can be added on top of the sand to make a desert habitat. Even land turtles should be provided with a shallow container of water. The depth of the water in the container should never be more than the height to the front edge of the turtle's carapace. The sides of the container should be gradually sloping so that the turtle can easily get out of the water. No material or substratum is needed in the bottom of an aquatic turtle habitat.

Small lizards, such as geckos and anoles, that are 5 to 6 inches long can be kept in vivariums 12 inches wide, 12 inches long, and 16 inches high. A vivarium of this size can keep one male and two females. The vivarium should also contain rocks, plants, tree limbs, or branches to give these active lizards things to climb on.

A ratio of length to width to height of 1:2:2 is sometimes used in determining the size of a vivarium for active tree- or wall-climbing lizards. In planning for terrestrial lizards, a ratio of 2:1:1 is sometimes used. Larger geckos and more active lizards need a larger vivarium. A vivarium 60 inches long, 24 inches wide, and 24 inches high may be needed to keep several spiny lizards. Common iguanas need a vivarium as large as 80 inches by 60 inches by 60 inches for one male and two females; smaller vivariums would be needed if only one animal is to be housed.

The most commonly used material for vivariums is glass aquariums; these are easily obtained from pet stores and discount stores. They should be fitted with an appropriate cover to keep the animals from escaping. Covers can be purchased or constructed from wood and wire mesh screen. Plexiglas can also be used to construct vivariums, with holes drilled to provide ventilation. Glass and Plexiglas are definitely needed for aquatic, semiaquatic, or rainforest habitats. Wood and wire mesh screen can be used to construct cages for dry habitats. In providing vivariums for reptiles, the environment should be duplicated as close as possible to the original; however, consideration should also be given to the ease with which the vivarium can be cleaned and maintained. Very elaborate vivariums can be used but are not always necessary. See Figure 18–72.

Water in aquatic and semiaquatic vivariums should be changed every three to seven days to prevent the buildup of bacteria and harmful wastes. An alternative to changing the water is to install a circulation pump and filter system, along with an aeration system, to help maintain water quality. See Figure 18–73.

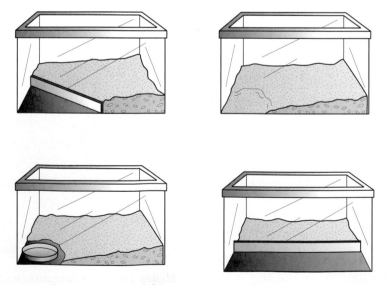

Figure 18–72 The most commonly used material for vivariums is glass aquariums. The substratum can completely cover the bottom, be partitioned into dry and wet areas, or a water bowl can be used to provide a wet area.

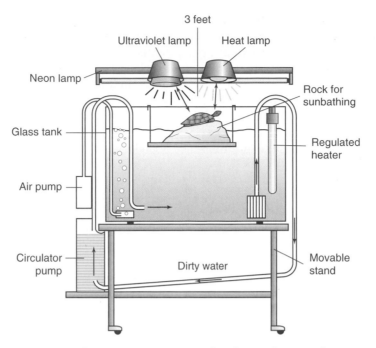

Figure 18–73 *Water in semiaquatic and aquatic systems must be changed every three to seven days, or a circulation pump and filter system, along with an aeration system, needs to be used.*

Reptiles are **ectotherms,** meaning they cannot generate their own body temperature and take on the temperature of their environment. Because reptiles are unable to generate their own body heat, a heat source must be provided. An incandescent light with a reflector shield makes a good light as well as a good heat source. The wattage of the bulb used depends on the amount of heat needed and the distance the light is placed above the vivarium. Reptiles bask until their body temperatures reach an optimal level, then they seek shaded areas or areas of lower temperatures. A thermometer in the vivarium is important so the temperature can be monitored and maintained between 65 and 85°F. The temperature under the heat lamp may need to be as high as 100°F. To duplicate normal habitats, a timer may need to be added to the heat lamp so that the lamp is turned off at night, thereby maintaining higher temperatures during the day and cooler temperatures at night. Relative humidity should be maintained from 50 to 70 percent and can be monitored with a hygrometer. Again, these are general guidelines and will vary with different species. See Figure 18–74.

Sunlight is important not only as a heat source, but also to absorb the ultraviolet (UV) rays that are needed in calcium metabolism, formation of pigment, and vitamin D synthesis. A fluorescent lamp should be used to provide the needed ultraviolet rays. The fluorescent lamp should *not* be placed on top of a piece of glass because the glass will filter out UV rays.

Vivariums for aquatic and semiaquatic species need a method of heating the water, which should be maintained at about 75 to 85°F; aquarium heaters can be used. To prevent reptiles from damaging the heating unit, it should be placed between a couple of clay bricks.

If additional heating is needed, a heating cable or a heating mat can be used. Heating cables run through the substratum in the bottom of the vivarium. A piece of wire mesh screen placed over the cables will prevent reptiles from digging up the cables and possibly causing burns. Heating mats can be placed under the vivarium to provide heat to the bottom of the vivarium. Artificial rocks with heating elements in them are also available to provide additional heat.

Figure 18–74 *Reptiles seek areas where the temperature feels good to them. A thermometer should be used to monitor temperatures.*

With the use of more than one heat source, the animals can move to the area they find most desirable. Remember that reptiles lie or bask in the warm areas until their body temperature reaches an optimal level, then they seek cooler areas.

Lizards that originate in tropical rainforests may need to have sprinkler or misting systems to duplicate the daily rainfall of their original habitat; these sprinklers need to be on a timer. Heat and humidity requirements for this type of habitat will differ considerably from other types of habitats.

Before leaving this discussion, it is important to talk about electricity and its safe use.

1. Always purchase electrical equipment that has been UL approved.

2. Always read the instructions before installing and using a piece of electrical equipment.

3. Make sure all heat lamps are placed so that they will not be sprayed or splashed with water.

4. Water heaters should be approved for underwater use.

5. Do not open the vivarium, attempt to handle animals, or make any changes to the vivarium until the electricity is turned off.

6. All electricity should flow through a switch box with appropriate fuses or circuit breakers.

7. All circuits should be equipped with Ground Fault Interrupters (GFIs); the GFIs will disconnect the power when a dangerous condition exists.

FEEDING

Turtles

Aquatic turtles are primarily carnivorous or omnivorous. In captivity, aquatic turtles soon learn to eat pieces of meat, liver, and canned dog or cat food; dry cat or dog food can also be used. Turtles should be offered food every day.

Land turtles are primarily herbivorous, feeding on grasses and plants. In captivity, they feed on canned dog or cat food. Along with the canned food, they should be given pieces of spinach, lettuce, grapes, alfalfa, and clover. They may also eat ground meat. Strawberries, cantaloupe, and other fruits can be added to their diet. Several different kinds of food should be tried to see what the turtle will eat.

Snakes

All snakes are carnivorous. Small snakes in captivity readily feed on earthworms, mealworms, crickets, and other insects. Larger snakes feed on baby mice and rats. If available, they may also feed on frogs and toads. Big snakes feed on full-grown mice, rats, baby chicks, guinea pigs, and rabbits. In feeding snakes live food such as rats and mice, one should observe the snake closely. If the snake does not eat the rodent within a few minutes, the rodent should be removed from the snake's vivarium. A rat or mouse left in with a snake can bite and injure the snake.

Snakes may learn to feed on canned dog and cat food, raw hamburger, or pieces of meat or fish. All foods should be supplemented with vitamins and minerals.

If several types of food have been offered to a snake and it has not eaten in four weeks or more, the snake may need to be force-fed. On small snakes, one can force open their mouth and place a piece of food in it; hopefully, the snake will swallow the food. If it is stressed or frightened, it may spit the food out. For larger snakes, one can use forceps and place a piece of meat in the snake's mouth and then try massaging the meat down into the snake's stomach. Liquids may also be tried; one can beat an egg into a liquid and place a plastic tube down the throat of the snake and into its stomach. The stomach is usually about one-third the length of the snake. The beaten egg should be placed in a plastic squeeze bottle and gradually squeezed into the tube. If the food successfully makes it to the snake's stomach, the snake should be returned to its vivarium. It should not be handled any more than necessary because it may regurgitate. Snakes that are about to molt do not eat. When the eyes cloud over, a snake is getting close to molting.

Lizards

Feeding lizards in captivity is fairly easy. Most are insect eaters. Lizards feed on crickets, as well as mealworms and earthworms, which are readily available from bait shops. Iguanas and larger lizards eat mealworms, earthworms, lettuce, flower blossoms, fresh fruit, vegetables, ground meat, and dog or cat food.

If a reptile captured from the wild refuses to eat, one may want to consider returning it to the wild instead of letting it die from starvation. If it is a purchased animal, one may want to enlist the help of a veterinarian.

HANDLING

Turtles

Most turtles found in pet shops can easily be handled. Small turtles are picked up by the shell with the fingers and thumb and placed in the hand. Larger turtles may take both hands, placing the fingers under the lower shell (plastron) and the thumbs on the top shell (carapace). One should not be too alarmed if a hissing sound is heard because this is usually a rush of air being expelled from the lungs as the turtle withdraws its head and limbs. Very few problems will be encountered with turtles in the wild; however, snapping turtles and some soft-shelled turtles have long necks and can reach around and bite. Turtles that are going to be transported can be placed in a cardboard box.

Snakes

Snakes that have not been handled will be frightened, and their reaction is to bite in self-defense. In handling a snake the first few times, a pair of soft, leather gloves should be worn. The bite of a small snake is not very painful, but sharp teeth can tear the skin as the hand is pulled back in a reflex action. Once the snake has been handled a few times and does not struggle or attempt to bite, the gloves will not be needed.

When picking up a small snake, one should grasp it firmly just behind the head. A small snake usually coils up around the hand and allows itself to be lifted. Larger snakes may need their body supported and lifted with the other hand. The snake should be held close to the handler's body; most snakes enjoy being held close because of the warmth generated. See Figure 18–75.

When attempting to remove a snake from its cage, the handler should let the snake know he or she is there and let the snake come to him or her. The snake may not like an intrusion and may defend its area. A snake should never be grabbed by its tail; most snakes will not like being held this way, and it is possible to cause injury and break off part of the tail.

When transporting a snake, one should place it in a cloth bag; a pillowcase works well. A pillowcase is porous and allows the snake to have air. If the weather is cool, the bag can be placed in a Styrofoam box or cooler. Holes must be punched or drilled into the box to allow air circulation. If necessary, a hot-water bottle can be placed in the box to provide warmth. The snake must not be allowed to get chilled.

Lizards

The temperament of lizards may vary considerably. Some lizards may tame easily and allow themselves to be handled, whereas others may remain wary and aggressive forever. Lizards that are obtained at a young age are probably the easiest to work with. Small lizards can be picked up by grasping them with the thumb and forefinger behind the head and holding the animal around the body with the rest of the hand. Larger lizards can be held in the same way, but the other hand should be used to support the rear of the lizard. The large iguanas, tegus, and monitors can be picked up using one hand to grasp the animal around the neck and the other hand to support the mid- and rear section of the animal. The tail can be held close to the body with the arm and elbow. See Figures 18–76 and 18–77.

The bite of small lizards most likely will not cause any pain or injury; they are usually not strong enough to even break the skin. Larger lizards, however, are capable of causing serious bites; some have powerful jaws and may clamp down and not let go. The handler should wear gloves to pick up animals that may possibly bite. Some lizards are capable of inflicting serious scratches from sharp and powerful claws. Any bites and scratches should be cleaned with soap and water and treated by a doctor.

One should never pick a lizard up by the tail; lizards are capable of dropping or losing their tails as a defense mechanism. The predator is left holding the tail while the lizard heads for cover. The lizard will grow or regenerate another tail, but the new tail usually lacks the coloration and pattern of the original.

Figure 18–75 Small snakes usually coil up around the hand when being lifted. Photo by Isabelle Francais.

Figure 18–76 Small lizards can easily be picked up with and held in one hand. Photo by Isabelle Francais.

Figure 18–77 *Some lizards are capable of causing serious bites as well as inflicting serious scratches with their sharp and powerful claws. Photo by Isabelle Francais.*

DISEASES AND AILMENTS

When obtaining a reptile, one should make sure the reptile is healthy. Many reptiles are imported and have been subjected to stressful conditions, extreme changes in temperature, unsanitary conditions, and lack of food or water. The reptile should be observed carefully, making sure it has been eating or is eating. An animal that has been eating and is well nourished will be filled out, and the ribs, vertebrae, and pelvic bones will be covered and not show noticeably. The animal's eyes should be open, and the animal should be alert to its surroundings. Its mouth should be closed; any frothing or foaming at the mouth, around the nose, or running at the eyes indicates an unhealthy animal. If the animal shows bite wounds or marks of other injuries, these should be healed and scarred over. One should avoid purchasing animals with open wounds, wounds that are seeping, or wounds that do not appear to be healing. Turtles that are sick usually drag the rear of their shell on the ground instead of carrying the shell horizontally off the ground. Aquatic turtles that are sick usually lean to one side when in the water. See Figure 18–78. One should look carefully for parasites; reptiles can be infected with ticks and mites. Animals that appear to be heavily infested should be avoided. Also one should check to see what the policy of the pet shop is on returning sick animals or if they have a replacement policy for animals that may die after leaving the shop.

Figure 18–78 *A healthy land turtle walks with its shell parallel to the ground, whereas a sick turtle drags the rear of its shell on the ground. An aquatic turtle should swim level in the water. If it swims at an angle, it may be sick.*

Reptiles are surprisingly free of diseases and ailments. If one starts out with a healthy, active reptile, chances are very good that one can avoid disease problems in the future. It is important to keep the vivarium clean and sanitary. Feces should be removed as soon as they dry, and any leftover food materials should be removed before they spoil. One must make sure the reptiles have a supply of fresh water.

Bacterial Diseases

Salmonella. *Salmonella* is a bacteria that may appear normally in the digestive tract of a reptile, especially in turtles. A normal population of *Salmonella* does not cause any problems, but when populations get out of control, prolonged diarrhea and anorexia may occur; this diarrhea is often watery, green, and foul smelling. *Salmonella* bacteria can be transmitted from reptiles to humans; children are especially susceptible. It is important to thoroughly wash the hands after handling reptiles and cleaning vivariums. If *Salmonella* infection is suspected, one should consult a veterinarian, who will probably treat the condition with antibiotics.

Aeromonas and Pseudomonas. *Aeromonas* and *Pseudomonas* are two types of bacteria that cause digestive infections, which can lead to poor nutrition and stress. Animals with digestive infections have diarrhea and appear anorexic. If not treated, further problems in the form of respiratory ailments and pneumonia can develop. Symptoms of respiratory ailments are difficulty breathing, blocked nostrils, and nasal discharge; bubbles may be seen around the nostrils when the animal exhales. The animal may hold its mouth open in an attempt to breathe easier.

Treatment includes making sure the animal has a clean, dry, warm vivarium and the use of antibiotics.

Pathogenic Amoebas. Pathogenic amoebas, *Entamoeba invadens,* may invade the digestive system and cause watery, slimy diarrhea and anorexia. Treatment with metronidazole by a veterinarian is recommended.

Environmental Control. Dirty water, dusty conditions, cold temperatures, and drafty conditions may cause irritation to the eyes of reptiles and lead to swollen eyelids. Antibiotic eye ointment applied between the eye and eyelid normally treats the condition; however, the conditions that originally caused the problem must be corrected.

Mouth Rot. Mouth rot is a fungal infection associated with open wounds or sores on the mouth of snakes. Snakes that rub their mouths and snouts up against the side of the vivarium or on decorations in the vivarium are likely to become infected. Hydrogen peroxide can be used to clean the sore, and then treatment with a liquid solution of sulfamethazine is recommended.

Shedding

Snakes and lizards in good health shed their skin several times during the year; snakes normally shed their skin in one piece. The first sign that a snake is getting close to shedding is that the covering over the eye becomes cloudy. The shedding starts at the snout. The snake rubs the snout on a rock or piece of wood to get the shedding process started. Lizards normally shed their skin in patches. The tokay gecko pulls the skin off in one piece and then eats it. This is normal, and the skin is considered a good source of protein. The process of shedding should not take more than a couple of days. A reptile that does not shed properly or completely may have a nutritional or health problem. Sometimes, bacteria and other disease organisms grow under the dead skin that has not shed; this can lead to serious infections. Moisture is usually needed for the skin to shed. Snakes may lie in their water container prior to shedding; for larger snakes, this may be impossible to do. One may want to place the snake in a damp burlap bag for a couple of days, or mist the snake occasionally with water.

Turtles also shed their skin. Small pieces of old skin come off constantly; however, if there is a continuous shedding or peeling of the skin from the same area, a bacteria or fungus may be causing a problem. One should make sure the vivarium is not too damp or wet, provide the turtle with drier conditions, and place a fluorescent lamp over the vivarium. The infected area can be treated with an iodine solution. One must make sure the turtle is eating and has a good nutritional diet. A visit to the veterinarian may be in order so that the turtle can be treated with antibiotics.

External Parasites

Mites and Ticks. Mites and ticks are external parasites that often attach themselves to reptiles. One should not attempt to pull ticks directly out of the skin; this may break the mouthparts off the tick and leave them in the skin, which can lead to an infection. Ticks are best removed by applying alcohol or petroleum jelly to the body of the tick; the tick will relax and back out of the wound in an attempt to get air. The tick can then be removed using a pair of tweezers. Individual mites are so small that they may not be noticed, but when an animal becomes heavily infested, the mites can be seen crawling around. Heavy infestations of mites can cause anemia, anorexia, depression, stress, listlessness, lack of appetite, and even death. If mites are visible or are suspected, a veterinarian should be consulted. Several methods are recommended for controlling mites, and a veterinarian can prescribe the best method. Both the reptile and its environment must be treated. Before new reptiles are introduced into a vivarium, they should be put in a quarantine vivarium and observed for ticks and mites.

Internal Parasites

Roundworms and Tapeworms. Roundworms and tapeworms are internal parasites that can occasionally become a problem. Reptiles caught in the wild almost always carry these worms. Normally, they are not a problem, but improper diet, stress, and illness may cause an increase in the number of worms. Symptoms are anemia, listlessness, loss of appetite, anorexia, and possibly death. Several worm medicines are available to control these parasites; one may want to check with a veterinarian.

Vitamin Deficiencies

Vitamin deficiencies occasionally occur in captive reptiles. Untreated nutritional deficiencies may lead to deformed shells on turtles, respiratory diseases, and pneumonia. Proper use of heat lamps, fluorescent lamps, and vitamin and mineral supplements should prevent deficiencies.

REPRODUCTION

Turtles

Turtles are oviparous (lay eggs that hatch later). The female deposits eggs in a nest cavity that she digs in the soil. The eggs are covered and then left unat-

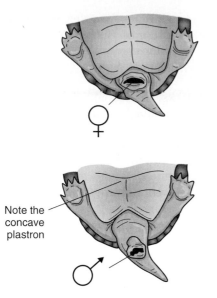

Note the concave plastron

Figure 18–79   Determining the sex of turtles. The plastron of the male may be slightly concave. The tail of the male is thicker and longer than that of the female. The cloacal opening is located further down the tail on the male.

tended. The eggs of turtles vary considerably in size, shape, shell texture, and number laid. Number laid may vary from one to three to as many as 200. Some species lay eggs with soft leathery shells, whereas others lay hard-shelled eggs. The incubation period also varies considerably; the eggs of soft-shelled turtles hatch in about thirty days, whereas the eggs of some land tortoises may take up to eighteen months. The average incubation period is sixty to ninety days. Incubation time depends on the temperature of the soil. When the young hatch, they are on their own, relying on inherited instincts for survival.

It is almost impossible to tell the sex of young turtles; as they age, some differences begin to appear. On some species, the plastron of the male becomes somewhat concave. This may aid in the male's ability to mount the female; the tail of the male becomes thicker and longer than that of the female. The cloacal opening is located further down the tail on the male. The cloaca of reptiles, amphibians, and birds is a cavity into which both the intestinal and genitourinary tracts empty. See Figure 18–79.

During mating, the male climbs onto the carapace of the female. The claws on the front limbs are used to grasp onto the front edge of the female's carapace. The large domed shell of the box turtle presents some problems for the male during mating; it is impossible for the male to hold onto the female's carapace. To complete copulation, the male falls backward off the female's carapace.

Snakes

Snakes may be oviparous, ovoviviparous, or viviparous (see Glossary). Most are oviparous, laying their eggs in shallow holes covered with a thin layer of soil. Some snakes lay their eggs under or around rocks, in hollow logs, or around the stumps of trees. It is important that the eggs not get too hot during the day or too cool at night; temperature is important in the development of the embryos. Snakes that are ovoviviparous retain the eggs within their bodies until they hatch and then give birth to the young. The young receive no nourishment from the female, only the nourishment contained in the egg. The garter snake is an example of an ovoviviparous species. In a few species, the young are retained in a preplacental sac that allows for some exchange of oxygen and nutrients. This is referred to as **viviparous.** Pit vipers are an example of a viviparous species.

Most egg-laying snakes leave the eggs unattended after laying them. Pythons, however, coil themselves around their eggs and brood them until they hatch. Incubation periods vary with different species. Egg-laying snakes lay eggs from thirty to eighty-five days after mating, with the eggs incubating for forty to ninety days. The gestation period for live-bearing snakes varies from 90 to 150 days. The number of eggs varies from one to as many as sixty. Live young varies from two to as many as 100. The young free themselves from the eggs with the aid of a sharp egg tooth on the tip of their snout and slice open a hole in the egg through which it can emerge. Most young are on their own as soon as they are born or hatch. Pythons may stay near the mother for two or three days before they move off on their own. Some species of rattlesnakes may stay with their young and defend them for several days, but most snakes show no interest in their young; some may even eat their own young.

The sex of snakes may be difficult to determine; very little difference exists between males and females. Females usually have a shorter tail. Many species of snakes still have spurs or claws near the base of the tail; these are what remains of the legs of their ancestors. These spurs or claws are usually longer in males than females. Females may also be heavier and longer than males.

In male snakes, the cloaca is modified to form a double penis, or **hemipenes.** The hemipenes often contains recurved spurs that aid the male in holding onto the female's cloaca during mating. During mating, the male and female usually entwine themselves around each other. The male moves forward until their two cloacas are together, and the male then inserts one or both hemipenes. The female may store the sperm in the oviduct for several months before fertilizing eggs.

Lizards and Iguanas

Lizards and iguanas may be oviparous, ovoviviparous, or viviparous; most species are oviparous. The females lay eggs in holes they dig in the soil; they may dig several holes before finding one that meets their approval. The time necessary for young to hatch varies considerably. For the smaller anoles and geckos, hatching takes place in thirty days or less, but for the larger monitor lizards, it may take as many as 120 days. Incubation periods vary depending on environmental factors. Anoles usually lay a single egg, whereas some species of the iguana family lay as many as fifty soft-shelled eggs. Geckos differ from other lizards because they lay two hard-shelled eggs, which are attached by an adhesive substance to cracks in tree bark.

Most of the young lizards have a sharp egg tooth on the end of their snout that they use to slice open the egg. Most lizard females do not stay around after the eggs are laid, so the young are on their own as soon as they emerge.

Many species of lizards show definite differences between the males and females. This is referred to as **dimorphism.** Determining sex in species with definite dimorphism is fairly easy; however, this characteristic may not develop until the lizards reach sexual maturity. Some differences may be observed to help determine the sex of young lizards. The male hemipenes, when not being used for copulation, is withdrawn into

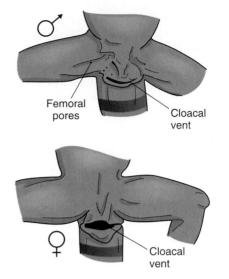

Figure 18–81 **Male agamas and most species of the family Iguanidae have thigh or femoral pores.**

the cloacal vent. This gives the male a larger or swollen appearance to the area of the cloaca and tail base; the female has a thinner tail base. Male anoles have two large scales behind the cloaca. Male geckos have anal or preanal pores forward of the cloaca. Male agamas and some iguanas have femoral pores on the inside of the rear thighs. See Figures 18–80 and 18–81.

During the period prior to mating, males may take on bright body colors, display brightly colored dewlaps, or engage in a series of ritual head bobbing or body movements to entice a female into mating. If the male finds an interested female, he may grasp or bite the female by the neck in an effort to hold onto her. The male then moves his cloaca as close as possible to the cloaca of the female. When the two cloacas are together, one of the male's hemipenes is inserted into the female's cloaca. The sperm flows into the female's cloaca, where it then enters the oviducts. In some species, the sperm can be retained or stored for long periods before fertilization takes place. In some cases, several fertilizations of eggs may occur from only one mating.

In species that bear live young, the mother usually shows no mothering instincts; the young are left to fend for themselves as soon as they are born.

Crocodilians

The reproduction habits of crocodilians are similar to that of lizards, except that crocodilians usually mate in the water. All crocodilians lay hard-shelled eggs, which are usually deposited in nests. The nests may be cavities that are dug into the soil. The eggs are laid

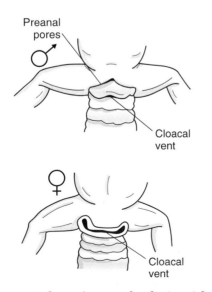

Figure 18–80 **Male geckos can be distinguished from females by the presence of anal or preanal pores.**

into the cavity and then covered with soil. Nests may also be built into mounds from soil and vegetation, with the eggs being deposited in the center of the mound.

Female crocodiles lay between sixteen and 100 eggs, alligators and caimans lay between twenty and thirty eggs, and gharials lay about forty eggs. The number of eggs laid depends on the age and size of the female. Crocodilian females stay within the vicinity of the nest to guard it against predators. After about ninety days, the young inside the eggs start making high-pitched sounds that alert the female. The female begins to dig the top away from the nest so the young can emerge as they hatch; some species even gather up the young in their mouth and carry them to the water.

Environment

To get reptiles to mate and reproduce in captivity, the environment that the species came from must be duplicated as closely as possible. This may involve a series of heat lamps, ultraviolet lights, water heaters, timers, thermostats, and spray equipment.

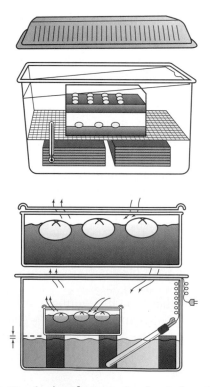

Figure 18–82 *An incubator using heated water to provide warmth and humidity. The eggs are placed in peat moss or vermiculite and then placed in a small container. The eggs should be covered one-half to two-thirds. Place the container with the eggs on a rack inside the incubator. Temperature should be maintained at 85°F with a heater that has a thermostat.*

Reptilian females in captivity should be provided with a medium in which to lay their eggs. It is important to know the habits of the species so that proper facilities can be provided. Normally, a tray or box filled with damp sand is sufficient.

The eggs of reptile species that do not **brood** (stay with and incubate their eggs) can be removed after the female has completed laying her clutch of eggs and then artificially incubated.

To artificially incubate eggs, they should be placed in a shallow container that has been filled with an incubation medium; this can be sterile peat moss or vermiculite. It should be of enough thickness so that it covers the eggs; an equal weight of water should be added to the medium. The medium should be damp but not overly wet. The eggs should be placed in the medium so that they are covered from one-half to two-thirds. One should try to place the eggs in the same position they were in originally; unlike bird eggs, they should not be turned or rotated. See Figure 18–82.

A cover with several small holes in it should be placed over the container; the holes allow oxygen to get to the eggs. The container should be placed in an incubator at about 85°F and the eggs observed daily. Most reptile eggs have a soft, leathery shell; if there is too much moisture, a white, fluffy mold may grow on the eggs. This mold should be wiped off the eggs, and some moisture allowed to evaporate. If a small indentation appears on the eggs, this is because moisture is lacking, and more water should be added to the medium.

As the young hatch, they can be moved from the incubator and placed in a nursery cage. If proper conditions exist, the young should begin feeding shortly after hatching. Proper knowledge of the species is important so that proper food is provided for the young. Several different kinds of food may need to be tried before the young will feed.

Lizards usually feed on insects such as crickets, flies, mealworms, bee moths, and small worms. Snakes are considered carnivorous; small snakes feed on earthworms, mealworms, crickets, and other insects. Larger snakes feed on baby mice, baby rats, and frogs. Turtles feed on ground meat and dog and cat food.

SUMMARY

Reptiles are cold-blooded vertebrates that possess lungs and breathe atmospheric air. They also have a bony skeleton, scales, or horny plates covering the body, and a heart that has two auricles and in most species one ventricle.

There are approximately 6,500 species of reptiles belonging in four different orders: Testudines (called Chelonia in some references), which are made up of the turtles, tortoises, and terrapins; Serpentes, which includes snakes, pythons, and boas; Squamata, which includes iguanas and lizards; and Crocodilia, which includes crocodiles, alligators, caimans, and gharials.

Present-day reptiles basically have three body types. The first have long bodies and clearly defined tails like the crocodiles and lizards. The second have long bodies that taper into tails like the snakes, and the third have short, thick bodies encased in shells like the turtles and tortoises.

The skin of reptiles has a horny surface layer. This surface layer must shed regularly to allow for the continued growth of the reptile. Snakes and some lizards cannot close their eyes. Instead of eyelids, they have a transparent layer, or brille, that permanently covers the eye. When the animal sheds its skin, this covering is also shed and is replaced by a new covering.

Habitats for reptiles generally fall into one of four types: terrestrial habitats for reptiles that live on land, semiaquatic habitats for reptiles that live on land or in trees but also spend part of their time in the water, aquatic habitats for reptiles that spend their time primarily in the water, and arboreal habitats for reptiles that live in trees. To be successful at raising reptiles in captivity, it is important to duplicate the animal's original habitat as closely as possible.

Reptiles are ectotherms; they are unable to generate their own body heat and take on the temperature of their environment. Because they cannot generate their own body heat, a heat source must be provided.

Most reptiles tame easily and allow themselves to be handled. Handlers should wear gloves the first few times as a matter of caution.

Reptiles are generally healthy animals. One should observe a reptile carefully when considering obtaining one. If the animal appears healthy, if it is eating, and if it does not appear to have any parasites, chances are good the animal will survive for a long time in captivity.

Turtles are oviparous. They lay eggs that hatch after an incubation period. Snakes may be oviparous, ovoviviparous, or viviparous. Snakes that are ovoviviparous retain the eggs within their bodies until they hatch, then give birth to the young. The young receive no nourishment from the female, only the nourishment contained in the egg. In a few species, the young are retained in a preplacenta sac that allows for an exchange of oxygen and nutrients; this is referred to as viviparous. Lizards and iguanas may be oviparous, ovoviviparous, or viviparous.

Crocodilians are oviparous and usually mate in the water.

DISCUSSION QUESTIONS

1. What is a reptile?

2. What are the four orders of reptiles?

3. What are the characteristics of turtles, tortoises, and terrapins?

4. What are the characteristics of lizards?

5. What are the characteristics of snakes?

6. What is the difference between pythons and boas?

7. What are the characteristics of crocodiles, alligators, caimans, and gharials?

8. Why do reptiles shed their skin?

9. What are scutes? What are the different types of scutes?

10. Discuss the terms *oviparous, ovoviviparous,* and *viviparous*. What would be some advantages and disadvantages of each?

11. What are the four types of habitats that reptiles live in?

12. Discuss the types of housing and equipment that would provide the four habitats required by reptiles in captivity.

13. What is an ecotherm?

14. Discuss the types of foods needed to maintain reptiles in captivity.

15. What are the signs of good health to look for when obtaining a reptile?

16. What are some signs of ill health?

SUGGESTED ACTIVITIES

1. Visit local pet shops amd make a list of the different species of reptiles that are available. Using the library, compile as much information as you can about the species.

2. Obtain some aquariums or use wood and wire mesh screen to construct the four types of habitats for reptiles. Obtain reptiles from the pet shop or capture some from the wild. Place them in the habitats and observe their habits.

3. Research the different orders, families, genus groups, and species of reptiles. Why are they classified as they are?

ADDITIONAL RESOURCES

1. Pet Support
   www.petsupport.com/REPTILES/index.html

2. Pets Warehouse
   www.petswarehouse.com/Reptile.htm

3. THE EMBL REPTILE DATABASE
   www.embl-heidelberg.de/~uetz/
   LivingReptiles.html

4. Welcome to the Amphibian Embryology
   Tutorial
   http://worms.zoology.wisc.edu/frogs/
   welcome.html

5. Florida Museum of Natural History
   www.flmnh.ufl.edu/natsci/herpetology/
   herpetology.htm#Top

6. The World-Wide Web Virtual Library:
   Herpetology
   http://cmgm.stanford.edu/~meisen/herp

7. Jason's Snakes and Reptiles
   www.snakesandreptiles.com/index.html

8. The ACCESS INDIANA Teaching and
   Learning Center
   http://tlc.ai.org/index.htm

chapter 19

Birds

OBJECTIVES

After reading this chapter, one should be able to:

- describe the characteristics of birds.
- compare and contrast the different orders of birds discussed.
- describe methods used in handling and feeding birds.
- describe the various housing and equipment needs of birds in captivity.
- describe the various diseases and ailments of birds.

TERMS TO KNOW

aviary	filoplume feathers	mantle
cere	flight feathers	papilla
clutch	grit	powder-down feathers
contour feathers	isthmus	preen
coverts	lores	scalloped feathers
crown	lutinos	scapulars
down feathers	mandibles	sternum

CHARACTERISTICS

Birds are two-legged, egg-laying, warm-blooded animals with feathers and wings, and they are believed to have evolved from prehistoric reptiles. Evidence for this is threefold: birds have scales on their legs and feet similar to those on reptiles, birds are egg-laying like many species of reptiles, and there are bird-like fossils in existence. The first bird is believed to be *Archaeopteryx;* the fossil of this bird-like reptile was found in the Bavarian area of Germany. See Figure 1–4. *Archaeopteryx* lived approximately 150 million years ago during the Jurassic period and was about the size of a pigeon. Its forelimbs formed short wings that were covered with feathers. *Archaeopteryx,* unlike modern birds, had teeth, a long tail, and three claw-like digits on the wings. Modern birds appeared about 80 million years ago during the Cretaceous period; by the Cenozic Era, about 70 million years ago, they were very similar to birds today. See Figure 19–1.

Modern birds have thin skin that consists of two layers: the outer layer, or epidermis, and the inner layer, or dermis. From the dermis emerge the bird's feathers,

be used, and milk may be added. Because their diet consists of liquids, their feces are also soft and of a liquid consistency. More effort and time is needed to provide proper feeding and care for these birds; proper sanitation is very important. Most members of this family come from Australia, New Guinea, Indonesia, and Polynesia. Many are rare and not always available in pet stores. They may be found only in the wild, in zoos, or in aviaries.

The Rainbow Lory. The rainbow lory, *Trichoglossus haematodus,* is one of the more common species. The forehead and crown of the bird are black. Yellow patches are found on each side of the neck and are connected by a band across the back of the neck. The chin is black; the feathers on the chest are red with black margins; and the back, wings, and belly are green. The thighs are yellow with green margins. The bill is orange, and the feet and legs are gray. See Figure 19–3.

The Subfamily Cacatuinae

The subfamily Cacatuinae is commonly referred to as cockatoos, and there are six genus groups. Members of this subfamily are easily recognized by the crest or tuft of feathers on the top of the head. Cockatoos are found in Australia and on islands of the South Pacific. Cockatoos and related species make excellent pets. Most are beautiful birds, tame easily, are considered intelligent, and learn to mimic words and sounds. Most cockatoos breed in captivity, although this may present a challenge.

Figure 19–3   Rainbow lory. Photo by Isabelle Francais.

The Palm Cockatoo. The genus *Probosciger* consists of one species—the palm cockatoo, *Probosciger aterrimus.* This is a completely gray-black cockatoo that reaches a length of 30 inches. It has a large beak and long crest feathers. It feeds on seeds and nuts. This species is rare in captivity.

The Black Cockatoos. The genus *Calyptorhynchus* consists of four species. They range in size from 23 to 28 inches in length. Because the primary color is black, they are referred to as black cockatoos. In the wild, they feed on grubs found in the bark and wood of trees. In captivity, they can be fed on seeds and nuts. Adult males have black beaks; the females have light-colored or white beaks. These large birds need large cages to be able to spread their wings and tails. Members of the genus group are the red-tailed black cockatoo, *C. magnificus;* the glossy black cockatoo, *C. lathami;* the white-tailed black cockatoo, *C. baudinii;* and the yellow-tailed black cockatoo, *C. funereus.*

The Gang-Gang Cockatoo. The genus group *Callocephalon* consists of a single species and two subspecies. *C. fimbriatum* is commonly referred to as the gang-gang cockatoo. It is about 14 inches long. The primary body color is gray, and the head is red or scarlet-colored. The crest of red feathers is curly and appears to be in disarray. The gang-gang cockatoo is rare in captivity.

The Greater Sulphur-Crested Cockatoo. The genus group *Cacatua* consists of about ten species and represents some of the most popular cockatoos. The greater sulphur-crested cockatoo, *C. galerita,* is the most common. See Figure 19–4. The bird is white with a bright yellow crest. The feet and beak are black. Males have dark brown to black eyes, and the females have a reddish tinge to their black eyes. These birds become very tame and affectionate and like a lot of attention. There are five subspecies.

The Lesser Sulphur-Crested Cockatoo. The lesser sulphur-crested cockatoo, *C. sulphurea,* is similar to the greater sulphur-crested cockatoo except it is smaller in size. They are about 13 inches long. Five subspecies are recognized.

The Umbrella-Crested Cockatoo. The umbrella-crested cockatoo, *C. alba,* is also referred to as the white-crested or greater white-crested cockatoo. This bird is white with a slight tinge of yellow under the wings and tail.

The Moluccan Cockatoo. The Moluccan cockatoo, *C. moluccensis,* is about 20 inches long. The primary

Figure 19–4   Greater sulphur-crested cockatoo. Photo by Isabelle Francais.

Figure 19–5   Moluccan cockatoo. Photo by Isabelle Francais.

body color is a salmon-pink underlayer of feathers with white outer feathers. The crest is salmon-pink, and the beak is black. The eyes of the male are black, and the female's eyes are black with a reddish tinge. See Figure 19–5.

The Philippine Cockatoo.   The Philippine cockatoo, *C. haematuropygia,* is a small cockatoo that is about 13 inches long. It is reddish to pink with a short, white crest.

Leadbeater's Cockatoo.   Leadbeater's cockatoo, *C. leadbeateri,* is one of the most common of the cockatoos. It is about 15 inches long, and the primary body color is white. The underside of the wings, abdomen, neck, chin, and sides of the face are light pink. The crest feathers have bands of red and yellow with white tips. This cockatoo makes an excellent pet and is a good breeder in captivity. Two subspecies are recognized.

The Bare-Eyed Cockatoo.   The bare-eyed cockatoo or Corella cockatoo, *C. sanguinea,* is considered one of the most intelligent of all birds. It is white with a short, white crest. Light yellow feathers are found un-

der the wings and tail, and a distinctive blue skin patch is found around and extends below the eye. The beak is white.

The Slender-Billed Cockatoo.   The slender-billed cockatoo, *C. tenuirostris,* is very similar to the bare-eyed cockatoo except it has a very long, slender upper beak.

Ducorps' Cockatoo and Goffin's Cockatoo.   Two other species are the Ducorps' cockatoo, *C. ducorps,* and Goffin's cockatoo, *C. goffini.* Both species are white with tinges of red and pink. They are about 11 inches long and are rare in captivity.

The Rose-Breasted Cockatoo.   The genus group *Eolophus* consists of one species and three subspecies. The rose-breasted cockatoo, *E. roseicapillus,* is another of the more popular species of cockatoo. The primary body color is gray, and the face, breast, and abdomen are light pink. The crest is also light pink. These birds are about 15 inches long. They may also be referred to as roseate cockatoos and Galah cockatoos. This species is considered mischievous and aggressive, but they do make excellent pets.

The Cockatiel. The genus *Nymphicus* consists of only one species, the cockatiel, *Nymphicus hollandicus.* Cockatiels are native to Australia but have become one of the most popular pet birds in the United States. These birds are about 12 inches long with the tail being about the same length as the body. They make very affectionate companions. Cockatiels are common in pet shops and are fairly reasonable in price. Although there is only one species, several color varieties are available. The gray or wild-colored cockatiel is the most readily available. White cockatiels are primarily white but have light yellow and peach markings; their eyes are red. White cockatiels with black eyes are also available. Cockatiels with primarily yellow coloring are referred to as **lutinos**. Pearled cockatiels are light gray in color with white or yellow spots over the body and wings. Bordered cockatiels are similar to the pearled variety but have dark margins on the feathers. The pied cockatiels or harlequin variety have symmetrical pied markings. The cinnamon cockatiel is a brownish color with the feathers becoming yellow at the tips, giving the bird a cinnamon appearance. See Figure 19–6.

Adult male cockatiels have bright yellow faces and crests with a bright orange patch on the cheeks. Females and immature birds have light yellow faces

Figure 19–6 Cockatiel. Photo by Isabelle Francais.

and crests. The cheek patches are a duller shade of orange, and yellow patches or markings may appear on the underside of the tail.

Cockatiels are considered ideal for beginning bird fanciers and for youngsters. They are easy to raise, breed readily in captivity, and make affectionate pets.

The Subfamily Psittacinae

The subfamily Psittacinae consists of more than 330 species, including macaws, conures, parrots, parakeets, parrotlets, and other birds. All members of the subfamily have four toes; the two center toes face forward, and the first and fourth toes point backward. The bills of these birds are large and are used in climbing, clasping, and feeding and for defense. Their heads are large in comparison to the rest of their bodies; they have short necks and a large, fleshy tongue.

Macaws. Macaws are large, beautiful birds found from Mexico to Paraguay and Southern Brazil. They are hardy, long-lived birds that may live for seventy-five years or more. Macaws have large beaks, long tails, and bare areas around the eyes and face. The beak of the macaw is the largest of the parrot family. The power in the bird's beak is tremendous. They could easily cut off a human finger if they chose to do so; however, the birds are gentle and affectionate. They are considered highly intelligent, are very playful, and have an average ability to mimic the human voice. The macaw's natural voice is very loud and harsh. There are three genus groups of macaws.

The Hyacinth Macaw. The genus group *Anodorhynchus* consists of three species. The hyacinth macaw, *A. hyacinthus,* is one of the most beautiful macaws. The primary body color is a deep hyacinth blue. Yellow skin patches are around the eyes and on the lower mandible. The beak and feet are black. The hyacinth macaw may reach a length of more than 34 inches. These birds are rare and are not readily available; prices may range as high as $9,000. See Figure 19–7. Two other species in this genus group are the glaucous macaw, *A. glaucus,* and the Lears macaw, *A. leari.* Both are smaller versions of the hyacinth macaw; however, they lack the bright, deep blue sheen and are only about 30 inches long.

The Blue and Gold Macaw. The genus group *Ara* consists of about thirteen different species. The blue and gold macaw, *A. ararauna,* is one of the more popular and commonly seen macaws. These macaws may reach a length of more than 30 inches. The neck and abdomen are a bright golden-yellow, and feathers on the crown are greenish-blue. Bright blue feathers run

The primary color is green, with the head usually being a darker green than the body. A yellow collar on the lower back part of the neck is the distinguishing characteristic. There is a yellow patch of feathers above the nostrils. A few red and blue feathers are found on the wings. Smaller macaws like the yellow-caped macaw are sometimes referred to as dwarf macaws or mini-macaws. These smaller macaws are usually lower priced than the larger macaws. Yellow-caped macaws are considered good talkers. Severe macaw, *A. severa,* is another of the smaller macaws; it may reach a length of up to 20 inches. The primary color is green, with a patch of dark brown above the beak. The crown is blue; lines of dark feathers appear on a large, white, bare cheek patch. The beak and feet are black. There are a few blue wing feathers, and red appears on the underside of the wings and tail. The bird is considered a fairly good talker and makes an affectionate pet. A second species, *A. severa castaneifrons,* is a little longer.

Illiger's Macaw. Illiger's macaw, *A. maracana,* is also a small macaw; it reaches a length of about 15 inches. The primary color is a dull olive green. There is a red-orange patch of feathers above the nostrils; the crown and head are blue-green. Some red feathers appear on the abdomen and wings. The upper tail

Figure 19–7 Hyacinth macaw. Photo by Isabelle Francais.

from the back of the head down the back; the wing feathers and tail feathers are blue. The underside of the blue tail feathers are tinged with yellow. The blue and gold macaw has a white cheek patch with rows of tiny black feathers. There is a black collar or bib under the lower mandible; the beak and feet are black. This species is much lower in price than the hyacinth macaw but still commands prices from $600 to $1,000. The blue and gold macaw is considered more alert and more intelligent than other macaws. It is also very curious and mischievous. See Figure 19–8.

The Scarlet Macaw. The scarlet macaw, *A. macao,* is sometimes referred to as the red and yellow macaw. The primary body color is a bright red or scarlet. The wings are red, but bright blue feathers and yellow feathers tinged with green are also found on the wings. The cheek patch is white with no rows of tiny feathers, and the upper mandible is white; the lower mandible is black. This macaw reaches a length of about 30 inches. It is one of the most commonly seen macaws but may command prices of more than $1,000.

The Yellow-Caped Macaw. The yellow-caped macaw (or yellow-collared macaw), *A. auricollis,* is a small parrot that reaches a length of about 16 inches.

Figure 19–8 Blue and gold macaw. Photo by Isabelle Francais.

feathers show a reddish tinge, and a yellow tinge is found on the underside of the feathers. The beak is black.

The Red-Bellied Macaw. The red-bellied macaw, *A. manilata,* is another of the small macaws and reaches a length of about 15 inches. The primary color is a dull olive green. The forehead and crown are black, changing to blue-green on the nape of the neck. The long flight feathers are blue with black tips. The upper side of the tail is tinged with red, and the underside is tinged with yellow. The beak is black, becoming lighter at the tip.

The Noble Macaw. The noble macaw, *A. nobilis nobilis,* is the smallest of the macaws. It reaches a length of about 14 inches. The primary color is a dull green. The forehead and crown show a bluish tinge. Some red appears on the wings, and the flight feathers are tinged with black. The underside of the wings are tinged in red, and the underside of the tail is tinged with yellow. A bare, white skin area exists between the eyes and the beak. Two other subspecies are recognized: *A. nobilis longipennis* and *A. nobilis cumanensis.*

The Red-Crowned Macaw. The red-crowned or red-fronted macaw, *A. rubrogenys,* is about 17 inches long. The primary color is a dull olive green. A bright red-orange patch appears on the forehead and crown. This bright red-orange also appears on the shoulders, **scapulars**, the small feathers that cover the base area,

Figure 19–9 Sun conure. Note that the long flight feathers have been trimmed on this bird. Photo by Isabelle Francais.

and the outer tips of the wing. The flight feathers are a dull blue; the underside of the tail is tinted gray. The upperside of the tail is tinged with blue and yellow. The feathers on the thighs are reddish-colored. There is a bare skin patch on the cheeks with lines of black feathers.

Other Macaws. Coulon's macaw, *A. couloni;* the great green macaw, *A. ambigua;* and Wagler's macaw, *A. caninde,* are three other species of rare macaws.

The Spix Macaw. The third genus group is *Cyanopsitta.* One species, the Spix macaw, *C. spixii,* is known. This is among the most rare species of macaws. Perhaps as few as twenty birds are known to exist in captivity; there may not be any left in the wild. The distinctive characteristic of this macaw is a head completely covered with feathers; there is no bare skin patch around the eyes or on the cheek like the other macaws. The primary color is a blue-green. The head is gray, and the beak is black. This macaw reaches a length of about 22 inches. Its tail is longer than that of other species of macaws.

Conures. Conures are about the size of the dwarf or mini-macaws. They are found in the wild from Mexico throughout Central America and most of South America. They differ from the macaws because they do not have a bare skin area around the eyes or cheek area; all macaws have a bare area to some extent. Conures are long and slender in shape and have large, tapered tails. They also have large heads and large, powerful beaks. Conures make very loud and harsh sounds. There are eight genus groups of conures; several are rare and unavailable in the pet market.

The Queen of Bavaria Conure. The genus group *Aratinga* consists of eighteen species; seven species are discussed here. The Queen of Bavaria conure, *A. guarouba,* is also referred to as the golden or yellow; conure. The primary color is a bright golden-yellow; the long flight feathers are green. The beak and feet are pink, and it has orange eyes. It reaches a length of about 14 inches. This beautiful bird is rare and very expensive.

The Sun Conure. The sun conure (or yellow conure), *A. solstitialis,* is considered by many to be the most beautiful and brilliantly colored of the conures. The primary color is a bright yellow-orange. The head shows more orange, especially around the eyes and in the cheeks. The wings are yellow with green and blue flight feathers. The eyes, beak, and feet are black. The sun conure is about 12 inches long. Although considered rare, it is available in the pet market. See Figure 19–9.

The Janday Conure. The Janday conure, *A. jandaya,* is one of the more popular conures. The head, neck, throat, and chest are a vivid mixture of red-orange; the entire head is yellow. The wings are primarily green, but the longer flight feathers are blue. The tail feathers are an olive green shading to blue and ending with black tips.

The Petz' Conure. The Petz' conure (or half-moon conure), *A. canicularis eburnirostrum,* is usually sold in the United States as a dwarf parrot; it is one of the most popular birds sold. The primary color is green; the upper parts are usually a darker green than the lower parts. The flight feathers are blue and green, and there is a bare yellow ring around the eye. There is an orange band on the forehead, and the crown is a dull blue. The Petz' conure becomes very tame and makes an excellent pet, except for its loud, harsh sound.

Wagler's Conure. The Wagler's conure (or red-fronted conure), *A. wagleri,* is a somewhat larger conure that reaches a length of about 14 inches. It is all green except for red on the forehead and crown. Some streaks of red appear on the throat; the beak is yellow.

The Cherry-Headed Conure. The cherry-headed (or red-headed) conure, *A. erythrogenys,* is about 13 inches long. The primary color is a bright green. The front part of the head is red, and red covers the eye and cheek area as well. There is a large, yellow ring around the eyes; some red also appears on the thighs. See Figure 19–10.

The Brown-Eared Conure. Eight subspecies of the *A. pertinax* are recognized. The common names of the various subspecies are derived from variations in color patterns. The primary color is a dull green, and they are not considered very attractive. The most common of the subspecies is the brown-eared conure, *A. pertinax ocularis.* It has brown on the sides of the face, cheeks, and throat. Other subspecies are the brown-throated conure, *A. p. aeruginosa;* the Margarita brown-throated conure, *A. p. margaritensis;* the Tortuga conure, *A. p. tortugensis;* the yellow-cheeked conure, *A. p. chrysophrys;* the Aruba conure, *A. p. arubensis;* the St. Thomas conure, *A. p. pertinax;* and the Bonaire conure, *A. p. xanthogenia.*

The Cactus Conure. The cactus conure, *A. cactorum,* is a small bird that is about 11 inches long. The forehead and crown are brown; the throat and chest are brown to olive brown. Green tinges the cheeks, and the wings are green. Flight and tail feathers are gray-blue. There is a white ring around the eye; the beak is yellow.

Figure 19–10 Cherry-headed (or red-headed) conure. Photo by Isabelle Francais.

The Golden-Crowned Conure. The golden-crowned conure, *A. aurea,* is about 10 inches long. There is a bright orange patch above the nostrils and on the forehead followed by blue on the crown, changing to green on the head. The throat and chest are yellow, tinged with green. The upper parts of the wings and tail are green with yellow on the underside of the wings and tail; the beak is black.

The Nanday Conure. The genus *Nandayus* consists of one species—the Nanday conure, *N. nenday.* The primary color is green, and the head is black. The flight feathers and the tail feathers are a blue-black. It has bright red thighs; the bill, legs, and feet are black.

The remaining genus groups of conures, *Leptosittaca, Cyanoliseus, Pyrrhura, Ognorhynchus, Microsittace,* and *Enicognathus,* are rare in captivity and are not discussed in this text.

Parrots. There are nineteen species of what are commonly referred to as the true parrots.

The African Gray Parrot. The genus group *Psittacus* consists of only one species—the African gray parrot, *P. erithacus erithacus.* The African gray parrots are native to Africa, where they may be found from the Ivory Coast to Angola, and inland to Kenya and

Tanzania; they are also found on islands near the Gold Coast. The primary color is gray. Various shadings of black and gray are found on the bird. Many of the feathers are tipped with white; the tail feathers are bright red. The beak is black, and the feet are a dusty black color. See Figure 19–11.

The African gray parrot is a beautiful bird, and many people consider it the best talker of all birds, with a voice most closely resembling a human's. They are also considered very alert and intelligent, and they make affectionate pets. The African gray parrot reaches a length of about 13 inches.

The Timneh Parrot. The Timneh parrot, *P. e. timneh,* is a smaller version of the African gray. It is a darker gray, and the tail is brown. They are long-lived birds, with a normal life span of twenty to twenty-five years, but they may live for up to seventy years. They sell from $600 to more than $1,000.

The Yellow-Headed Amazon. The genus *Amazona* is the largest genus group, consisting of about twenty-seven species. Most are considered loud and are not favorable pets; only the more common species are discussed. The Mexican double yellow head, *Amazona ochrocephala oratrix,* is also referred to as

Figure 19–11 ***African gray parrots. Photo by Isabelle Francais.***

Levaillant's Amazon or yellow-headed Amazon. It is one of the most popular of the Amazon parrots. The primary color is a bright green. The adult has a yellow head and neck; red appears on the shoulders, wings, and near the base of the tail feathers. They reach a length of 14 inches. These parrots are considered excellent talkers.

The Yellow-Naped Amazon. The yellow-naped Amazon, *A. o. auropalliata,* is primarily green. There is a yellow patch above the upper beak and a yellow band on the nape of the neck. The parrot has a bare eye ring. The tail feathers are green with a reddish tinge near the base and a yellow tinge near the tips of the feathers; some red and blue appear on the flight feathers. The beak is gray, becoming darker at the tip. It is considered a good talker.

The Panama Amazon. The Panama Amazon, *A. o. panamensis,* is an excellent talker. It is very similar to the yellow-naped Amazon but does not have the yellow band on the nape of the neck. The beak is yellow with a dark tip. Both parrots are about 14 inches long.

The Yellow-Fronted Amazon. The yellow-fronted Amazon, *A. o. ochrocephala,* resembles the Panama Amazon. The yellow patch above the upper beak is larger and extends up into the crown. The eyes are orange with bare eye rings. It is considered a good talker. It is sometimes referred to as the Colombian Amazon or single yellow head.

The Green-Cheeked Amazon. The green-cheeked Amazon, *A. viridigenalis,* is also referred to as the Mexican Red Head. The primary color is green. The flight feathers are red and blue; the head is red, and the red descends down to the eyes. A blue-green strip runs from the eye backward and down the back of the head; the cheeks are green. The bill is a yellow-white color. This parrot is about 13 inches long. See Figure 19–12.

The Finsch's Amazon. Finsch's Amazon, *A. finschii,* is also referred to as the lilac-crowned Amazon. The primary color is a dull green. There is a band of bright purple or plum above the upper bill connecting the eyes; the color becomes lighter as it moves up into the crown and top of the head. The cheeks are a bright green; the feathers on the chest and abdomen have dark margins. The Finsch's Amazon is about 13 inches long. It is considered a fairly good talker and makes a good pet.

The Spectacled Amazon. The spectacled Amazon, *A. albifrons,* is a small parrot that reaches a length of

Figure 19–12   Green-cheeked Amazon. Photo by Isabelle Francais.

about 11 inches. It is sometimes referred to as the red, white, and blue parrot because those three colors are found on the head. A red patch above the upper beak extends back to encircle the eyes; above this and extending up into the crown is white changing to blue. As with most of the Amazon parrots, the flight feathers are blue and green. The beak is yellow-white. Two subspecies are recognized: the lesser white-fronted Amazon, *A. a. nana,* and the Sonoran spectacled Amazon, *A. a. saltuensis.*

The Yellow-Lored Amazon.   The yellow-lored Amazon, *A. xantholora,* is very similar to the spectacled Amazon except the patch above the bill and between the eyes is yellow. It has a black cheek patch, and there is a red ring around the eye.

The Yellow-Cheeked Amazon.   The yellow-cheeked Amazon, *A. autumnalis,* is about 14 inches long. Like all Amazon parrots, the primary color is green. A bright red patch above the upper beak connects the eyes; the forehead and crown are light purple to lilac, changing to light blue on the head and back down the back of the head. The cheeks are bright yellow. The beak is yellow-white with darker edges and tip. It is considered a good talker. Three subspecies are recognized: Salvin's Amazon, *A. a. salvini;* Diademed Amazon, *A. a. diadema;* and Lesson's Amazon, *A. a. lilacina.*

The Blue-Crowned Amazon.   Four subspecies of *Amazona farinosa* are known. The blue-crowned Amazon, *A. farinosa guatemalae,* is considered the most colorful of the subspecies. The forehead and crown are a bright blue fading to a lighter blue down the back of the head and nape of the neck. All four species are about 16 inches long.

The Blue-Fronted Amazon.   The blue-fronted Amazon, *A. aestiva aestiva,* is primarily green. A blue patch above the upper bill extends into the forehead; the head has a mixture of blue and yellow feathers. The cheeks are yellow, and the flight feathers are red. The blue-fronted Amazon is considered an excellent talker. A subspecies, the yellow-winged Amazon, *A. a. xanthopteryx,* is very similar to the blue-fronted Amazon except that yellow appears on the wings, and there is a yellow band on the crown above the blue. The feathers on the neck, chest, and back have dark margins. See Figure 19–13.

The Orange-Winged Amazon.   The orange-winged Amazon, *A. amazonica amazonica,* is similar to the blue-fronted Amazon except that the blue on the forehead is a brighter color, and there is a tinge of orange in the yellow cheeks. There is a yellow circle on the crown. A subspecies, the Tobago orange-winged Amazon, is very similar.

Figure 19–13   Blue-fronted Amazon. Photo by Isabelle Francais.

The Festive Amazon.   The festive Amazon, *A. festiva festiva,* has a bright red rump. When the wings are folded, the red does not show. A light purple band is located above the upper bill, and the cheeks, throat, and crown are tinged with blue. A subspecies, Bodin's Amazon, *A. f. bodini,* is similar in appearance.

The San Domingo Amazon.   The San Domingo Amazon, *A. ventralis,* is about 12 inches long. It is also referred to as Salle's or the white-headed Amazon because it has a white forehead.

The Red-Vented Parrot.   There are eight species of the *Pionus* genus of parrots and about twenty-eight subspecies. The red-vented parrot, *P. menstruus,* is about 11 inches long. It is also referred to as the blue-headed parrot. The primary color is green. The head is blue, and the blue extends downward onto the neck and upper part of the breast. The vent area and tail **coverts** are bright red. The red-vented parrot makes a quiet and affectionate pet but is not considered a good talker.

The White-Crowned Parrot.   The white-crowned (or white-capped) parrot, *P. senilis,* is about 10 inches long. White appears above the upper beak and extends upward onto the forehead and crown; white also appears on the throat area. The chest and abdomen are a dull blue, and bronze coloring appears on the green wings. There is a bare ring around the orange eyes. See Figure 19–14.

Figure 19–14   White-crowned (or white-capped) parrot. Photo by Isabelle Francais.

The Bronze-Winged Parrot.   The bronze-winged parrot, *P. chalcopterus,* is about 11 inches long. The crown, cheeks, back of the neck, tail, and upper breast are dark blue-green. The wing feathers are bronze; specks of pink appear on the throat and breast. The vent area is red, and the bill is yellow-white.

The White-Bellied Caique.   The *Pionites* genus of parrots is commonly referred to as caiques. Caiques are considered very intelligent and can be amusing comics. They are about 10 inches long. The white-bellied caique, *P. leucogaster leucogaster,* is the most common of the caiques. It is also referred to as the white-breasted caique. The entire head of the bird is orange-yellow down to and including the neck; the breast area is white. The wings and tail are green, and there is a white ring around the eyes. The thighs of the bird are green; the yellow-thighed caique, *P. l. xanthomeria,* is similar to the white-bellied caique except that it has yellow thighs.

The Black-Headed Caique.   The black-headed caique, *P. melanocephala melanocephala,* has a black forehead and crown. Blue feathers appear on both sides of the upper beak. Below the black cap, the feathers around the neck are yellow tinged with some white; the thighs are yellow.

The Pallid Caique.   The pallid caique, *P. melanocephala pallida,* is similar to the black-headed caique except the thigh color is an apricot orange.

Hawk-Headed Parrot.   The genus *Deroptyus* contains two species, and they are referred to as hawk-headed parrots. The hawk-headed parrots are about 14 inches long and have a broad, square tail that is about 4 inches long. The forehead and crown are brown with streaks of light brown to white. The feathers from the top of the head and down the back of the neck are longer than normal; these feathers are maroon with bright blue tips. When the bird is excited or threatened, these long feathers on the back of the head and neck become erect, presenting a threatening pose. The breast of the bird also has maroon feathers with blue tips. The wings and rump are green. The upper tail feathers are a mixture of blue, green, and brown. The base of the tail feathers on males is red. The female does not have the red on her tail feathers. The beak is black. The birds are considered affectionate and make excellent pets. When feeding hawk-headed parrots, it is necessary to add fruits to their diet of nuts and seeds. Two species are recognized: the Guiana hawk-headed parrot, *Deroptyus accipitrinus accipitrinus,* and the Brazilian hawk-headed parrot, *D. a. fuscifrons.*

The Red-Capped Parrot. There are six species of the genus *Pionopsitta.* They are small parrots about 7 to 9 inches long. Considered rare, they are not readily available in the United States. One species, the red-capped or pileated parrot, *Pionopsitta pileata,* is the most common. It is primarily green with red on the cheeks and wings. The primary flight feathers are blue.

The Brown-Necked Parrot. The genus group *Poicephalus* consists of nine species. They range in size from 8 to 13 inches. They are native to the African continent. The brown-necked parrot, *P. robustus fuscicollis,* is primarily green with red on the thighs, outer edges of the wings, and shoulders. The forehead of the male is brown, and the forehead of the female is red. The feathers on the head and neck are silver, tinged with brown. Two other subspecies are recognized. All are rare and not readily available in the pet market.

Jardine's Parrot. Jardine's parrot, *P. gulielmi gulielmi,* is also referred to as a red-headed or red-crowned parrot. It is primarily green with a red crown, thighs, and shoulders. There are dark, almost black feathers on the wings and tail. The upper bill is yellow-white with a black tip; the lower bill is black. The orange-crowned parrot, *P. g. fantiensis,* is very similar to the Jardine's parrot except the markings are orange.

The Yellow-Fronted Parrot. The yellow-fronted parrot, *P. flavifrons flavifrons,* is primarily green with bright yellow on the top of the head and facial area. The tail and flight feathers are a dark green tinged with bronze; the eyes are orange.

The Senegal Parrot. The Senegal parrot, *P. senegalus,* is one of the most common of the African parrots. The primary color is green; the head is gray, and they have yellow eyes. Immature Senegal parrots have gray eyes. The breast feathers are tinged with yellow. There are three subspecies of the Senegal parrot: the scarlet-bellied Senegal parrot, *P. s. versteri;* the orange-bellied Senegal parrot, *P. s. mesotypus;* and the Senegal parrot, *P. s. senegalus.* Hand-reared Senegal parrots can be affectionate and make good pets. See Figure 19–15.

Brown's Parrot. Brown's parrot, *P. meyeri,* is also referred to as Meyer's parrot. The primary color of this parrot is brown. Yellow appears on the wings; the lower breast and abdomen are green. Some blue appears on the rump and tail feathers; the eyes are red. The beak is black, and there are six subspecies.

Figure 19–15   Senegal parrot. Photo by Isabelle Francais.

The Orange-Bellied Parrot. The orange-bellied parrot, *P. rufiventris,* is also referred to as the Abyssinian or red-breasted parrot. The upper parts of this parrot are gray; the underparts are blue-green. The throat is tinged with orange changing to bright orange on the breast.

Grand Eclectus Parrot. The genus *Eclectus* consists of one species and ten subspecies. These beautiful parrots are native to Australia, New Guinea, and the islands of the South Pacific. Eclectus parrots are about 12 to 14 inches long. Unlike most of the other parrots, the males and females show sexual dimorphism. The males are bright green with red patches on the sides of the breast. The upper side of the wings is green, and the underside is red; the upper side of the tail is green with blue and white tips, and the underside is a blue-black with yellow tips. The lower mandible is black, and the upper mandible is yellow-white with a pink-orange base. The female has a bright red head and breast; the red changes to a bright blue down the neck. The lower breast and abdomen are blue. The upper part of the wings is maroon while the underside is a mix of blue, maroon, and green. The upper side of the tail is maroon, changing to orange at the tips. The beak is all black and is long,

Figure 19–16   Grand eclectus parrot. The female is the more brightly-colored of the two. She has a bright red head and breast. The red changes to a bright blue down the neck. The male is green with red patches on the sides of the breast. The male is the least colored of the two. Photo by Isabelle Francais.

the beak is a black band that goes down the throat and then spreads out and back, making a black band below the cheek. There is a light blue band above the black, and a light rose band below the blue. The tail and flight feathers show some yellow; some blue tinting may appear on the nape of the neck and down the back. The beak is red; an orange ring surrounds the eye. Females are a duller green and lack the black, blue, and rose-colored rings. Young ringnecks are dull in color, and it may take two to three years before the males develop their beautiful color markings. Indian ringneck parakeets make excellent pets and are good talkers; ringnecks are being bred in blue and yellow forms. Ringnecks are fairly common and may be found in the pet markets for $150 to $500. See Figure 19–17. An African subspecies, the African ringneck, *P. k. krameri,* is smaller and lighter in color.

The Alexander Ringneck Parakeet.   The Alexander ringneck parakeet, *P. eupatria nidalensis,* is a larger version of the Indian ringneck. The species is about 20 inches long; several subspecies are recognized.

The Moustache Parakeet.   The moustache parakeet, *P. alexandri fasciata,* is also referred to as the banded

with the upper mandible showing more curve from the base to the tip than the upper mandible of other parrots. The feathers of both the male and female almost resemble fur. Eclectus parrots are considered very affectionate, make excellent pets, and are very good talkers. There are ten subspecies of the Grand Eclectus parrot, *Eclectus roratus.* Eclectus parrots are fairly rare and command prices from $600 to more than $1,000. See Figure 19–16.

There are several species of the genus group *Tanygnathus.* Most are rare and are seldom seen in captivity. The great-billed or large-billed parrot, *T. megalorhynchos,* may be kept in some breeding aviaries but seldom reaches the pet market. The primary color is green, and the beak is a bright red.

Parakeets.   The genus *Psittacula* is made up of the larger parakeets; the males all have red beaks.

The Indian Ringneck Parakeet.   The Indian ringneck parakeet, *P. krameri manillensis,* is one of the most beautiful of the larger parakeets. The birds are about 17 inches long with a long, tapering tail that makes up about half the body length. The primary color is pastel green. Males have a narrow, black band above the nostrils that connects the eyes; below

Figure 19–17   Indian ringneck parakeet. This is a lutino or yellow-colored ringneck. Photo by Isabelle Francais.

parakeet. It has a black band across the forehead connecting the eyes, and a black patch below the lower mandible that includes the chin and throat area; most of the body is green. The head is blue-gray, and the breast is reddish brown with tinges of violet; the beak is red. Seven subspecies are recognized. See Figure 19–18.

The Derbyan Parakeet. The Derbyan parakeet, *P. derbiana,* is similar to the moustache parakeet but is about 20 inches long. The primary color is green. A black band across the forehead connects the eyes, and a black patch is on the chin and throat; the head is blue-gray. The wings are green with a yellow wing patch. The tail is 7 to 9 inches long. The upper mandible of the male is red with yellow edges and tips; the lower mandible is black with yellow edges. The beak of the female is black. The female is not as colorful as the male.

The Plum Head Parakeet. The plum head parakeet, *P. cyanocephala,* is a beautiful bird, with the head being a plum or light purple color. A black band extends down from the lower mandible, curving backward and toward the nape of the neck. The primary body color is yellow-green. The wings are green with

Figure 19–18 Moustache parakeet. Photo by Isabelle Francais.

a maroon wing patch; the beak is yellow-white. Females have gray heads. The plum head parakeet is about 12 inches long. The slaty-headed parakeet, *P. himalayana himalayana,* resembles the female plum head parakeet.

The Malabar Parakeet. The Malabar (or blue-winged) parakeet, *P. columboides,* is another beautiful bird. The male is primarily gray. A black band completely circles the neck; a blue band is below the black band and changes to a light green. The wings and tail feathers are dark blue with a yellow tinge to the tips; yellow also appears on the vent and underside of the tail. Males have a red beak, and there is a tinge of green around the eyes. Females have a black head and do not have the collar.

The Barraband Parakeet. The genus group *Polytelis* is made up of three species. The Barraband parakeet, *P. swainsonii,* is a very beautiful bird with a long, tapering tail. The bird reaches an overall length of about 16 inches. The forehead, crown, chin, and upper throat area are bright yellow. A bright red patch appears on the lower throat area; the bill is yellow-white. The top of the head and down the back are green; the breast and abdomen are yellow-green. The wings are green with blue flight feathers; the tail is green with tinges of pink and black. The female does not have the red and yellow markings, and the overall colors are not as bright. They are considered very affectionate and make excellent pets.

The Rock Pebbler. The rock pebbler, *P. anthopeplus,* is about 17 inches long. The primary color is olive yellow, with the upper parts being darker than the underparts. There is a green wing patch, and the flight feathers are dark blue; some of the shorter secondary flight feathers are red. The upper tail feathers are shades of dark blue, olive, and black; the underside of the tail is tinged with red. The beak is red. The female is more of an olive green with duller shades of blue and red.

The Princess Alexandra Parakeet. The Princess Alexandra parakeet, *P. alexandrae,* is larger than the other two birds in this genus group. It is also referred to as the Princess of Wales, Princess parakeet, and Queen Alexandra's parakeet. The male is 18 inches long; the coloring is a mixture of soft pastels. The crown is blue, changing to bright green on the top of the head; this color changes to an olive green down the back. The wings and tail feathers are olive green; the throat is a reddish orange. The breast and abdomen are a light yellowish red. The female is duller and has shorter tail feathers.

The Crimson Rosella. The genus group *Platycercus* contains nine species of parakeets; most are referred to as rosellas. Only four species are common in the United States. Rosellas are recognized by a black shoulder patch and scalloped feathers on the back, which are black with the outer edges being of a different color. The crimson rosella, *P. elegans elegans,* is about 14 inches long. The head, neck, and underparts are crimson red. The upper side of the tail is dark blue and green, and the underside is light blue; the cheek feathers are blue. The scalloped feathers on the back are black with green and red margins. The rump is red, and there is a violet-blue coloring to the wing feathers.

The Red Rosella. The red rosella, *P. eximius eximius,* is smaller than the crimson rosella. The head, neck, and throat are bright red, which descends into the breast area in a V-shape; there is a white patch on the chin. The breast and abdomen are yellow. The scalloped feathers down the back are black with yellow margins. The flight feathers are dark blue; secondary flight feathers are lighter blue, and there is a red wing patch. The upper side of the tail feathers is greenish blue, and the underside is light blue; the vent is red. The male is a very bright color with the patterns very distinct, whereas the female is duller in color with less distinct color patterns.

The Stanley (or Western) Rosella. The Stanley (or Western) rosella, *P. icterotis icterotis,* is about 11 inches long. The head and underparts of the male are red; there is a yellow cheek and throat patch. The scalloped feathers on the back are black with green margins. The upper wing and tail feathers are blue, green, and black. The undersides of the wings and tail are green with a blue tinge.

The Blue (or Mealy) Rosella. The blue (or mealy) rosella, *P. adscitus adscitus,* is about 12 inches long. The head is yellow with tinges of white; there is a white chin and throat patch with a narrow, light blue band below the white. The scalloped feathers on the back and wings are black with yellow margins; the flight feathers are blue. The breast is yellow tinged with blue; the vent is red. The females are duller in color.

The Elegant Parakeet. The genus group *Neophema* is made up of seven species commonly referred to as grass parakeets. They are about 8 to 9 inches long. The elegant parakeet, *N. elegans elegans,* is primarily olive yellow in color with yellow shading on the face, abdomen, and underside of the tail. The flight feathers are dark blue. Females are duller in color. The other six species are less known and not easily obtained in the United States.

The Quaker Parakeet. The genus *Myiopsitta* contains one species—the Quaker parakeet, *M. monachus.* The primary color is gray. A green area extends from the eye back to the nape of the neck and down to the shoulders and back. The wings and tail are green; the undersides of the wings and tail are gray. The vent and thighs are green; the gray feathers on the breast show light gray margins. The bill is yellow-white. The Quaker parakeet is one of only a few parrot-like birds that builds a nest. In the wild, they live in colonies, and the nests may resemble large apartment complexes. The birds are about 12 inches long.

The Red Rump Parakeet. The genus group *Psephotus* is made up of five species; this group is fairly rare and not usually available in pet shops. The red rump parakeet, *P. haematonotus,* is primarily green with a red rump and yellow on the chest and lower abdomen. The females are duller in color. Although not one of the most attractive birds, the red rump parakeet makes an excellent pet. See Figure 19–19.

The Many Color Parakeet. The many color parakeet, *P. varius,* is about 11 inches long. The primary color is green. There is a yellow patch on the forehead, a red patch on the nape of the neck, and a yellow patch on the shoulder. The long flight feathers are blue, and there is a red patch on the green wings. The upper

Figure 19–19 ***Red rump parakeet. Photo by Isabelle Francais.***

side of the tail is green, and the underside is yellow. Flecks of red appear on the green breast and the abdomen, and the vent area is red. The female is duller in appearance.

The Budgerigar. The genus *Melopsittacus* is made up of one species—the budgerigar, *M. undulatus.* The "budgie" is the most popular pet bird in the world. This small bird is about 7 inches long. Most people refer to the budgie as a parakeet. The budgerigar is native to Australia, where the aborigines used them for food. The term *budgerigar* in Aborigine means "good bird" or "good food." The wild budgerigar is primarily a yellowish green. Brown lines are found on the head feathers and the feathers on the nape of the neck. These lines get wider on the back and wings; the underside is pale green. The throat is pale yellow with several small, blue spots and larger black markings; numerous color mutations are now available. The bill and legs are brown. Budgerigars are ground feeders and, in captivity, drop to the floor of cages to feed. They learn to talk if properly taught. It may be difficult to determine the sex of young birds. Adult males have a bright blue **cere** (the fleshy portion of the beak that surrounds the nostrils), and on adult females, the cere is pink, light blue, or brown. Budgerigars are easy to tame, can be taught to talk, are easy to care for, make excellent companions, and are fairly inexpensive. Budgerigars normally live for about 4 to 6 years. See Figure 19–20.

Figure 19–21   Peach-faced (or rosy-faced) lovebird. Photo by Isabelle Francais.

Lovebirds. The genus group *Agapornis* are referred to as lovebirds and are native to Africa. There are nine species; three species are fairly common. Lovebirds are very hardy and long-lived. They make excellent pets and tame readily if obtained at a young age; older birds may become aggressive. Determining the sex of lovebirds is difficult.

The Peach-Faced Lovebird. The peach-faced lovebird (or rosy-faced lovebird), *A. roseicollis,* is the largest species of this group. The primary color is a soft, pastel green. Rose or peach color makes up the forehead and crown, encircles the eyes, and extends down to the chin, cheeks, and throat. The wings and the long, central tail feathers are dark green; the upper tail coverts are blue. The beak is yellow. See Figure 19–21.

The Black-Masked Lovebird. The black-masked lovebird, *A. personata,* is another very popular species. The head is black with a white ring around the eye. A yellow-orange patch appears on the throat; a faint yellow band encircles the neck. The rest of the body is primarily green. The wings and central tail feathers are a darker green; the underside of the tail is yellow-green. The beak is a bright red. The blue-masked lovebird is a mutation of the black-masked lovebird; the color pattern is the same except the green is replaced with blue, and the yellow-orange is replaced with an off-white to gray. The beak is light pink.

Figure 19–20   Budgerigar. Photo by Isabelle Francais.

Fischer's Lovebird. Fischer's lovebird, *A. fischeri,* has similar markings to the black-masked lovebird. The head is bright orange-red; the beak is bright red, and there is a white ring around the eye. The rest of the body is primarily green. The rump, tail feathers, and flight feathers are tinged with blue.

The Nyassaland Lovebird. The Nyassaland lovebird, *A. lilianae,* resembles the Fischer's lovebird but is smaller in size; it is about 5 to 6 inches long. The Nyassaland is considered a more pleasant bird than the other lovebirds. Its voice is less harsh, and it is not as aggressive.

Parrotlets. The genus group *Forpus* are native to South America. They are referred to as parrotlets. They are very similar to lovebirds but are not as attractive. They are not as popular because they are not as colorful and not as easily tamed as the lovebirds.

Hanging Parakeets. The genus group *Loriculus* is made up of a group of small, agile birds referred to as hanging parrots or hanging parakeets. Unlike other birds, they hang upside down when roosting; they have long claws that interlock to prevent them from falling. They are 5 to 6 inches long and have short tails. They feed on nuts and nectar. The blue-crowned hanging parakeet, *L. galgalus,* is the most common species. It is primarily green with a bright blue crown and black beak. A red patch covers the rump and descends down into the upper tail coverts. The male has a red and yellow throat patch.

CLASSIFICATION— WOODPECKERS (TOUCANS)

The order Piciformes consists of woodpeckers and related birds; there are six families in the order. Most species in this order are strictly arboreal, rarely descending to the ground. They feed on fruits, vegetation, and insects. Only the toucans are kept as pets.

MAJOR FAMILY— WOODPECKERS (TOUCANS)

The Family Ramphastitidae

Toucans belong to the family Ramphastitidae and are only occasionally available in pet shops. There are forty species of toucans, the most common being the Toco toucan, *Ramphastos toco;* Swainson's toucan, *Ramphastos swainsonii;* and the channel-billed toucan, *Ramphastos vitellinus.* These are large birds that have about the same sized body as a macaw. The most distinctive feature of the toucan is its large bill, which can be almost as long as the bird's body.

The bill, despite its enormous size, is light in weight because of a honeycomb of fibers within the solid outer shell. The exact purpose for such a large bill is not fully understood. The bird feeds primarily on fruits but will also feed on insects and lizards. Toucans are native to the tropical forests of Central and South America; the birds roost in holes in trees. Its tail has an unusual muscular and bony attachment that allows the tail to be folded over the bird's back. With the tail folded over the bird's back, the bird can fit more easily into roosting holes; toucans are usually very noisy. Prices for birds that are available may run more than $2,500.

The Toco Toucan. The Toco toucan is primarily black with a large white throat and chest patch. The bill is yellow with a black band around the bill where it joins the head; the bill may have bright orange and blue markings. They have blue surrounding their large, black eyes.

Swainson's Toucan. Swainson's toucan is also black but has a yellow chin and chest patch. The upper mandible of the bill is yellow, with brown on the lower part of the upper mandible; the lower mandible is brown, and the tip of the upper mandible is black. There is a green circle around the large black eyes.

The Channel-Billed Toucan. The channel-billed toucan is black with a white throat patch. There is a band of orange below the white; it has a blue band surrounding its black eyes. The upper part of the upper mandible is yellow; the lower part of the upper mandible and the lower mandible are brown. There is a white band around the bill where it joins the head. The upper tail coverts are yellow and orange.

CLASSIFICATION— PERCHING BIRDS

The order Passeriformes includes the perching birds. Of the approximately 9,000 species of birds, 5,100 are classified as perching birds; these are further classified into sixty-five families. Perching birds have three toes that point forward and one toe that points backward; the toes are not webbed. Perching birds are also noted for their pleasing songs. Of the sixty-five families, only seven are discussed in this text.

MAJOR FAMILIES—PERCHING BIRDS

The Family Zosteropidae

The family Zosteropidae consists of eighty-five species classified into eleven genus groups; the family is commonly referred to as white eyes and related species. They are native to Africa, Asia, New Guinea, and Australia. White eyes are small, only about 2½ inches long; the eyes have a ring around them. They feed primarily on nectar from flowers. In captivity, they feed on liquid nectars, fruits, peanut butter, cakes, mynah meal, or mynah pellets; canned milk should also be provided with the nectar. Some species in this family are the gray-breasted white eye, *Zosterops lateralis;* cinnamon white eye, *Hypocryptadius cinnamomeus;* and the Norfolk white-throated white eye, *Zosterops albogularis.*

The Family Emberizidae

The family Emberizidae is represented by one subfamily; the subfamily Emberizinae is commonly referred to as buntings. There are 281 species classified into sixty-nine genus groups. Many buntings are native to the United States and cannot legally be captured or kept in cages. Two species that may be available, however, are the rainbow bunting and the red-headed bunting, *Emberiza melanocephala.*

The Rainbow Bunting. The rainbow bunting is native to Mexico. This bird is about 5 inches long. The upper parts are blue, and the underparts are yellow; some orange specks appear on the chest. The forehead and crown are bright green; there is a yellow ring around the dark eyes. Females are dull in comparison to males.

The Red-Headed Bunting. The red-headed bunting is native to India; it is about 7 inches long, and the primary color is brown. The head is a rich brown, and the color changes to a lighter brown down the nape of the neck and into the mantle. The wings are brown with lighter brown edges to the feathers. The tail is brown with tinges of gray and black. The underparts of the bird are dull yellow. Females are lighter in color with tinges of green and black.

The Family Sturnidae

The family Sturnidae is commonly referred to as starlings. There are 106 species classified into twenty-two genus groups.

The Talking Mynah. One member of this family is the popular talking mynah, also known as the Hill mynah or Indian grackle, *Gracula religiosa.* The mynah bird is native to Asia and India. It is primarily black. There is a white patch on its wings; the bill is orange. A bare, yellow skin patch extends from below the eye back toward the nape of the neck; its legs are yellow. It feeds primarily on fruits. The mynah bird is noted for its ability to mimic the human voice and other sounds.

Keeping mynah birds can be more time consuming than keeping other species of birds because of their diet. They consume fruits, and because of this, their cage needs to be cleaned at least once a day. Mynah birds command prices from about $300 to $500.

The Family Ploceidae

The family Ploceidae consists of three subfamilies, eighteen genus groups, and 143 species. The subfamily Viduinae are referred to as whydahs. There are nine species of whydahs in the genus group *Vidua.* Whydahs are native to areas of Africa south of the Sahara desert, where they are also called widow birds. Whydahs go through a period of about six to eight months when they have bright colors, and another period when their colors fade to duller shades. The diet of these birds consists primarily of seeds.

Several species are rare and unavailable in the pet market. Species of whydah that may be available are the paradise whydah, *Vidua paradisaea,* and the pintailed whydah, *Vidua macroura.*

The Paradise Whydah. The paradise whydah is primarily black with two long, black central tail feathers that may reach 14 inches in length. A reddish brown band encircles the neck, and the color extends down onto the chest. Two black bands extend from the crown back down the back of the head. Females are duller in appearance.

The Pintailed Whydah. The pintailed whydah is smaller than the paradise whydah. Its central tail feathers are only about 9 inches long and are very narrow. The primary color is white; the crown, wings, and tail are black. There is a white band across the flight feathers; the bill is red.

The Red-Collared Whydah. The red-collared whydah has a body that is about 4 inches long, and a tail that is 6 to 8 inches long. The primary body color is black. Brown shadings appear on the margins of the wing feathers, and a red-orange band encircles the neck; the bill is black. Several subspecies of the red-collared whydah may also be available in the pet market. Among these are the red-shouldered whydah, the yellow-backed whydah, the white-winged whydah, and the yellow-shouldered whydah.

The Orange Weaver. The weavers make up the subfamily Ploceinae. There are ninety-four species classified into seven genus groups; all are native to Africa. Weavers also go through color phases. The orange weaver, *Ploceus aurantius,* is bright orange. The head, chest, and lower abdomen are black; red tinges are found on the nape of the neck. The wings, tail, and rump have a brownish tinge.

Masked Weaver. Masked weavers are occasionally available in the pet market. Most are black with yellow markings on the face; their wings are brown with lighter margins. Among the masked weavers are the little-masked weaver, the half-masked or vitelline weaver, and the rofous-necked weaver.

The Red-Billed Weaver. The red-billed weaver is primarily light red with various shades of red on the head, neck, chest, and abdomen. A black patch covers the face, throat, and forehead; the beak, feet, and legs are red. They are very aggressive and should not be caged with other small birds. Two other species of weavers are the Baya weaver and the scaly crowned weaver.

The Family Pycnonotidae

The family Pycnonotidae consists of 118 species of bulbuls classified into sixteen genus groups. These small birds are native to Africa, the Middle East, India, Southern Asia, the Far East, Java, and Borneo. They are soft-billed birds that feed on nectar, fruits, and mynah pellets. Species that may be available are the black bulbul, *Hypsipetes madagascariensis;* the black-headed bulbul, *Pycnonotus xanthopygus;* the brown-eared bulbul, *Hypsipetesam aurotis;* the common bulbul, *Pycnonotus barbatus;* the finch-billed bulbul of the genus *Spizixos;* the pale olive-green bulbul, *Phyllastrephus fulviventris,* the red-vented bulbul, *Pycnonotus cafer;* and the red-whiskered bulbul, *Pycnonotus jocosus.*

The Family Fringillidae

The family Fringillidae consists of several canaries, finches, siskins, and grosbeaks. Family members have nine instead of ten large, primary feathers in each wing; they also have twelve large tail feathers. Females of the family build a cup-like nest and incubate the eggs. There are three subfamilies.

The subfamily Fringillinae consists of one genus group, *Fringilla,* and three species. The three species are the brambling, *F. montifringilla;* the Canary Islands chaffinch, *F. teydea;* and the chaffinch, *F. coelebs.*

The subfamily Drepanidinae consists of fifteen genus groups and twenty-eight species. All species are native to the Hawaiian Islands and are either endangered, threatened, or thought to be extinct.

The subfamily Carduelinae consists of 122 species in seventeen genus groups. Species in the subfamily are native to North and South America, Europe, Asia, and Africa. A few species are available in the pet market.

The Common Canary. One species that is very important as a pet is the common canary, *Serinus canarius domesticus.* The early history of the canary is not known, but it is believed they were native to the Canary Islands and other islands off the northwestern coast of North Africa. Spanish sailors and soldiers are believed to be responsible for bringing the first canaries to Europe as early as the 1500s. The Spanish controlled the sale and distribution of canaries for about 100 years. As the birds became available to the French, Dutch, German, and English birdkeepers, more research and development took place. Several new varieties were developed. Today, more than twelve different varieties in more than fifty different color combinations are available. Development in the breeding of canaries was concentrated into three areas: the physical appearance of the bird, the coloration, and the singing ability.

The Gloster Fancy Canaries. A physical development that has become popular is the "crested" canary. The feathers on the top of the head fan out to form a crest or fringe of feathers that hangs out over the eyes and beak of the bird. These birds are known as the Gloster fancy canaries, and two types exist: One type is known as the "corona" and has the crest, and the other is known as a "consort" and has normal head feathering. When mating Gloster fancy canaries, it is important to mate a corona with a consort. If two corona types are mated, 25 percent of the offspring will die in the shell or shortly after hatching.

Red-Factor Canaries. The most important color varieties are those carrying the red color factor. Birds that carry red are usually more expensive; the more red, the more expensive they are. These birds are known as red-factor canaries.

The Roller Canary. Canaries are also popular because of their singing ability. The best known for its singing ability is the roller canary. Only the male canary sings, and the songs differ among birds.

Varieties of canaries that may be available include the lizard canary, Yorkshire canary, clear fife canary, Norwich canary, cinnamon canary, frilled canary, Belgian canary, Scotch fancy canary, and Lancashire canary.

The Family Estrildidae

The family Estrildidae consists of grass finches, waxbills, and mannikins. They are separated from the family Fringillinae because they have ten large primary feathers. There are approximately 124 species in twenty-seven genus groups. These are primarily ground-feeding birds, and their main food consists of seeds.

The Zebra Finch. The zebra finch, *Poephila quttata,* is found wild over most of the continent of Australia, where it is as common as the sparrow is in the United States. They are active during most of the day and are very curious. Zebra finches are social, getting along well with other birds and people. They are considered very hardy and may live for several years. They are probably the most widely kept and bred finch in captivity.

The Wild (or Normal) Finch. The wild (or normal) finch has an orange beak and orange cheek patch. A white patch extends from the top front of the eye downward to the chin; a narrow black "tear" marking extends downward from the eye. The forehead, crown, top of the head, and nape of the neck are gray. The upper chest is black and white striped with zebra-like markings; there is a black chest patch. The underside is white, and the tail is black with white bars. The sides below the wings are reddish-brown with white spots. The females lack the zebra mark-

Figure 19–23   Several Bengalese (or society) finches. Photo by Isabelle Francais.

ings, cheek markings, and side marking; they are also lighter in color. See Figure 19–22.

Several other color patterns are recognized. Among them are the white, fawn, dominant silver or dilute normal, dominant cream or dilute fawn, pied, chestnut flanked white, penguin, recessive silver or dilute normal, recessive cream or dilute fawn, and yellow-beaked varieties.

The Bengalese (or Society) Finch. The Bengalese (or society) finches, *Lonchura striata domestica,* are probably derived from the white-backed munia, *Lonchura striata.* These small birds are about 4 inches long. The white-backed munia is found in India and Ceylon. It is primarily brown with very dark, almost black, head and breast; the rump and underside are white. See Figure 19–23.

There are about eight different varieties of the domesticated Bengalese or society finch. The categories are self chocolate, pied chocolate, dilute chocolate and white, white with brown eyes, self fawn, pied fawn, dilute fawn and white, and white with ruby eyes.

The Owl Finch. The owl finch, *Poephila bichenovii,* is about 4 inches long. It is basically a gray-brown color with various shades of gray-brown on the body, wings, and tail. A dark ring surrounds the face similar to that of a barn owl.

The Black-Throated Finch. The black-throated finch, *Poephila cincta cincta,* is primarily a cinnamon-brown color. The forehead, crown, cheeks, and nape of the neck are gray. The beak is black, and a black patch extends down from the lower beak to the chin and throat area; the wings are brown. The tail feathers are black, and the upper tail coverts are

Figure 19–22   Wild (or normal) finch. Photo by Isabelle Francais.

white. The eyes are brown, and the area in front of the eye (lores) is black; the legs and feet are flesh-colored. The black-throated finch can be aggressive to other birds and may need to be kept separated. They are about 4 inches long.

The Long-Tailed Finch. The long-tailed finch, *Poephila acuticauda acuticauda,* is very similar to the black-throated finch except it is more slender and is about 7 inches long. The beak is black, and there is a black chin and throat patch; the lores are also black.

The Double-Bar Finch. The double-bar finch, *Poephila bichenovii bichenovii,* is primarily dark brown. The lores, the area around the eyes, cheeks, chin, and throat are white with a black band bordering the white. The chest is white, and another black band runs across the lower chest; the abdomen is white tinged with yellow and brown. The wings are black with white spots, and the tail is black. The beak is blue-gray, the eyes are brown, and the legs and feet are gray.

The Shafttail Finch. The shafttail finch, *Poephila acuticaudata,* is primarily brownish gray. The forehead, crown, head, and nape of the neck are gray. There is a large black chin and throat patch; the lores are black. The beak is orange, and the wing and tail feathers are brown. Some white and black may appear on the rump and tail. It is difficult to distinguish sex of males and females.

The Parson Finch. The Parson finch, *Poephila cincta,* is very similar to the shafttail finch except it has a black beak and the tail is shorter. In addition to their normal diet of seeds, they also feed on mealworms.

The Masked Finch. The masked finch, *Poephila personata personata,* is primarily a light cinnamon-brown color. The beak is yellow. A triangular-shaped, black patch extends from the eyes forward to the beak. There is also a black chin and throat patch; a large black patch is also located on the flanks. The tail is black and the underside of the tail is white; the legs and feet are red. The masked finch is about 5 inches long.

The Diamond Firetail Finch. The diamond firetail finch, *Emblema guttata,* is primarily brown. The beak is red; and the lores are black. The forehead, crown, top of the head, and nape of the neck are gray. There is a white chin and throat patch. A broad, black band extends across the upper chest and down the sides. The lower chest and abdomen are white. The wings are brown, and the long tail feathers are black; the upper tail coverts are red. The eyes are brown, and there is an orange ring around the eyes.

The feet and legs are gray. The diamond firetail finch, sometimes referred to as a diamond sparrow, is about 5 inches long.

The Painted Finch. The painted finch, *Emblema picta,* is only about 4 inches long. The beak is long and pointed; the upper part of the beak is black with a red tip, and the lower beak is red with blue at the base. The forehead, lores, chin, throat, and area around the eyes are scarlet. An irregular scarlet patch runs down the center of the chest and abdomen. The crown, nape of the neck, and upper parts of the bird are brown. The wing feathers and tail feathers are brown; the underparts of the bird are black with numerous white spots. The feet and legs are flesh-colored.

The Red-Browed Finch. The red-browed finch, *Aegintha temporalis,* is about 5 inches long and has many different colors, although the primary color is gray. The beak is red with black. A red patch extends from the beak backward above and past the eye. The forehead, crown, nape of neck, cheeks, chin, throat, and lower portions of the bird are gray. The wings are an olive yellow, and the underside near the vent is white. The tail feathers near the base of the tail are red. The long tail feathers are gray. The feet and legs are flesh-colored.

The Plumb-Headed Finch. The plum-headed finch, *Aidemosyne modesta,* is about 4 inches long. The primary color is brown. The feathers on the back, shoulders, wings, and tail are dark brown with white markings. The feathers on the underside of the bird are lighter brown with white stripes or bars. The long tail feathers are black, with the two outer tail feathers having white tips. The forehead, crown, and top of the head are red to violet (plum-colored). Ear coverts, cheek, and chin feathers are white with slight brown markings. The eyes are black with a white ring around them. The upper beak is black, and the lower beak is black with white at the base. The feet and legs are flesh-colored.

The Chestnut-Breasted Finch. The chestnut-breasted finch, also called a chestnut-breasted mannikin, *Lonchura castaneothorax,* is about 5 inches long. The colors of this finch are various shades of brown. The beak is blue-gray. The forehead, crown, and nape of the neck are gray-brown with darker brown markings; the lores, area around the eyes, chin, and throat area are dark brown. The upper chest is a chestnut color; a black band runs below this and up to the shoulders. The areas below the black band and the abdomen are straw yellow; the wings and tail feathers are light brown. The feet and legs are gray-brown. Other related species are the pictorella finch, *Lon-*

chura pectoralis, and the yellow-rumped finch, *Lonchura flaviprymna.*

The Gouldian Finch. The Gouldian finch, *Chloebia gouldiae,* is recognized in three color forms. The three forms are distinguished by the black, red, or yellow coloration on the bird's head. Young finches are a dull green above and gray below. The young begin to molt at about three months of age and complete the process at six months. The adults are very brightly colored birds that reach a length of about 6 inches. See Figure 19–24.

The red-headed form of the Gouldian finch is primarily a grass-green color on the upper body, including the upper wings; the primary wing feathers and outer secondary feathers are brown. The rump and upper tail coverts are bright blue; the long tail feathers are black. The lores, forehead, crown, cheeks, and ear coverts are red. A narrow, black band borders the red and extends down and widens into a black chin and throat patch; a narrow, blue band borders the black. The chest area is light purple, and the lower chest and abdomen are yellow. The beak is gray-white with a red tip. The feet and legs are yellow. The black-headed form is very similar to the red-headed form, except that the lores, forehead, crown, cheeks, chin, and throat area are black. The yellow-headed form is very similar to the other two forms except the lores, forehead, crown, cheek, chin, and throat are yellow.

Some color variations in the Gouldian finch have been developed; some may have white chests, whereas others have a blue upper body and wings. All three forms breed indiscriminately. Female Gouldian finches usually have the same color patterns but are duller in appearance.

Temperature is very important in the survival of Gouldians. Gouldian finches should be kept at minimum temperature of 72°F.

The Crimson Finch. The crimson finch (or crimson waxbill), *Neochmia phaeton,* is primarily brown. The beak is red with white at the base of the lower mandible. The forehead, crown, back of neck, and shoulders are gray-brown, turning to a purple-red down the back; the upper tail coverts are crimson. The lower chest, abdomen, and under tail coverts are black. The lores, cheeks, ear coverts, and chest are crimson with white spots; the feet and legs are yellow. Females have more brown on them.

The Star Finch. The star finch (or rufus finch), *Bathilda ruficauda,* is primarily olive brown on the upper parts with darker brown on the wings; the beak is red. The lores, forehead, and cheeks are crimson; the crown and nape of the neck are light yellow and white. The feathers from the throat down the chest

Figure 19–24   Gouldian finch (red-headed). Photo by Isabelle Francais.

and abdomen are yellow with white spots. The tail feathers are brown with a crimson tinge; the feet and legs are yellow. The star finch is about 4 inches long.

The Cordonbleu Finch. The cordonbleu, *Uraeginthus bengalus,* is native to Africa. This finch is about 4 inches long. The upper parts are brown, and the lores, chin, cheek, chest, and abdomen are blue. The male has red cheek patches. Other related species are the blue-breasted cordonbleu, *U. angolensis,* and the blue-capped cordonbleu, *U. cyano-cephala.* Both are very similar, except the male blue-breasted cordonbleu lacks the red cheek patch, and the blue-capped cordonbleu has more blue, especially on the crown.

The Blue-Faced Parrot Finch. The *Erythrura* genus of finches are commonly referred to as parrot finches. All are native to countries and islands of the Southwest Pacific area. Among the species that sometimes become available is the blue-faced parrot finch, *Erythrura trichroa.* The primary color is green. The beak is black, and the lores, forehead, and cheeks are blue. The tail feathers are red, and the feet and legs are light brown. The blue-faced parrot finch is about 4 inches long.

The Red-Headed Parrot Finch. The red-headed parrot finch, *Erythrura psittacea,* is primarily green. The forehead, crown, cheeks, chin, and throat are red. The tail feathers are red, and the beak is black; the feet and legs are brown.

The Pintailed Nonpareil. The pintailed nonpareil, *Erythrura prasina,* is primarily green on the upper parts of the body. The forehead, chin, throat, and cheeks are blue. The rump and upper tail coverts are red; the center tail feathers are long and form a point. The chest is red, and the lower chest and abdomen are a duller shade of red. The beak is black, and the feet and legs are light brown. The pintailed nonpareil is about 5 inches long.

Other *Erythrura* finches that may be found in aviaries are the blue-green parrot finch, *E. tricolor;* the short-tailed parrot finch, *E. cyaneovirens;* the Mindanao (or many-colored) parrot finch, *E. coloria;* Kleinschmidt's (or the pink-billed) parrot finch, *E. kleinschmidti;* the green-faced parrot finch, *E. viridifacies;* the Papua parrot finch, *E. papuana;* and the green-tailed (or Bamboo) parrot finch, *E. hyperythra.*

The Red-Eared Waxbill. The genus group *Estrilda* is referred to as waxbills. They are small, nervous, alert, and very active birds; most are native to Africa. The red-eared waxbill, *E. troglodytes,* is one of the most popular. The primary color is light brown with a light red tinge; the undersides are darker brown. The beak is red, and the red color extends back to and around the eyes. The tail and rump area are black. The red-eared waxbill is about 3½ inches long.

The Orange-Cheeked Waxbill. The orange-cheeked waxbill, *E. melpoda,* is primarily brown. The beak is red, and there is a large orange cheek patch. The rump is red, the tail is black, and there is orange on the lower abdomen. The orange-cheeked waxbill is about 3½ inches long.

The Lavender Waxbill. The lavender waxbill, *E. caerulescens,* is primarily a pearl-gray color except for a red rump and tail. White spots are scattered on the flanks and sides. The lavender waxbill is about 3½ inches long.

The Gold-Breasted Waxbill. The gold-breasted waxbill, *Amandava subflava,* is only about 3 inches long. The upper parts are primarily brown. The beak is red, the head and throat are red, the rump is orange, and the tail is black. The chest of the males is orange.

The Melba Finch. The melba finch (or waxbill), *Pytilia melba,* is larger than the other waxbills; it is about 5 inches long. The primary color is gray. The beak is red, and the head and throat are red. The lower crown is gray, changing to dark olive green down the shoulders, back, and rump. The upper tail coverts are red, and the wings are olive green. The throat and chest is an olive yellow with scattered white spots. The melba finch and other members of the *Pytilia* are insectivorous and should be supplied with a variety of small insects.

HOUSING AND EQUIPMENT

A wide variety of cages are available at pet shops and variety stores. Equipment used for the large parrot-type birds must be constructed of very heavy materials; wood and plastics cannot be used for parrot cages because they will shortly be destroyed. Heavily constructed metal cages must be used. The cage should be inspected for sharp points or edges that can cut or injure the bird's tongue or feet. Secure latches or locks should be used; these birds are very curious and ingenious and may soon learn to open their cage and escape.

If one choses to construct a cage for macaws, the wire size should be ½-inch by 3-inch wire mesh screen in 12- to 14-gauge wire. Smaller mesh wire may be used for outside cages if mice, rats, or other birds are a problem. Finches should have rectangular cages to allow for long horizontal flight. Flying in a circular pattern is unnatural to the finch and may cause undue stress.

Most new cages are equipped with plastic perches. These are hard and may be uncomfortable for many birds. If birds refuse to perch, one should suspect something is wrong with the perch. Size is also important. Wood makes the best perches. Parrots will soon destroy wood perches, and the perch will have to be replaced. Wood does give the parrot the opportunity to exercise its beak, which helps keep the beak trim and the bird busy. Limbs and tree branches can also be used for perches; one must make sure the limbs or branches are clean and free of molds or pesticides. If the perch tapers from end to end, this will give the bird the opportunity to find the most comfortable location. The ideal perch size for parrots is a 1-inch square perch. Budgerigars prefer an oval-shaped perch about ½ inch in diameter. Finches and canaries like round perches about ½ inch in diameter. See Figure 19–25.

Water containers should be made of glass, ceramic materials, or stainless steel. These materials are hard and are easy to clean. Gravity-type waterers are excellent for birds; they hang on the outside of the cage, and a metal tube extends into the cage. The water is released on demand when the bird touches the end of the tube.

The cage should be placed in an area of the room that is free from drafts, where constant temperature is maintained, not in direct sunlight, and

Parrot/2.5 cm
(1 in) square

Finch/1.25 cm
(0.5 in) round

Budgie/1.25 cm
(0.5 in) oval

Figure 19–25   Ideal perch size and shape for birds.

away from poisonous plants or other hazards the bird can reach.

To prevent caged birds from becoming bored, toys may need to be provided for them. The only thing that can be put in with large parrot-type birds is stainless-steel chains with bells; anything else will be destroyed. Mirrors, chains with bells, and ladders can be put in with smaller birds such as budgerigars, canaries, and finches. See Table 19–1.

When selecting cages, the following points should be kept in mind:

1. The cage should be big enough for the bird or birds.

2. The cage should be easy to clean.

3. The cage should be large enough to provide room for feeders, waterers, nest boxes, and other items.

4. The cage should be affordable.

5. The cage should be made of stainlees steel or anodized aluminum or be chrome-plated. These materials will not rust and are easy to design. Materials that rust present two hazards to birds: (a) the rust may be ingested and damage the linings of the digestive system and (b) the rust pieces may contain lead or other poisonous materials.

Feed containers for the large parrot-type birds must be constructed of sturdy materials. Glass, ceramic, or stainless steel are preferred; plastic material will be destroyed. Plastic feeders are fine for the smaller birds, however.

FEEDING

Most bird diets consist of seeds, except for the fruit- and nectar-consuming birds. Seeds consumed by birds are of two types: cereal seeds and oil seeds. Cereal seeds contain a higher proportion of carbohydrates compared with oil; examples of cereal seeds are canary seed, millet, corn, and dehusked oat kernels. Oil seeds are high in fat and low in carbohydrates; exam-

	Table 19–1 Minimum Cage Size for Caged Birds	
	CAGED PAIR	**SINGLE BIRD**
SPECIES	**Length × Width × Height**	**Length × Width × Height**
Canaries	18″ × 10″ × 10″	*
Cockatiels	4′ × 2′ × 3′	26″ × 20″ × 20″
Cockatoos	4′ × 4′ × 3.5′	3′ × 2′ × 3.5′
Conures	4′ × 4′ × 4′	4′ × 3′ × 4′
Finches	2′ × 2′ × 2′	12″ × 12″ × 12″
Lovebirds	4′ × 4′ × 4′	*
Macaws	6′ × 6′ × 6′	3′ × 2′ × 3.5′
Budgerigars	24″ × 14″ × 8″	*
Mynah birds	(⁺) 6′ × 3′ × 3′	6′ × 3′ × 3′
Amazon Parrots	4′ × 3′ × 4′	3′ × 2′ × 2′
African Gray	4′ × 3′ × 4′	3′ × 2′ × 2′

*These birds prefer the company of other birds and should not be caged singly.

†The cage should be at least this size for a pair.

ples of oil seeds are sunflower seeds, peanuts, safflower, pine nuts, rape, maw, niger, and linseed. Both cereal and oil seeds can be purchased separately and mixed together to provide variety. Most people, however, probably prefer to buy premixed rations. When selecting seeds for a mix, one must make sure the seed is clean and free of dust and dirt. The seed should also be dry and free of molds; both shelled and unshelled peanuts should be examined for mold. Peanuts are attacked by a mold that produces strong toxins that can cause liver damage; fed over a period of time, they can cause the death of a bird.

Soaked seeds may be fed to young birds that are having trouble breaking hard seeds, and to birds during breeding and molting seasons. The seeds should be soaked in warm water for about 24 hours; this will stimulate the germination of the seeds. The chemical changes in the seeds during germination increase the protein content. When feeding soaked seeds, it is important that the water be poured off and that the seeds be washed thoroughly with tap water. Soaked seeds provide an ideal medium for the growth of molds and fungi; birds must not be allowed to consume seeds that have mold and fungi on them. Any seeds not consumed by the birds in a few hours should be removed and thrown away. The container that the seeds were in should be washed thoroughly before feeding more soaked seeds.

Birds can also be fed green plant materials. Chickweed, dandelion leaves, carrot tops, kale, and spinach are good choices that can be fed. One must be careful not to feed too much green plant material because it may cause diarrhea; this is especially true if the bird is not used to green plant material. Green material must be thoroughly washed to remove any residue of pesticides or other chemicals. Lettuce should be avoided because it is not as nutritious as other types of greens. If green foods are taken directly from the refrigerator, they should be allowed to warm to room temperature before being offered.

Birds cannot grind up their food. They can break the seeds open with their beaks, but the actual grinding of the food takes place in the ventriculus. This is accomplished with the aid of **grit** that must be supplied in their diet. Grit is available from pet stores in two forms: soluble and insoluble. The soluble form is usually soluble oyster shell that breaks down and serves as a source of minerals. The insoluble form is usually crushed granite; the insoluble grit provides the primary base for which the food material is rubbed and worked against to grind the food up. Grit should be available to the birds at all times.

Cuttlefish bone should be provided as a source of calcium; cuttlefish are marine mollusks. Cuttlefish bone can be obtained from pet supply stores and clipped to the side of the bird's cage. Female birds need a supply of calcium for the making of eggshells. Some small birds may have trouble breaking or chewing off pieces of the cuttlefish bone; pieces may have to be cut or shaved off for them.

Mynah birds will not eat seeds and do not need grit or cuttlefish in their diet. Mynah birds are fed special softbill pellets or foods that are available from pet supply stores. Some of these foods can be fed directly from the package, and some need to be mixed with water before feeding. Mynah birds also need fruit in their diet; apple slices, grapes, orange slices, and banana slices are the most common. Dried fruits can also be used, provided they are soaked and rinsed off before feeding. Mynah birds also like live food such as mealworms.

Lories and lorikeets feed on nectar and pollen; nectar is usually available from pet supply stores. Nectar foods are usually powders that need to be mixed with water and provided to the birds in special feeders. Lories and lorikeets also eat fruit.

Hand-reared birds are in great demand because they are usually tamer and more easily handled. Birds that are hand-reared need to be provided a brooder or heating pad so that the birds can be kept warm. A spoon bent up on the sides makes an ideal tool for hand-feeding young birds. Hand-feeding is very time consuming; newly hatched birds need to be fed every hour and a half. As the birds approach weaning, they still need to be fed every 3 or 4 hours. Feeding is required from early morning to late evening. Foods commonly used are dry baby cereals, fruits, and canned baby food. These foods are mixed with water in a blender and then heated; the food must be warm and fairly runny. Supplements such as calcium and vitamins can be added. As the birds get older, they can be gradually weaned from the liquid diet to seeds.

TRAINING AND HANDLING

After bringing a new bird home, it should be left alone for a day or two to get acquainted with its new surroundings. After a couple of days, begin trying to offer the bird a treat through the bars of the cage. The bird may refuse the treat, but one should be patient and continue at regular intervals to offer a treat. One should talk softly to the bird, and soon the bird will take the treats. The next step is to open the door of the cage and offer the treat through the open door. One should move slowly and talk calmly to the bird. The bird may retreat, but again, one must be patient and the bird will soon be taking the treats. Next, a stick or perch should be introduced into the cage and the bird encouraged to step up onto this new perch. The perch should be pressed up against the bird's chest just above its legs. Parrots use their beaks to aid in climbing, so one should not be alarmed if the bird grasps the perch with its beak. The feet should step up on the perch next. After the bird becomes accustomed to climbing on a perch held into the cage, one can slide a hand or finger (depending on the size of the bird) in under the stick and attempt to get the bird to perch on the hand or finger. While attempting this, one should offer the bird a treat to distract it and continue to talk calmly to the bird. A leather glove may be needed when working with the large parrot-type birds. Once the bird feels comfortable perched on the hand, one can try to remove the bird from its cage while perched on the hand. The bird may jump from the hand at first, but one should keep trying; this process may take several days. Some birds are more adventurous than others. Patience is important. The next step is to try to get the bird to use a perch outside its cage or to perch on one's arm or shoulder.

When working with birds outside their cages, it is important to be aware of possible hazards. All windows must be closed, as should blinds or curtains. Aquariums and open containers of water should be covered. The bird should not be left unsupervised, especially where electrical wires and other hazards may exist. Other pets, especially cats, should be kept out of the room when the bird is outside its cage.

Clipping a bird's wings will restrict its ability to fly. If a bird is going to be allowed out, the wings will need to be clipped to prevent it from escaping. When clipping the wings, one should cut across the primary and secondary flight feathers, leaving the two outermost primary feathers uncut; otherwise, the clipped wings will be obvious. One should cut just above the base of the shaft, being careful not to cut down to the base of the feathers because this could cause serious bleeding. This is a painless procedure and will not harm the bird. The bird may be a little withdrawn for awhile if it is not used to this procedure. See Figure 19–26.

Budgerigars, cockatiels, parrots, macaws, cockatoos, and mynah birds can learn to talk. They actually mimic or imitate human words, the calls and sounds of other birds, and other sounds that they hear. Trying to teach a bird to talk should begin after it becomes tame and accustomed to its handler and surroundings. Toys, mirrors, food, and any other distractions should be removed while working with the bird. Young birds are the easiest to teach, and males usually learn more words and learn quicker than females, although this is not always true. One should develop a routine in the training. It is helpful to try to give the bird its lesson at the same time each day. Lessons should be only about 15 minutes in length; the bird may become bored if lessons are longer. Short words and phrases are easiest

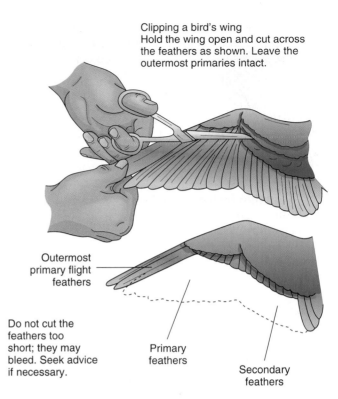

Clipping a bird's wing
Hold the wing open and cut across the feathers as shown. Leave the outermost primaries intact.

Outermost primary flight feathers

Do not cut the feathers too short; they may bleed. Seek advice if necessary.

Primary feathers

Secondary feathers

Figure 19–26   The correct way to clip a bird's wing.

to mimic. One should slowly repeat several times the word or phrase he or she wants the bird to mimic. Only one person in a family should be trying to teach the bird to talk. Birds may learn to talk for females and children more quickly. Patience is important. Some birds may never learn to talk.

DISEASES AND AILMENTS

Pet birds are usually very healthy because they do not come in contact with parasites or disease organisms that birds in the wild may encounter. However, pet birds are susceptible to diseases, and owners should be alert for the following signs that might indicate a problem:

1. Normally, adult birds sleep on one leg. If a bird is observed sleeping on both legs, it could indicate the bird is uncomfortable or is ailing.

2. Fluffed feathers could indicate that the bird is trying to retain body heat; the bird may be chilled.

3. Runny feces could indicate a digestive ailment.

4. Lack of activity or failure to fly may indicate an ailment.

5. Eyes that are continually closed or have discharges are signs of colds and other ailments.

6. Irregular or difficulty breathing, wheezing, or noisy breathing could indicate a respiratory ailment.

7. Loss of appetite or failure to eat may indicate an ailment.

At the first sign of illness, a bird should be provided with warmth, rest, food, water, and possibly medication. The cage should be moved to a warmer area or a cage provided in which a warmer temperature can be maintained. The cage temperature for an ill bird should be maintained at 85 to 90°F. The additional heat can be provided with a heating pad under the cage or a light bulb placed near the cage. Two or three perches should also be provided. The bird will move to the area where it finds the most comfortable temperature. The cage should be partially covered with a towel or sheet to prevent drafts. Some quick-energy liquids such as sugar water, honey water, or orange juice may prove beneficial.

Proper management is important in preventing diseases and ailments.This begins with selecting a healthy bird from a reliable source. Many ailments occur when the birds are first taken home. Make sure the bird is placed in a dry, warm, draft-free area. It is important that the bird be subjected to as little stress as possible; one should make sure there are no animals around to bother the bird. No new birds should be introduced in with other birds until they have gone through a quarantine and observation period. Sanitation is extremely important. The bird should receive fresh food and clean, fresh water. The cage and perches must be kept clean.

Internal Parasites

Internal parasites are rarely seen in pet birds because they are not exposed to wild birds or to the ground where exposure can occur. Occasionally, internal parasites may occur, especially in wild birds or in aviaries that do not maintain good sanitary practices and rodent control.

Roundworms. Roundworms can be diagnosed by observing the feces. Roundworms in the feces appear as long, thin, white round worms. Infection occurs by ingestion of the worm eggs from contaminated feces, soil, or food. Infected birds show blockage of the intestine, poor plumage, loss of weight, and diarrhea. Several treatments can be used to control roundworms.

Capillaria Worms. Capillaria worms (threadworms) appear in the feces as small, microscopic, thread-like worms. Infection occurs when birds ingest contaminated feces, soil, or feed. Infected birds may have decreased appetite, lose weight, regurgitate frequently, show poor plumage, and have diarrhea. Treatment with some drugs may be effective in controlling capillaria worms.

Tapeworms. Tapeworms can be diagnosed by observing the small, rice-like, white segments in the feces of infected birds. Birds become infected by ingesting intermediate hosts, such as houseflies, fleas, ticks, or earthworms. Infection with tapeworms is slight if proper cleaning and sanitation practices are observed. Piperazine, nicotine sulfate, and kamal powder have been suggested in the control of tapeworms.

Trichomoniasis. Trichomoniasis is a disease caused by the flagellated protozoan *Trichomonas gallinae* and is spread through contaminated feces. The disease has primarily been observed in pigeons and in birds in outdoor aviaries. Birds indoors may become infected if infected birds are introduced into the cage. The protozoan lives in the head sinuses, mouth, throat, esophagus, and other organs. The first signs of the disease are small, yellowish lesions on the linings of the mouth and throat. These lesions, which are cheese-like in appearance, enlarge rapidly and may block the throat of the bird and prevent it from closing its mouth. Breathing may be difficult, and loss of weight may also occur. Watery discharge may occur from the eyes, and blindness can occur in advanced stages. Death usually occurs within eight or ten days. Metronidazole and dimetridazole treatment may be effective in suppressing the disease.

External Parasites

Red Mites. External parasites such as red mites, *Dermanyssus gallinae,* feed on the blood of infected birds. The mites usually feed at night and then move to cracks in the cage and nest box. Heavy infestations can cause anemia and even death. Mites, although small, can be seen with the naked eye, and they appear as tiny red specks after feeding on the bird's blood. Birds infected with mites are restless and scratch and pick at their feathers. Mites can be spread to caged birds by introducing infected birds, or birds in outdoor aviaries can be infected from wild birds visiting the aviary. Adult birds may be dusted with pyrethrum powder. Young birds in nest boxes should be dipped in appropriate insecticides. Infected nest boxes should be destroyed or thoroughly disinfected. Cages and equipment should also be thoroughly cleaned and disinfected.

Feather Mite. The feather mite, *Knemidocoptes laevis,* should be considered in the case of feathers that look chewed, and where no other cause is known; severe infestation can cause a bird to lose its feathers.

Feather mites bury into and feed on the feather quill just below the surface of the skin. Infestation usually occurs on the beak or neck first and spreads over the entire body. Restlessness, severe scratching, feather picking, and skin irritation are signs of infection. The mites may be seen as small, gray-colored moving specks. Unlike red mites, they are on the bird both day and night. Cages and equipment should be treated with nicotine sulfate, coumaphos, or malathion. Birds can be sprayed with mite sprays that are available at most pet shops.

Scaly Face and Leg Mites. Scaly face and leg mites, *Knemidocoptes pilae,* and related species are mites that may infect budgerigars, lovebirds, and canaries. Scaly face mites tunnel into the soft tissues around the beak, ceres, and face. Scaly leg mites tunnel under the scales on the legs. These mites live their entire life cycle on the bird and can cause severe irritation. White, scaly deposits are the first signs; these deposits then become thickened, enlarged, and encrusted. Early diagnosis and treatment will prevent these unsightly deposits. Application of vaseline or mineral oil will loosen the deposits and plug the air holes used by the mites; this will suffocate and kill the mites. Cages, perches, and equipment should be thoroughly cleaned.

Bacterial Diseases

Parrot Fever. Bacterial diseases such as psittacosis, also called parrot fever or chlamydiosis, are caused by the bacteria *Chlamydia psittaci.* The disease is also referred to as ornithosis when found in other birds such as turkeys and pigeons. Feces and contaminated food and water are the primary source of infection; the bacteria can also be inhaled into the respiratory system. When infection occurs, the bacteria travel to the liver and spleen, where they multiply. Symptoms may include nasal discharges, closing of the eyes, listlessness, loss of appetite, loss of weight, greenish-colored diarrhea, and labored breathing. It is possible for birds to be carriers of the disease and show no signs. Pet birds should be purchased from reliable, disease-free sources. If the disease source is not known, birds should be treated with chlortetracycline-impregnated seed for a minimum of twenty-one days. Psittacosis can also be transmitted to humans. Symptoms are flu-like and pneumonia-like with fever, headache, and coughing. Treatment with antibiotics is usually effective.

Pullorum Disease. Pullorum disease is caused by the bacteria *Salmonella pullorum.* Transmission is usually by direct contact with infected birds. Infection may cause a high mortality in young birds. Infected birds do not eat, appear sleepy, and have a white pasting of feces around the vent. Furazolidone in the feed is one of the recommended treatments.

Colibacillosis. Colibacillosis is caused by the bacteria *Escherichia coli.* This bacteria normally exists as part of the intestinal flora of all animals. Infection from *E. coli* usually is associated with another disease. One form of colibacillosis is air sac disease where infection produces severe, cheese-like inflammations over the heart, liver, and air sacs. Colibacillosis of the navel or omphalitis causes a reddening, swelling, mushiness, or wetness around the navel. Young birds seven to ten days old are infected. Chronic infections of the oviduct, salpingitis, is a result of bacteria entering the oviduct. Infection causes swelling of the oviduct and a foul-smelling discharge. Many infections of the legs and wing joints, mouth, and respiratory tract are associated with *E. coli* bacteria. Prevention is probably more effective than treatment. Sanitation and cleanliness are of utmost importance.

Pasteurellosis. Pasteurellosis, or fowl cholera, is caused by the bacteria *Pasteurellosis multocida.* Symptoms may include joint, sinus, nasal, middle ear, or skull infections. Fever, depression, anorexia, ruffed feathers, mucous discharges, and difficulty breathing are more visible signs of the disease. Disease organisms may be spread by contaminated feed and equipment. Some of the sulfa-based drugs, tetracycline and penicillin, may be effective.

Bumblefoot. Bumblefoot is a painful ailment of the feet of birds and is usually associated with staphylococcal infections. The feet and joints become hot and swollen with a thick, grayish white fluid. The bird may not want to walk or may have trouble bending its toes or clasping onto the perch. Treatment with antibiotics is recommended. Suitable perches and sanitation of cages is important in preventing bumblefoot.

Tuberculosis. Tuberculosis is a disease caused by the bacteria *Mycobacterium avium.* The disease usually shows no outward signs; there is just a slow weight loss and deterioration of the bird's condition. Internal organs have hard, rounded gray to yellow pea-sized nodules. These nodules increase in size and number as the disease progresses. Regular tuberculin testing is recommended in breeding aviaries.

Viral Diseases

Psittacine Beak and Feather Disease. Psittacine beak and feather disease syndrome is a viral disease that attacks the immune system. Symptoms usually become evident at the first molt; new feathers may not emerge or may be deformed and soon break off. The beak and nails may be soft, overgrown, and lose their pigment. There is no cure for the disease. Treatment to control secondary diseases and supplements of vitamins and minerals may prolong the life of the bird. This disease is also referred to as French molt.

Newcastle Disease. Newcastle disease is a rapidly spreading viral disease of birds. Respiratory difficulty such as wheezing may be one of the signs. Nervous ailments such as tremors, wing droop, partial or complete leg paralysis, or twisted necks may also develop among birds that survive the disease. Some birds may serve as carriers of the disease. Domestic poultry are especially susceptible, and there may be 100 percent mortality of infected birds. Imported birds serve as the main source of possible infections; these birds need to be quarantined and tested at licensed bird import stations. Birds that test positive should be destroyed. Vaccination is the recommended method of prevention.

Pacheco's Parrot Disease. Pacheco's parrot disease is a herpesvirus infection that primarily infects parrots, conures, cockatoos, and macaws. Symptoms are not specific; birds may become lethargic, regurgitate, and have diarrhea. They may have an increased thirst and eventually recover. Other birds may show signs of weakness shortly before dying. No specific treatment is recommended; however, the bird should be kept warm and given plenty of liquids, vitamin C supplements, and antibiotics. Acyclovir is an antiherpes viral drug that may prove effective.

Nutritional Problems

Goiter. A deficiency of iodine in the diet of birds may lead to goiter. Symptoms of goiter are a swelling of the thyroid glands in the neck; the swelling can cause pressure on the windpipe and interfere with breathing. Budgerigars seem to have an especially high requirement for iodine; iodine blocks are available at pet supply stores.

Rickets. A deficiency of calcium or phosphorous, an imbalance of calcium and phosphorus, or a deficiency of vitamin D_3 can cause a deterioration or softening of bone. This condition is called osteomalacia, or rickets. Symptoms are lameness, a stiff-legged gait, or a constant resting in a squatting position. There is usually a decrease in growth. The beak can easily be bent from side to side, and leg and wing bones become soft, bowed, or twisted. Oyster shell or coarse limestone in the diet usually prevents rickets from occurring. Supplements of vitamin D_3 can also be provided.

Candidiasis. Candidiasis is a condition caused by a deficiency of vitamin A. An early sign is a bird seen playing with its food but not consuming any. As the condition progresses, white fungus-like spots appear in the mouth. These spots spread throughout the digestive system if the deficiency is not corrected. Vitamin supplements are available at pet stores to treat the condition; young birds are most commonly affected.

Obesity. Obesity can be a problem with caged birds. Obese birds are usually lethargic and less active. Feeding too many sunflower seeds, which are high in fat content, can lead to obesity. A change of diet and more exercise can correct the problem.

Other Problems

Overgrown Claws. Overgrown claws may cause a bird to get caught on its cage and injure itself. The claws can easily be clipped with nail clippers designed for pets. One must be careful not to cut into the blood vessel that runs down the center of the claw; this vessel appears as a pinkish streak.

Feather Plucking. Feather plucking may become a problem in some birds; the reason for the problem is not always apparent. Improper diet, lack of bathing facilities, boredom, and a frustrated desire to breed may be possible causes. Many times, getting the bird a companion is an effective cure.

Birds that live indoors need to receive regular bathing or spraying; this encourages birds to **preen** themselves. Preening is the bird's natural cleaning and trimming of its feathers using its beak. Regular preening reduces the chances of feather plucking. Regular bathing and spraying reduces feather dust and dirt. Small birds such as budgerigars, canaries, finches, mynahs, lories, and cockatiels should be provided with a container of water to bathe in. This container should be fairly heavy to prevent the bird from tipping or upsetting the container. The water container may be placed in the cage every few days for about 30 minutes at a time. Larger birds such as parrots and cockatoos can be sprayed with a plant sprayer that produces a fine mist. One should spray the mist above the bird so that the mist settles down. The bird may be startled at first but will soon spread its wings and enjoy it. The bird should not be completely saturated and should receive a gentle spray two or three times a week.

REPRODUCTION

The reproductive organs of a male bird are the testes, ductus deferens, and the cloaca. See Figure 19–27. The testes of the male, which produce the sperm, remain in the abdominal cavity and do not descend into a scrotum. The sperm is released from the testes and enter the ductus deferens, small tubes that lead to the cloaca. The male bird does not have a copulatory organ. The sperm is released from the ductus deferens into the cloaca. During mating, the cloaca of the male and female are joined together, and the sperm is transferred from the cloaca of the male to the female. The dorsal wall of the male's cloaca is lined with small finger-like projections called **papilla,** which aid in transferring the sperm to the female's cloaca. The sperm make their way up the oviduct of the female and are stored in the folds of the oviduct. The sperm of birds may be viable for up to thirty days, which is longer than the sperm of mammals; however, newer sperm is probably more active. When the egg yolk is released by the ovary, it drops down into the infundibulum. As the yolk moves into the oviduct, the walls are stretched, and the sperm are released to fertilize the yolk. After fertilization, the egg yolk moves down through the oviduct. The thick, white albumen is secreted around the yolk. Two shell membranes are added in the **isthmus.** The egg then enters the uterus, where the eggshell is formed. The eggshell takes about 20 hours to form; the egg moves to the vagina and is then laid. See Figures 19–28 and 19–29.

After the egg is laid, the embryo continues to grow inside the egg. This period of growth is referred to as the incubation period. The period varies with the species but averages around twenty-one days.

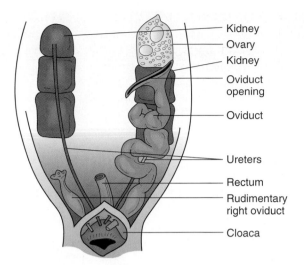

Figure 19–28   Urogenital system of a female bird. For more detail, see Figure 19–29.

Proper temperature and humidity are especially important in the incubation of eggs.

Some species of pet birds are easily bred in captivity, especially the smaller birds like budgerigars, finches, and canaries. Larger species may be more difficult to breed; the African gray parrots are especially difficult.

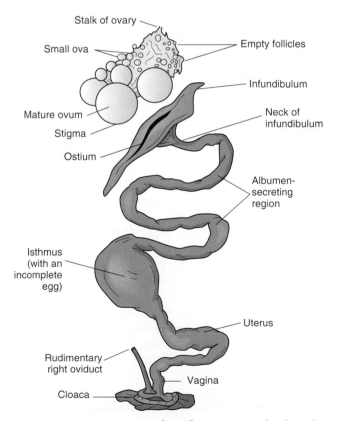

Figure 19–29   Ovaries and oviduct system of a female bird.

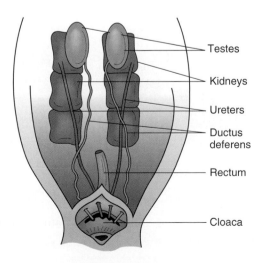

Figure 19–27   Urogenital system of a male bird.

Figure 19–30   Young society finches. Photo by Isabelle Francais.

The first problem that may be encountered in reproducing birds is getting both a male and female of a species. Determining the sex of some species is difficult because no dimorphism exists. Sex can be surgically determined by making a small incision in the side of the bird. Because of the high price paid for some birds, this procedure may be a wise investment.

Canaries and finches are good choices for beginning bird fanciers who want to breed their birds, because these birds breed and reproduce easily in captivity. Most species show various degrees of sexual dimorphism, and pairs can more easily be selected. Breeding cages should be provided for these small birds and equipped with a wood or preferably a wire divider. The male is then placed on one side and the female on the other side. The divider should be removed when the male begins to call to the female and begins to feed her through the wire divider. A small wire, wood, plastic, or ceramic open-topped container should be provided for the bird to make a nest in. The container should be hung about halfway up the side of the cage. Straw, grass, cotton, feathers, or dry peat moss can be provided for nesting materials. The female usually lays one egg per day until a clutch of two to six eggs are laid; incubation lasts about fourteen days. The female feeds the young. A soft food is needed so that she can feed them. Commercial foods are available, or bread and milk or the yolk of a hard-boiled egg and milk can be supplied in a cup. The young develop rapidly; their eyes open in about seven days. At about four weeks,

they can be separated from their parents. Soft foods can then be provided for the young; seeds should be cracked and soaked for 24 hours to soften them. See Figure 19–30.

It is fairly easy to distinguish sex of budgerigars. The males have blue ceres, and the females have pink or brown ceres. If a pair are introduced and they fight, they need to be separated for a couple of weeks and introduced again. Budgerigars may breed more successfully if two or more pairs are kept in a cage. A nest box should be placed in the upper third of the cage and should be about 7 inches square and 9 inches high. The nest box should have a 1½-inch hole placed about two-thirds the way up the side, and a wooden perch should be placed on the outside. Budgerigars usually do not use nesting materials. Rather, a concave depression should be made in the floor of the nesting box to hold the eggs together in a clutch. Budgerigars lay four to six eggs, one egg being laid every two days; the incubation period usually is seventeen to eighteen days. The young are fed a colostrum-like milk that is produced in the proventriculus, or gizzard of the female; this milk is then regurgitated up and fed to the young. After a few days, the mother mixes the feeding of milk with other soft foods. After two weeks, the male helps with the feeding duties. Soft foods like those fed to finches and canaries should also be provided. The young budgerigars' eyes open at eight to ten days. They can be separated from their parents at five to six weeks of age. See Figure 19–31.

It is difficult to determine the sex of lovebirds; surgically sexed birds will prevent disappointment at trying to mate same-sexed birds. A pair of birds of the

Figure 19–31   A young budgerigar. Photo by Isabelle Francais.

same sex will preen each other, and everything may appear to be going well. One bird may even try to mount the other.

A nest box similar to those used for budgerigars can be used for lovebirds. Straw, grass, small tree branches, and sticks can be provided for nesting materials. Some green materials may be provided to add moisture necessary for incubation. The female lays four to seven eggs in a clutch; the incubation period varies from eighteen to twenty-four days. The young begin to venture from the nest at about five weeks of age and are weaned at seven weeks. Both parents assist in feeding the young.

Male cockatiels usually have brighter orange or yellow about their head and cheeks than the females. A nest box approximately 9 inches by 9 inches by 15 inches high should be provided. A hole 3 inches in diameter and a wooden perch should be provided. Wood chips and straw can be provided for nesting materials; however, some cockatiels may not use nesting materials. A concave depression should be made at the bottom of the nest box. The female lays three to six eggs in a clutch, with one egg laid every other day; the incubation period varies from eighteen to twenty days. Both parents incubate the eggs and feed the young. At four to five weeks, the young leave the nest. They are ready to be separated from their parents at seven to eight weeks. See Figure 19–32.

The size of the nest box for the true parakeets will vary depending on the species. The smaller parakeets can use a nest box similar in size to those used for budgerigars; the larger parakeets such as the ringneck parakeets and rosettas need larger nest boxes. Boxes up to 13 inches square and 3 feet deep may be required; a hole 4 inches in diameter should also be provided. The box should then be filled up to the hole with pine shavings. The birds will remove the shavings until the desired level is reached. Those interested in breeding true parakeets should obtain more specific information for their species.

Most parrots use holes in the ground or in trees to make nests in the wild. In captivity, a wooden or metal nest box can be used. Metal might be preferred because parrots can destroy wooden nest boxes. Metal barrels or garbage cans serve as suitable nesting boxes. A wooden perch should be attached to the outside, and a hole 6 inches in diameter for an entrance should be made 3 to 12 inches from the top. Another hole with a lid should be added so that the process of incubation and hatching may be observed. The nest box should be hung high up in the corner of the bird's cage. A large cage or **aviary** is needed to provide plenty of room for breeding large birds. Wood shavings and peat moss make suitable nesting materials. Two to four eggs are usually laid per clutch, and the incubation period is twenty-four to thirty days. Both the male and female take turns incubating the eggs. When the young hatch, they are completely blind and helpless. At about eight to ten weeks of age, the young leave the nest; however, the parents continue to feed the young for another three to seven weeks. The young are weaned and on their own at eleven to seventeen weeks of age. See Figure 19–33.

Figure 19–32 A young cockatiel. Photo by Isabelle Francais.

Figure 19–33 Young macaw parents. Photo by Isabelle Francais.

SUMMARY

Birds are two-legged, egg-laying, warm-blooded animals with feathers and wings, and they are believed to have evolved from prehistoric reptiles. The first bird is believed to be *Archaeopteryx*, which lived approximately 150 million years ago during the Jurassic period. Modern birds appeared about 80 million years ago during the Cretaceous period, and by the Cenozic Era some 70 million years ago, they were very similar to the birds today.

Modern birds are classified into twenty-eight orders containing 170 families and 9,000 species. Three orders are important as pets. The parrots, cockatoos, macaws, lovebirds, lories, lorikeets, budgerigars, and parakeets belong to the order Psittaciformes. Perching birds, which include the finches, waxbills, buntings, canaries, siskins, and grosbeaks belong to the order Passeriformes. Toucans belong to the order Piciformes, which also include the woodpeckers and other related birds.

Cages for birds are available in a wide variety of sizes and styles. Cages should be selected depending on the species of bird, ease of cleaning, affordability, and the materials used in its construction.

Most birds' diet consists of seeds. Seeds consumed by birds are of two types: cereal seeds and oil seeds. Cereal seeds contain a higher proportion of carbohydrates compared with oil. Oil seeds are high in fat and low in carbohydrates. Nectar- and fruit-consuming birds are an exception because their diets consist of pollen, nectar, fruits, and other soft foods. Some green plant materials can be fed to birds but should be limited because they can cause digestive problems and diarrhea. Seed-eating birds must also receive grit in their diet. Grit is important in helping grind up the food material when it reaches the bird's gizzard. All female birds should receive ample amounts of calcium because it is important in the formation of egg shells.

Budgerigars, cockatiels, parrots, macaws, cockatoos, and mynah birds can learn to mimic the human voice and other sounds they hear. In teaching a bird to mimic the human voice or to learn other tricks, patience is very important; some birds may never learn to mimic sounds or learn tricks, but they still make excellent companions.

Pet birds are usually very healthy because they do not come in contact with parasites or disease organisms that birds in the wild may encounter. However, pet birds are susceptible to diseases, and owners should be alert for signs that might indicate a problem.

DISCUSSION QUESTIONS

1. What is a bird?

2. Why do many people believe that birds evolved from prehistoric reptiles?

3. What are the types of feathers found on birds, and what are their purposes?

4. What are the three orders of birds discussed, and what are the main characteristics of them?

5. Name three nutritional ailments of birds, and list the symptoms of each.

6. What are the guidelines to follow when placing a cage in a room?

7. What are some important factors to remember when selecting a cage?

8. Describe how a cage for large parrot-type birds should be constructed.

9. What are two types of seeds consumed by birds, and what are some examples of each?

10. Why is grit important in the diet of most birds? Which birds do not need grit?

11. Describe the steps in getting a bird to feel comfortable with its handler and the steps in training.

12. What are some signs that might indicate poor health in a bird?

13. What are some external parasites of birds, and what are the symptoms that might indicate a bird has external parasites?

14. What are some internal parasites of birds, and what are the symptoms that might indicate a bird has internal parasites?

15. What are some bacterial diseases of birds, the symptoms, and the treatment for them?

16. What are some viral diseases of birds, the symptoms, and the treatment for them?

SUGGESTED ACTIVITIES

1. Visit a local pet shop and make a list of the different species of birds that are available. Using the local library, compile as much information as you can about the species.

2. Research the different orders, families, genus groups, and species of birds. Why are they classified as they are?

3. Contact government agencies and compile a current list of endangered, threatened, and extinct species of birds.

ADDITIONAL RESOURCES

1. Electronic Zoo/NetVet Bird Sites
   http://netvet.wustl.edu/birds.htm

2. Avian Health
   http://www.avianweb.com

3. Birds Of A Feather – Bird Links
   www.boaf.com/birdlinks.htm

4. Exotic Birds Page
   www.flash.net/~flycolor/birds.htm

5. Dr. Jungle's Exotic Pets of the World
   www.exotictropicals.com/encyclo/encyclo.htm

6. The John Rice Reef Page
   www.globaldialog.com/~jrice/reef.html

OBJECTIVES

After reading this chapter, one should be able to:

■ describe the characteristics of fish.

■ compare the three classes of fish.

■ describe and setup a freshwater and marine aquarium.

■ describe the feeding habits of fish.

■ describe the common diseases of fish.

TERMS TO KNOW

anal fin	community aquariums	pelvic fins
adipose fin	dorsal fin	protrusive
anterior	gonopodium	shoals
aquarists	labyrinthine chamber	spawning
barbels	neuromasts	species aquarium
brackish	pectoral fins	symbiosis
caudal fin	peduncle	vent

CHARACTERISTICS

Evidence of the early development of fish is found in fossils dating back to the Ordovician period 425 to 500 million years ago. These early fish were called Ostracoderms. They were slow, bottom-dwelling animals covered with thick, bony plates and scales. Ostracoderms had very poorly developed fins and did not have jaws. They are also believed to be the first animals to have a backbone. Ostracoderms became extinct about 250 million years ago.

From the Ostracoderms evolved two groups of fish with movable jaws. One group was called Placoderms and first appeared about 395 million years ago. They had thick, bony plates and paired fins. Their upper jaw was fused to the skull, but the lower jaw was hinged and movable. This group became extinct about 345 million years ago, but from them evolved the class Chondrichthyes. Today, Chondrichthyes are represented by the sharks, rays, skates, and other fish with cartilaginous skeletons.

The other group of jawed fish to evolve from Ostracoderms was called Acanthodians. It first appeared about 410 million years ago. From Acanthodians evolved the class Osteichthyes; these were fish with bony skeletons. Osteichthyes are divided into two subclasses: Sarcopterygii and Actinopterygii. Sarcopterygii are fish with fleshy fins or lobed fins, and their skeletons are made up of part cartilage and part bone. Today, only two orders exist: Dipteriformes, which is made up of lungfish, and Coelacanthiformes, which contains only one living species, *Latemeria chalumnae* (see Figure 1–2). It is believed that one extinct order of Coelacanthiformes, the Phipidistia, crawled out of the water about 390 million years ago to become the first amphibians. The subclass Actinopterygii is represented today by more than 20,000 species in thirty-four different orders.

Fish are cold-blooded vertebrates that breathe with gills and move with the aid of fins. They are the most numerous vertebrates, with more than 30,000 species. The smallest fish is the Philippine island goby, *Pandaka pygmaea,* which is only ⅓ to ½ inch long. The largest is the whale shark, *Rhincodon typus,* which grows to about 50 feet and weighs several tons.

ANATOMY

Most fish are covered with scales, which are thin, bony plates that overlap each other and provide protection. See Figures 20–1, 20–2, and 20–3. The scales develop from and are embedded in a pocket of the dermis. The exposed part of the scale is covered with a thin layer of epidermis. The skin of fish contains glands that produce a slimy mucus, which makes the fish slippery and provides protection from bacteria. The skin contains chromatophores, pigment cells that give the fish its colors. The colors a fish has usually allows it to blend with its surroundings, and most fish are able to change their color if necessary. Sensory receptors are also contained in the skin.

The scales on modern fish are of four types: ctenoid, cycloid, gamoid, and placoid. Most aquarium fishes have either ctenoid or cycloid scales. Ctenoid scales have serrations on the edges and rough

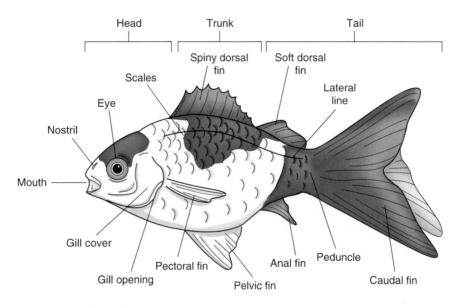

Figure 20–1   External anatomy of a fish.

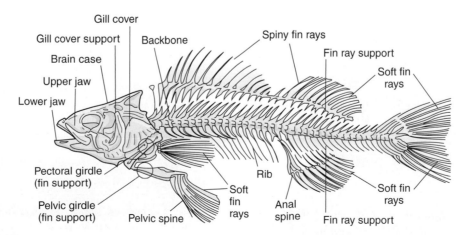

Figure 20–2   Skeletal structure of a fish.

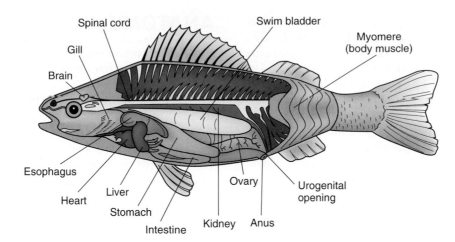

Figure 20–3 Internal anatomy of a fish.

surfaces; fish that have these scales feel rough to the touch. Cycloid scales have smooth surfaces and edges that make the fish feel smooth and slick.

Fins are movable structures that aid the fish in swimming and maintaining its balance. Most modern bony fish have rayed fins. These fins consist of a web of skin supported by bone or cartilage rods called rays. Fish may have sharp, spiny, or soft rays. These fins are very flexible.

Most fish have at least one vertical **dorsal fin** along their back, a single **anal fin** on the underside near the tail, and a single **caudal fin** or tail fin. Some species of fish have an **adipose fin;** a small, fleshy fin located on the back between the dorsal and caudal fin. One set of identical paired fins, which is found on the side just behind the head, is called the **pectoral fins,** and one set called the **pelvic fins** is found below and just behind the pectoral fins.

Fish breathe through organs called gills. Water is drawn in through the mouth by constant opening and closing, forced back into the pharynx, and out through the gills. The dissolved oxygen in the water is taken

into the blood, and carbon dioxide is released into the water because of the differences in the concentration of gases between the water and the blood. The blood comes in close contact with the water in thin-walled filaments on the gills; the oxygen and carbon dioxide are exchanged through the walls of the filaments. The gills in bony fish are protected by a cartilaginous gill cover. See Figures 20–4, 20–5, and 20–6.

All fish must maintain proper levels of salt and water in their bodies. Water flows from areas of weak salt solution to areas of strong solution in an effort to dilute the salt solution (osmosis). In freshwater fish, the body salts are a higher concentration than the salt in the water around them; therefore, water is drawn into their bodies. Freshwater fish have no need to drink extra water; their kidneys retain the salts and eliminate large amounts of water. Saltwater fish are surrounded by water that has a higher salt content than their bodies. Water is drawn from their bodies, so the fish must drink water to keep from dehydrating. The excess salts are eliminated in small amounts of very concentrated urine.

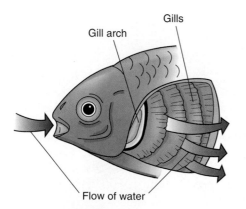

*Figure 20–4 **Fish breathe through organs called gills. Water is drawn in through the mouth and forced back into the pharynx and out through the gills by the constant opening and closing of the mouth.***

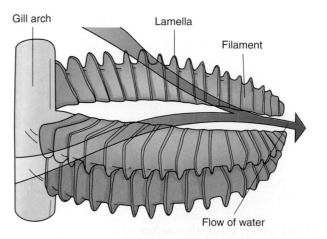

*Figure 20–5 **Each gil consists of many filaments.***

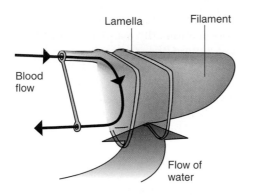

Figure labels: Lamella, Filament, Blood flow, Flow of water

Figure 20–6 Each filament consists of lamella. Lamella are thin walled and contain the blood vessels through which oxygen and carbon dioxide are exchanged.

Most bony fish have a swim bladder in their abdominal cavity. The swim bladder is filled with gases produced by the blood, which enables the fish to maintain a particular depth. A few bony fish are able to breathe atmospheric air because their swim bladder is supplied with blood vessels, and this allows the swim bladder to function like a lung. This is an important adaptation for fish living in swamps and other waters that are poorly oxygenated.

A few species of fish are able to use atmospheric oxygen because part of their intestines are modified to allow for oxygen intake. These fish rise to the surface and gulp air into their mouths. The air is then swallowed into the digestive system, where it is taken into the blood. Anabantoid fish have special rosette-shaped plates in a **labyrinthine chamber** behind their gills. These rosette-shaped plates are supplied with numerous blood vessels that absorb oxygen from the atmospheric air they inhale through their mouth.

The fish's heart is a long, folded organ consisting of only two chambers: the atrium (or auricle) and the ventricle. The blood is pumped by the ventricle through the arteries to the gills. In the capillaries of the gills, the blood receives oxygen in exchange for carbon dioxide, and the blood then continues on through the arteries to other organs and parts of the body. As the blood flows through other organs and the other parts of the body, nutrients are picked up and later exchanged for waste products. Oxygen is exchanged for carbon dioxide, and waste products are eliminated. After these exchanges, the blood returns through veins to the atrium of the heart. From the atrium, the blood flows to the ventricle, where the circulation of the blood began, and the cycle is repeated. The blood makes a simple, single circuit in the fish's circulation system.

Most fish are able to react to changes in water pressure, temperature, currents, and sounds because of a lateral line running along their side from behind the gill cover to the base of the tail. A series of pressure-sensitive cells called **neuromasts** are contained in tubes along the lateral line. Pores opening into the tubes bring the cells in contact with water. See Figure 20–1.

The eyes of fish are similar to those of other vertebrates, but differ in a couple of ways. The fish's eye has a spherical lens that focuses by moving within the eyeball, not by changing the curvature of the lens. Fish do not have eyelids; the eye is kept moist by the flow of water. The size of the eye usually depends on the amount of light reaching the eye. Fish that live in shallow, brightly lit waters usually have small eyes, whereas fish living in dimly lit water have larger eyes.

All fish have inner ears. Species that have swim bladders have a more acute sense of hearing because the bladder acts as a resonator and amplifies the sound. Bony fish are sensitive to sounds in the range of 100 to 10,000 cycles per second.

Fish have taste buds in their mouth, on their lips, and on their body and fins. Their sense of smell and taste is highly developed. Some fish have taste buds on their **barbels,** which are whisker-like projections around their mouth.

Fish occupy different levels of an aquarium; some occupy the top, middle, or bottom levels. This usually gives an indication of the type of diet and feeding habits of the fish. The structure of their mouth also gives an indication of what level the fish will occupy. Fish that occupy the top level feed on insects that fall into the water; their mouths are usually upturned to enable them to feed easier at the surface. Some top-level fish feed on other small fish that also occupy the top level. Most top-level fish have long, streamlined bodies and can propel themselves through the water rapidly to catch their prey.

Some middle-water fish are found in flowing waters. These fish are active swimmers and have streamlined bodies. Some middle-water swimmers come from slower waters, live in shoals, and feed among the reeds and other vegetation. Fish that feed among the reeds usually have narrow bodies that enable them to easily move through the reeds. Fish species that feed on leafy vegetation usually have long, thin bodies with small fins that enable them to move among the leaves and to conceal themselves. Middle-water swimmers usually have small mouths and are straight forward, neither upward nor downward turned.

Bottom-dwellers live on the bottom of streams, rivers, and lakes. Their mouths are downward turned or underslung, which enables them to pick up food from the bottom areas. Bottom-feeders are equipped with barbels to assist them in feeling their way along the bottom and to help them locate food.

Digestive System

The digestive system of fish varies depending on the type of food consumed. Fish that are predators have teeth on the front of the jaws, the roof of the mouth, and throat teeth just in front of the esophagus. Predator fish usually consume their prey whole, and the teeth function primarily to clasp and move their prey into place for easier swallowing, normally head first. Sharks and piranha have cutting teeth for biting pieces from their prey. Some fish have rasp-like teeth to scrape plant and animal growth from rocks. Some species do not have teeth in their jaws but have teeth only in their pharynx to grind food material before it enters the esophagus. From the esophagus, food enters the stomach, which also varies considerably depending on the diet. The stomach is a tube-like organ, and food is digested here and then enters the intestine. Undigested material leaves the body through the anus. Lung fish, sharks, and rays have a cloaca through which the waste material passes.

GENERAL CLASSIFICATION

Kingdom — Animalia
Phylum — Chordata
Superclass — Pisces (all fish)
Class — Agnatha or Cyclostomata (jawless fish, possessing sucking or filter-feeding mouths)
Class — Chondrichthyes (fish with cartilaginous skeletons)
Class — Osteichthyes (fish with bony skeletons)
  Subclass — Actinopterygii (ray-finned fish)
    Order — Cypriniformes (characins, gymnotid eels, loaches, minnows, suckers)
    Order — Siluriformes (catfish)
    Order — Atheriniformes (flying fish, half beaks, killifish, needle fish)
    Order — Perciformes (cardinal fish, glassfish, and related fish)
    Order — Tetraodontiformes (box fish, ocean sunfish, puffers, triggerfish)
    Order — Scorpaeniformes (scorpion fish, sculpins)
    Order — Gasterosteiformes (pipefish, sea horses, sticklebacks, trumpetfish)
    Order — Mormyriformes (mormyrids)
    Order — Osteoglossiformes (bony tongues, freshwater butterfly fish, mooneyes)

The subclass Actinopterygil includes all ray-finned fish; they are divided into thirty-four orders. Nine of these orders have species that are common in the aquarium trade and are listed and discussed in this text.

CLASSIFICATION—CYPRINIFORMES

The order Cypriniformes is made up of three suborders, nineteen families, and 3,500 species. Eight families are discussed here.
  Suborder — Characoidei
    Family — Characidae (tetras and characins)
    Family — Gasteropelecidae (hatchet fish)
    Family — Anostomidae (headstanders)
    Family — Hemiodontidae (pencil fish)
    Family — Citharinidae (moon fish)
  Suborder — Cyprinoidei
    Family — Cyprinidae (minnows and carps)
    Family — Gyrinocheilidae (algae eaters)
    Family — Cobitidae (loaches)

MAJOR FAMILIES— SUBORDER CHARACOIDEI

The Family Characidae

The family Characidae is a large family containing 1,300 species. See Figure 20–7. About 1,000 species are found in Central America and South America; the remaining species are found in Africa. Most species are brightly colored and have a narrow dorsal fin and small adipose fin. They inhabit shallow, slow-moving rivers of the rainforest and live among sand and ground shoal areas. The waters are usually soft and slightly acidic. Most species are omnivorous and consume all types of foods; a few species are carnivorous. Most members of this family are sociable and do well in aquariums with other fish species, or community aquariums. Common members of the family include the following:

cardinal tetra, *Cheirodon axelrodi*

neon tetra, *Paracheirodon innesi*

silver-tipped tetra, *Hemigrammus nanus*

red phantom tetra, *Megalamphodus sweglesi*

rummy or red-nosed tetra, *Hemigrammus rhodustomus*

glow-light tetra, *Hemigrammus erythrozonus*

head and taillight tetra, *Hemigrammus ocellifer*

bloodfin, *Aphyocharax rubripinnis*

bleeding-heart tetra, *Hyphessobrycon erythrostigma*

black tetra, *Gymnocorymbus ternetzi*

lemon tetra, *Hyphessobrycon pulchripinnis*

rosy or rosaceous tetra, *Hyphessobrycon rosaceus*

jewel tetra, *Hyphessobrycon callistus*

x-ray tetra or x-ray fish, *Pristella maxillaris*

emperor tetra, *Nematobrycon palmeri*

Congo tetra, *Phenacogrammus interruptus*

striped Pyrrhulina, *Pyrrhulina vittata*

penguin fish, *Thayeria obliquua*

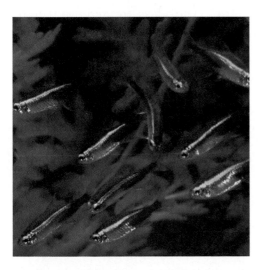

Figure 20–7   Neon tetra. Photo by Aaron Norman.

All of these are small fish varying in length from 1½ to 3½ inches. Water temperature for these species should be kept at about 75°F plus or minus 3°. They consume all types of foods.

The Blind Cave Fish.   The blind cave fish, *Astyanax mexicanus,* is native to the underground cave waters of Mexico; it grows to about 3½ inches long. Its body is primarily flesh-colored, and the fins are colorless. The young usually have eyes, but as they mature, skin grows over them and they become nonfunctional. This species does best in an aquarium with its own kind, which is referred to as a species aquarium. Although they are blind, they use their sense of smell to seek out and consume food.

The Spraying Characin. The spraying or splashing characin, *Copella arnoldi,* is unusual among aquarium fish because of its spawning habits; the female deposits her eggs on leaves, laying them 1 to 2½ inches above the water. She leaps out of the water and deposits her eggs, and the male immediately leaps out of the water to fertilize them. The male constantly splashes water on the eggs to keep them from drying out. When the young hatch, they drop into the water. In an aquarium, characins deposit their eggs on the cover glass or aquarium lid.

The Bucktoothed Tetra. The bucktoothed tetra (or Exodon), *Exodon paradoxus,* grows to about 5 inches in length. This species of tetra is very aggressive and should be kept in thickly planted aquariums with fish that are able to protect themselves. This species is gold-colored with reddish orange fins. There are two large, black spots; one on the side and one on the peduncle of the tail (part in front of the caudal fin). See Figure 20–1.

Schreitmuller's Metynnis. Schreitmuller's Metynnis, *Metynnis schreitmulleri,* grows to about 5½ inches long; it has a disc-shaped silver body. There is a reddish tinge to the head, back, and fins. During breeding season, the females have a dark red edge to the anal fin, and the male's is black. There are several species of *Metynnis,* and they are difficult to identify: All species have a broad-based adipose fin. They feed on vegetation and fruit; these species are native to South America.

The Glass (or Red-Eyed) Tetra. The glass (or red-eyed) tetra, *Moenkhausia oligolepis,* grows to slightly less than 5 inches in length. It is silver gray with dark edges to the scales on the upper body; the upper part of the eye is red. There is a yellow band around the peduncle and a black band at the base of the caudal fin.

The Red Piranha. The red piranha (or Natterer's piranha), *Serrasalmus nattereri,* is the most widespread of the piranha species found in South America; it grows to about 12 inches in length. The red piranha is a disc-shaped, muscular, and very powerful fish. Its diet consists of young fish, lean meat, meat-based flake foods, and insects. The primary color is steel gray; the back is a darker blue-gray color, and the underside is red. There are numerous black spots on the body. The most unusual feature of this species is its large, sharp teeth. Piranhas must be kept in an aquarium by themselves. Water temperature should be maintained at about 78°F. The red piranha is classified in some references in the separate family Serrasalmidae.

The Family Gasteropelecidae

The family Gasteropelecidae are commonly referred to as hatchet fish. See Figure 20–8. They have distinctive hatchet-shaped, deep, narrow bodies and are small fish, 1½ to 2½ inches long. Their pectoral fins are elongated and used to fly short distances across the water's surface. Aquariums must be fitted with a cover lid to prevent hatchet fish from flying out. Water temperature must be maintained at about 79°F, plus or minus 6°. Hatchet fish are native to South America. They are top-feeders and must be provided with floating

foods, including insects and dried foods. Several common species of hatchet fish are listed as follows:

> marbled hatchet fish, *Carnegiella strigata*
>
> common hatchet fish, *Gasteropelecus sternicla*
>
> silver hatchet fish, *Gasteropelecus levis*
>
> black-winged hatchet fish, *Carnegiella marthae*

Figure 20–8   Marbled hatchet fish. Photo by Aaron Norman.

The Family Anostomidae

The family Anostomidae are commonly referred to as headstanders; they are native to South America. See Figure 20–9. Headstanders reach a length of 3 to 5 inches. During rest, these fish position themselves vertically with their head down. They are found in shallow water among plants and rocks. Headstanders are middle- to bottom-dwelling fish and do well in community aquariums. The water should be soft, slightly acidic, filtered through peat, and maintained at 79°F. Members of the family are good jumpers, so the aquarium should be covered. They feed on all types of food, including lettuce, boiled spinach, and soaked oat flakes. Some common species of headstanders are listed as follows:

> marbled headstander, *Abramites hypselonotus*
>
> striped headstander, *Anostomus anostomus*
>
> spotted headstander, *Chilodus punctatus*
>
> banded Leporinus, *Leporinus fasciatus*
>
> nite tetra, flagtailed Prochilodus, *Prochilodus insignis*

Figure 20–9   Striped headstander. Photo by Aaron Norman.

The Family Hemiodontidae

The family Hemiodontidae (also called Lebiasinidae) is commonly referred to as pencil fish. See Figure 20–10. These are $1\frac{1}{2}$ to $7\frac{3}{4}$ inches in length and are slim, with a protrusive mouth. They are native to South and Central America. This family is distinguished from others in the suborder Characoidei by not having teeth in its lower jaw. Pencil fish have small, pointed mouths; some have an adipose fin, and some swim at an oblique angle. All pencil fish take on different coloration at night than during the day. They seek the upper and middle level of a community aquarium, where they feed on the surface. The water in the aquarium should be soft, slightly acidic, filtered through peat, and 76°F, plus or minus 6°. Some common aquarium species are listed as follows:

Figure 20–10   Golden or one-lined pencil fish. Photo by Aaron Norman.

golden pencil fish or one-lined pencil fish, *Nannostomus unifasciatus*

dwarf pencil fish, *Nannostomus marginatus*

three-lined pencil fish, *Nannostomus trifasciatus*

half-lined Hemiodus, silver Homiodus, flying swallow, *Hemiodopsis* or *Hemiodus semitaeniatus*

tube-mouthed pencil fish, *Nannostomus eques*

common pencil fish, *Nannostomus espei*

half-banded Pyrrhulina, *Pyrrhulina laeta*

Rio Meta Pyrrhulina, *Pyrrhulina metae*

The Family Citharinidae

The family Citharinidae is sometimes referred to as moon fish and is native to Africa. There are about seventy different species with a wide diversity of size and shape; four species are common in the aquarium trade.

The Six-Banded Distichodus. The six-banded Distichodus, *Distichodus sexfasciatus,* gets its name from the six or seven vertical, dark bars on the side of the fish. The primary color is orange with silver or gold iridescence. The adipose fin is white with a dark edge, the other fins have a reddish tinge, and the caudal fin is bright red with black edges. The eyes are large and set in a rather small, pointed head. It does well in a community aquarium, but because of its large size, up to 10 inches, it may be aggressive to smaller fish. Plants cannot be maintained in an aquarium with Distichodus because the fish eat the young, tender shoots. They feed on all types of food, including boiled lettuce, spinach leaves, and soaked oat flakes. Water should be maintained at 78°F, plus or minus 3°. See Figure 20–11.

Figure 20–11   Six-banded distichodus. Photo by Aaron Norman.

The One-Striped African Characin.   The one-striped African characin, *Nannaethiops unitaeniatus,* grows to about 2½ inches. The upper parts of the fish are brown and the lower parts yellow, with a white belly. A dark stripe runs from the mouth through the eye to the base of the caudal fin; above this is a golden iridescent stripe. This species does well in a community aquarium. Water temperature should be maintained at 78°F. They feed on all types of foods. See Figure 20–12.

Figure 20–12   One-striped African characin. Photo by Aaron Norman.

The Pike Characin.   The pike characin, *Phago maculatus,* grows to about 6 inches. It is a very thin, narrow fish with long beak-like jaws. Both upper and lower jaws have two rows of teeth. The caudal fin is large and deeply forked; other fins are small. The pike characin is light brown with a yellow tinge. A dark brown stripe runs from the eye back through the center of the caudal fin. There are dark, horizontal bars on the caudal fin. This very aggressive fish should be kept only in a species aquarium. Its diet consists of insects and other fish. Water temperature should be maintained at 81°F, plus or minus 2°.

The African Redfin.   The African redfin, *Neolebias ansorgii,* grows to about 1½ inches. This is a very colorful species; its primary color is dark blue with a white underside and red fins. There is an overall blue-gold iridescence to its color. This species occupies the lower levels of an aquarium and does best in shoals (groups of fish) of its own kind in a species aquarium. It feeds on all types of food. Water temperature should be maintained at 78°F, plus or minus 5°.

MAJOR FAMILIES— SUBORDER CYPRINOIDEI

The Family Cyprinidae

The family Cyprinidae is a very large family consisting of more than 1,500 species; they are commonly referred to as carps and minnows. See Figures 20–13, 20–14, and 20–15. This family of fish does not have teeth in its jaws, but uses teeth in its throat (pharyngeal teeth) to grind and break up food. Some have one or two barbels at the corners of the mouth; none have an adipose fin. Most occupy the lower levels of an aquarium and do well in community aquariums. Water temperatures should be maintained at 75°F, plus or minus 2°. These fish feed on all types of food. A few species of Cyprinidae are coldwater fish. Most of the tropical cyprinids are classified into three main genus groups, *Barbs, Rasbora,* and *Brachydanio,* and are commonly referred to as barbs, rasboras, and danios. All are colorful favorites of the aquarium trade. Most grow from 2 to 4 inches in length.

Figure 20–13   Tiger barb. Photo by Aaron Norman.

Some common species are listed as follows:

    Arulius barb or longfin barb, *Barbus arulius*

    rosy barb, *Barbus conchonius*

    Cuming's barb, *Barbus cumingi*

    striped barb or zebra barb, *Barbus fasciatus*

    black ruby barb or purple-headed barb, *Barbus nigrofasciatus*

    checker or island barb, *Barbus oligolepis*

    dwarf or pygmy barb, *Barbus phutunio*

    golden barb, *Barbus schuberti*

    tiger barb, *Barbus tetrazona*

    cherry barb, *Barbus titteya*

    red-tailed Rasbora, *Rasbora borapetensis*

    Harlequin fish or red Rasbora, *Rasbora heteromorpha*

Figure 20–14   Red-striped or glow-light rasbora. Photo by Aaron Norman.

pygmy or spotted Rasbora, *Rasbora maculata*

red-striped or glow-light Rasbora, *Rasbora pauciperforata*

scissor-tail or three-lined Rasbora, *Rasbora trilineata*

pearl danio, *Brachydanio albolineatus*

leopard danio, *Brachydanio frankei*

blue or Kerr's danio, *Brachydanio kerri*

spotted or dwarf danio, *Brachydanio nigrofasciatus*

zebra danio or zebra fish, *Brachydanio rerio*

Figure 20–15   Zebra danio. Photo by Aaron Norman.

The Tinfoil Barb.   The tinfoil barb, also called the goldfoil or Schwanenfeld's barb, *Barbus schwanenfeldi,* is a large member of this genus group that grows to 12 inches. It is native to Thailand, Malaysia, and Indonesia. The primary body color is silver with gold or blue iridescence. The fins are red, the dorsal fin has a black tip, and the caudal fin has black edges. Two pair of barbels are on the upper jaw. It does well in a community aquarium but should be kept only with other fish similar in size. Water temperature should be maintained at 72°F, plus or minus 5°. Its diet consists of vegetation, and these fish should be fed lettuce leaves along with other types of food.

The Spanner Barb.   The spanner barb (or T-barb), *Barbus lateristriga,* is another large barb that grows to 7 inches. It is native to Thailand, Malaysia, and Indonesia. The primary color is greenish gold, darker on the back and getting lighter down the sides, with silver on the underside. Two large, dark, vertical bars run from the back to the belly, and one dark, horizontal bar runs through the peduncle to the caudal fin. This community fish occupies all levels; however, it should be kept with fish similar in size. Water temperature should be maintained at 71°F, plus or minus 5°. It consumes all types of food, including plant material.

The Clown Barb.   The clown barb (or Everett's barb), *Barbus everetti,* is native to Singapore and Borneo; it grows to 6 inches. The primary color is reddish brown with an orange tinge and silver or gold iridescence; the fins are light red. It does well in a community aquarium but should be kept with other fish similar in size. Water temperature should be maintained at 79°F, plus or minus 2°. It consumes all types of food, including plant material.

The Giant Danio.   The giant danio, *Danio malabaricus,* is found in Southwest India and Sri Lanka; it grows to 4 inches. The back color is bluish green and the underside light pink. The sides of the fish are blue, and there are three or four yellow, horizontal stripes running across the **anterior** half. The front part of the fish has irregular vertical stripes and spots. All the fins except the colorless pectoral fin have a reddish tinge. This species does well in a community aquarium but may show aggression toward smaller fish. Water temperature should be maintained at 72°F, plus or minus 3°. It occupies the upper level of an aquarium and feeds on all types of foods. See Figure 20–16.

Figure 20–16   Giant danio. Photo by Aaron Norman.

The White Cloud Mountain Minnow. The white cloud mountain minnow, *Tanichthys albonubes,* is native to China and grows to 1¾ inches. See Figure 20–17. This very hardy fish tolerates a wide range of aquarium temperatures from 61 to 72°F. The back is olive brown with green iridescence; the underside is white. A thin, dark blue stripe runs the length of the body and is bordered above by a golden iridescent stripe and below by a reddish band. The fins are reddish with silver-blue tips. It does well in a community aquarium and feeds on all types of foods.

Figure 20–17   White cloud mountain minnow. Photo by Aaron Norman.

The Dadio. The dadio, *Laubuca dadiburjori,* is native to Indonesia, Burma, Malaysia, Sri Lanka, and Thailand, where it is found in slow-moving waters. It grows to 1¾ inches. The primary color is golden-brown. A blue stripe runs the length of the body; the fins are yellow. The pectoral fins are well developed and positioned so that rapid movement of these fins propels the fish or helps it "fly" across the surface of the water to capture flying insects. Dadios spawn in pools of water on the upper surface of aquatic plants. This species does well in community tanks where it occupies the upper levels. A cover glass or lid is needed. Water temperatures should be maintained at 75°F. It feeds on all types of food.

The Flying Fox. The flying fox, *Epalzeorhynchus kallopterus,* is native to Borneo, Indonesia, Sumatra, and Java. It is found in creeks with heavy vegetation. This species has an elongated body and grows to 5½ inches. The mouth faces downward and has two pair of barbels. The back is olive green, and the underside is white. A yellow stripe runs from the top of the snout, above the eye, and down to the peduncle. A larger black stripe runs below this from the snout, through the eye, the length of the body, and through the caudal fin. The fins are reddish brown with yellow tips. This species does well in community aquariums and occupies the lower levels; it should be provided with plenty of vegetation and rocks to hide among. Water temperature should be maintained at 76°F, plus or minus 5°. It feeds on all types of food, including plant material. See Figure 20–18.

Figure 20–18   Flying fox. Photo by Aaron Norman.

The Red-Tailed Shark. The red-tailed shark, *Epalzeorhynchus bicolor,* is native to Thailand. The dorsal fin is well developed, and gives this fish a shark-like appearance; it grows to 5 inches. It is completely black except for a red caudal fin. The mouth faces slightly downward, and there are two pair of barbels. This species does better in a species aquarium. It can be introduced into a community aquarium but may be aggressive to smaller fish and to its own species. This fish prefers the middle and lower levels of an

aquarium and should be provided with some vegetation and rocks. Water temperature should be maintained at 75°F, plus or minus 5°. Water should be soft, slightly acidic, and filtered through peat. It feeds on all types of food, including plant material. See Figure 20–19.

Figure 20–19   Red-tailed shark. Photo by Aaron Norman.

The Goldfish.   The cold-water Cyprinids that receive the most attention are the goldfish, *Carassius auratus.* Goldfish have been kept and developed by the Chinese for centuries; today, there are several varieties. Among them are the common goldfish, Bristal shubunkin, comet fantail, veiltail, moor, oranda, celestial, lionhead, bubble-eye, and pompon. The common goldfish, comet, and Bristal shubunkin have a single caudal fin. Fantails, veiltails, moor, and oranda have twin caudal fins and a single dorsal fin; the lionhead, celestial, pompom, and bubble-eye have twin caudal fins and no dorsal fin. Most goldfish grow from 3½ to 8 inches long. See Figures 20–20 and 20–21.

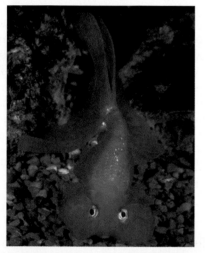

Figure 20–20   Bubble-eye variety of goldfish. Photo by Aaron Norman.

Goldfish occupy all levels of an aquarium. The single-tail varieties are the easiest to keep; the twin tails are more delicate and require more care. Water temperature should be maintained between 32 and 68°F for the single tails, and 46 to 68°F for the twin tails. They feed on all types of foods. The type of water is not critical, but it must be kept clean.

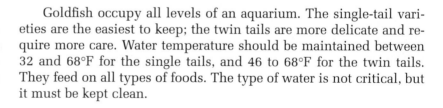

Figure 20–21   Oranda variety of goldfish. Photo by Aaron Norman.

Koi.   Koi are a species of carp, *Cyprinus carpio.* They originated in Japan, where they have been kept and developed for centuries. There are three main varieties: single-colored, two-colored, and three-colored koi. They are also divided by the type of scales they have. Those that have a few large scales are called Doitsu, those with pinecone scales are called Matsuba, those with gold metallic speckled scales are called Kin-rin, and those with silver metallic speckled scales are called

Gin-rin. The single-colored koi are yellow or orange. The two-colored koi are either white with red markings, called Kohaku, or gold and silver, called Hariwaki. The three-colored koi are light blue with orange and black markings, called Asagi; white with red and black markings, called Taisho Sanke; or black with red and white markings, called Showa Sante. Koi grow to 3 feet and can be worth thousands of dollars. Small koi are usually kept indoors and then moved to outdoor pools when they reach 5 inches; they occupy all levels. Water temperature should be maintained between 32 and 68°F. The water must be well filtered. They consume all types of foods.

The Family Gyrinocheilidae

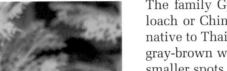

The family Gyrinocheilidae contains a single species, the sucking loach or Chinese algae eater, *Gyrinocheilus aymonieri.* This fish is native to Thailand and grows to 10 inches. Its primary body color is gray-brown with darker blotches along its sides and back. There are smaller spots on the head and caudal fin. It has large, fleshy lips on the underside of the snout and can cling to vegetation, rocks, and to the glass sides of an aquarium. It feeds primarily on algae and other vegetation but eats other foods. They do well in community aquariums. Water temperature should be maintained at 75°F, plus or minus 5°. See Figure 20–22.

Figure 20–22 Chinese algae eater or sucking loach. Photo by Aaron Norman.

The Family Cobitidae

The family Cobitidae is commonly referred to as loaches. These are bottom-dwelling fish native to the waters of Indonesia. They are distinguished by their flat underside, mouths on the lower side of their snout, and three pairs of barbels. They are nocturnal and hide among vegetation and rocks during the day. There are several species, which grow from 2¼ to 12 inches. Most do best in aquariums with their own species. Loaches feed on all types of foods. Water temperature should be maintained at 80°F, plus or minus 5°.

The clown loach, *Botia macracantha,* is the most colorful and popular; the primary color is bright orange. See Figure 20–23. Three large, black bands encircle the body. The first band covers the top of the head, the eyes, and the lower jaw; the second encircles the body in front of the dorsal fin; and the third encircles the peduncle and includes part of the dorsal and anal fins. The front portion of the dorsal and anal fins are orange; the other fins are red. The clown loach does well in community aquariums.

Figure 20–23 Clown loach. Photo by Aaron Norman.

Some other species of loaches are listed as follows:

skunk Botia, Hora's loach, or mouse loach, *Botia horae*

Pakistani loach or reticulated loach, *Botia lohachata*

banded loach, *Botia hymenophysa*

orange-finned loach, *Botia modesta*

chain or dwarf loach, *Botia sidthimunki*

zebra loach, *Botia striata*

Kuhli or Coolie loach, *Acanthophthalmus kuhli*

Nichol's loach, *Noemacheilus nicholsi*

CLASSIFICATION—SILURIFORMES

The order Siluriformes is made up of twenty-four families and more than 2,500 species; they are commonly referred to as catfish. Most species are covered with skin but lack scales. Some species are covered with bony plates; most species have an adipose fin. The pectoral and dorsal fins have stiff, sharp rays or spines. Their bodies are usually flattened on the underside, and they have barbels; these characteristics aid in their bottom-dwelling nocturnal habits. Six families are discussed here:

Family Siluridae (glass catfish)

Family Schilbeidae (three-striped glass catfish)

Family Mochokidae (upside-down catfish)

Family Pimelodidae (unarmored catfish)

Family Callichthyidae (armored catfish)

Family Loricariidae (sucker catfish)

MAJOR FAMILIES—CATFISH

The Family Siluridae

The family Siluridae contains one species commonly found in the aquarium trade, the glass catfish, *Kryptopterus bicirrhis*. This species is native to Indonesia. It is transparent, with the backbone and internal organs clearly visible. It has a very long anal fin, only one ray to the dorsal fin, and no adipose fin; the caudal fin is forked. There is one pair of barbels on the upper jaw that can be extended outward. The glass catfish grows to 4 inches in length and should be kept in small shoals; as a single specimen, it does not do well. This species does well in a community aquarium where it occupies the upper level. The glass catfish swims at an oblique angle with its tail down. Water temperature should be maintained at 75°F, plus or minus 5°. It feeds on all types of foods. See Figure 20–24.

Figure 20–24   Glass catfish. Photo by Aaron Norman.

The family Schilbeidae also contains one species of glass catfish that is common in the aquarium trade—the three-striped glass catfish, *Eutropiellus debauwi*. See Figure 20–25. This species from Central Africa is blue-gray with three dark, horizontal stripes running behind the gill to cover the full length of the body. The center

stripe runs through the center of the caudal fin. The upper and lower stripes end at the peduncle and appear as spots on the caudal fin. The anal fin is very long, and there is a small dorsal fin towards the front part of the back; it also has an adipose fin. This species grows to 3 inches and does well in a community aquarium where it occupies the upper level and swims obliquely with its tail down. The three-striped glass catfish should be kept in shoals; a single specimen does not do well. It feeds on all types of foods. Water temperature should be maintained at 76°F, plus or minus 3°.

Figure 20–25   Three-striped glass catfish. Photo by Aaron Norman.

The Family Mochokidae

The family Mochokidae contains one species common in the aquarium trade—the upside-down catfish, *Synodontis nigriventris.* It is native to Central Africa and grows to 3 inches. The primary color is gold-brown with numerous dark spots. It has a very large adipose fin; the fins are colorless with dark spots. The upside-down catfish has three pair of barbels: one smooth pair on the upper jaw and two feathered pair on the lower jaw. Its upside-down habit enables it to feed on the underside of leaves and rocks. It feeds primarily on algae but takes other foods from the surface of the water. This species does well in a community aquarium and should be provided with vegetation and rocks. Water temperature should be maintained at 77°F, plus or minus 4°.

The Family Pimelodidae

The family Pimelodidae is referred to as unarmored catfish. They are covered with skin but have no scales or bony plates. Two species, both from South America, are common in the aquarium trade: the spotted Pimelodella, *Pimelodella* or *Pimelodus pictus,* and the graceful Pimelodella, *Pimelodella gracilis.* The spotted Pimelodella grows to 4½ inches in length. It is silvery-white with numerous black spots along its body and caudal fin. It also has three pair of very long barbels. See Figure 20–26. The graceful Pimelodella grows to more than 6½ inches. Females are almost colorless with a dark stripe running horizontally from the gill cover to the caudal fin; the males are a blue-black iridescent color. The graceful Pimelodella has three long pairs of barbels; the one pair on the upper jaw are extremely long. Both species do well in community aquariums provided they have vegetation and rocks to hide among during the day; they become active at night. They occupy the middle and lower levels of the aquarium. Water temperature should be maintained at 72°F, plus or minus 5°. They prefer soft, slightly acidic water. Both species are primarily carnivorous, feeding on insects, worms, and meat-based flaked foods.

Figure 20–26   Spotted pimelodella. Photo by Aaron Norman.

The Family Callichthyidae

The family Callichthyidae is commonly referred to as armored catfish and is native to Central and South America. See Figure 20–27. They have two rows of overlapping bony plates or scutes. They are small, ranging in size from 2½ inches to 4 inches. Their mouths are on the underside of the snout, and they have three pairs of barbels. They are primarily scavengers, feeding on the bottom of streams usually at night. These species can be seen in an aquarium dashing to the surface, where they gulp in air that is later absorbed in their digestive tract. They do well in a community aquarium if provided with vegetation and rocks to hide among; single fish do not do well. Water temperature should be maintained at 73°F, plus or minus 6°.

Figure 20–27   Peppered or salt-and-pepper Corydoras catfish. Photo by Aaron Norman.

Some common species of armored catfish are listed as follows:

short-bodied catfish, *Brochis splendens*

bronze catfish, *Corydoras aeneus*

elegant catfish, *Corydoras elegans*

dwarf or pygmy Corydoras, *Corydoras hastatus*

leopard Corydoras, *Corydoras julii*

black-spotted Corydoras, *Corydoras melanistius*

peppered Corydoras, *Corydoras paleatus*

swartz's Corydoras, *Corydoras schwartzi*

Myer's Corydoras, *Corydoras myersi*

reticulated Corydoras, *Corydoras reticulatus*

Albino Corydoras, albino species of *C. paleatus* and *C. aeneus*

armored catfish, *Callichthys callichthys*

Port Hoplo or Atipa, *Hoplosternum thoracatum*

striped-tail catfish, *Dianema urostriata striata*

The Family Loricariidae

The family Loricariidae is commonly referred to as sucker catfish. These fish have three or four rows of bony plates or scutes and have underslung mouths designed for sucking or clinging onto vegetation and rocks found in the fast-moving mountain streams of South America. The whiptail catfish, *Loricaria filamentosa,* grows to about 10 inches in length. It has an elongated body with a very elongated peduncle; the upper lobe of the caudal tail is elongated into an appendage. The primary color is gray-brown to yellow-brown with several dark markings and spots. The front rays of the dorsal, pelvic, and pectoral fins form stiff, sharp spines; it does not have an adipose fin. It prefers water temperature of 75°F, plus or minus 3°. See Figure 20–28.

Figure 20–28   Whiptail catfish. Photo by Aaron Norman.

The golden Otocinclus, *Otocinclus affinis,* is a small sucker-mouth catfish, growing only to a little over 1½ inches. See Figure 20–29. The primary color is golden-brown; a broad black stripe runs from the snout to the base of the tail. The fins are colorless except for black markings on the caudal fin; there is no adipose fin. The underside of the fish is silver-white. It prefers a water temperature of 73°F, plus or minus 6°. The sucker catfish, *Plecostomus punctactus,* grows to more than 12 inches. The primary color is dark brown to black with dark spots. On the lighter fish, five oblique bars can be seen on the side. The fins are brown to black with dark spots, and there is an adipose fin. The fins are large, with the front rays forming stiff, sharp spines. It also prefers water temperatures of 75°F, plus or minus 6°. All three of these suckermouth catfish do well in a community aquarium. They should be provided with vegetation and rocks to hide among during the day, although the plants may be rooted up. They feed on all types of food, including algae and other plant material.

Figure 20–29   Sucker catfish. Photo by Aaron Norman.

CLASSIFICATION—ATHERINIFORMES

The order Atheriniformes is made up of three suborders and fifteen families. Five families are discussed here.

    Suborder — Exocoetoidei
        Family — Exocoetidae (halfbeaks and flying fish)
    Suborder — Cyprinodontoidei
        Family — Cyprinodontidae (killifish or egg-laying tooth carps)
        Family — Anablepidae (four-eyed fish)
        Family — Poeciliidae (live-bearers or viviparous top minnows)
    Suborder — Atherinoidei
        Family — Atherinidae (rainbow fish or silversides)

MAJOR FAMILIES—SUBORDER EXOCOETOIDEI

The Family Exocoetidae

The family Exocoetidae or Hemiramphidae contains one species that is common in the aquarium trade, the halfbeak or wrestling halfbeak, *Dermogenys pusillus.* This fish is native to the waters of Thailand, Malaya, Sumatra, and Java. It is found in fresh and brackish waters (waters where fresh and salt waters mix). The halfbeak grows to almost 3 inches; its distinguishing characteristic is that the lower jaw protrudes far beyond the upper jaw. This fish is greenish brown, and the underside is greenish yellow. The caudal fin is rounded, and the dorsal and anal fins are opposite each other, giving this fish a pike-like appearance. The male has a red spot on the lower front of the dorsal fin; the fins are greenish yellow. The fish does well in a community aquarium and occupies the upper levels. It feeds on insects and floating foods. Water temperature should be maintained at 73°F, plus or minus 5°. See Figure 20–30.

Figure 20–30   Celebes halfbeak. Photo by Aaron Norman.

MAJOR FAMILIES—
SUBORDER CYPRINODONTOIDEI

The Family Cyprinodontidae

The family Cyprinodontidae is referred to as killifish or egg-laying tooth carps. See Figures 20–31 and 20–32. They are native to Africa, Asia, Europe, South America, Central America, and North America. They are small fish, reaching lengths of 1½ to 4 inches. They have elongated bodies, their heads are usually flattened, and their mouths are protrusive to pick up food. They are brightly colored; the males are usually more colorful than the females. The fins of the males are also larger, elongated, and more colorful. Some species lay their eggs on vegetation, whereas others lay their eggs in the mud on the bottom of stream beds. Those species that lay their eggs in the mud are referred to as annual fish because after their eggs are laid, the stream or pond dries up, and the parent fish die; the next year when the rains come, the streams and ponds are filled and the eggs hatch. The young grow quickly and become sexually mature in a single season. The eggs have very hard shells and are able to withstand the weeks or months that the dry period may last. Most species are very aggressive and should be raised as a pair in an aquarium without other fish. Water temperature should be maintained at about 75°F, plus or minus 5°. The water should be soft, slightly acidic, and filtered through peat. These fish are usually not kept in large aquariums because of their reproductive habits and their short life spans.

Figure 20–31   Golden lyretail or golden lyretailed panchax. Photo by Aaron Norman.

Figure 20–32   American flag fish. Photo by Aaron Norman.

Some examples of the killifishes are listed as follows:

   lyretail or lyretailed panchax, *Aphyosemion australe*
   red lyretail, *Aphyosemion bivittatum*
   steel blue Aphyosemion, *Aphyosemion gardneri*
   blue gularis, *Aphyosemion sjoestedti*
   lampeyed panchax or lampeye, *Aplocheilichthys macrophthalmus*
   plumed lyretail, *Aphyosemion filamentosum*
   sabrefin or sicklefin killi, *Austrofundulus dolichopterus*
   striped panchax, *Aplocheilus lineatus*

Ceylon killifish, *Aplocheilus dayi*

blue panchax, *Aplocheilus panchax*

black-finned pearl fish, *Cynolebias nigripinnis*

American flag fish, *Jordanella floridae*

Rachov's nothobranch, *Nothobranchius rachovi*

Medata or rice fish, *Oryzias latipes*

Playfair's panchax, *Pachypanchax playfairi*

clown killi or rocket panchax, *Pseudoepiplatys annulatus*

six-barred Epiplatys, *Epiplatys sexfasciatus*

featherfin panchax, *Pterolebius longipinnis*

red Aphyosemion, *Roloffia occidentalis*

green panchax or dwarf panchax, *Aplocheilus blocki*

The Family Anablepidae

The family Anablepidae is a small family consisting of only one genus group and two species. They are referred to as four-eyed fish, *Anableps anableps,* and inhabit waters of Mexico, Central America, and Northern South America. See Figure 20–33. This fish has an elongated body, a flat head, and large, protruding eyes. It swims just below the surface, and the large, protruding eyes are divided by the water's surface, which allows the fish to see above and below the water at the same time. It grows to 8 inches in length; its body is primarily blue-gray with a yellowish tinge to the underside. The dorsal fin is set far back on the body, and the caudal tail is rounded. Males can be distinguished from females by their prominent **gonopodium,** a modification of the anal fin into a tube-shaped organ through which sperm packets pass and enter the oviduct of the female. The sperm packets break open and release sperm to fertilize the eggs. Four-eyed fish give birth to live young twice a year. They do well in community tanks but may be aggressive to smaller fish. Water temperature should be maintained at 75°F, plus or minus 2°, and 5 teaspoons of salt should be added for every 2½ gallons of water.

Figure 20–33 Four-eyed fish. Photo by Aaron Norman.

The family Poeciliidae consists of live-bearers, viviparous top minnows, or live-bearing tooth carps. See Figure 20–34. This family of fish give birth to live young. The males are usually smaller than the females but have much brighter colors and striking patterns. The anal fin of the males is modified into a gonopodium. Sperm that is passed to the female may be stored in the folds of the oviduct, so further pregnancies can occur from one mating. The fertilized eggs develop within the female, and she gives birth about twenty to sixty days after fertilization, depending on the species. Live-bearers do well in community aquariums, but the young must be protected from the adults or they will become food. Water temperature varies with the species, but if maintained at 79°F, plus or minus 5°, it should be sufficient for most species. Water should be medium-hard and slightly alkaline. Some species do best if salt is added to the water. Live-bearers consume all types of foods, including plant material. Beginning **aquarists** may want to check an additional reference before raising live-bearers. Some of the common live bearers are listed as follows:

Piketop minnow, *Belonesox belizanus*

Eastern mosquito, *Gambusia affinis*

Girardinus, *Girardinus metallicus*

pseudo helleri, *Heterandria bimaculata*

merry widow, *Phallichthys amates*

guppy or millions fish, *Poecilia reticulata*

blue limia or black-bellied limia, *Poecilia melanogaster*

humpbacked limia, *Poecilia nigrofasciata*

Cuban limia, *Poecilia vittata*

sailfin molly, *Poecilia latipinna*

black molly, *Poecilia hybrid*

blue-eyed Priapella, *Priapella intermedia*

swordtail, *Xiphophorus helleri*

platy, *Xiphophorus maculatus*

Variatus platy, *Xiphophorus variatus*

Montezuma swordtail, *Xiphophorus montezumae*

Mexican sailfin molly, *Poecilia velifera*

Figure 20–34   Black sailfin molly. Photo by Aaron Norman.

The Family Atherinidae

The family Atherinidae is referred to as rainbowfish and silver-sides; they are found in the shallow coastal waters of Australia and Madagascar. See Figure 20–35. They are distinguished from other fish because they lack a lateral line; their pelvic fins are midway back on the underside, and they have two dorsal fins. The fins are positioned so that the top part of the fish is almost identical to the bottom half. Four species are found in the aquarium trade: the dwarf rainbowfish, *Melanotaenia maccullochi;* the red-tailed rainbowfish, *Melanotaenia nigrans;* the Madagascar rainbowfish, *Bedotia geayi;* and the Celebes sailfish, *Telmatherina ladigesi.* These fish do well in community aquariums, although plenty of vegetation should be provided. Water should be medium-hard and maintained at about 75°F, plus or minus 3°. Some salt may need to be added to the water for them to do best.

Figure 20–35   Madagascar rainbowfish. Photo by Aaron Norman.

CLASSIFICATION—PERCIFORMES

The order Perciformes consists of more than 6000 species in sixteen suborders and 160 families. Included in this order are both marine and freshwater species; nineteen families are discussed here. Species in this order do not have their swim bladder connected to an open duct to the throat. The dorsal, anal, and pelvic fins usually have spines. The dorsal fin consists of an anterior (front) part that has spines and a posterior (rear) part that has soft rays.

Suborder — Percoidei
    Family — Apogonidae (cardinalfish)
    Family — Centropomidae (snooks, robalos, and glassfish)
    Family — Grammidae (basslets)
    Family — Nandidae (leaf fish)
    Family — Cichlidae (cichlids)
    Family — Pomacentridae (damsel fish, anemone fish)
    Family — Monodactylidae (fingerfish)
    Family — Toxotidae (archer fish)
    Family — Platacidae (bat fish)
    Family — Scatophagidae (scats)
    Family — Chaetodontidae (butterfly fish)
    Family — Labridae (wrasses)
Suborder — Blennioidei
    Family — Acanthuridae (surgeons, tangs, and unicorn fish)
    Family — Zanclidae (Moorish Idol)
Suborder — Mastacembeloidei
    Family — Mastacembelidae (spiny eels)
Suborder — Anabantoidei
    Family — Anabantidae (climbing perch)
    Family — Helostomatidae (kissing gourami)
    Family — Belontiidae (fighting fish and paradise fish)
    Family — Osphronemidae (gourami)

MAJOR FAMILIES— SUBORDER PERCOIDEI

The Family Apogonidae

The family Apogonidae is a group of small, narrow fish less than 4 inches in length. They have short, compact bodies, and the peduncle is unusually large. These species are mouth-brooders, and the male incubates the eggs in his mouth. Most species are nocturnal and hide among the vegetation and rocks during the day. Water temperature should be maintained at 77°F, plus or minus 2°.

The Pajama cardinalfish has an interesting color pattern. The part of the fish in front of the dorsal and pectoral fins is yellow and white; next there is a large, black band from top to bottom, and the back half of the fish is white with brown spots. The anterior dorsal fin and pelvic fins are yellow with red on the front edges. The posterior dorsal fin, pectoral fins, and caudal fin are colorless. Two other species are the barred cardinalfish, *Apogon binotatus,* and the sawcheek cardinalfish, *Apogon quadrisquametus.*

The Family Centropomidae

The family Centropomidae is a group of marine fish found in the waters near India, Thailand, and Burma. See Figure 20–36. One species found in the aquarium trade is the Indian glassfish, *Chanda ranga.* This narrow-bodied species grows to about 3 inches long. It is transparent, and the skeleton and internal organs are visible; in light, it shows a gold, yellow, and green iridescence. It is a peaceful fish and

does well in a community tank. Water temperature should be maintained at 71°F, plus or minus 6°. The water should be hard and slightly alkaline. Three to six teaspoons of salt should be added for every 2½ gallons of water.

Figure 20–36   Indian glassfish. Photo by Aaron Norman.

The Family Grammidae

The family Grammidae consists of small fish found around the coral reefs of the Indo-Pacific and Caribbean. See Figure 20–37. One species found in the aquarium trade is the royal Gramma or fairy basslet, *Gramma loreto,* which is native to the Caribbean. It grows to about 3 inches. The head and front half of the fish are bright pinkish purple, and the back half is bright yellow. There is a black spot on the dorsal fin and an oblique, narrow yellow stripe running through the eye from the mouth. This species adapts well to a community aquarium and should have some rocks and coral to hide around. It occupies all levels of the aquarium and feeds on chopped foods and live foods. Water temperature should be maintained at 77°F, plus or minus 2°.

Figure 20–37   Royal Gramma or Fairy basslet. Photo by Aaron Norman.

The Family Nandidae

The family Nandidae is a small family made up of very aggressive predators. The South American leaf fish, *Monocirrhus polyacanthus,* is a very narrow fish, brown and yellow in color, and resembles a leaf. Its head is pointed, the forehead slightly concave, and it has a protrusive mouth. The lower jaw is elongated and appears to have a worm-like growth. It moves slowly among vegetation in search of prey. The leaf fish is able to change its coloration so it blends in with changes in the vegetation; it must be kept in a species aquarium. Water temperature should be maintained at 79°F, plus or minus 4°. The Schomburgk's leaf fish, *Polycentrus punctatus,* is a narrow, short-bodied fish with a short peduncle. See Figure 20–38. It is another predator that slowly stalks its prey. It is usually gray-blue with numerous dark markings but can change its coloration to blend into its surroundings. It must be kept in a species tank. Water temperature should be maintained at 76°F, plus or minus 3°.

Figure 20–38   Schomburgk's leaf fish. Photo by Aaron Norman.

The Badis, *Badis badis,* is another fish in this family, but it is more peaceful and can be kept in a community aquarium. See Figure 20–39. Its general color is reddish brown with blue and red spots; the dorsal fin has several dark markings. It prefers the middle and lower levels of an aquarium. Water temperature should be maintained at 81°F, plus or minus 2°.

Figure 20–39   Badis. Photo by Aaron Norman.

The Family Cichlidae

The family Cichlidae, commonly referred to as cichlids, is a large family of heavy-bodied fish. Most cichlids are native to South America and Africa; there are more than 700 species. They have one pair of nostrils instead of two like most other fish. The dorsal fin has a long base, and the first three rays form spines. The anal fin is short, and the first three rays are also spined; the caudal fin is usually rounded. There are two main patterns of spawning. Some species lay their eggs on rocks, leaves, or logs, or in the holes dug by the male. Other species, known as shelter breeders, lay their eggs in spawning areas out in the open that have been cleared. Spawning sites are usually defended by both parents. They tend the nest until the young are hatched, and the female may stay with them for a short time; a few species are mouth-brooders.

The Oscar, *Astronotus ocellatus,* is one of the most popular and easily recognized cichlids. In an adequate aquarium, it can grow to 14 inches in length. The coloration varies considerably, but it is usually gray with black, vertical blotches along the sides; some have reddish orange scales along the lower sides. The one constant in coloration is a black spot encircled with a reddish orange ring on the peduncle. Oscars are aggressive and should be kept in a species aquarium; they occupy all levels. Rooted plants are not recommended because they will dig them up. Water temperature should be maintained at 75°F, plus or minus 3°. They consume all types of food, including chopped meat, live minnows, and small fish.

The angelfish, *Pterophyllum scalare,* is another popular aquarium fish from South America. See Figure 20–40. It is a narrow, disc-shaped cichlid with large fins. The dorsal, pelvic, and anal fins are usually elongated. It grows to about 6 inches. Angels are usually silver with brown to black markings; numerous patterns are available. It is very difficult to determine the sex of angels. They are peaceful fish that do well in community aquariums. Water temperature should be maintained at 79°F, plus or minus 7°. They consume all types of foods.

Figure 20–40   Angelfish. Photo by Aaron Norman.

Other cichlids available in the aquarium trade are listed as follows:

Port acara, *Aequidens portalegrensis*

keyhole cichlid, *Cleithacara maronii* (See Figure 20–41.)

Figure 20–41   Keyhole cichlid. Photo by Aaron Norman.

blue acara, *Aequidens pulcher*

Agassiz's dwarf cichlid, *Apistogramma agassizi*

yellow dwarf cichlid, *Apistogramma reitzigi*

Jack Demsey, *Cichlasoma biocellatum*

Festivuum, *Cichlasoma festivum*

Rio Grande perch, *Cichlasoma cyanoguttatum*

firemouth cichlid, *Cichlasoma meeki*

zebra cichlid, *Cichlasoma nigrofasciatum*

pike cichlid, *Grenicichla lepidota*

orange chromide, *Etroplus maculatus*

green chromide, *Etroplus suratensis*

earth-eaters, *Geophagus jurupari*

jewel cichlid, *Hemichromis bimaculatus*

golden Julie, *Julidochromis ornatus*

Egyptian mouth-brooder, *Hemihaplochromis multicolor*

Fuelleborn's cichlid, *Labeotropheus fuelleborni*

Trewavas's cichlid, *Labeotropheus trewavasae*

Golden-eyed dwarf cichlid, *Nannacara anomala*

Kribensis, *Pelvicachromis pulcher*

Malawi blue cichlid, *Pseudotropheus zebra*

Mozambique mouth-brooder, *Sarotherdon mossambicus*

brown discus, *Symphysodon aequifasciata* (See Figure 20–42.)

Figure 20–42   Brown discus. Photo by Aaron Norman.

blue discus, *Symphysodon aequifasciata haraldi*

discus, *Symphysodon discus*

waroo, *Uaru amphiacanthoides*

The Family Pomacentridae

The family Pomacentridae includes the damsel fish, anemone fish, and sergeant majors. These are marine fish of the coral reefs of tropical areas. They are brightly colored active fish, and most are territorial and aggressive.

Clown Fish.   One group, the clown fish, is also called anemone fish because they live in and around the tentacles of the sea anemone. See Figure 20–43. Most fish would be fatally stung by the tentacles, but clown fish are immune and enjoy the safety of the sea anemone.

Clown fish provide bits of food to the sea anemone in exchange for safety. These small fish do best if sea anemones are also kept in the aquarium; this mutually beneficial relationship is known as **symbiosis.** Common species are the tomato or fire clown fish, *Amphiprion ephippium frenatus,* and the common clown fish, *Amphiprion ocellaris.* The tomato or fire clown fish is tomato red in color with a black blotch on its side. A white vertical stripe runs down the head just behind the eye; a narrow black stripe runs along the edge of the white stripe. The common clown fish is orange with three white bands encircling the body; the white bands are edged with black. The fins are orange, edged in black. If more than one clown fish is kept in an aquarium, plenty of space should be provided, and each should have its own sea anemone.

Figure 20–43   Common clown fish. Photo by Aaron Norman.

The damsel fish and sergeant majors are related to the clown fish but do not live among the sea anemone; they prefer more open areas and seek protection among the coral branches. The yellow-backed damsel or black-footed sergeant major, *Abudefduf melanopus,* grows to about 2¾ inches in length. The primary body color is gray-turquoise, but the back is yellow from the upper lip through the eye to the back of the dorsal fin. The top and bottom of the caudal fin are yellow, and the pelvic and anal fins have black edges.

The Yellow-Tailed Blue Damsel.   The yellow-tailed blue damsel, *Abudefduf parasema*, is about 4 inches long. The body is blue, and the peduncle and caudal fin are yellow.

The Sergeant Major.   The sergeant major, *Abudefduf saxatitis* grows to 7 inches. See Figure 20–44. It is silver-blue with a yellow tinge to the upper body. There are six dark, vertical bands on the side.

Figure 20–44   Sergeant major. Photo by Aaron Norman.

The Humbug Damsel.   The humbug damsel, *Dascyllus aruanus,* is silver-white with three black bands encircling the body. It grows to 4½ inches.

The Domino Damsel. The domino damsel, *Dascyllus trimaculatus,* is all black except for a white spot on each side by the middle of the dorsal fin and one on the forehead. See Figure 20–45. It grows to 4¾ inches in length. The blue puller, *Chromis coeruleus,* is blue-green. It grows to 5 inches. All members of this family prefer water temperatures of 77°F, plus or minus 2°.

Figure 20–45   Domino damsel. Photo by Aaron Norman.

The Family Monodactylidae

The family Monodactylidae is referred to as fingerfish. One species of fingerfish, *Mondactylus argenteus,* is native to the brackish coastal waters of Africa and India. This tall, disc-shaped fish grows to about 9 inches. The dorsal and anal fins are long-based, large, and appear almost identical in shape; the fish is silver. The dorsal fin is tinged with yellow, and the anal fin has a black front edge. Young can be kept in hard fresh water, but as they grow, salt should be added and the amount gradually increased. Water temperature should be maintained at 75°F, plus or minus 3°. They feed on all types of food.

The Family Toxotidae

The family Toxotidae contains one genus group and six species. They are called archerfish and are found on the brackish coastal waters of Southeast Asia and the Western Pacific. See Figure 20–46. Archerfish shoot drops of water at insects to knock them into the water. The head is pointed; the tongue and roof of the mouth form a tube through which water can be forced out in a straight stream. One species of archerfish, *Toxotes jaculator,* grows to about 10 inches. The fish is silver with black marks and blotches on its side; the back is greenish brown. The dorsal fin is black with yellow markings, and the anal fin is black. The edges of the caudal fin are tinged with greenish yellow. Salt should be added to the aquarium water, and the temperature should be maintained at 80°F, plus or minus 3°. The archerfish feeds on insects and worms.

Figure 20–46   Archerfish. Photo by Aaron Norman.

The Family Platacidae

The family Platacidae is referred to as bat fish and is native to the marine waters of India and the Western Pacific. See Figure 20–47. These large, disc-shaped fish have tall, wide-based dorsal and anal fins. They grow to 20 inches in length. The round bat fish, *Platax orbicularis,* is easy to keep in marine aquariums. It prefers tall aquariums, and because it grows quickly, a large aquarium should be made available. The round bat fish is light brown with two dark, vertical bars crossing the body. One bar runs from the top of the head down

Figure 20–47 ***Round batfish. Photo by Aaron Norman.***

through the eye to the bottom of the lower jaw. The second bar runs from just in front of the dorsal fin down to just behind the pelvic fins. The pelvic fins are dark brown; the leading edge and trailing edge of the brown dorsal fin is black. The base of the anal fin is brown, and the rest is black. The base of the caudal fin is brown; the rest is colorless. Bat fish are peaceful fish but are easily frightened. Coral and rocks should be provided for them to hide among. Water temperature should be maintained at 77°F, plus or minus 2°.

The Family Scatophagidae

The family Scatophagidae consists of a small number of species referred to as scats. See Figure 20–48. They are native to the coastal waters of the Indo-Pacific. They are narrow, almost square-shaped fish, and their heads are short. The Argus fish, *Scatophagus argus,* grows to about 12 inches. The young live in fresh or brackish water, but the adults live in coastal sea areas. They are silver-brown with dark spots on the body. The fins are colorless on the young; on older individuals, the fins become yellow-brown with dark spots. They feed on all types of foods. Water temperature should be maintained at 76°F, plus or minus 6°.

Figure 20–48 ***Argus fish. Photo by Aaron Norman.***

The Family Chaetodontidae

The family Chaetodontidae is the butterfly fish and angelfish that live in the marine waters of the Indo-Pacific region. See Figure 20–49. They are brightly colored, attractive fish that live among the coral reefs. Most of them have small, elongated snouts that they use to pick and scavenge among the crevices of the coral for food. They are territorial, and fish of the same species will fight. Angelfish can be distinguished from butterfly fish because they have a backward-pointing spine on the gill cover. Butterfly fish and angelfish can be fed a variety of fresh meat, frozen brine shrimp, flake foods, and live foods. Water temperature should be maintained at 77°F, plus or minus 2°. It is impossible to describe the coloration of these fish.

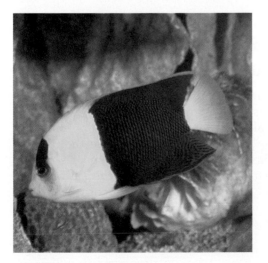

Figure 20–49 ***Black and gold angelfish, Oriole angelfish, two-colored angelfish, bicolored angelfish, or bicolored cherub. Photo by Aaron Norman.***

Common species are listed as follows:

threadfin butterfly fish, *Chaetodon auriga*

Pakistani butterfly fish or collared coral fish, *Chaetodon collare* copperbanded butterfly fish, beaked coral fish, or long-nosed butterfly fish, *Chelmon rostratus*

long-nosed coral fish or forceps fish, *Forcipiger longirostris*

wimple fish, pennant coral fish, Poor Man's Moorish Idol, or featherfin bullfish, *Heniochus acumineatus*

black and gold angelfish, Oriole angelfish, two-colored angelfish, or bicolor cherub, *Centropyge bicolor* (See Figure 20–49)

yellow-faced angelfish, blue-faced angelfish, *Euxiphipops xanthometapon*

rock beauty, *Holacanthus tricolor*

blue ring angelfish, *Pomacanthus annularis*

black or grey angelfish, *Pomacanthus arcuatus*

imperial or emperor angelfish, *Pomacanthus imperator*

The Family Labridae

The family Labridae is referred to as wrasses, hogfish, or cleaner fish. See Figure 20–50. They are found in temperate and tropical marine areas. Wrasses are found in a wide range of colors and patterns; most species change color as they mature. They live among the coral and feed on parasites they pick off the skin of other fish. They lie on the bottom or bury themselves into the sand at night. Common species are listed as follows:

Cuban hogfish, *Bodianus pulchellus*

clown labrid, *Coris angulata*

bird wrasse, *Gomphosus coeruleus*

cleaner wrasse, *Labroides dimidiatus*

green wrasse, *Thalassoma lunare*

Figure 20–50   Cuban hogfish. Photo by Aaron Norman.

MAJOR FAMILIES— SUBORDER BLENNIOIDEI

The Family Acanthuridae

The family Acanthuridae is made up of several species of brightly colored fish referred to as tangs, surgeons, or unicorn fish. See Figure 20–51. They are oval-shaped fish with bony, scalpel-like projections on the peduncle. These projections can give the unwary prey a painful wound. Three of the more common species are the powder-blue surgeon, *Acanthurus leucosternon;* the sailfin tang, *Zebrasoma veliferum;* and the smooth-headed unicorn fish, *Naso lituratus*. All of these fish are native to the Indo-Pacific oceans. It is recommended that they receive vegetative-type foods in their diet.

Figure 20–51   Smooth-headed unicorn fish. Photo by Aaron Norman.

The Family Zanclidae

The family Zanclidae contains only one species—the Moorish Idol, *Zanclus canescens* or *Zanclus cornutu S.* See Figure 20–52. The scientific name can be found both ways in different reference books. It is closely related to the tangs and surgeons. The coloration is a series of yellow, white, and black vertical bands on the body and includes the fins. The dorsal fin is elongated into a long, trailing fin. This species is not very hardy and is difficult for inexperienced marine aquarists to keep. It grows to 9 inches in length.

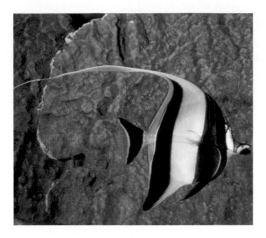

Figure 20–52   Moorish idol. Photo by Aaron Norman.

MAJOR FAMILIES— SUBORDER MASTACEMBELOIDEI

The Family Mastacembelidae

The family Mastacembelidae consists of the spiny eels. There are about fifty species native to the freshwaters from Africa to Southeast Asia. They are eel-like in appearance, with the dorsal, caudal, and anal fins merging into one continuous fin. There are several spiny rays at the front part of the dorsal fin. The upper jaw and snout are elongated and extend past the lower jaw.

The species *Macrognathus aculeatus* grows to about 14 inches. It is mainly brown. Four or five dark brown spots with light brown lines encircling the spots are located on the dorsal fin. The species *Mastacembelus argus* grows to about 10 inches. It is dark brown with numerous light brown spots and markings. Both species should be kept in species aquariums. Spiny eels occupy the middle to the lower levels and may bury themselves into the sand or gravel on the bottom with only the heads sticking out. Water temperature should be maintained at 77°F, plus or minus 2°. They feed primarily on worms and insects.

MAJOR FAMILIES— SUBORDER ANABANTOIDEI

The suborder Anabantoidei, order Perciformes, is a group of small- to medium-sized freshwater fish referred to as labyrinth fish. They are native to Africa, Southeast Asia, the Philippines, and Indonesia. All members of this suborder have a labyrinth organ that lies in the gill cavity and allows the fish to breath atmospheric air. This is an advantage for the fish that live in the poorly oxygenated tropical waters. In most species, the male builds a floating nest made from saliva and air bubbles. As the eggs are released from the female, the male fertilizes them, and the eggs float up under the bubble nest. The male then guards the nest and may even guard the young fry. There are four families in this suborder: Anabantidae, Helostomatidae, Belontidae, and Osphronemidae.

The Family Anabantidae

The family Anabantidae consists of about twenty species. Included in the family is the climbing perch, *Anabas testudines.* This species has an elongated body with large scales and a large mouth. The pectoral fins are well developed, and the fish can climb out of shallow pools and crawl across the ground to another body of water. Its general color is grayish brown, and the fins are tinged with yellow. It grows to 10 inches in length and is easy to keep. Water temperature should be maintained at 80°F, plus or minus 5°.

The Family Helostomatidae

The family Helostomatidae contains only one species—the kissing gourami, *Helostoma temmincki.* See Figure 20–53. This narrow, oval fish grows to 12 inches in length. The main characteristic of the species is its habit of extending its lips and "kissing." It is silver with a greenish or yellowish tinge to the fins. Kissing gourami consume all types of food material but are primarily vegetarians. They do well in a community aquarium. Water temperature should be maintained at 79°F, plus or minus 4°.

Figure 20–53   Kissing gourami. Photo by Aaron Norman.

The Family Belontidae

The family Belontidae consists of about fifty species, including the fighting fish, and a few gourami and paradise fish. See Figure 20–54. The Siamese fighting fish, *Betta splendens,* is a popular freshwater labyrinth species. It has an elongated body with long, flowing dorsal, caudal, anal, and pelvic fins. In the wild, they are reddish brown with blue-green iridescence. There are numerous red, green, and blue dots along the sides. By selective breeding, they are now available in green, blue, red, and violet. The males are very aggressive and must never be put together. One or two females and one male can exist in a community aquarium; however, they will probably do better and display more intense coloration in a species-only aquarium. They feed on all types of food. Water temperature should be maintained at 80°F, plus or minus 3°.

Figure 20–54   Male Siamese fighting fish. Photo by Aaron Norman.

Several species of gourami are in this family, including the honey gourami, *Colisa chuna;* dwarf gourami, *Colisa lalia;* banded gourami, *Colisa fasciata;* pearl gourami, *Trichogaster leeri;* and the three-spot gourami, *Trichogaster trichopterus.* See Figure 20–55. These gourami have pelvic fins that are elongated into long filaments. They do well in community aquariums. Water temperatures should be maintained at 77°F, plus or minus 5°. They feed on all types of foods.

Two species of paradise fish are commonly found in the aquarium trade: the paradise fish, *Macropodus opercularis,* and the brown-spike-tailed paradise fish, *Macropodus cupanus dayi.* The coloration on the paradise fish, *M. opercularis,* varies considerably;

the color is normally reddish brown with blue-green vertical bars. Coloration on some individuals may be primarily blue with reddish brown markings and stripes on the caudal fins. *M. opercularis* is aggressive and should be kept in a species aquarium. *M. cupanus dayi* is primarily reddish brown, with two dark, horizontal bars on the sides. The dorsal and anal fins are reddish brown with blue edges. The caudal tail has red elongated center rays, and the outer rays are reddish brown with blue edges. *M. cupanus dayi* can be kept in a community aquarium but may be aggressive to smaller fish. Water temperature for *M. opercularis* should be maintained at 67°F, plus or minus 8°; *M. cupanus dayi* should be kept at 71°F, plus or minus 7°. Both species consume all types of foods.

Figure 20–55   Three-spot gourami. Photo by Aaron Norman.

The Family Osphronemidae

The family Osphronemidae consists of one species—the gourami, *Osphronemus goramy.* It grows to 24 inches in length and thus is suited for the aquarist that has a good-sized aquarium. It is brownish gray with silver iridescence. It has a small head, and the lower lip projects forward in older fish. The pelvic fins are elongated and thread-like. The gourami consumes all types of food, but it is primarily a vegetarian. Lettuce and soaked oat flakes can provide the needed plant material to their diet. Water temperature should be maintained at 73°F, plus or minus 6°.

CLASSIFICATION—TETRAODONTIFORMES

The family Balistidae, order Tetraodontiformes, is made up of a group of marine species referred to as triggerfish. See Figure 20–56. They have three dorsal spines, with the second locking the first into an upright position. They are found throughout the world. As with other species of colorful marine fish, it is almost impossible to describe all the various colorations, patterns, and shapes of these unique fish. Common species are the undulate or orange-green triggerfish, *Balistapus undulatus,* and the Picasso or white-barred triggerfish, *Rhinecanthus aculeatus.* Both species have strong jaws and sharp teeth. They can be kept in community aquariums but may show aggression to smaller fish. Water temperature should be maintained at 75°F, plus or minus 2°. They feed on chopped meat, shellfish, and meat-based flake foods.

Figure 20–56   Picasso or white-barred triggerfish. Photo by Aaron Norman.

CLASSIFICATION—SCORPAENIFORMES

The family Scorpaenidae, order Scorpaeniformes, is made up of the lion fish, also called the dragon fish or turkey fish, *Pterois volitans.* See Figure 20–57. This species grows to about 14 inches in length. It is native to the Indo-Pacific and Red seas. Its primary color is reddish-

brown with numerous white bands encircling the body. The fin rays are long, and the spiny ray is poisonous. It feeds on live fish or chopped meats and has a voracious appetite. It can be kept in a community aquarium with other fish that are equal in size. Water temperature should be maintained at 77°F, plus or minus 2°.

Figure 20–57   Lion fish, dragon fish, or turkey fish. Photo by Aaron Norman.

CLASSIFICATION—GASTEROSTEOIDEI

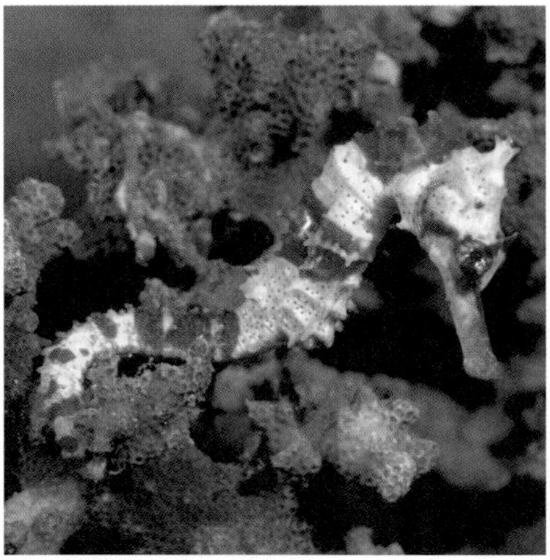

The family Syngnathidae, order Gasterosteoidei, contains the sea horses. See Figure 20–58. The yellow or oceanic sea horse, *Hippocampus kuda,* grows to about 2 inches in length, and it is native to the Indo-Pacific marine waters. Sea horses may vary in color from black to yellow. They have a hard body covering and no caudal or anal fins; they swim in a vertical position. Sea horses have a prehensile tail that they can use to anchor themselves to plants and coral. They are fairly difficult to keep and should be kept in species aquariums. Water temperature should be maintained at 77°F. They feed on live foods.

Figure 20–58   Golden or yellow sea horse. Photo by Aaron Norman.

CLASSIFICATION—MORMYRIFORMES

The family Mormyridae, order Mormyriformes, contains 130 species of fish referred to as mormyrs and elephant-snout fish. The long-nosed elephant-snout fish, *Gnathonemus petersi,* is occasionally seen in the aquarium trade. See Figure 20–59. It is native to the dark, turbid fresh waters of Africa. It grows to 9 inches in length. Elephant-snout fish are dark brown-black, with a violet iridescence. There are two yellowish white irregular, vertical stripes running between the single dorsal and anal fins. The edges of the dorsal, anal, and caudal fins are edged with white. The lower lip is elongated into a tool that is used to dig for food on the bottom of streams. They also emit an electrical field around their bodies that enables them to move around in the darkness. They feed on insects, insect larvae, and worms. They will adapt to a community aquarium provided it has some dark secluded places to hide. Water temperature should be maintained at 75°F.

Figure 20–59   Long-nosed elephant-snout fish. Photo by Aaron

CLASSIFICATION— OSTEOGLOSSIFORMES

The family Notopteridae, order Osteoglossiformes, has six species consisting of fish with long anal fins and reduced or no dorsal fins. See Figure 20–60. Several species of knife fish exist, and there are some discrepancies in their scientific name. They move about by the undulating motion of their long anal fin. Members of the family are commonly referred to as knife fish. The African knife fish, *Xenomystus nigri,* grows to about 8 inches in length. It is primarily light brown with a lighter underside. It occasionally rises to the surface to swallow atmospheric air. The African knife fish is peaceful and can be kept in a community aquarium with other peaceful fish. Darkened areas and plenty of vegetation should be provided. Water temperature should be maintained at 79°F, plus or minus 4°.

Figure 20–60   Knife fish. Photo by Aaron Norman.

AQUARIUM TANKS AND EQUIPMENT

Modern aquarium tanks are available in all sizes and shapes. The glass used in tank construction can be bonded together with a sealant to form a strong leak-proof container. These all-glass aquariums are the cheapest to purchase. Framed tanks are available for additional cost; the frames are usually for decoration and add very little to the structural strength of the aquarium. Plastic tanks are available on the market but must be cleaned carefully to prevent scratching the surfaces; these tanks are rather inexpensive.

The size of tank will depend on the amount of money a person wants to invest in an aquarium, and the species he or she plans to purchase. The larger the aquarium, the more money it will cost for the tank and accessory equipment. Most people will want to start small and decide later whether they want to get more involved and invest more money.

To determine the number of fish that can be put in a tank, one should multiply the length by the width to determine the square inches of surface area. A rule of thumb is to have no more than 1 inch of fish for every 10 square inches of surface area in a tropical freshwater aquarium; 1 inch of fish for every 30 square inches of surface area in a cold freshwater aquarium, and 1 inch of fish for every 48 square inches in a marine aquarium. An aquarium tank 30 inches long by 12 inches wide has 360 square inches of surface area. Dividing the 360 square inches by 10 indicates that the tank will hold 36 inches of tropical freshwater fish. That could be three 12-inch fish, eighteen 2-inch fish, or twelve 3-inch fish. One should also allow for rocks, coral, or branches that are in the aquarium.

The aquarium should be situated in a draft free area and out of direct sunlight. Aquariums placed in direct sunlight have an excessive growth of algae, and the tank may overheat during the summer and cause an oxygen deficiency. Aquariums situated in drafty areas may have fluctuations in the water temperature. The surface should be level to prevent cracking the glass. A thin piece of styrofoam can be placed under the aquarium to allow for an uneven surface.

The water in an aquarium must be kept clean and contain adequate amounts of dissolved oxygen; aquariums use filters and air pumps to accomplish this. Air pumps introduce compressed air (oxygen) into the water and at the same time disperse carbon dioxide (aeration). As the air pump introduces air into the water, air bubbles rise to the surface. Air is usually forced through airstones (fused, porous glass) to break up the flow into minute bubbles. As the bubbles rise, they pull the water with them and cause the water in the tank to circulate. This circulation brings water from the bottom of the tank to the surface, where aeration (the exchange of oxygen and carbon dioxide) can take place. The constant circulation of the water also helps equalize the temperature throughout the aquarium. See Figure 20–61.

The purpose of the filter is to remove solid waste and uneaten food materials from the water. Filter systems work in one of three ways: through the mechanical removal of waste and food materials, by chemical removal of dissolved materials, and by biological filtration or conversion of harmful substances into harmless ones.

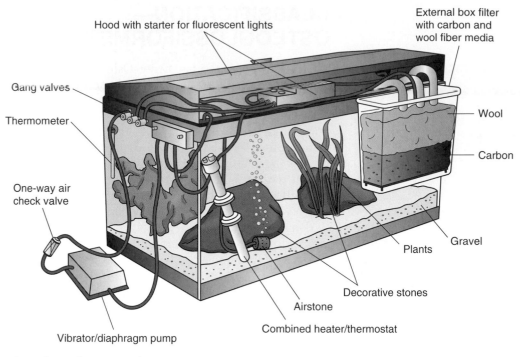

Figure 20–61   A modern glass aquarium system.

Older aquariums used internal mechanical filters driven by airstones. See Figure 20–62. As the air rose, this system pulled water through the media in the box. The media separated the waste and food materials out of the water, and the water was returned to the tank. Newer models still used the air-driven uplift principle, but instead use external box filters or internal sponge filters. See Figure 20–63.

Most aquariums today use power filters with an electric motor connected to an impeller by a magnetic device through the wall of the filter box. By not housing a drive shaft through the filter box, there is no chance for leaks in the system. Power filters increase water flow through the media and remove more waste materials. It is important that the connection of the return hose to the aquarium is secure. If the hose is loose, large amounts of water can be pumped onto the floor, and the entire tank could possibly be emptied. Undergravel filters and internal and external canister filters are now available with the use of electric power pumps. See Figure 20–64.

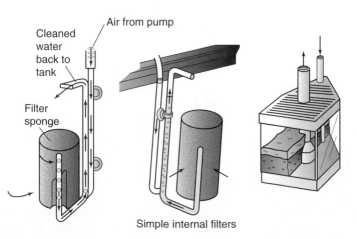

Figure 20–62   Internal filters. An internal filter can consist of a sponge or plastic box containing wool and carbon media. Dirty water enters the hole in the sponge or holes in the plastic box, where suspended materials are trapped in the filter media. The air pulls the dirty water into the filter and forces the clean water out through the tube.

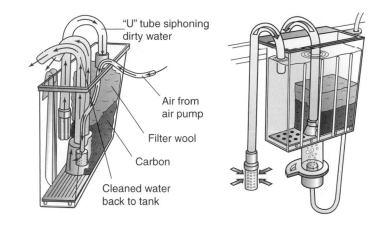

External box filters

Figure 20–63 *An external box filter hangs on the back of the aquarium. The dirty water is fed into the filter by a siphon tube. Air from the pump enters the bottom of the filter. As the air rises, it pulls the dirty water up the siphon tube and through the filter media. The filtered, clean water flows back into the aquarium.*

Chemical filtration is accomplished with the use of activated charcoal; the charcoal soaks up dissolved minerals and chemicals. Activated charcoal can be placed in box filters, or return water can flow through the charcoal. The charcoal must be replaced periodically as it reaches a point where it can no longer absorb dissolved material.

Aeration is achieved by the use of electrically driven air pumps; pumps are either a vibrator-diaphragm type or a rotary-vane type. Pumps are available in several sizes; the size needed depends on the size of the aquarium, the number of airstone filters, and other equipment used.

Biological filters neutralize toxic substances, especially ammonia excreted from fish during respiration and produced from decaying waste and food materials. In this system, a slotted plastic plate is placed in the bottom of the aquarium. See Figure 20–65. A 2- or 3-inch layer of gravel is placed over the slotted plate; the gravel should be 0.125 inches in diameter. Gravel that contains large particles allows food to fall into the spaces around the gravel and decay; too-small particles block the action of the air pump. Marine aquariums should have coral sand placed above the slotted plate. In some aquariums, peat should be placed over the slotted plate. The slotted plate has an uplift tube on the corner; this tube should be placed at the back of the aquarium. When the tank is filled, an air line with an aerator attached is placed down the uplift tube; this line is connected to an air pump. Air is pumped down the line and through the airstone. As the air bubbles rise in the uplift tube, water is drawn through the gravel, up through the uplift tubes, and back into the aquarium at the surface. The gravel layer acts as a filter for suspended particles in the water. After several hours of aeration, colonies of aerobic bacteria (bacteria that use oxygen) begin to grow in the gravel; this process is called nitrification. *Nitrosomonas* bacteria convert ammonia into nitrites, and then *Nitrobacter* bacteria convert the nitrites into nitrates. Nitrates may inhibit the growth of fish, but they are not nearly as toxic as ammonia. Nitrites and nitrates may be utilized by plants in the aquarium.

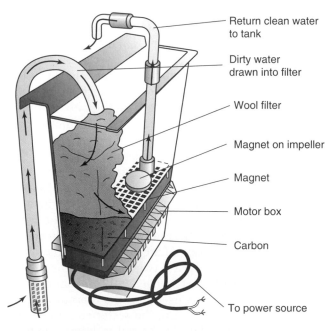

Return clean water
to tank

Dirty water
drawn into filter

Wool filter

Magnet on impeller

Magnet

Motor box

Carbon

To power source

Figure 20–64 *A motorized or power filter operates the same as an external filter except that a motorized pump is used to draw the dirty water through the filter media, and the filtered, clean water is pumped back into the aquarium.*

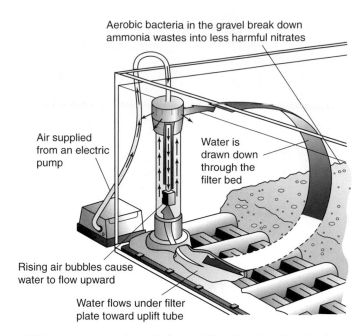

Aerobic bacteria in the gravel break down
ammonia wastes into less harmful nitrates

Air supplied
from an electric
pump

Water is
drawn down
through the
filter bed

Rising air bubbles cause
water to flow upward

Water flows under filter
plate toward uplift tube

Figure 20–65　An undergravel filter system. A slotted plastic filter is placed at the bottom of the aquarium, and a layer of ground is placed on the filter. Air is pumped through an airstone. As the air rises in the tube, it lifts water upward inside the tube. As the water rises, it pulls the dirty water through the gravel, where suspended materials are filtered, and aerobic bacteria convert the harmful ammonia wastes into less harmful nitrites and then nitrates.

Several factors contribute to an increase in ammonia and nitrites:

■ Increase in waste material and uneaten food on the bottom of the aquarium

■ Dirty filters

■ Failure to change water at regular intervals (One-fourth to one-fifth of the water should be changed every three to four weeks.)

■ Overfeeding that leads to an accumulation of uneaten food

■ Overpopulation of fish

Nitrite levels can be measured with test kits. It is very important to remove nitrites from tapwater used in marine aquariums; many marine species are sensitive to nitrites.

In order for the aerobic bacteria to flourish, the pH and oxygen levels must be maintained; pH is a measure of acidity or alkalinity. The nitrification process is most efficient at a pH of 9. The pH scale reads from 0, which is very acidic, to 14, which is very alkaline; 7.0 is neutral. Most freshwater tropical fish prefer slightly acid conditions. The pH of an aquarium is easily measured with a pH test kit.

A protein skimmer can be added to marine systems to eliminate dissolved proteins and other waste products that pollute and discolor the water; they op-

erate on the airlift water principle. As the air bubbles reach the surface of the airlift tube, the foaming residue that contains the proteins and other wastes are collected in a chamber; these residues settle into a yellow liquid that can be poured out.

Tropical aquariums must have water heaters to heat the water and thermostats to maintain water temperature. Heaters are of two types. One type is a heater that is placed in the aquarium and controlled by a separate thermostat; the electrical connection between the heater and the thermostat is on the outside of the aquarium. The second type is a condensed heater and thermostat; the heater and the thermostat are connected together inside a glass tube. This tube is placed into the aquarium, and the temperature is controlled by an adjustment knob at the top of the tube. Heaters are available in several standard sizes. Approximately 10 watts of power for each $1\frac{1}{2}$ gallons of water is required. A 30-gallon aquarium would need a heater of approximately 250 watts. Large aquariums may have better distribution of heat if two heaters are used. It is important to use a heater that is of a recommended size. If a large heater is used and the thermostat unit fails while the heater is on, the water could become overheated. If a small heater is used and the room temperature should drop, the heater might not produce enough heat to maintain the water temperature.

Another internal type of heater uses a heating cable buried under the gravel. The advantage is that fish do not come in contact with the heater and cannot be burned or dislodge it.

Several types of thermometers are available for monitoring water temperature. Liquid crystal thermometers in the form of an adhesive strip stuck to the outside of the aquarium are very popular. Mechanical dial thermometers can also be used and attached to the outside of the aquarium. Conventional float-type thermometers that float or are attached to the inside of the aquarium with a suction cup are also available.

Water hardness is another factor that needs to be considered. Hardness is caused by dissolved magnesium and calcium salts. Hardness is of two types: temporary and permanent. Temporary hardness can be removed by boiling the water. Permanent hardness needs to be removed by chemical means or distillation. The hardness can also be diluted by removing some of the hard water and adding soft water.

Hardness can be measured in several different ways. A German scale measures as degrees of hardness, °DH; and the Clark or English scale measures as degrees of hardness, °H. Hardness can also be measured as ppm (parts per million of either $CaCo_3$ or CaO); and as gpg (grains per gallon). Conversion factors are as follows:

1 °DH on the German scale is 17.9 ppm of $CaCo_3$

1 °H on the English scale is 14.3 ppm of CaO

gpg $CaCo_3$ = ppm $CaCo_3$ × 0.058 ppm

Test kits are available to measure hardness as parts per million of calcium carbonate (ppm of $CaCo_3$). To avoid confusion with the various scales, we use the ppm scale: 17.9–89.5 ppm of $CaCo_3$ is considered very soft, 89.5–179 soft, 179–358 medium-hard, 358–537 hard, and above 537 ppm is considered very hard water.

Chlorine in tapwater may be a concern to aquarists; however, it is easily removed by aerating the water for 12 to 24 hours. Commercial dechlorinators or activated charcoal filters are also effective.

The water in marine aquariums needs to be measured for the amount of salt (density); this is done with a hydrometer. This instrument floats higher or lower in the water depending on the density of the water. The density or specific gravity of the water can be read on a scale located on the glass tube; the recommended specific gravity is 1.020 to 1.026. Water should be tested at the recommended operating temperature. Salt or marine mixes can be added to freshwater to provide a marine water environment.

Lighting in an aquarium is necessary for visual display and for plant growth. Plants need light for photosynthesis, which in turn uses carbon dioxide and releases oxygen. The amount of light depends on the kind and number of plants, the height of the aquarium, the type of water, and the thickness of the glass cover. Lighting can be provided with tungsten filament bulbs or fluorescent bulbs; fluorescent bulbs are available in different colors and can add to the attractiveness of an aquarium. Fluorescent bulbs operate at cooler temperatures and do not increase the water temperature. A standard level of fluorescent lighting is approximately 0.10 to 0.14 watts per square inch of surface area. Lighting should be provided 10 to 15 hours per day.

Decorative items may be added to an aquarium to make it more attractive; these also provide shelter and hiding places for aquarium inhabitants. Decorations added should make the aquarium look as natural as the original environment of the fish. In freshwater aquariums, rocks and wood pieces are often added; in marine aquariums, corals, shells, rocks, and wood may be used. Wood pieces should be boiled several times to kill bacteria and resins that may pollute the aquarium. Many manufactured look-alike pieces are now available for aquarium use as well. Petrified woods, bamboo canes, and pieces of slate can also be used.

Plants can be added for attractivenes and to provide a means of using carbon dioxide while releasing oxygen. Some common floating plants for tropical aquariums include water lettuce, *Pistia stratiotes,* and duckweed, *Lemna minora.* Rooted plants include Japanese dwarf rush, *Acorus gramineus;* Madagascar lace plant, *Aponogeton madagascariensis;* baby's tears, *Bacopa monnieria;* fanwort, *Cabomba aquatica;* hornwart, *Ceratophyllum demersum;* water trumpet, *Cryptocoryne blassii;* Amazon sword plant, *Echinodorus grandiflorus;* water weed, *Egeria densa;* hairgrass, *Eleocharis acicularis;* water thyme or Canadian pondweed, *Elodea canadensis;* water star, *Hygrophila polysperma;* water wisteria, *Hygrophila difformis;* Ambulia, *Limnophilia aquatica;* Ludwigia, *Ludwigia repens;* Java fern, *Micosorium pteropus;* water milfoil, *Myriophyllum aquaticum;* giant Indian water star, *Nomaphila stricta;* European curly pondweed, *Potamogeton crispus;* arrowhead, *Sagittaria graminea;* giant eel grass, *Vallisneria gigantea;* and Java moss, *Vesicularia dubyana.*

Japanese dwarf rush, baby's tears, hornwort, water weed, hairgrass, water thyme, duckweed, Ludwigia, water milfoil, European curly pondweed, arrowhead, giant eel grass, and Java moss can be used in cold-water aquariums in addition to willow moss, *Fontinalis antipyretica.* These plants should be planted to a depth where the crown of the plant is level to or just above the gravel. Plants may be fertilized with liquid or tablet plant foods specially formulated for aquarium use. Plastic plants are also available and add the same attractiveness but do not filtrate the aquarium.

FEEDING

Fish consume all types of foods and should be given a variety so that they do not become bored. Fish should be fed two or three times a day and should receive only what they can consume in 3 to 5 minutes.

Flake foods are ideal for small fish up to 4 or 5 inches long. These foods are produced from the meat of fish, fish eggs, wheat, and vegetables. They usually contain additional vitamins and minerals. These food materials are also processed into pellet form for larger fish; floating fish sticks are good for large top-feeding fish, while small pellets that sink slowly are best for middle-feeders, and sinking tablets are appropriate for bottom-feeders.

Several live foods are also available; among them are small crustaceans called water fleas, *Daphnia.* Brine shrimp, *Artemia,* are small shrimp that live in salt lakes or brackish waters. These shrimp should be rinsed off and may be fed to both small and large fish. River shrimp and bloodworms are available for still larger fish. Earthworms, flies, maggots, wood lice, caterpillars, crickets, and grasshoppers are other live foods that can be used. One problem with feeding live foods is the parasites and diseases that they may carry can be harmful to fish.

Freeze-dried and frozen foods are basically the same as live foods except they are in a safe form. Some freeze-dried or frozen foods available are *Mysis* shrimp, Pacific shrimp, tubiflex worms, krill, and plankton.

Carnivorous fish can be fed minced or chopped meat; common foods include beef heart, liver, raw fish meat, and shellfish meat. Pieces of turkey and chicken can also be used. Meat pieces can also be thinly sliced and frozen; these can then be broken and fed. Some of the larger carnivorous fish may need to be fed other fish; small goldfish may be used for this purpose.

Many fish also require vegetative material in their diets. Chopped or shredded lettuce, chopped spinach leaves, canned peas, wheat germ, and oat flakes can be used. Any vegetables not consumed within 8 hours should be removed,

Marine species of invertebrates are of four types:

1. Those that feed on plankton that is filtered from the water, including stone and horny coral, tubeworms, bivalves, some species of sea cucumbers, and crustaceans. Commercially prepared plankton foods and frozen foods are available for the plankton feeders.

2. Those that feed on plant material, including sea urchins, mollusks, and sea slugs. A diet of lettuce and spinach will hopefully prevent them from feeding on aquarium plants.

3. Those that are carnivorous, including crabs, starfish, sea anemones, shrimp, and lobsters. They can be fed small pieces of crab, fish, shrimp, and flake foods. Sea anemones should be fed only when their tentacles are out (in bloom). Drop the food onto their tentacles.

4. Those that are scavengers, including sea cucumbers. They feed on debris and uneaten foods on the bottom of the aquarium.

DISEASES AND AILMENTS

Aquarists need to be able to recognize symptoms of diseases in their fish as soon as they begin to appear. Early recognition and treatment increase the chances for recovery. Some signs that may indicate a disease problem include the following:

1. Unusual behavior such as swimming movements

2. A fish leaning to the side, floating to the surface, or sinking to the bottom

3. Fins not extended and laying flat against the body

4. Caudal fins closed or rolled

5. A fish shying away from the shoal

6. A fish slower in its movements than others

7. Respiration faster and deeper than normal

8. Fish up on the surface gasping for air

9. Fish that are rubbing against decorations on the bottom

10. Fish that are not eating

11. Fish that are thin with sunken sides

12. Abnormally large or swollen belly

13. Abnormal color

14. Frayed fins

15. Malformation of the back and spinal column

16. Cloudiness of the eyes

17. Scales sticking out away from the body instead of lying flat

Parasites

White Spot.   White spot, or "Ich," is caused by the parasitic organisms *Ichthyophthiriis multifiliis.* This is one of the most common diseases of aquarium fish; the organism is often present in fish but usually does not cause any problems. When the fish is stressed, the parasite multiplies rapidly, causing numerous white spots on the body and fins; the parasite lives under the epidermis. When the parasite matures, it breaks through the wall of the epidermis and swims about. The mature parasite eventually settles on a stone or plant and forms a cyst (resting stage). During this resting stage, the cell divides into 1,000 to 2,000 ciliospores (reproductive spores). These break from the cyst and swim freely in the water. When they make contact with a fish, they burrow into the epidermis and mature in about four days. The parasites living under the skin and the ciliospores in the cysts cannot be destroyed by chemicals; however, the free-swimming adult parasite and ciliospores can be destroyed with chemicals added to the water. Chemicals are readily available at most pet stores with aquarium products. The organism *Cryptocaryon irritans* causes the same type of problem in marine fish.

Slime Disease. Slime disease is caused by parasites that attack the skin of fish. The fish's body produces large amounts of mucus or slime in reaction to the irritation caused by the parasites. The heavy mucus produced dulls the colors of the fish, weakens the fish, and may cause death if the parasites attack the gills. Parasites of the genus *Chilodonella* may cause slime disease in fresh water; *Cyclochaeta (Trichodina)* and *Costia (Ichthybodo)* in all types of water; and, *Brooklynella* in marine water. Chemical water treatments are available for the treatment of slime disease; however, some of the parasites may be resistant. Infected fish should be removed from the aquarium and put in a water-salt solution. A formalin bath solution should follow if the water-salt solution does not succeed.

Hole-in-the-Head. Hole-in-the-head disease is caused by the parasitic organism *Hexamita.* These organisms live under the skin and in the muscle tissues of the host fish. As the organisms multiply, the tissues break down, the skin breaks open, and the decayed tissue breaks away, leaving "holes" in the body. These large, open sores or ulcers are most common around the head, along the lateral line, and at the base of the dorsal fin. The disease may be linked to vitamin deficiencies, and supplements may aid in healing. Metronidazole or dimetridazole added to the water is the recommended treatment.

Velvet Disease. The organism *Oodinium* causes a condition called velvet disease in freshwater fish and coral disease in marine fish. *Oodinium* is a member of a group of parasites known as dino flagellates, which are related to algae and contain chlorophyll. The or-

ganisms penetrate the skin cells and feed on them. Both diseases appear as a velvet-like gold coating of dust on the sides of the fish. The gills are usually affected, and the fish will have rapid movement of the gills. Copper-based treatments can be added to the water. Some marine invertebrates cannot tolerate copper-based treatments, however, and the affected fish should be removed and treated in a hospital aquarium.

White Fungus Growth. A fungus organism called *Saprolegnia* causes white fungus growths on the fins, mouth, eyes, and gills. These fungus growths sometimes appear as secondary infections in fish already infected with spot or bacterial diseases. Fish should be removed to a hospital aquarium where the water has been treated with malachite green oxalate or medicinal methylene blue. Fish may also be treated directly by applying malachite green oxalate, povidone-iodine, or mercurochrome to the affected area. Treated fish should be kept in a hospital aquarium until the fungus disappears. Some commercial treatments are available at pet stores with fish supplies.

Flukes. Two species of flukes that can cause problems are *Dactylogyrus* and *Gyrodactylus.* Resulting damage from both can be fatal depending on the degree of infestation. Flukes are tiny worm-like parasites that attach themselves to the gills and body of fish. They have numerous hooks on the rear part of their body and a sucking mouth. A single fluke has both male and female reproductive organs (hermaphroditism). *Dactylogyrus* permanently attach themselves to the gills. Fish affected with *Dactylogyrus* will have rapidly moving gills, swim at the surface, and pant heavily trying to obtain more oxygen. Their gills will be covered with slime, and parts of the gills may be eaten away. Fish may also scrape their bodies up against objects trying to relieve the irritation. *Gyrodactylus* primarily attach themselves to the body of the fish. The infected fish will also scrape up against objects; their color will fade; the fish will produce more slime in response to the irritation; and the fins may also become ragged and eaten away. Commercial treatments are available at pet stores with fish supplies. Infected fish should be removed and treated in a hospital aquarium; salt and formalin baths are recommended. The aquarium that the fish were removed from should be thoroughly cleaned before returning fish to it.

Anchor Worms. Anchor worms are members of the class Crustacea, phylum Arthropoda. Two species, *Lernaea cyprinacea* and *Lernaea carassii,* are found regularly on goldfish; other species of tropical fish are rarely affected. Only the females of *Lernaea* species are parasitic. They have a chitinous adhesive organ around their mouth and will burrow under the scales and attach themselves to the muscle tissue. Two whitish-colored egg sacs can be seen trailing from the parasite. Commercial water treatments are available. The aquarist may need to remove the fish from the aquarium, remove the parasite with tweezers, and treat the wound with iodine and mercurochrome.

Fish Lice. *Argulus foliaceus* is a species of fish lice that fish can be infected with from feeding on infected plankton. These flat, oval-shaped crustaceans attach themselves to the body of the fish with sucking mouth parts (proboscis); they pierce the skin with a spine and discharge a poison. This poison can be fatal to young fish. Fish should be removed, and the lice should be removed with tweezers. The fish should then be placed in a hospital tank to which a commercially available treatment has been added. The original tank should be thoroughly cleaned before returning fish to it.

Bacterial Disease

Finrot. Finrot is a condition especially prevalent in brightly colored fish or those with long, trailing fins like the black mollies. The edges of the fins start to lose their color, the tissue between the rays begins to break down, and the fins become ragged. The fins become shorter as the condition persists. The disease is believed to be caused by a bacteria, along with a contributing factor such as a vitamin-deficient diet, poor water conditions, and fin nipping by other fish. Commercially prepared treatments are available; treatment of marine species may be more difficult.

Mouth Fungus. Mouth fungus is a misleading name because the disease is actually caused by a bacteria, *Chondrococcus columnaris.* Symptoms appear as white, tufty material around the mouth and white patches on the skin. Severely affected fish may be unable to close their mouths and will have respiratory trouble. Commercially prepared treatments are available. If the condition persists, the infected fish may need to be removed, placed in a hospital aquarium, and treated with antibiotics.

Neon Disease. Neon disease is a disease that primarily affects the neon tetra but may also affect other species of tropical fish. The disease is caused by the parasite *Plistophora hyphessobryconis.* The organism lives in the tissue of the fish's body, where it multiplies and produces spores. These spores are released when the fish dies or when parts of the fish slough off. The spores later hatch, and the newly hatched organisms can infect other fish. The first symptoms of the disease are gray to white discolorations on the fish's body. These areas usually enlarge and bulge outward. After a period of time, the fish dies; no treatment or cure is available.

Tuberculosis. Tuberculosis of fish is caused by the bacteria *Mycobacterium;* cold-water species are usually the most affected. Once the bacteria has invaded the fish, it moves to various tissues and organs of the body; the symptoms can be varied. The fish usually become slow, move off to an area of the aquarium by themselves, and die. Usually, there are discolored areas or bulges. The back may be disfigured because of deformities to the spinal cord. All weak or diseased fish should be removed from the aquarium and destroyed; other fish should be moved to a hospital aquarium. The aquarium, gravel, decorations, and equipment should be thoroughly cleaned and disinfected before fish are returned to the aquarium.

Pseudomonas and Aeromonas. Bacterial infections from *Pseudomonas* and *Aeromonas* species sometimes occur in cold-water fish. These bacteria infect only fish that have already been weakened by other disease conditions or old age. Infected fish have skin lesions or ulcers, and their bellies may be swollen as a result of an accumulation of fluid. This swelling causes the scales to stand out from the body; swelling behind the eyes may also cause the eyes to protrude.

Environmental Control. Fish living in an aquarium that has a lack of oxygen and an excess supply of carbon dioxide may show signs of disease. If single fish are seen swimming near the surface, this is normal; if almost all of the fish are swimming at the surface, it may be a sign of insufficient oxygen. One-fourth to one-half of the water in the aquarium should be replaced immediately, and the cause, the lack of oxygen, should be corrected.

Aquariums with an incorrect water pH can cause respiratory distress and excess mucus production in fish. This excess mucus clouds the color of the fish. Partial replacement of the water and the addition of chemicals can correct the pH; any changes should be gradual.

After a period of time, the chemical filtration abilities of an aquarium are reduced, and a buildup of ammonia occurs. The first sign of this buildup is a cloudiness or murky appearance of the water; there is also a buildup of anerobic bacteria. Anaerobic bacteria produce foul-smelling poisonous gases that are harmful to fish. A characteristic sign of gravel filtration system pollution is a black color to the gravel, which is caused by the buildup of iron sulphite produced by the interaction of hydrogen sulphite and iron compounds in the gravel. The hydrogen sulphite is produced by the anerobic bacteria; at this point, the aquarium should be cleaned. Fish living in a polluted aquarium show signs of respiratory distress and eventually die.

REPRODUCTION

Fish reproduce by means of fertilization of the female's eggs by the sperm (milt) from the male. This may take place outside the female's body as the eggs are being expelled, or as the milt is being deposited in the vent (the urogenital opening) of the female and fertilization takes place internally. When fertilization takes place internally, the eggs are retained in the body of the female, and she gives birth to live, free-swimming young. The young may be nourished within her body (viviparous), or the eggs may hatch within her body (ovoviviparous), with the young then expelled.

Egg-Layers

Most aquarium fish are egg-laying species. The eggs are expelled from the female and then fertilized by the male during a reproduction ritual called spawning. Egg-laying fish fall into one of five groups.

Egg-Scatterers. Egg-scatterers are fish that lay their eggs in a haphazard manner on the floor of an aquarium. Some species lay adhesive eggs that stick to the gravel on the aquarium floor or on decorations and plant materials. Other species lay nonadhesive eggs. There is no parental care of the eggs or young; once the eggs are laid, they are forgotten. In many cases, steps must be taken to prevent the parents from eating the eggs and the newly hatched fry. To reduce the number of eggs eaten by the parents, several steps can be taken:

1. Cut down on the amount of light; this encourages spawning and increases the number of eggs produced.

2. Have aquarium plants so dense that adult fish will have difficulty swimming among the plants to find eggs and newly hatched fry.

3. Place one or two layers of glass marbles on the floor of the aquarium; the eggs will settle among them, and the adults will not be able to get to the eggs.

4. Place spawning grass, or Spanish moss, or man-made grass on the aquarium floor; this material can be placed over layers of marbles also.

5. Drape a nylon curtain in the tank; the adults can swim above the curtain, and as the eggs are laid, they will fall through the curtain to the floor of the aquarium.

6. Use a large spawning tank that allows the eggs to go undiscovered and the newly-hatched fry to hide or evade the adults.

Examples of egg-scatterers are the barbs, danios, some species of Rasbora, some species of characins, and goldfish.

Egg-Buryers. Egg buryers are fish that lay their eggs in the mud of rivers and ponds; these are the killifish and other annual fish that live in areas where the rivers and ponds dry up. The adults lay their eggs in the mud and die when the river and ponds dry up. The eggs survive in the mud, and when the next rains come, the young hatch. In aquariums, annual fish lay their eggs in the material on the floor of the aquarium. Two or three inches of peat is the ideal material. Because the water does not dry up, they live for more than one year. These fish may also lay eggs in dense plant growth or spawning grass. The eggs can be removed with tweezers and stored in moist peat. One should take the peat containing the eggs, squeeze out the water, place the peat in plastic bags, and seal them. After two or three months, the peat containing the eggs should be placed in water so that the eggs will hatch.

Egg-Depositors. Egg-depositors have complex spawning routines and are excellent parents; these fish usually select their own mate. They clean off a nesting site where the female deposits her eggs and the male fertilizes them. The parents usually take turns guarding the nest; they constantly fan the nest and eggs with their fins to keep them clean from dirt or silt and dust that may settle on them. When the young fry hatch, the parents keep watch for several days to protect them from predators. Most of the cichlids fit into this group, as well as the corydoras, catfish, sunfish, and some Rasboras.

Mouth-Brooders. Mouth-brooders are fish that carry their eggs in their mouth until they hatch, after which the young may continue to be carried in the parent's mouth until they are ready to fend for themselves. In some species, the young, after leaving the parent's mouth, may return if threatened. After the eggs deposited by the female are fertilized by the male, the female, and in some species the male, goes around and picks up the eggs by mouth. During this incubation period of two to three weeks, the parent does not eat. Some species of cichlids and some species of *Betta* are mouth-brooders.

Nest-Builders. The nest-builders construct a nest in which the eggs are deposited; it is usually constructed by the male. This nest may be a bubble nest on the surface made from saliva-blown bubbles or may be prepared from materials found on the floor of the aquarium. After the eggs are fertilized, the female should be removed because the male will become aggressive towards her; the male then guards the nest. Gouramis, some sunfish, some cichlids, a few catfish, and the *Betta splendens* are nest-builders.

Live-Bearers

In live-bearing species, the male's anal fin is modified so that he can deposit milt into the vent of the female. The modification is called a gonopodium and is a characteristic that helps identify the male of the species. Live-bearers may give birth to twenty or as many as 200 young fry. Females store the milt in their oviducts for several months, thereby several broods can be produced from one mating. In viviparous species, the young are nourished from the female's bloodstream. In ovoviviparous species, the young are nourished by the yolk sac. Guppies, platies, swordtails, mollies, and half beaks are common live-bearing species.

In preparing to breed fish, several conditions should be observed. The temperature of the water should be increased 2 or 3°F; this stimulates the fish into breeding. Filtration and healthy water conditions are vital to successful breeding and reproduction of fish; however, a system that provides too much up-lift or suction may draw the young fry into the gravel on the bottom, and if the aeration is too strong, the young can be injured. An air-operated sponge filter and a gentle flow of air from an airstone provide adequate conditions.

Young fry of live-bearers can begin taking food immediately; fry of egg-laying fish survive the first few days from the nutrients in the egg sac. Commercially prepared foods are available in liquid and powdered form for both live-bearing and egg-laying fry. Young fry should also receive brine shrimp. These are available from aquarium supply stores and can be hatched from eggs also. After a few days, the young can be fed dried foods that have been mashed into small pieces.

SUMMARY

Evidence of the early development of fish is found in fossils dating back to the Ordovician Period 425 to 500 million years ago. These early fish were called Ostracoderms. From the Ostracoderms evolved two groups of fish: the Placoderms and Acanthodians. From Placoderms evolved our present-day sharks, rays, skates, and other fish with cartilaginous skeletons. From the Acanthodians evolved the Osteichthyes, represented today by fish with bony skeletons.

Fish are cold-blooded vertebrates that breathe with gills and move with the aid of fins. They are the most numerous vertebrates, with more than 30,000 species; these are divided into three main classes and thirty-four orders. Most fish are covered with scales; these are thin, bony plates that overlap each other and provide protection. Fins are movable structures that aid the fish in swimming and maintaining balance. Gills are organs through which fish obtain oxygen from the water.

Fish have a long, folded heart that consists of only two chambers: the atrium and the ventricle. Most species have a lateral line that is composed of a series of pressure-sensitive cells, called neuromasts, which enable the fish to react to changes in water pressure, temperature, currents, and sounds.

The teeth and digestive systems of fish vary depending on the type of food consumed. Some fish are predators and consume other fish or animals; others feed primarily on plant material. The eyes of fish are similar to other vertebrates but have spherical lenses that focus by moving within the eyeball, not by changing the curvature of the lens. Fish do not have eyelids; the eye is kept moist by the flow of water.

The senses of smell and taste in fish are highly developed; they not only have taste buds in their mouth and on their lips, but also on their body and fins. Some fish also have taste buds on barbels around their mouth.

All fish have inner ears. Species that have swim bladders have a more acute sense of hearing because the bladder acts as a resonator and amplifies the sound.

Modern aquarium tanks are available in many sizes and shapes. The size of the aquarium depends primarily on the species kept and the amount of money the aquarist is able to spend. The larger the aquarium, the more money that will be needed for accessories.

The water in an aquarium must be kept clean and contain adequate amounts of dissolved oxygen; filters and air pumps are used to accomplish this. Filter systems work in one of three ways: through the mechanical removal of waste and food materials, by chemical removal of dissolved materials, and by the biological filtration or convection of harmful substances into harmless ones.

Fish consume all types of foods and should be given a variety so that they do not become bored. Fish should be fed two or three times a day and should receive what they can consume in 3 to 5 minutes. Common foods for aquarium fish are commercially prepared flake foods and pellets; live foods such as earthworms, flies, maggots, wood lice, caterpillars, small crustaceans, and brine shrimp; and freeze-dried and frozen foods. Carnivorous fish may be fed minced or chopped beef hearts, liver, raw fish, and shellfish. Many species of fish also eat vegetable and plant material.

Aquarists need to be able to recognize symptoms of diseases in their fish as soon as they appear. Some common diseases of fish are white spot, or "Ich"; slime disease; hole-in-the-head; and velvet disease of freshwater fish; and coral disease of marine fish. Parasitic fungus, flukes, worms, and lice may also cause ailments in fish; bacterial infections are also common. Maintaining water cleanliness and quality is important in controlling many of these diseases and ailments.

Fish reproduce by means of fertilization of the female's eggs by the milt from the male. This fertilization can take place after the eggs are expelled from the female's body, or the milt can be deposited in the vent of the female and fertilization takes place internally. When fertilization takes place internally, the eggs are retained in the body of the female, and she gives birth to live, free-swimming young. Egg-laying fish fall into one of five groups: egg-scatterers, egg-buryers, egg-depositors, mouth-brooders, and nest-builders.

In preparing to breed fish, several conditions should be observed; water cleanliness, quality, and temperature are extremely important in successful breeding.

DISCUSSION QUESTIONS

1. Describe the characteristics of fish.
2. What are the three main classes of fish, and what are their important characteristics?
3. Describe the general characteristics of the twelve orders that are commonly found in the aquarium trade.
4. What are the three types of filter systems, and how does each one work?
5. Describe the nitrogen cycle that takes place in an aquarium.
6. What is a power filter, and what is the advantage of using one?
7. What are some common foods that fish consume?
8. What are the five types of egg-laying fish, and what are their habits?
9. What are signs that might indicate a fish has a disease or other ailment?
10. What are common diseases caused by parasites, and what are the symptoms and treatment for each?
11. What are two species of flukes that attack aquarium fish, and what are the symptoms and treatments for each?
12. What causes finrot, and how can it be contolled?

SUGGESTED ACTIVITIES

1. Visit the local pet shop and make a list of the different species of fish available. Using the local library, compile as much information as you can about the different species.
2. Set up an aquarium and put some common species of aquarium fish in it. Observe and record their habits.
3. Set up an aquarium with live-bearers and try raising some young fish. Observe and record their daily habits.
4. Species of fish such as guppies can be used to conduct color and genetic studies. Raise young guppies, and select and mate them for differences in color and fins. Are some characteristics dominant?
5. Obtain test kits and sample the aquarium water for dissolved oxygen, ammonia, nitrites, pH, hardness, and carbon dioxide. Keep a daily or weekly record. What changes occur and why?
6. Determine the species, number, and size that can be raised in an aquarium 30 inches long, 19 inches high and 12 inches wide. What is the ideal temperature for the species you selected?

ADDITIONAL RESOURCES

1. Aquasite
www.aquasite.com/index.shtml
2. Fish Link Central
www.fishlinkcentral.com
3. Aqualink Aquaria Service
www.aqualink.com
4. Tropical Fish
www.tropicalfishcentre.co.uk
5. Discus Breeders' Web site
www.aquaworldnet.com/dbws.shtml
6. Freshwater Information Service & Homepage (FISH)
www.worldzone.net/family/aquaria
7. Fins: Fish Information Service
www.actwin.com/fish/index.cgi

 Glossary

abomasum: the fourth compartment of a ruminant's stomach, sometimes referred to as the "true stomach" because it functions very similarly to the stomach in monogastric animals.

abscess: a swelling caused by the accumulation of pus from an infection.

absorption: the taking in of fluids.

adipose fin: a small, fleshy fin located on the back and between the dorsal and caudal fins on certain species of fish.

adrenal glands: glands located near the front of the kidneys that produce hormones involved in sexual activity and metabolism.

aeration: the exchange of carbon dioxide in water for oxygen.

aerobic bacteria: bacteria that live in the presence of oxygen.

agile: graceful, quick, and easy movement.

agouti: color pattern where one color is streaked or intermingled with another color.

airstones: fused, porous glass devices used in aquariums to oxygenate the water.

albino: an animal that lacks color pigmentation. Albino animals are white with red eyes.

albumen: the thick, white matter that surrounds the yolk of an egg.

algae: a group of single-celled aquatic plants that have chlorophyll.

allergy: a hypersensitivity to animal hair, dust, and other allergens.

aloof: one who holds his or her head high in a proud, confident manner.

amiable: mild, easygoing.

amino acids: a compound that contains both the amino (NH_2) group and the carboxyl (COOH) group; considered the building blocks of proteins.

amniotic sac: a sac or thin membrane that surrounds the embryo and contains a watery amniotic fluid.

amphibians: members of the class Amphibia, which includes frogs, toads, and newts. They live part of their lives in water and part on land.

amphibious: being able to live both on land and in the water.

amphiumas: a form of amphibian commonly known as Congo eels. They have long, slender bodies with tapered heads; have two pairs of tiny weak limbs with one, two, or three toes on each limb; and are found in muddy, swampy areas of the Southern United States.

anaerobic bacteria: bacteria that live in the absence of oxygen.

anal fin: a fin located on the underside of a fish between the vent and the caudal fin.

anal pores: openings located just forward of the vent on certain species of geckos. They can be used to identify the male of the species.

anal sacs: glands located at the four o'clock and eight o'clock position on either side of the dog's anus. These glands produce a substance that allows the dog to mark its territory when it defecates. These glands may sometimes become infected or impacted.

analgesics: substances used to make an animal insensitive to pain without losing consciousness.

anatomist: person who studies the structure and form of plants and animals.

anemia: a condition in which the blood is lacking in red blood cells; may result from a lack of iron and copper in the blood.

anesthetics: substances used to produce a lack of awareness or sensitivity.

angora: a type of long fur on cats and rabbits.

animal breeder: person who arranges for the mating of animals, cares for the mother and the young, and trains or sells them when they reach the right age. They often specialize in one breed such as German Shepherd dogs or Siamese cats.

animal groomer: person who bathes pets and keeps them free of ticks, fleas, and other pests. Groomers brush the pets, trim their hair, and may cut and trim their nails.

animal health technician: see *veterinary technician*.

animal husbandry: the management and care of domestic animals.

animal hygiene: the management of animals so that good health is maintained.

animal rights: the position that animals should not be exploited; i.e., not used for food, clothing, entertainment,

384

medical research, or product testing. This also includes the use of animals in rodeos, zoos, circuses, and even as pets.

animal shelters: the places where abandoned and unwanted animals are taken for care or to be euthanized.

animal trainer: person who teaches animals to obey commands, to compete in shows or races, or to perform tricks to entertain audiences. They also teach dogs to protect lives or property, or act as guides for the visual- or audio-impaired, and to hunt or track.

animal welfare: the position that animals should be treated humanely; i.e., properly cared for with adequate housing, good nutrition, disease prevention and treatment, responsible care, humane handling, and humane euthanasia or slaughter.

Animalia: the animal kingdom, the division of life forms that includes all living and extinct animals. The animal kingdom includes the invertebrates and vertebrates.

annual fish: fish that lay their eggs in the mud of streams and ponds that are drying and then die. When the rains come, the eggs hatch and the young grow quickly to adults.

anterior: the front of an animal.

antibodies: substances produced by an animal's body in response to antigens. Antibodies neutralize the effects of microorganisms and the toxins produced by microorganisms.

anticoagulant infusion: the process of slowly introducing anti–blood-clotting solution into the veins of patients having heart surgery.

anticonvulsant: a substance that controls or prevents convulsions.

antifungal agent: a substance that kills fungi.

antiseptic: substance applied to a wound to prevent the growth of microorganisms.

antivivisectionists: those who are against the use of living animals for experimentation and study.

apathetic: showing no interest.

aplastic anemia: a disease common to female ferrets that occurs as a result of not being bred. The female ferret stays in heat if not bred, and the continued production of the hormone estrogen causes a decline in the production of red and white blood cells and the blood clotting factor in the bone marrow.

aquarist: a person who raises fish in an aquarium.

aquarium: a water tank or container used to raise fish or other aquatic animals.

aquatic: living in a water habitat.

arachnids: eight-legged arthropods of the class Arachnida, which includes spiders, scorpions, mites, and ticks.

arboreal: living in trees.

aristocratic: holding the head high and having a certain air of nobility.

arrogant: excessively proud to the point of being overbearing and hard to handle.

arthritis-like: symptoms similar to arthritis; i.e. inflammation of the joints, stiffness, and difficulty in moving.

arthropods: invertebrate animals such as insects, arachnids, and crustaceans that have jointed bodies and legs and an external skeletal structure.

ascarids: intestinal parasitic worms; the common roundworm is an ascarid.

ascorbic acid: vitamin C.

assimilation: conversion of food material into a form that can be absorbed.

assistant laboratory animal technician: a person who assists laboratory animal technicians in feeding and caring for animals used in laboratories. The lowest level of certification for laboratory animal care workers.

atria: part of an animal's heart that receives blood from the veins and pumps blood to the ventricle.

atrophy: when a part of the body decreases in size or wastes away.

Aves: the class for birds; warm-blooded vertebrate animals having two legs, wings, and feathers and that lay eggs.

aviary: a place where large numbers of birds are kept.

avidin: a protein in egg white that combines with biotin in the intestinal tract to make the biotin unavailable to the animal.

awn fluff: a soft, wavy wool with a guard hair tip found between the underwool and the awn hair of Giant Angora rabbits.

awn hair: strong, straight hair (guard hair) protruding above the awn fluff of Giant Angora rabbits.

bacteria: single-celled organisms that lack chlorophyll and reproduce by fission. They live in colonies and can cause diseases in animals.

barbels: the whisker-like projections on the mouth of several species of fish. These projections help the fish feel its way along the bottom.

bias: a view or belief that is slanted or a prejudice.

bib: distinct marking under the chin and on the upper chest.

binomial nomenclature: Linnaeus's system of naming species, which used the Latin language. Each species has a name composed of two words: first, the genus that is written with the first letter capitalized, and second, the species that is written with a small first letter.

biochemical reaction: chemical reactions that take place in the cells of plants and animals and include respiration, digestion, and assimilation.

biochemist: person who studies the chemical processes of living organisms.

biologist: person who studies the origin, development, anatomy, function, distribution, and other basic principles of living organisms. The biologist may also be called a biological scientist or life scientist.

biology: the study of living organisms.

biophysicist: person who studies the atomic structure and the electrical and mechanical energy of cells and organisms.

biotechnology: industries that combine the sciences of biology and engineering to help or aid humans.

black plague: disease that killed more than one-fourth of the population during the 1300s.

bold: showing no hesitation or fear.

botanist: person who studies plants.

brackish: a mixture of fresh and salt water found where rivers and streams enter the salt waters and also found in some inland swamp areas that receive some salt water.

brawny: very muscular and strong.

breeder: one who breeds animals.

brille: a transparent layer that covers the eye of a snake.

broad-spectrum antibiotics: antibiotics that are effective against a large range of diseases.

broken: white-furred animal with a spotted or colored pattern.

brood: the incubation of eggs where the female keeps them warm. Also a group of young produced by a female.

buck: male rabbit.

butterfly nose: a dark nose marking on some breeds of rabbits; covers the nose, mouth, and whisker bed.

caecilians: class Amphibia, order Gymnophiona; worm-like amphibians that do not have limbs and are usually without sight.

caiman: class Reptilia, order Crocodilia, family Alligatoridae. Reptiles found in Central and South America that grow up to 15 feet; smaller species may be found in the pet trade.

calico: also known as tortoiseshell and is made up of three colors: black, orange, and white.

camouflage: the body coloring of an animal that allows it to hide or conceal itself in its environment.

canines: sharp, pointed teeth on either side of the jaws.

cannibalism: practice of eating one's own kind.

caped: a type of rat that is white with a darker color about the head, ears, and neck.

capybara: *Hydrochoerus hydrochaeris,* found in South America, is the largest of the rodents, growing to 4 feet in length and weighing up to 120 pounds.

carapace: the hard, protective upper shell on a turtle, tortoise, or terrapin; it also forms part of the vertebrae and ribs.

carbohydrates: compounds such as sugars and starches composed of carbon, hydrogen, and oxygen that are needed by an animal for energy.

Carboniferous period: geologic time period from 280 to 345 million years ago that saw the growth of trees, ferns, and conifers as well as an increase in amphibians and the appearance of the first reptiles.

carboxylation: the combining of organic acids with carbon dioxide to form carboxylic acid.

carnivorous: a meat-eating animal.

carrier: an animal that appears healthy but carries disease-causing organisms that can infect other animals.

cartilaginous: species of fish that has a skeleton wholly or partly composed of cartilage.

casque: a structure on the top of the head that resembles a helmet.

catheter: a tube that is placed internally so that fluids can be injected or withdrawn.

caudal fin: the tail fin of a fish.

cavy: another name for a guinea pig.

cecum: where the small intestine joins the large intestine. It serves no purpose in most animals, but in horses and rabbits, bacterial digestion of roughages takes place here.

Cenozoic era: the geologic period of time from the present to 70 million years ago during which placental animals, carnivores, large mammals, and humans appeared.

central scutes: the scutes located down the center of the shell on turtles, tortoises, and terrapins.

cere: the fleshy portion of the beak that surrounds the nostrils of birds.

chatty: very talkative; a cat that makes a considerable amount of vocal noise.

chemistry: the study of various substances and how they react with each other.

cherts: sections of rocks that show fossilized remains of blue-green algae.

chimaera: class Chondrichthyes, order Chimaerae. A group of fish with cartilaginous skeletons and having their upper jaw fused to the skull.

chitinous: material that forms the hard outer covering on insects and crustaceans.

Chordata: a phylum with four distinctive characteristics: embryonic notochord, which is usually replaced by the vertebra; a dorsal tubular nerve cord running down the back side; pharyngeal or throat area gill slits; and a rear area tail.

chromatophores: pigment cells in the skin of animals that give the animals their color.

ciliospores: reproductive spores of certain species of parasites.

civet-like animal: an animal that has a long body and short legs similar to the African civet cat.

clan: colony structure.

class: a division of the classification of life forms beneath phylum and above order.

classical LD50 test: determines the dosage required to kill 50 percent of a population of test animals within a specified period.

climax: period of greatest abundance of a particular group of wild animals.

cloaca: a common chamber where the digestive and urinary tracts empty, and the reproductive tracts of birds, reptiles, amphibians, and fish are located.

clutch: a nest of eggs or a group of young birds.

coagulation: clotting of the blood.

coastal scutes: scutes located between the central and marginal scutes on turtles, tortoises, and terrapins. See also *scutes.*

cochlea: the organ within the ear that enables an animal to detect sounds.

cold-blooded: animals that cannot regulate their internal body temperature. They take on the temperature of their surrounding environment.

collar: a band or stripe around the neck of an animal.

colony mating: placing several females in with one male for mating. This method is practiced in hamsters, mice, rats, and guinea pigs.

colony structure: living in groups for protection against enemies, finding and storing food, and sharing the duties of raising the young.

colorpoint: the color of the nose, ears, feet, and tail of the animal. See also *points.*

colostrum: first milk produced by the mother. It is rich in proteins and minerals and contains antibodies.

community aquarium: an aquarium that has more than one species of fish.

composed: does not become overexcited, anxious, or hurried.

conception: the point when the sperm fertilizes the egg.

condo: an open multilevel habitat without walls for mice.

confinement systems: animals are confined to cages or pens in partially enclosed or totally enclosed buildings so that production improves through closer control of the environment.

conformation: the size, shape, and form based on what an ideal animal of the species should look like.

congenial: easygoing and cooperative.

congenital: present at birth.

conjunctiva: the mucous membrane lining the inner surface of the eyelids and covering the front part of the eyeball.

conjunctivitis: inflammation of the membrane lining the inner surface of the eyelids.

consignee: the person to whom animals are being delivered.

consignor: the person who ships animals to be delivered.

constitution: the physical makeup of an animal.

consumer: the person who buys products or goods for his or her own use, not for a business.

contagious: a disease that can be spread from one animal to another by direct or indirect contact.

contour feathers: the feathers that give the bird its outward form.

coprophagy: the consuming of fecal material.

copulation: the act of mating.

cornea: the transparent covering over the eyeball.

corneal ulceration: a tear or break in the cornea of the eye that becomes infected.

coronavirus: a disease that causes severe vomiting and diarrhea in dogs and cats; is usually spread through contaminated feces.

cough suppressant: a substance that checks, restrains, or subdues a cough.

coverts: the smaller feathers that cover the base of the quills of the wing and tail feathers.

crepuscular: animals that are active during the twilight hours.

crest: the tuft of hair or feathers on the top of the head.

Cretaceous period: geologic period 135 to 70 million years ago during which the dinosaurs and marine reptiles reached their greatest abundance followed by their extinction.

crown: the top of the head.

crude fiber: substance composed mostly of nondigestible bulk or roughage.

ctenoid scales: scales on certain fish that have serrations on the edges that makes the fish feel rough.

cud: regurgitated food not fully digested in the rumen. The animal chews on this and swallows it again for further digestion.

cycloid scales: scales on certain fish that have smooth edges that make the fish feel smooth and slick.

cyst: a capsule formed around an organism that is in a dormant or spore stage. Also a sac-like structure or pocket on an animal that is filled with fluid or pus.

decarboxylation: the breakdown of carboxylic acid.

defecate: the movement of solid waste from the anus.

degenerate: to decrease in size or waste away.

demodectic mites: class Arachnida, order Acari. Mites found on the skin of many animals can cause demodicosis if present in large numbers.

demodicosis: skin disease caused by large numbers of demodectic mites; also called demodectic mange.

deoxyribonucleic acid: see *DNA.*

dermal: relating to the dermis.

dermis: the sensitive layer of skin that lies below the epidermis.

devilish: overplayful; tending to get into trouble.

Devonian period: geologic period about 370 million years ago, during which the first amphibians appeared and freshwater fish became abundant.

dew-drop valve: a water valve fitted on bottles to provide water on demand to small animals.

dewlap: loose fold of skin under the chin or hanging from the throat.

dexterous: skillful in use of hands or body.

digestion: the breakdown of food materials and the absorption of nutrients.

dignified: an excellent-appearing animal that shows beauty of form and grace.

dilate: to widen, expand, or increase in size.

dimorphism: the male and female of a species having different physical characteristics so that they can be identified.

disc-shaped: circular or round in form.

disposition: how an animal acts under various circumstances.

dissection: cutting an animal into pieces or parts so that it can be studied or examined.

diurnal: animals that are active during the daylight hours.

DNA (deoxyribonucleic acid): genetic material that carries the hereditary codes.

docile: tame, easygoing, and obedient.

doe: female rabbit.

dog trainer: person who teaches dogs to hunt or track, to obey signals or commands, to guard lives or property, to run races, or to lead the blind.

domesticated: animals that have adapted to a life around humans.

dominant gene: in a pair of genes, the one that controls the particular genetic characteristic.

dorsal: the topside or back of an animal.

dorsal fin: fin located on the back of a fish.

down feathers: small, soft feathers located under the contour feathers that conserve body heat.

Draize eye test: measures the effects of products such as cosmetics on the unprotected eyes of restrained rabbits. Rabbits do not have tear ducts to wash the products from their eyes.

ductus deferens: small tubes in the male bird that lead from the testes to the cloaca.

dull: the lack of usual shine or luster.

duodenum: the first section of the small intestine.

ecologist: person who studies the relationship of plants and animals to the environment.

ecoterrorism: terrorism or crimes committed under the disguise of saving nature.

ectoparasite: an external parasite that attacks the outside of an animal's body.

ectotherm: an animal that is cold-blooded.

egg-buryers: fish that lay their eggs in the mud of rivers and ponds.

egg-depositors: fish that prepare a nest, deposit their eggs, fertilize them, and guard the eggs until they hatch.

egg-scatterers: fish that lay their eggs in a haphazard manner on the bottom of streams, rivers, and lakes.

embryo: the early developing stages of a vertebrate animal before it is born or hatched.

embryologist: person who studies the development of animals from egg to birth.

encephalitis: an inflammation of the brain.

endoparasite: an internal parasite that attacks the inside of an animal's body.

enteritis: an inflammation of the intestinal tract.

enterprising: hardworking, determined, and aggressive in its work.

entropion: a condition in which the eyeball sinks into its socket or in which there is a spasm of the eyelid due to discomfort. A congenital defect or a result of an injury.

enzyme: complex protein substance produced by the body and necessary for biochemical reactions to occur.

Eocene epoch: geologic period 40 to 60 million years ago, during which placental mammals spread and increased.

epidermal: relating to the epidermis.

epidermis: the layer of skin that lies above the dermis.

epinephrine: a substance made from the adrenal glands of cattle used to relieve symptoms of hay fever, asthma, and some allergies in humans. It is also used as a heart stimulant, as a muscle relaxant, and to constrict blood vessels.

epithelial cells: the tissue that covers the inside of the intestines.

epoch: geologic division of time less than a period.

era: refers to one of the five major geologic divisions of time.

estivation: a deep sleep-like state triggered by excessively warm temperatures.

estrogen: a female sex hormone that promotes estrus and is responsible for the development of secondary sex characteristics.

estrus: female's heat period or season.

ethical: belief or ideal that is basically good or bad; dealing with what is good and bad.

euthanasia: to induce the death of an animal quickly and painlessly.

evulsion: a tear or pulled-out wound.

exoskeleton: the hard, external protective shell of a turtle, tortoise, or terrapin.

expectoration: to expel material from the throat or lungs by coughing or spitting.

external parasite: an ectoparasite.

extroverted: not shy; forward and aggressive.

factory farming: practice of keeping chickens in cages and veal calves in small crates.

fad: a passing craze; interest for a short time.

family: a division of the classification of life forms beneath order and before genus.

fats: nutrient made up of the same chemical elements as carbohydrates but in different combinations. Fats contain 2.25 times as much energy as an equivalent amount of carbohydrates and proteins.

fatty acids: the saturated and unsaturated monocarboxylic acids in the form of glycerides in fats and oils.

feces: solid waste material discharged through an animal's anus.

feisty: nervous, excitable, and easily agitated or irritated.

femoral pores: openings located on the inside of the rear legs of agamas and some species of iguana. They can be used to determine sex.

fennecs: a small, fox-like animal native to Africa.

feral: wild.

ferreting: the practice of using ferrets to drive rabbits and rats from burrows.

fetal development: the natural and normal growth of the fetus within the uterus.

fetus: the unborn or unhatched young of a vertebrate.

filoplume feathers: short, hair-like feathers.

fish: used when referring to several species of fish.

flight feathers: the longer contour feathers that extend beyond the body and that are used in flight.

fly-back fur: animal fur that quickly returns to its normal position after the animal is stroked.

formalin: a solution of water, formaldehyde, and a small amount of methanol.

fossils: hardened or petrified remains of plant or animal from previous periods of history.

freshwater: water habitat that is neither brackish nor marine.

Fungi: plant kingdom including the molds, yeasts, and fungi that obtain their food by absorption and that lack chlorophyll.

fungi: a group of plants that lack chlorophyll and obtain their nourishment from other plants and animals. Includes molds, mildews, smuts, mushrooms, and certain bacteria.

fungus disease: disease that is caused by a fungus.

gaiety: being cheerful, happy, or merry.

gait: the way an animal moves.

gamoid scales: scales on gar and birches. These scales are very thick, heavy, and similar to scales on ancient species of fish.

gastrointestinal tract: the stomach and intestines.

geneticist: person who specializes in the study of heredity and how characteristics vary in all forms of life.

genetics: the study of heredity and how and what characteristics are passed from generation to generation.

genus: a division of the classification of life forms beneath family and before species.

gestation: period of time from conception until birth of young.

gloves: white paws.

glucagon: a substance produced from cattle pancreas that increases the sugar content in blood and is used to counteract insulin shock in humans.

goiter: a swelling of the thyroid glands in the neck area. The ailment is caused by a lack of iodine in the diet.

gonopodium: a modification to the anal fin of live-bearing male fish that aids in transferring sperm to the vent and cloaca of the female.

gram-negative bacteria: bacteria that will not hold the purple dye used in Gram's method of identifying bacteria.

gram-positive bacteria: bacteria that will hold the purple dye used in Gram's method of identifying bacteria.

gregarious: associates in groups with others of their own kind; not solitary.

grit: gravel-like material fed to birds that accumulates in the gizzard and aids in the grinding of food material.

grotzen: the center back strip of a chinchilla pelt.

guard hairs: the longer, coarse hairs that form a protective coat above the shorter underfur of an animal. The guard hair protects the underfur from rain and cold.

handler: person who holds and commands trained animals during a show, sporting match, or hunt.

hardness: the soluble minerals in water.

haw: another term for nictitating membrane.

heat period: the estrus of females.

heel: a command for the dog to walk close to the side of the handler, usually the left side.

hemipenes: the modified cloaca in male reptiles of certain species used to transfer sperm to the cloaca of the female.

hemoglobin: red-colored, iron-containing protein component of blood that is responsible for carrying oxygen to the body tissues.

hemorrhage: uncontrolled bleeding.

herbicide: a chemical substance used to destroy plants.

herbivorous: animals that feed on plant material.

hertz: a unit of frequency; one cycle per second.

hibernation: going into a resting or dormant state during winter.

hob: a male ferret.

hookworms: bloodsucking parasitic worms that have hooks near their mouth so they can attach themselves to the intestinal lining of animals.

hormones: substances that are produced in one part of the body, but influence cells in another part of the body. Hormones may influence growth.

host: an animal from which a parasitic organism derives nourishment.

humane: kind, gentle, merciful, sympathetic.

humane shelters: places where unwanted or abandoned animals are taken for care or to be euthanized.

humanize: to believe that animals have the same rights as humans and to treat them as humans.

husbandry: care and management of domestic animals.

hutch: house used to keep rabbits.

hutch burn: a chapped or burned condition around the external genital area of female rabbits caused by urine-soaked bedding.

hypomagnesemia: a condition that is caused by low blood magnesium.

ileum: the last section of the small intestine. It is located between the jejunum and the large intestine.

immune gamma globulin: antibodies.

immunoglobulins: protein substances that are produced in response to antigens and act as antibodies.

immunosuppressive therapy: therapy used to correct an immune system problem.

immune system: the body system that prevents the development of disease organisms and counteracts their effects.

immunize: to give an animal protection against disease-causing organisms.

impaction: a condition in which the small intestines, cecum, or large intestines become tightly packed with food material or feces.

impetuous: abrupt, hasty, and rushing headlong into work without much efficiency.

in vitro: refers to in glass or in laboratory containers. Cultured skin can be grown in vitro in laboratories.

inbreeding: breeding closely related animals.

incisor: front teeth between the canines in either jaw.

incorrigible: unable to be controlled or corrected.

incubation: the period of time in which the embryo is developing within the egg. Also the act of setting on or brooding an egg.

independent: wanting to have its own way and not wanting to be controlled.

indigenous: native to a particular region.

indiscriminately: the mating habits of animals where different species, types, or varieties mate and produce young.

indolent: being lazy or inactive.

infectious diseases: diseases that are caused by microorganisms such as viruses, bacteria, protozoa, and fungi that are capable of invading and growing in living tissues.

infundibulum: the funnel-shaped opening to the oviduct in which the egg yolk drops after being released from the ovary.

inherent: belief or ideal that is by nature right or wrong; a belief or ideal that we are born with; a deep-seated belief or ideal.

inherited: passed on to offspring from parents or ancestors.

inherited defects: problems, ailments, or deformities that are passed on to offspring from parents or ancestors.

inhumane: cruel and savage.

inquisitive: wanting to check things out or wanting to investigate things.

insecticide: a chemical substance that is used to destroy insects.

insectivorous: animals that feed primarily on insects.

insulin: a substance produced from cattle pancreas and used to treat people with diabetes.

intensive operations: farming operations in which the farmer or manager tries to increase output through better breeding, feeding, and management.

intermediate host: a host that a parasitic organism lives on or in during an immature stage.

internal parasite: an endoparasite.

intradermal: within the skin.

intramuscular: within the muscle.

intravenous: entering by way of the vein.

invertebrates: members of the animal kingdom without backbones, such as arthropods and worms.

inviolate sanctuaries: a place where animals are protected and no hunting is allowed.

iridescence: colors that have a brilliant shine or glitter.

isotopes: atomic forms of elements.

isthmus: the area in a bird's oviduct where the two shell membranes are added to surround the yolk and albumen.

Jacobson's organ: smell and taste receptors located in the roof of the mouth of cats and reptiles.

jaundice: abnormal yellowing of the skin, tissues, eyes, and urine due to bile pigments building up in the blood as a result of liver disease or excessive breakdown of red blood cells.

jejunum: the second section of the small intestine. It is located between the duodenum and ileum.

jill: a female ferret.

jovial: good-natured, happy, easygoing.

Jurassic period: geologic period 135 to 180 million years ago, during which the first birds and mammals appeared and the dinosaurs became abundant.

keen: very sensitive, alert, and intense.

kennel attendant: person who feeds and waters animals, keeps their quarters clean, exercises animals, and keeps records on the animal's health, feeding, and breeding.

keratin: a sulfur-containing fibrous protein that is necessary for the growth of hair, wool, fur, and horn-like epidermal tissues.

keratitis: an inflammation of the cornea.

kilohertz: a unit of frequency; 1,000 cycles per second.

kindling: giving birth to a litter of rabbits.

kingdom: one of the five major areas in which all life forms are classified; i.e., Monera, Protista, Plantae, Fungi, and Animalia.

kinked: twisted or curled fur.

kits: the name given to young ferrets.

laboratory animal care workers: persons who provide food and water, clean cages, check for illness, and keep records on each animal.

laboratory animal technicians: the second level of certification of laboratory animal care workers. They work with scientists and carry out their instructions.

laboratory animal technologists: the highest level of certification of laboratory animal care workers. They work with scientists and carry out their instructions.

labyrinthine chamber: a chamber behind the gills of certain species of fish that consists of rosette-shaped plates supplied with numerous blood vessels that absorb oxygen from the atmosphere.

labyrinthine organ: rosette-shaped plates within the labyrinthine chamber.

laceration: a wound, tear, or cut that has rough edges.

lactating: producing milk.

lamellae: special adaptations on the underside of the toes of most geckos that allow them to grip onto smooth surfaces.

lateral line: a line running horizontally on the side of most fish comprised of pressure-sensitive cells called neuromasts. See also *neuromasts*.

lethal gene: a recessive gene that, if carried by both parents, causes death of the offspring.

lethargy: lack of energy, sluggish.

lice: class Insecta, phylum Arthropoda. Small wingless insects belonging to the order Mallophaga (biting lice) and Anaplura (sucking lice). They can cause great irritation and loss of blood.

life scientist: biologist.

limber: very agile and flexible.

litter: multiple offspring produced at one birth.

live-bearers: fish that give birth to live young.

locomotor problems: problems involving the coordination of movement.

longy: a Manx cat with a complete tail.

lores: the area between the eye and the bill of a bird.

luster: shine by reflected light.

lutinos: birds, especially cockatiels, that are primarily yellow in color.

macrominerals: minerals needed in the largest quantity and most likely to be lacking in the feed supply; includes calcium, phosphorus, potassium, sodium, sulfur, chlorine, and magnesium. Also called major minerals.

malocclusion: a condition in which the upper and lower teeth grow to the extent that the animal cannot eat. The condition results because the teeth do not wear down properly, or it may be an inherited condition in which the teeth do not come together properly.

Mammalia: vertebrates that possess mammary glands, have a body that is more or less covered with hair, and possess a well-developed brain.

mandibles: both the lower and upper segments of a bird's bill; the lower jaws of mammals.

mane: the hair from the head down the back of the neck.

mantle: the feathers on the back of a bird.

marginal scutes: the scutes located around the outer edge on the shell of turtles, tortoises, and terrapins.

marine: saltwater habitats.

maxilla: the upper jaws of mammals.

meek: showing a lack of aggression and weakness.

meningitis: inflammation of the membranes covering the brain and spinal cord.

metabolism: the physical and chemical processes that take place in an animal's body that produce the energy necessary for the body's activities.

metamorphosis: a series of physical changes that an animal goes through from birth to adult.

microfilariae: the very small larvae form of the filament nematodes, *Filaria,* that causes heartworm disease in dogs.

microminerals: minerals that are needed in trace amounts; includes iron, iodine, copper, cobalt, fluorine, manganese, zinc, molybdenum, and selenium. Also called trace minerals.

milt: the sperm-containing substance of male fish.

mimic: ability to imitate sounds or words.

minerals: inorganic compounds needed in small amounts by an animal's body for the growth of bones, teeth, and tissues and to control various chemical reactions. They are components of the ash or noncombustible part of the ration that is left after the material has been burned.

mites: class Arachnida, order Acari, phylum Arthropoda; small animals that have four pairs of legs. They can infest larger animals by burrowing into the skin and feeding on the blood.

mitt: the feet or paws of a ferret.

molds: a fungus that produces a soft, woolly growth usually on damp, decaying materials.

molt: the shedding of feathers, skin, or shells to allow for the internal growth of an animal.

Monera: the kingdom including the bacterial organisms that lack a true nucleus in the cell and reproduce by fission.

monk: a man of a religious order who lives in a monastery.

monogamous: pair bond may last throughout the life of the animals; having only one mate at a time.

monogastric: having a single stomach.

moral: a belief or ideal that is basically right or wrong; morals are dealing with right and wrong.

mouth-brooders: fish that carry their eggs in their mouth until they hatch.

mutant: an individual or species of animal produced by a change in the genetic structure or hereditary material. See also *sport.*

mutation: a sudden change in an inherited trait; a departure from the parent type. Mutant animals are also called sports.

nape: the back of the neck.

nature: the mental and emotional qualities of an animal.

nest box: a box that animals use to give birth, lay eggs in, and raise their young.

nest-builders: fish that construct a nest in which their eggs are deposited.

neuromasts: a series of pressure-sensitive cells along the lateral line of fish.

neuter: the surgical procedure performed on male animals to prevent reproduction by the removal of the testicles.

neutered: castrated or spayed animal.

newts: class Amphibia, order Caudata; small semiaquatic amphibians similar to salamanders.

nictitating membrane: the third eyelid of certain animals that moves diagonally under the eyelid to help lubricate the cornea. It is also referred to as the haw.

night feces: the fecal material consumed by rabbits. See *coprophagy.*

nitrification: nitrogen cycle.

nitrogen cycle: the process by which ammonia is converted into nitrites and nitrates.

nitrogen free extracts (N.F.E.s): the easily and completely digested sugars and starches.

nocturnal: sleeping during the day and being active at night.

noncontagious: a disease or ailment that cannot be spread to another animal by contact.

noninfectious diseases: diseases caused by physical injuries and genetic defects; diseases that are not contagious.

nonruminant animals: single-stomached, or monogastric, animals. All of the small animals discussed in this book are nonruminants.

normal: fur type that is regular or the most common.

nose break: the curve the nose makes where it connects with the skull.

notochord: a flexible cord of cells in the embryo stage of growth that eventually becomes the vertebra.

nutrient: a single class of foods or group of food materials of the same general chemical composition, which aids in the support of animal life.

nutrition: receiving a proper and balanced ration so that an animal can grow, maintain its body, reproduce, and produce what we expect from it.

nutritionist: person who studies how food is used and changed into energy.

obesity: excessive accumulation of fat in the body.

oblique: running at approximately a 45-degree angle.

observant: quick to see things and attentive.

obstinate: determined to have its own way; stubborn.

occlusion: teeth on both jaws coming together correctly.

olfactory mucosa: an area of the nasal lining that picks up smells from the air as the animal breathes.

olm: class Amphibia, order Caudata (*Proteus anguinus*). A dull, white animal with red feathery, external gills. It is about 1 foot long, blind, and found in underground lakes, streams, and rivers in parts of Europe.

omasum: the third compartment of a ruminant's stomach. It removes large amounts of water from food as it moves from the rumen.

omnivorous: animals that feed on both animal and plant material.

oocyst: an egg that has not fully matured.

ophthalmic ointments: a salve or cream used in the eyes to relieve pain or irritation.

oral mucosa: the mucous membrane of the mouth.

order: a division of the classification of life forms beneath class and above family.

Ordovician period: geologic period about 420 million years ago, during which the first land plants were believed to have become established. During this time invertebrates, such as arthropods and worms, appeared on land.

osmosis: the flow of a substance from an area of high concentration to an area of lower concentration through a semipermeable membrane.

Osteichthyes: the bony fish, a large group of vertebrate animals that live in the water and have permanent gills for breathing, fins, and a body covered with scales.

osteomalacia: a disease in which the bones of the body become soft and weak.

osteoporosis: a condition in which the bones become porous and brittle.

otitis: an inflammation or infection of the ear.

oviparous: animals that lay eggs.

ovoviviparous: animals that retain eggs within their bodies until they hatch and then give birth to live young.

ovulation: release of egg cells from the ovaries into the oviduct.

ozone: a gas found in a layer of the atmosphere. It filters out ultraviolet radiation from the sun and prevents heat loss from the earth.

palatable: a food material that will be eaten and digested by an animal.

palate bone: the bone of the roof of the mouth.

Paleocene epoch: geologic period 60 to 70 million years ago, during which placental mammals appeared.

palpating: a method of using the hands to externally examine internal structures for physical diagnosis; e.g., pregnancy or tumors.

papilla: small projections on the dorsal wall of a male's cloaca that aids in transferring sperm to the female's cloaca.

papillae: knobs on the tongue that help hold food or prey; some carry taste buds.

paralytic: a partial or complete loss of motion and sensation in a part of the body.

parasite: organism that lives on or within another organism and derives nourishment from the host.

pathogenic organisms: organisms capable of causing disease.

pathologist: a person who studies the effects of diseases on plants and animals.

peccaries: small nocturnal mammals native to Central and South America that are related to pigs.

pectoral fins: fins located on each side of a fish behind the gill opening; they correspond to the forelimbs on other animals.

pedigree: a record of an animal's ancestry.

peduncle: the section of a fish's tail from the main body to the caudal fin.

pelvic fins: a pair of fins located toward the anterior underside of a fish; they correspond to the rear limbs on other animals.

penicillin: an antibiotic produced from molds.

periodontal disease: a disease affecting the tissues that surround the teeth.

peripheral vision: able to see in a wide circle or a large surrounding area.

Permian period: the geologic time period from 230 to 280 million years ago, which saw an increase and spread of reptiles and the extinction of many marine invertebrates.

perosis: an ailment of birds caused by malnutrition. Symptoms include the shortening and thickening of the bones in the limbs and enlargement of the hock.

pesticide: a chemical substance used to destroy pests.

pet care counselor: a person employed by pet shops to counsel, guide, and recommend the correct pet for the customer.

pet care workers: general workers who perform several tasks in animal hospitals, boarding kennels, animal shelters, pet grooming parlors, pet training schools, and pet shops.

pet groomer: person who combs, cuts, trims, and shapes the fur of all types of dogs and cats.

pet shops: specialized stores in which consumers can purchase a pet and a full line of pet products.

pH: the acidity or alkalinity of a substance. A pH of less than 7 is acidic, and a pH of greater than 7 is alkaline.

pharmaceutical: medicinal drug industry.

pharmacologist: a person who studies the effects of drugs and other substances such as poisons and dusts on living organisms.

pharyngeal: the area of the pharynx.

pharyngeal teeth: teeth located in the pharynx of certain fish species that are used to grind food material.

philosophers: persons who attempt to understand the values, reality, and truths about life.

photosynthesizing: the formation of carbohydrates from carbon dioxide and water in the chlorophyll containing cells of plants.

phylum: a division of the classification of life forms directly beneath the kingdom.

physiologist: a person who studies the life functions (growth, respiration, and reproduction) of plants and animals.

piebald: mice with spots, patches, or broken patterns.

pied markings: markings of two or more colors that appear as blotches of color all over the body.

pika: a small, short-eared animal related to rabbits. Order Lagomorpha, family Ochotonidae. They are found in Asia and western North America.

pinworms: small parasitic worms of the family Oxyuridae that infect the intestines of animals.

placenta: sac that contains a newborn; must be removed before the baby animal can breathe.

placental: having a vascular organ that unites the fetus to the uterus. The developing fetus receives nourishment through this organ.

placental mammals: mammals having a placenta through which the embryo and fetus are nourished while in the uterus.

placental membrane: same as placental sac.

placid: calm and quiet.

placoid scales: scales found on sharks and rays that are pointed and similar to teeth.

Plantae: plant kingdom, including multicellular photosynthesizing organisms, higher plants, and multicellular algae.

plastron: lower part of the shell of turtles, tortoises, and terrapins.

Pleistocene epoch: the geologic period of time that occurred from 500,000 years to 3 million years ago; during this time, early humans appeared.

Pointer: a breed of dog that gets its name from the stance it takes in the presence of game.

points: nose, ears, tail, and feet of an animal.

poliomyelitis: a disease caused by a virus that results in paralysis of the body and a decrease in muscle size, often with permanent disability and deformity.

polygamous: having more than one mate at a time.

posterior: the rear part of an animal.

pound seizures: dogs and cats from animal pounds and shelters are sold to laboratory animal dealers and to research and educational facilities for use in animal experimentation.

powder-down feathers: feathers that have tips that break down as they mature and then release a talc-like powder that provides waterproofing and luster to the feathers.

preanal pores: openings located just forward of the vent on certain species of geckos. They can be used to identify the male of the species. Also called anal pores.

precocious: showing premature development.

predator: an animal that seeks out smaller or weaker animals for food.

preen: the process a bird uses to trim and clean its feathers with its bill.

preferential: have a greater liking for or use of an element.

prehensile: an animal's tail that is adapted for grasping or wrapping around objects.

preplacenta sac: sac through which some oxygen and nutrient exchange takes place, but not to the extent of a placenta in mammals.

prime fur: a pelt from a chinchilla that has completely shed and been replaced by new fur.

priming line: new hair that comes up through old hair as a chinchilla sheds. This new hair forms a distinct line called the priming line.

proboscis: the long, tubular sucking mouth parts on certain species of fish.

proestrus: period preceding a dog's estrus.

progeny: offspring.

prolapse: an organ of an animal's body that falls out of its normal position. The anus, vagina, and uterus, for example, may turn inside out and extend out from the body opening.

proteins: compounds made up of amino acids that supply materials needed for an animal to produce muscle tissue, hair, hooves, horns, skin, and various internal organs.

GLOSSARY

Protista: kingdom including the single-celled or microscopic animals and single-celled algae.

protozoa: microscopic, single-celled animals.

protrusive: species of fish that can protrude their mouth outward to aid them in capturing prey.

proventriculus: the glandular stomach of a bird located between the crop and gizzard.

provoked: to arouse or incite an animal to bite or attack.

puppy mills: places where dogs are bred without regard to the pedigrees, defective traits, and abnormalities. The primary purpose of the breeder is to produce puppies for the pet trade. Also called puppy factories.

purebred breeder: a person who breeds animals from parent stock eligible for registration.

quarantine: to isolate animals while treating them for diseases or while determining if they have a disease or ailment.

queen: female cat.

rabbitry: place where domestic rabbits are kept.

rabies: a viral disease that affects the nervous system of warm-blooded animals.

random source: obtaining animals from animal shelters or humane shelters for use in laboratories.

rays: the hard, spiny structures that support the fins on fish.

recessive gene: in a pair of genes, the one that does not control the particular genetic characteristic; see *dominant gene.*

rectal prolapse: the protrusion of the rectum from the anus.

red nose: a condition in gerbils, usually caused by bacteria, that causes hair loss and red, swollen areas of the skin around the nose and muzzle.

regurgitated: not fully digested food material is brought back to the mouth from the stomach or rumen.

reptile: class of cold-blooded vertebrates having lungs, an entirely bony skeleton, a body covered with scales or horny plates, and a heart with two atria and usually a single ventricle.

Reptilia: the class that includes turtles, tortoises, terrapins, snakes, pythons, boas, iguanas, lizards, crocodiles, alligators, caiman, and gharials.

reserved: being shy, cautious, or restrained.

reservoir: an immune host.

respiration: the process by which oxygen is supplied to the cells and tissues in exchange for carbon dioxide; breathing.

reticulum: the second stomach compartment of ruminants often referred to as the "hardware stomach" because foreign bodies such as nails or pieces of wire can be held there for long periods of time without causing harm; also called the honeycomb because of its beehive interior appearance.

retractile: the claws of a cat that can be drawn back into the toes.

rex: the type of fur on the rex breed of rabbit. The guard hairs and the underfur are the same length, giving the fur a very soft, plush feel.

robust: a strong, rough, healthy-looking individual.

rodenticide: chemical substance used to kill rats and mice.

rodents: order Rodentia. Small, gnawing mammals that have a single pair of continuously growing incisors in both the upper and lower jaw.

roll-back fur: fur that gently returns to the original position after the animal is stroked.

roman nose: a nose that is shorter and more rounded than the nose of similar species.

rosettes: the circles or swirls of fur found on Abyssinian guinea pigs.

roundworms: round, unsegmented worms that occupy the intestinal tract of animals.

ruddy: color of a cat breed; an orange-brown color, ticked with dark brown or black.

rumen: the largest compartment of ruminants that makes up about 80 percent of the total capacity of the stomach.

ruminant: forage-consuming or multistomached animals; includes farm animals such as cattle, sheep, and goats, and zoo animals such as elk, deer, giraffes, buffalo, camels, and antelope.

rumpy: a Manx cat that has a dimple where the tail should be.

rustic looking: a rough, coarse-looking animal.

salamander: class Amphibia, order Caudata. Small, lizard-like amphibians lacking scales and breathing through gills during their larval stage.

sanitation: proper cleaning or sterilizing to prevent the growth of microorganisms.

saprophyte: a plant that lives on and receives its nourishment from dead or decaying organic material.

satin: fur with a very shiny appearance due to its transparent outer shell.

scalloped feathers: feathers that are black, with the outer margins or edges being of a different color.

scapulars: the small feathers that cover the base area on a bird's wing.

scurvy: a disease caused by the lack of vitamin C; a condition characterized by swollen, painful joints and bleeding gums.

scutes: the hard, tough scales or plates on the shell of turtles, tortoises, and terrapins.

selective breeding: selecting animals to breed that have certain traits or qualities.

self: one-color fur.

self-propagating cells: cells that reproduce when transferred to an artificial medium as in tissue culture.

semiaquatic: animals that live both in water and on land.

serene: calm and quiet.

serrations: the notches found on the rear edge of the carapace on certain species of turtles.

serum: the liquid portion of blood and other body fluids left after the solids have coagulated.

Setter: breed of bird dogs that take a half sitting stance when on point.

sheen: brightness, shininess.

shoal: a group or school of fish.

sirens: class Amphibia, order Caudata, family Sirenidae. Eel-like amphibians that live in freshwater areas of Southern and Central United States. They have feathery, external gill structures; small, weak front legs; and no rear legs.

skin irritancy test: measures the effects of products such as cosmetics on the skin of restrained rabbits. A patch of fur is shaved from the animal. Products are applied to the area and then observed for any reactions.

small animal breeder: a person who specializes in raising small animals for the pet trade or for use in laboratories.

solitary: prefers being alone; by itself.

solubility: capable of being dissolved. Vitamins are classified as fat soluble or water soluble. Fat-soluble vitamins can be stored by the body; water-soluble vitamins are stored in limited amounts.

sore hock: an ulceration to the foot pad caused by the animal's body weight pressing down on the rough wire of cages.

spawning: the act of depositing and the fertilization of eggs by certain species of fish.

spayed: a female animal that has had her ovaries removed by surgical procedure to prevent her from reproducing.

specieism: belief that any use of animals by humans reflects a bias or that humans are superior to animals.

species: a division of genus groups into individuals that are related.

species aquarium: an aquarium used to keep only one species of fish.

spermatophore: a capsule containing sperm deposited by the male.

spiracles: openings behind the eyes of sharks and rays through which water is drawn in and forced out the gills.

spirit: showing a lot of pep, vigor, and energy.

spirochete: a type of bacteria that is spiral shaped, is flexible, and does not have rigid cell walls. Bacteria of this type cause leptospirosis and syphilis in animals.

sport: an offspring that shows an unusual characteristic; a mutation. See also *throwbacks*.

spraying: the spraying of urine on objects by male cats to mark territory.

status animal: an animal regarded as a symbol of high social status.

status symbol: something that would be an indication of a person's wealth, position, or importance.

staunchness: strong, steady, steadfast while on point.

steady: staying on task.

steatitis: a deficiency disease, also called yellow fat disease, caused by feeding too much oily fish such as tuna.

sternum: the breastbone of an animal.

streptococcal bacteria: round-shaped bacteria that forms rows of cells in a chain-like arrangement.

stress: any condition that is not normal. Putting an animal in a situation or under conditions that it is not used to.

stromatolites: layered structures found in rock created through the activity of primitive algae or bacteria.

stumpy: a Manx cat with a short tail.

styptic: a substance used to stop bleeding by causing the blood to clot.

subclass: a division of the classification of life forms beneath class and above order. Life forms placed in certain classes are sometimes further divided into subclasses.

subcutaneous: beneath the skin; between the skin and muscle.

suborder: a division of the classification of life forms beneath order and above family. Life forms placed in certain orders are sometimes further divided into suborders.

subordinate: being submissive to. In a colony structure, there is one lead or dominant male; the others are subordinate or submissive to him.

subphylum: a division of the classification of life forms beneath phylum and above class. Life forms placed in certain phylums are sometimes further divided into subphylums.

substratum: the layer of gravel, peat, or soil that is placed in the bottom of an aquarium or vivarium.

suppressants: a drug that is used to keep under control or in check various conditions, ailments, or diseases.

sustenance: food or nourishment necessary to maintain life.

symbiosis: two different species living together in a mutually beneficial relationship.

synthesis: making or producing of a compound.

systemic: a substance that moves throughout all body systems.

systemic antibiotics: antibiotics such as tetracycline, gentamicin, tylosin, and neomycin.

tabby: cat with striped markings.

tapeworm: a parasitic segmented worm that lives in the intestines of a host animal. Segments are called proglottids.

tapirs: small, hoofed animals found in South America. They have four toes on each front foot and three toes on each rear foot. They have a short, movable trunk.

taxa: the seven categories of Linnaeus's classification system that include kingdom, phylum, class, order, family, genus, and species.

taxidermist: a person who skins, prepares, and mounts the skins of animals for display.

GLOSSARY

taxonomy: the science that is concerned with the naming and classification of organisms.

teased: masses of hair are untangled and straightened with a comb.

temperament: the emotional and mental qualities of an individual animal.

temperate zone: the mild climate zone between the Tropic of Cancer and the Arctic Circle and the zone between the Tropic of Capricorn and the Antarctic Circle.

terrapin: class Reptilia, order Testudines, family Emydidae, genus *Malaclemys*. Turtles found in fresh and brackish water that are considered excellent eating.

terrarium: a cage used for keeping land animals.

terrestrial: an animal that lives on land.

Tertiary period: the geologic time period from 3 million to 70 million years ago, during which the first placental animals appeared, spread, and increased in numbers. Monkeys, apes, whales, large carnivores, and grazing animals also appeared during this period.

testosterone: a male sex hormone that is responsible for the development of secondary sex characteristics.

theologians: persons who study the theory and teachings of religion.

therapy: methods used in the treatment of a disease, disorder, or condition. Also methods used to maintain good health.

thorough: describes a dog that works every bit of ground and cover.

thrombin: a substance made from cattle blood that aids in the clotting of blood and is used in skin grafting.

throwbacks: an offspring that shows an ancestor's characteristics that have not appeared in recent generations.

ticked coat: darker colors are found on the tips of each hair of the animal's coat.

ticking: a fur coat that has hairs with darker tips mixed in.

timid: showing lack of confidence, somewhat afraid.

tom: male cat.

tortoise shell: two colors producing a spotted or blotched pattern.

toxemia: a condition in which there is a buildup of toxins in the blood.

toxicology: the science that deals with poisons and their effects on an organism.

toxin: a poisonous substance produced by an organism.

trace minerals: see *microminerals.*

traits: characteristics or qualities of an animal that are usually inherited.

tranquil: quiet and peaceful.

tranquilizers: substances used to reduce mental anxiety or tension.

Triassic period: geologic time period from 180 to 230 million years ago, during which the dinosaurs appeared along with mammal-like reptiles.

trinomial nomenclature: adds a third word to the binomial nomenclature for an organism that has a subspecies name.

trophozoites: the growing stage of a protozoan parasite distinguished from the dormant or reproductive stage.

tympanum: a membrane that picks up vibrations and transfers them to the inner ear.

Tyzzer's disease: a disease primarily of rabbits caused by the bacterium *Bacillus piliformis*. The disease causes severe diarrhea and death among young rabbits.

udder development: the natural and normal growth of the mammary system of female animals.

ultrasonic: frequencies above what can be detected by the human ear; usually above 20,000 cycles per second.

ultraviolet: a portion of light rays that are beyond the visible violet spectrum because the wavelengths are shorter.

umbilical cord: cord that connects the fetus with the placenta to carry food to and remove waste from the fetus.

underfur: the very dense, soft, short fur that is beneath the longer, coarser guard hairs.

underwool: the short, dense, soft fur underneath the longer fur of Giant Angora rabbits.

undulating: moving in a wavy form or in a series of waves similar to the way snakes move.

unethical: a belief that something is going against one's moral principles and values.

unprovoked: an animal bite or attack in which the animal was not aroused or incited.

unthrifty: not thriving, flourishing, or growing vigorously; showing impaired performance.

uremic poisoning: a condition that occurs with severe kidney disease. Toxic substances that are normally removed from the blood by the kidneys build up and cause a toxic condition.

uterus: the organ in the female mammal in which the young is nourished and develops.

vain: appearing conceited, overconfident, or showing excessive pride.

vegetarian: person who lives on a diet primarily of vegetables, fruits, grains, and nuts.

veil: refers to the jet-black tips of the fur on black chinchilla mutations called Black Velvets.

vent: the external opening of the cloaca.

ventricle: part of the heart that receives blood from the atria and pumps blood into the arteries.

ventral: the underside or belly of an animal.

ventriculus: commonly called the gizzard, the largest organ of a bird's digestive system; it grinds and crushes food, which then enters the small intestine.

vermin: small, destructive animals such as wild mice and rats.

versatile: having a wide range of skills.

vertebrate: animals that have a segmented spinal column.

veterinarian: Doctors of Veterinary Medicine who treat and control animal injuries and diseases. They immunize healthy animals, inspect animals and meat used as food, perform surgery, set broken bones, establish diet and exercise routines, and prescribe medicines.

veterinary technician: a person who works under the supervision of the veterinarian to soothe and quiet animals under treatment, draw blood, insert catheters, and conduct laboratory tests.

vicious: overly aggressive, ferocious, or violent.

vigilant: always alert and watchful.

vigor: healthy growth; vitality; physical or mental strength.

viruses: infective agents that can cause diseases. They are only capable of growth and multiplying in living cells.

vitamins: organic compounds needed in small amounts by an animal body for various chemical reactions to occur.

vivariums: cages in which the environment is duplicated as close as possible to the natural environment of the species.

viviparous: animals that retain the young within their bodies in a preplacenta sac.

vivisection: research consisting of surgical operations and experiments to study the structure and function of organs.

vogue: something that is popular, fashionable, or in style at a given period of time.

waltzing mice: mice with a genetic defect that causes them to lose their sense of hearing and balance; they are unable to walk in a straight line and dance in circles.

wary: showing caution, being somewhat afraid.

wet tail: hamster enteritis; symptoms include wetness around the tail and rear area of the animal caused by runny diarrhea.

whelping: the act of dogs giving birth.

wholesale: selling animals in large numbers usually for resale.

witty: intelligent, clever, or quick to learn.

yeasts: a type of fungus that causes fermentation and is used in the making of alcoholic beverages and baking bread.

zoo administrators: persons who are employed at zoos throughout the country. There are zoo directors, animal curators, veterinarians, and resident zoologists. Their job is to provide services for the entertainment and education of the public.

zoologist: person who studies animals.

zoology: the study of animals.

zoonoses: diseases that can be transmitted from animals to humans; rabies is an example.

Colleges of Veterinary Medicine

U.S. COLLEGES OF VETERINARY MEDICINE ACCREDITED BY THE AMERICAN VETERINARY MEDICAL ASSOCIATION

Alabama: Auburn University
College of Veterinary Medicine
Auburn University, Alabama 36849
(334) 844-3691
http://www.vetmed.auburn.edu

Tuskegee University
School of Veterinary Medicine
Tuskegee, Alabama 36088
(334) 727-8173
http://svmc107.tusk.edu/Tu/
svmfront.html

California: University of California
School of Veterinary Medicine
Davis, California 95616-8734
(530) 752-1360
http://www.vetmed.ucdavis.edu

Colorado: Colorado State University
College of Veterinary Medicine and
Biomedical Sciences
Ft. Collins, Colorado 80523
(970) 491-7051
http://www.cvmbs.colostate.edu

Florida: University of Florida
College of Veterinary Medicine
Gainesville, Florida 32610-0125
(352) 392-4700
http://www.vetmed.ufl.edu

Georgia: University of Georgia
College of Veterinary Medicine
Athens, Georgia 30602
(706) 542-3461
http://www.vet.uga.edu

Illinois: University of Illinois
College of Veterinary Medicine
2001 South Lincoln
Urbana, Illinois 61801
(217) 333-2760
http://www.cvm.uiuc.edu

Indiana: Purdue University
School of Veterinary Medicine
1240 Lynn Hall
West Lafayette, Indiana 47907-1240
(317) 494-7607
http://www.vet.purdue.edu

Iowa: Iowa State University
College of Veterinary Medicine
Ames, Iowa 50011
(515) 294-1242
http://www.vetmed.iastate.edu

Kansas: Kansas State University
College of Veterinary Medicine
Manhattan, Kansas 66506
(913) 532-6011
http://www.vet.ksu.edu

Louisiana: Louisiana State University
School of Veterinary Medicine
Baton Rouge, Louisiana 70803
(504) 346-3295
http://www.vetmed.lsu.edu

Massachusetts: Tufts University
School of Veterinary Medicine
200 Westboro Road
North Grafton, Massachusetts 01536
(508) 839-5302
http://www.tufts.edu/vet

Michigan: Michigan State University
College of Veterinary Medicine
East Lansing, Michigan 48824-1314
(517) 355-6509
http://www.cvm.msu.edu

Minnesota: The University of Minnesota
College of Veterinary Medicine
St. Paul, Minnesota 55108
(612) 624-9227
http://www.cvm.umn.edu

Mississippi: Mississippi State University
College of Veterinary Medicine
Mississippi State, Mississippi 39762
(601) 325-3432
http://www.cvm.msstate.edu

Missouri: University of Missouri
College of Veterinary Medicine
Columbia, Missouri 65211
(573) 882-3877
http://www.cvm.missouri.edu

New York: Cornell University
College of Veterinary Medicine
Ithaca, New York 14853-6401
(607) 253-3700
http://www.vet.cornell.edu

North Carolina: North Carolina State University
College of Veterinary Medicine
4700 Hillsborough Street
Raleigh, North Carolina 27606
(919) 829-4200
http://www2.ncsu.edu/ncsu/cvm/
cvmhome.html

Ohio: The Ohio State University
College of Veterinary Medicine
Columbus, Ohio 43210
(614) 292-1171
http://www.vet.ohio-state.edu

Oklahoma: Oklahoma State University
College of Veterinary Medicine
Stillwater, Oklahoma 74078
(405) 744-6648
http://www.cvm.okstate.edu

Oregon: Oregon State University
College of Veterinary Medicine
Corvallis, Oregon 97331-4801
(503) 737-2141
http://www.vet.orst.edu

Pennsylvania: University of Pennsylvania
School of Veterinary Medicine
3800 Spruce Street
Philadelphia, Pennsylvania
19104-6044
(215) 898-5434
http://www.vet.upenn.edu

Tennessee: University of Tennessee
College of Veterinary Medicine
Knoxville, Tennessee 37901
(423) 974-7262
http://www.vet.utk.edu

Texas: Texas A&M University
College of Veterinary Medicine
College Station, Texas 77843-4461
(409) 845-5051
http://www.cvm.tamu.edu

Virginia: Virginia Tech and University of
Maryland
Virginia-Maryland Regional
College of Veterinary Medicine
Blacksburg, Virginia 24061-0442
(540) 231-7367
http://www.vetmed.vt.edu

Washington: Washington State University
College of Veterinary Medicine
Pullman, Washington 99164-7010
(509) 335-9515
http://www.vetmed.wsu.edu

Wisconsin: The University of Wisconsin-Madison
School of Veterinary Medicine
Madison, Wisconsin 53706
(608) 263-6716
http://www.vetmed.wisc.edu

Reprinted with permission of the American Veterinary
Medical Association.

APPENDIX B

Veterinary Technology Programs

AMERICAN VETERINARY MEDICAL ASSOCIATION PROGRAMS IN VETERINARY TECHNOLOGY

Programs Accredited by the AVMA Committee on Veterinary Technician Education and Activities (CVTEA)

1931 North Meacham Road, Suite 100, Schaumburg, Illinois 60173-4360 Phone (708) 925-8070

ALABAMA
*Snead State Community College
Veterinary Technology Program
Boaz, AL 35957
(205) 593-5120
(Glenn Sexton, DVM-Director)
2 years-Associate in Science
Initial Accreditation-Nov., 1979
FULL ACCREDITATION

CALIFORNIA
California State Polytechnic University
College of Agriculture
Animal Health Technology Program
3801 W. Temple Ave.
Pomona, CA 91768
(909) 869-2200
(Gerald Hackett, DVM-Coordinator)
4 years-Bachelor of Science
Initial Accreditation-April, 1996
PROVISIONAL ACCREDITATION

*Cosumnes River College
Veterinary Technology Program
8401 Center Pkwy.
Sacramento, CA 95823
(916) 688-7355
(Kathryn Graham, DVM-Director)
2 years-Associate in Science
Initial Accreditation-April, 1975
FULL ACCREDITATION

Foothill College
Animal Health Technology Program
12345 El Monte Rd.
Los Altos Hills, CA 94022
(415) 949-7203
FAX (415) 949-7375
(Karl Peter, DVM-Director)
2 years-Associate in Science
Initial Accreditation-April, 1977
Nov., 1982; Reaccredited Nov., 1986
FULL ACCREDITATION

Hartnell College
Animal Health Technology Program
156 Homestead Ave.
Salinas, CA 93901
(408) 755-6700
(Carol Thomas Kimbrough, AHT-Director)
2 years-Associate in Science
Initial Accreditation-April, 1981
FULL ACCREDITATION

Los Angeles Pierce College
Animal Health Technology Program
6201 Winnetka Ave.
Woodland Hills, CA 91371
(818) 347-0551
(Lisa Eshman, DVM-Director)
2 years-Associate in Science
Initial Accreditation-Dec., 1975; April, 1993
FULL ACCREDITATION

Mt. San Antonio College
Animal Health Technology Program
1100 N. Grand Ave.
Walnut, CA 91789
(909) 594-5611
(Kenneth Tudor, DVM-Director)
2 years-Associate in Science
Initial Accreditation-April, 1977
FULL ACCREDITATION

San Diego Mesa College
Animal Health Technology Program
7250 Mesa College Dr.
San Diego, CA 92111
(619) 627-2832
(Emmett Casey, Dean)
2 years-Associate in Science
Initial Accreditation-April, 1978
PROBATIONAL ACCREDITATION

*Yuba College
Veterinary Technology Program
2088 N. Beale Rd.
Marysville, CA 95901
(916) 741-6962
(Blaine Russell, DVM-Director)
2 years-Associate in Science
Initial Accreditation-April, 1978
FULL ACCREDITATION

COLORADO
*Colorado Mountain College
Veterinary Technology Program
Spring Valley Campus
3000 County Rd. 114
Glenwood Springs, CO 81601
(303) 945-7481
(Tom McBrayer-Assistant Dean)
2 years-Associate in Applied Science
Initial Accreditation-December, 1975
FULL ACCREDITATION

*Bel-Rea Institute of Animal Technology
1681 S. Dayton St.
Denver, CO 80231
(800) 950-8001
(Nolan Rucker, DVM-President)
2 years-Associate in Science
Initial Accreditation-April, 1975
FULL ACCREDITATION

Front Range Community College
Veterinary Research Technology Program
4616 S. Shields
Ft. Collins, CO 80526
(970) 204-0466
(Betsy Torgeson, DVM-Director)
2 years-Associate in Applied Science
Initial Accreditaiton-April, 1996
PROVISIONAL ACCREDITATION

CONNECTICUT
Quinnipiac College
Veterinary Technology Program
Mt. Carmel Ave.
Hamden, CT 06518
(203) 281-8958
(Steve Carleton, DVM-Director)
4 years-Bachelor of Science
Initial Accreditation-April, 1980
FULL ACCREDITATION

FLORIDA
St. Petersburg Junior College
Veterinary Technology Program
Box 13489
St. Petersburg, FL 33733
(813) 341-3652
(Guy Hancock, DVM-Director)
On-Campus Program
2 years-Associate in Science
Initial Accreditation-April, 1978
FULL ACCREDITATION
Distance Learning Program
Associate in Science
Initial Accreditation-November, 1995
PROVISIONAL ACCREDITATION

GEORGIA
*Fort Valley State College
Veterinary Technology Program
Fort Valley, GA 31030
(912) 825-6353
(Getter Huggins, PhD-Director)
2 years-Associate of Applied Science
Initial Accreditation-November, 1978
PROBATIONAL ACCREDITATION

NEBRASKA
*Nebraska College of Technical Agriculture
Veterinary Technology Program
Curtis, NE 69025
(308) 367-4124
(Ricky Sue Barnes Wach, DVM and
Barbara Berg AAS, VT-Co-Coordinators)
2 years-Associate in Technical Agriculture in
Veterinary Technology
Initial Accreditation-August, 1973
FULL ACCREDITATION

*Omaha College of Health Careers
Veterinary Technician Program
10845 Harney
Omaha, NE 68154
(402) 333-1400
(Travis Littledike, DVM-Director)
18 months-Associate in Applied Science
Initial Accreditation-Nov., 1978
FULL ACCREDITATION

NEW JERSEY
Camden County College
Animal Science Technology Program
P.O. Box 200
Blackwood, NJ 08012
(609) 227-7200
(Harriet Doolittle, VMD-Coordinator)
2 years-Associate in Applied Science
Initial Accreditation-Nov., 1978
FULL ACCREDITATION

NEW YORK
La Guardia Community College
The City University of New York
Veterinary Technology Program
31-10 Thomson Ave.
Long Island City, NY 11101
(718) 482-5764
(Douglas McBride, DVM-Director)
2 years-Associate in Applied Science
Initial Accreditation-April., 1983
FULL ACCREDITATION

Mercy College
Veterinary Technology Program
555 Broadway
Dobbs Ferry, NY 10522
(914) 693-4500
(Jack Burke, DVM-Director)
4 years-Bachelor of Science
Initial Accreditation-April, 1987
FULL ACCREDITATION

State University of New York
Agricultural & Technical College
Health Sciences & Paramedical Technologies
Veterinary Science Technology Program
Canton, NY 13617
(315) 306-7410
(Edward Gordon, DVM-Director)
2 years-Associate in Applied Science
Initial Accreditation-Nov., 1979
FULL ACCREDITATION

State University of New York
College of Technology
Veterinary Science Technology Program
Delhi, NY 13753
(607) 746-4385
(Foster Palmer, DVM-Acting Director)
2 years-Associate in Applied Science
Initial Accreditation-Dec., 1975
FULL ACCREDITATION

Suffolk Community College
Veterinary Science Technology Program
Western Campus
Crooked Hill Rd. Brentwood, NY 11717
(516) 434-6700
(Allen Jacobs, DVM-Director)
2 years-Associate in Applied Science
Initial Accreditation-April, 1996
PROVISIONAL ACCREDITATION

NORTH CAROLINA
*Central Carolina Community College
Veterinary Medical Technology Program
1105 Kelly Dr.
Sanford, NC 27330
(919) 775-5401
FAX (919) 775-1221
(Paul Porterfield, DVM-Director)
2 years-Associate in Applied Science
Initial Accreditation-Nov., 1974
FULL ACCREDITATION

NORTH DAKOTA
*North Dakota State University
Veterinary Technology Program
Department of Veterinary Science
Fargo, ND 58105
(701) 231-7511
FAX (701) 231-7514
Email tcolville@plains.nodak.edu
(Tom Colville, DVM-Director)
4 years-Bachelor of Science
Initial Accreditation-May, 1978
FULL ACCREDITATION

OHIO

Columbus State Community College
Veterinary Technology Program
550 E. Spring St.
Columbus, OH 43216
(614) 227-2569 or 2400
(H. Marie Suthers, DVM-Director)
2 years-Associate in Applied Science
Initial Accreditation-May, 1974
FULL ACCREDITATION

*UC Raymond Walters College
Veterinary Technology Program
P.O. Box 670571
Cincinnati, OH 45267-0571
(513) 558-5171
FAX (513) 371-9844
(David Bauman, DVM-Director)
2 years-Associate in Science
Initial Accreditation-April, 1975
FULL ACCREDITATION

Stautzenberger College
Veterinary Technology Program
5355 S. Wyck
Toledo, OH 43614
(419) 866-0261
(Bill Mewborn, DVM-Director)
2 years-Associate in Applied Science
Initial Accreditation-April, 1996
PROVISIONAL ACCREDITATION

OKLAHOMA

*Murray State College
Veterinary Technology Program
Tishomingo, OK 73460
(405) 371-2371
FAX (405) 371-9844
(Kay Helms, DVM-Director)
2 years-Associate in Applied Science
Initial Accreditation-Nov., 1980
FULL ACCREDITATION

OREGON

*Portland Community College
Veterinary Technology Program
P.O. Box 19000
Portland, OR 97219
(503) 244-7461
(Phillip Cochran, DVM-Director)
2 years-Associate in Applied Science
Initial Accreditation-April, 1989
FULL ACCREDITATION

PENNSYLVANIA

Harcum College
Veterinary Technology Program
Bryn Mawr, PA 19010
(610) 526-6055
FAX (610) 526-6031
(Nadine Hackman, VMD-Director)
6 semesters-Associate in Science
Initial Accreditation-April, 1976
FULL ACCREDITATION

*Johnson Technical Institute
Veterinary Science Technology Program
3427 N. Main Ave.
Scranton, PA 18505
(717) 342-6404
(Carl Reynolds, VMD-Director)
2 years-Associate in Specialized Technology
Initial Accreditation-November, 1995
PROVISIONAL ACCREDITATION

*Manor Junior College
Veterinary Technology Program
Fox Chase Rd. & Forest Ave.
Jenkintown, PA 19046
(215) 885-2360
(Joanna Bassert, VMD-Director))
2 years-Associate in Science
Initial Accreditation-April, 1992
PROBATIONAL ACCREDITATION

Wilson College
Veterinary Medical Technology Program
Chambersburg, PA 17201
(717) 264-4141
FAX (717) 264-1578
(William Hay, DVM-Director)
Bachelor of Science-College for Women
Associate in Science-Continuing Studies Division
Initial Accreditation-May, 1984
FULL ACCREDITATION

PUERTO RICO

University of Puerto Rico
Animal Health Technology Program
Medical Sciences Campus
P.O. Box 365067
San Juan, PR 00936-5067
(809) 758-2525, ext. 1051 or 1052
(Carlos Ortiz, AHT-Director)
4 years-Bachelor of Science
Initial Accreditation-April, 1996
FULL ACCREDITATION

SOUTH CAROLINA
*Tri-County Technical College
Veterinary Technology Program
P.O. Box 587
Pendleton, SC 29670
(803) 646-8361
Email Rmarshall@Al.Tricounty.TCC.SC.US
(Roseann Marshall, DVM-Director)
2 years-Associate in Health Science
Initial Accreditation-April, 1979
FULL ACCREDITATION

SOUTH DAKOTA
*National College Allied Health Division
Veterinary Technology Program
321 Kansas City St.
Rapid City, SD 57709
(800) 843-8892
(Peggy Behrens, DVM-Director)
8 quarters-Associate in Applied Science
Initial Accreditation-April, 1981
FULL ACCREDITATION

TENNESSEE
*Columbia State Community College
Veterinary Technology Program
Columbia, TN 38401
(615) 540-2722
(Boyce Wanamaker, DVM-Director)
2 years-Associate in Applied Science
Initial Accreditation-April, 1979
FULL ACCREDITATION

*Lincoln Memorial University
Veterinary Technology Program
Harrogate, TN 37752
(615) 869-6278
(Randall Evans, DVM-Director)
Initial Accreditation-November, 1987
FULL ACCREDITATION

TEXAS
Cedar Valley College
Veterinary Technology Program
3030 N. Dallas Ave.
Lancaster, TX 75134
(214) 372-8164
(David Wright, DVM-Director)
2 years-Associate in Applied Science
Initial Accreditation-November, 1978
FULL ACCREDITATION

*Midland College
Veterinary Technology Program
3600 N. Garfield
Midland, TX 79705
(915) 685-4619
FAX (915) 685 6431
(Kerry Coombs, DVM-Director)
2 years-Associate in Applied Science
Initial Accreditation-November 1992, April 1995
PROBATIONAL ACCREDITATION

Sul Ross State University
Range Animal Science Department
Veterinary Technology Program
Alpine, TX 79830
(915) 837-8205
FAX (915) 837-8409
(Paul Weyerts, DVM-Director)
2 years-Associate Degree
Initial Accreditation-November, 1977
FULL ACCREDITATION

*Tomball College
Veterinary Technology Program
30555 Tomball Pkwy.
Tomball, TX 77375-4036
(713) 351-3357
(George Younger, DVM-Coordinator)
2 years-Associate in Applied Science
Initial Accreditation-April, 1990
FULL ACCREDITATION

UTAH
*Brigham Young University
Veterinary Technology Program
Provo, UT 84602
(801) 378-4294
(Richard Thwaits, DVM-Director)
4 years-Bachelor in Animal Science
Initial Accreditation-April, 1983
FULL ACCREDITATION

VERMONT
Vermont Technical College
Veterinary Technology Program
Randolph Center, VT 05061
(802) 728-3391
(Amy Woodbury-St. Denis, DVM-Director)
2 years-Associate in Applied Science
Initial Accreditation-April, 1991
FULL ACCREDITATION

VIRGINIA
*Blue Ridge Community College
Veterinary Technology Program
Box 80
Weyers Cave, VA 24486
(540) 234-9261
(Stuart Porter, VMD-Director)
2 years-Associate in Science
Initial Accreditation-November, 1976
FULL ACCREDITATION

*Northern Virginia Community College
Veterinary Technology Program
Loudoun Campus
1000 Harry Flood Byrd Hwy.
Sterling, VA 20164-8699
(703) 450-2525
FAX (703) 450-2536
(Leslie Sinn, DVM-Head)
2 years-Associate in Applied Science
Initial Accreditation-April, 1980
FULL ACCREDITATION

WASHINGTON
Pierce College at Ft. Steilacoom
Veterinary Technology Program
9401 Farwest Dr., SW
Tacoma, WA 98498
(206) 964-6665
(Terry Teeple, DVM-Acting Director)
2 years-Associate in Animal Technology
Initial Accreditation-May, 1974
FULL ACCREDITATION

WEST VIRGINIA
Fairmont State College
Veterinary Technology Program
Fairmont, WV 26554
(304) 367-4303
(William Carpenter, DVM-Coordinator)
2 years-Associate in Applied Science
Initial Accreditation-April, 1981
FULL ACCREDITATION

WISCONSIN
*Madison Area Technical College
Veterinary Technician Program
3550 Anderson
Madison, WI 53704
(608) 246-6100
(Robert Taylor, DVM-Director)
2 years-Associate in Applied Science
Initial Accreditation-November, 1974
FULL ACCREDITATION

WYOMING
Eastern Wyoming College
Veterinary Technology Program
3200 W. "C"
St. Torrington, WY 82240
(800) 658-3195, ext. 8268
(Patti Sue Peterson, AHT-Director)
2 years-Associate in Applied Science
Initial Accreditation-April, 1976
FULL ACCREDITATION

*Student Chapter of the North American Veterinary
Technician Association (NAVTA).

APPENDIX B

CLASSIFICAITON OF ACCREDITATION

Programs evaluated by CVTEA will be placed in one of the following classifications with respect to the Essential Requirements listed in CVTEA's accreditation policies and procedures manual. The accreditation status of each accredited program will be published semiannually and made available upon request.

FULL ACCREDITATION—Those programs that meet or exceed the minimal requirements.

PROVISIONAL ACCREDITATION—Those programs that meet or exceed most, but not all, minimal requirements and that have not completed the entire program and placed graduates.

PROBATIONAL ACCREDITATION—Those programs that meet or exceed most, but not all, minimal requirements.

ACCREDITATION WITHHELD—Those programs that do not fall in any of the above categories. A program on Probational Accreditation that does not make annual progress toward correcting its deficiencies, or any program that has been on Probational or Provisional Accreditation for five successive years will be classified as Accreditation Withheld.

CANADA—For information regarding Canadian veterinary technology programs, contact:

Canadian Association of Animal Health Technologists & Technicians
Box 91
Grandora SK S0K 1V0
CANADA
(306) 329-8660
FAX (306) 283-4829

The following states do not have AVMA-accredited veterinary technology programs: Alaska, Arizona, Arkansas, Delaware, District of Columbia, Hawaii, Idaho, Maine, Montana, Nevada, New Hampshire, New Mexico, Rhode Island

Reprinted with permission of the American Veterinary Medical Association 1931 North Meacham Road, Suite 100, Schaumberg, Illinois 60173-4360.

Reported Cases of Lyme Disease

Reported Cases of Lyme Disease, by State, 1989–1998*

State	Region	1989	1990	1991	1992	1993	1994	1995	1996	1997	1998	TOTAL	1995 POP	INC 97	INC 98	ANN INC
ALABAMA	ESC	25	33	13	10	4	6	12	9	11	24	147	4.246	0.26	0.57	0.35
ALASKA	PAC	0	0	0	0	0	0	0	0	2	1	3	0.603	0.33	0.17	0.05
ARIZONA	MT	0	0	1	0	0	0	1	0	4	1	7	4.305	0.09	0.02	0.02
ARKANSAS	WSC	10	22	31	20	8	15	11	27	27	7	178	2.485	1.09	0.28	0.72
CALIFORNIA	PAC	250	345	265	231	134	68	84	64	154	149	1744	31.565	0.49	0.47	0.55
COLORADO	MT	1	0	1	0	0	1	0	0	0	6	9	3.748	0.00	0.16	0.02
CONNECTICUT	NE	774	704	1192	1760	1350	2030	1548	3104	2297	2969	17728	3.271	70.23	90.77	54.20
DELAWARE	SA	25	54	73	219	143	106	56	173	109	45	1003	0.717	15.20	6.28	13.99
DC	SA	0	5	5	3	2	9	3	3	10	8	48	0.555	1.80	1.44	0.87
FLORIDA	SA	6	7	35	24	30	28	17	55	56	68	326	14.184	0.39	0.48	0.23
GEORGIA	SA	715	161	25	48	44	127	14	1	9	5	1149	7.209	0.12	0.07	1.59
HAWAII	PAC	1	2	0	2	1	0	0	1	0	0	7	1.179	0.00	0.00	0.06
IDAHO	MT	42	1	2	2	2	3	0	2	4	7	65	1.166	0.34	0.60	0.56
ILLINOIS	ENC	79	30	51	41	19	24	18	10	13	9	294	11.790	0.11	0.08	0.25
INDIANA	ENC	8	15	16	22	32	19	19	32	33	69	265	5.797	0.57	1.19	0.46
IOWA	WNC	27	16	22	33	8	17	16	19	8	27	193	2.843	0.28	0.95	0.68
KANSAS	WNC	15	14	22	18	54	17	23	36	4	12	215	2.564	0.16	0.47	0.84
KENTUCKY	ESC	21	18	44	28	16	24	16	26	20	36	238	3.857	0.52	0.65	0.62
LOUISIANA	WSC	2	3	6	7	3	4	9	9	13	7	63	4.338	0.30	0.16	0.15
MAINE	NE	3	9	15	16	18	33	45	63	34	78	314	1.239	2.75	6.30	2.54
MARYLAND	SA	138	238	282	183	180	341	454	447	494	653	3410	5.039	9.80	12.96	6.77
MASSACHUSETTS	NE	129	117	265	223	148	247	189	321	291	782	2712	6.071	4.79	12.88	4.47
MICHIGAN	ENC	165	134	46	35	23	33	5	28	27	12	508	9.538	0.28	0.13	0.53
MINNESOTA	WNC	92	70	84	197	141	208	208	251	256	238	1745	4.615	5.55	5.16	3.78
MISSISSIPPI	ESC	7	7	8	0	0	0	17	24	37	6	96	2.696	1.00	0.22	0.36
MISSOURI	WNC	108	205	207	150	108	102	53	52	28	2	1015	5.319	0.53	0.04	1.91
MONTANA	MT	0	0	0	0	0	0	0	0	0	0	0	0.870	0.00	0.00	0.00
NEBRASKA	WNC	0	0	25	22	6	3	6	5	2	4	73	1.639	0.12	0.24	0.45
NEVADA	MT	7	2	5	1	5	1	6	2	2	6	37	1.533	0.13	0.39	0.24
NEW HAMPSHIRE	NE	3	4	38	44	15	30	28	47	39	46	294	1.148	3.40	4.01	2.56
NEW JERSEY	MA	680	1074	915	688	786	1533	1703	2190	2041	1818	13428	7.950	25.67	22.87	16.89
NEW MEXICO	MT	5	0	3	2	2	5	1	1	1	4	24	1.690	0.06	0.24	0.14
NEW YORK	MA	3224	3244	3944	3448	2818	5200	4438	5301	3327	4426	39370	18.191	18.29	24.33	21.64
NORTH CAROLINA	SA	61	87	73	67	86	77	84	66	34	63	698	7.202	0.47	0.87	0.97
NORTH DAKOTA	WNC	12	3	2	1	2	0	0	2	0	0	22	0.642	0.00	0.00	0.34

Continued.

Reported Cases of Lyme Disease, by State, 1989–1998*—cont'd

State	Region	1989	1990	1991	1992	1993	1994	1995	1996	1997	1998	TOTAL	1995 POP	INC 97	INC 98	ANN INC
OHIO	ENC	99	36	112	32	30	45	30	32	40	59	515	11.134	0.36	0.53	0.46
OKLAHOMA	WSC	16	13	29	27	19	99	63	42	45	10	363	3.275	1.37	0.31	1.11
OREGON	PAC	5	11	5	13	8	6	20	19	20	21	128	3.149	0.64	0.67	0.41
PENNSYLVANIA	MA	626	553	718	1173	1085	1438	1562	2814	2188	2713	14870	12.060	18.14	22.50	12.33
RHODE ISLAND	NE	415	101	142	275	272	471	345	534	442	720	3717	0.992	44.57	72.60	37.48
SOUTH CAROLINA	SA	18	7	10	2	9	7	17	9	3	7	89	3.667	0.08	0.19	0.24
SOUTH DAKOTA	WNC	3	2	10	1	0	0	0	0	1	0	8	0.730	0.14	0.00	0.11
TENNESSEE	ESC	30	28	35	31	20	13	28	24	45	45	299	5.247	0.86	0.86	0.57
TEXAS	WSC	82	44	57	113	48	56	77	97	60	20	654	18.801	0.32	0.11	0.35
UTAH	MT	3	1	2	6	2	3	1	1	1	0	20	1.958	0.05	0.00	0.10
VERMONT	NE	1	11	7	9	12	16	9	26	8	11	110	0.585	1.37	1.88	1.88
VIRGINIA	SA	54	129	151	123	95	131	55	57	67	73	935	6.615	1.01	1.10	1.41
WASHINGTON	PAC	33	30	7	14	9	4	10	18	11	7	143	5.448	0.20	0.13	0.26
WEST VIRGINIA	SA	15	11	43	14	50	29	26	12	10	13	223	1.825	0.55	0.71	1.22
WISCONSIN	ENC	762	337	424	525	401	409	369	396	480	657	4760	5.122	9.37	12.83	9.29
WYOMING	MT	6	5	11	5	9	5	4	3	3	1	52	0.479	0.63	0.21	1.09
U.S. TOTAL		8803	7943	9470	9908	8257	13043	11700	16455	12801	15934	114314	262.889	4.87	6.06	4.35

Courtesy of the Center For Disease Control and Prevention (CDC).

Index